CAMBRIDGE LIBRARY COLLECTION
Books of enduring scholarly value

Physical Sciences

From ancient times, humans have tried to understand the workings of the world around them. The roots of modern physical science go back to the very earliest mechanical devices such as levers and rollers, the mixing of paints and dyes, and the importance of the heavenly bodies in early religious observance and navigation. The physical sciences as we know them today began to emerge as independent academic subjects during the early modern period, in the work of Newton and other 'natural philosophers', and numerous sub-disciplines developed during the centuries that followed. This part of the Cambridge Library Collection is devoted to landmark publications in this area which will be of interest to historians of science concerned with individual scientists, particular discoveries, and advances in scientific method, or with the establishment and development of scientific institutions around the world.

Radiations from Radioactive Substances

Sir Ernest Rutherford (1871–1937) was a New Zealand-born physicist who has become known as the 'father of nuclear physics' for his discovery of the so-called planetary structure of atoms. He was awarded the Nobel Prize in Chemistry in 1908. His co-authors, James Chadwick and Charles D. Ellis also made significant discoveries in the field of nuclear physics, with Chadwick discovering the neutron particle in 1932. Research in nuclear physics in the 1930s had become focused on investigating the natures of alpha, beta and gamma radiation and their effects on matter and atomic structure. This volume provides a definitive account of the state of research into these types of radiation in 1930, explaining the theory and process behind inferring the structure of the atom and the structure of the nucleus. The text of this volume is taken from a 1951 reissue of the 1930 edition.

Cambridge University Press has long been a pioneer in the reissuing of out-of-print titles from its own backlist, producing digital reprints of books that are still sought after by scholars and students but could not be reprinted economically using traditional technology. The Cambridge Library Collection extends this activity to a wider range of books which are still of importance to researchers and professionals, either for the source material they contain, or as landmarks in the history of their academic discipline.

Drawing from the world-renowned collections in the Cambridge University Library, and guided by the advice of experts in each subject area, Cambridge University Press is using state-of-the-art scanning machines in its own Printing House to capture the content of each book selected for inclusion. The files are processed to give a consistently clear, crisp image, and the books finished to the high quality standard for which the Press is recognised around the world. The latest print-on-demand technology ensures that the books will remain available indefinitely, and that orders for single or multiple copies can quickly be supplied.

The Cambridge Library Collection will bring back to life books of enduring scholarly value (including out-of-copyright works originally issued by other publishers) across a wide range of disciplines in the humanities and social sciences and in science and technology.

Radiations from Radioactive Substances

Ernest Rutherford
James Chadwick
Charles Drummond Ellis

CAMBRIDGE
UNIVERSITY PRESS

CAMBRIDGE UNIVERSITY PRESS

Cambridge, New York, Melbourne, Madrid, Cape Town, Singapore,
São Paolo, Delhi, Dubai, Tokyo

Published in the United States of America by Cambridge University Press, New York

www.cambridge.org
Information on this title: www.cambridge.org/9781108009010

© in this compilation Cambridge University Press 2010

This edition first published 1951
This digitally printed version 2010

ISBN 978-1-108-00901-0 Paperback

RADIATIONS
FROM
RADIOACTIVE SUBSTANCES

RADIATIONS
FROM
RADIOACTIVE SUBSTANCES

by

SIR ERNEST RUTHERFORD, O.M., D.Sc., Ph.D., LL.D., F.R.S.

NOBEL LAUREATE

Cavendish Professor of Experimental Physics in the University of Cambridge

JAMES CHADWICK, Ph.D., F.R.S.

Fellow of Gonville and Caius College, Cambridge

and

C. D. ELLIS, Ph.D., F.R.S.

Fellow of Trinity College, Cambridge

Re-issue of the edition of 1930

CAMBRIDGE

AT THE UNIVERSITY PRESS

1951

PUBLISHED BY
THE SYNDICS OF THE CAMBRIDGE UNIVERSITY PRESS
London Office: Bentley House, N.W. I
American Branch: New York
Agents for Canada, India, and Pakistan: Macmillan

First Printed   1930
Reprinted with corrections 1951

First Printed in Great Britain at the University Press, Cambridge
Reprinted by Offset Litho by Bradford and Dickens

CONTENTS

CONTENTS

Appendix. Radioactivity of ordinary matter. The Radioactivity of Potassium and Rubidium. The Discovery of the Helium . . . Continued theory. The atomic media column. γ Activity of appendix. Preparation of Radioactive Sources. The Mass values of quantities of Radium.

Table of Radium. The α-ray Electron, their atomic Numbers and Atomic Weights. The Periodic Table of the Elements etc. Artificial Disintegration of Elements.

LIST OF PLATES

PREFACE

In 1904 I published through the Cambridge University Press a collected account of radioactive phenomena entitled *Radioactivity*, followed a year later by a revised and enlarged edition. In 1912 a new volume was issued, entitled *Radioactive Substances and their Radiations* (Cambridge University Press), which endeavoured to give a concise account of our knowledge of radioactivity within the compass of a single volume.

The issue of this book was sold out soon after the conclusion of the War, and I was unable in the press of other work to find time for the preparation of a revised edition.

Since the publication in 1912, there has been a very rapid growth of our knowledge of the transformations of radioactive substances and of the radiations which accompany these transformations. The literature has rapidly expanded and many thousands of new papers, dealing with various aspects of the subject, have been published. It was felt that any attempt to give a collected account of the researches on this subject along the lines of the 1912 edition would have necessitated a very bulky volume. In the meantime, the need for such a publication had been met by the appearance of several new books. Professors Stefan Meyer and Egon v. Schweidler published, in 1916, *Radioaktivität* (Teubner, Berlin), followed in 1927 by a second enlarged edition. This excellent volume, which gives references to all the literature on this subject, is of great value to the scientific student. In 1928, Professor K. W. F. Kohlrausch published a collected account of radioactive researches entitled *Radioaktivität*, which appeared as one of the volumes of the Wien-Harms *Handbuch der Experimental Physik*. This extensive work gives an excellent detailed survey of our knowledge of the subject. I should also refer to the very useful summary *Radioactivity*, 1925, revised and reprinted 1929, prepared by Professor A. F. Kovarik and Dr L. W. McKeehan as a *Bulletin* of the National Research Council of the National Academy of Sciences, Washington. This summarises in a compact form the more important additions to our knowledge of this subject since the year 1916.

With the exception of a few outstanding problems, the wonderful series of radioactive transformations of uranium, thorium, and actinium are now well understood and attention today tends to be

more and more concentrated on the study of the α, β, and γ rays
which accompany the transformations and of the effects produced by
these radiations in their passage through matter. This is a subject
not only of great scientific interest but of fundamental importance.
A detailed study of the nature and modes of emission of the radiations
spontaneously appearing during the disintegration of atoms promises
to give us information of great value on the structure of the nucleus
of the atom and of the energy changes involved in its transformation.

In addition, the bombardment of matter by swift α particles has
placed in our hands a powerful method for studying the artificial
transformation of the nuclei of a number of the ordinary elements.
Evidence is accumulating that by this method we are not only able
to cause a disintegration of a nucleus with loss of mass but also in
some cases to build up a nucleus of greater mass by capture of the
colliding α particle.

For these reasons, I have thought it desirable to confine this new
work mainly to an account of the radiations from active matter and
their application to physical problems. The way in which the long
series of radioactive transformations has been unravelled is only
referred to in an incidental manner.

It has been our object to give a concise and connected account of
our knowledge of the radiations and of the bearing of the results on
the problem of the structure of the nucleus. No attempt has been
made to include an account of all the numerous papers which have
appeared within the last decade, but it is hoped that the more
important and essential facts and theories have received adequate
treatment.

During the last two years there has been a vigorous attack on the
problem of nuclear structure based on the ideas of the wave mechanics.
In a time of such rapid advance, it is inevitable that a number of new
facts and theories which have been published while the book was
passing through the press could not be included. It is hoped that
such lacunae will be filled in a subsequent edition.

In the preparation of this volume, I have been fortunate in
obtaining the help of my colleagues Dr J. Chadwick, F.R.S., and
Dr C. D. Ellis, F.R.S., who have themselves made substantial contributions to our knowledge of the subject. While all three of us have
co-operated in the preparation of this book, I have been mainly responsible for the chapters dealing with the α rays, Dr Chadwick for
the chapters dealing with the scattering of α and β particles by matter

and the artificial disintegration of the elements, and Dr Ellis for the account of the β and γ rays.

We are indebted to Professors L. Meitner and C. T. R. Wilson, Dr Skobelzyn, M. Frilley, Mr Blackett, and the proprietors of the *Zeitschrift für Physik* for permission to publish some of the photographs in the plates; to the Council of the Royal Society and the proprietors of the *Philosophical Magazine* for the use of certain figure blocks, and to the Akademische Verlagsgesellschaft m.b.H., Leipzig, for Figs. 126, 128 and 130, which are based on corresponding figures in *Radioaktivität*, by K. W. F. Kohlrausch. We desire to express our thanks to Mr G. A. R. Crowe for preparation of many of the diagrams.

E. RUTHERFORD

CAVENDISH LABORATORY
CAMBRIDGE

October, 1930

CHAPTER I

RADIOACTIVE TRANSFORMATIONS

§1. In studying the history of the rapid progress in our knowledge of atomic physics during the past thirty years, one cannot fail to be impressed with the outstanding importance of three fundamental discoveries which followed one another in rapid succession at the close of the last century. We refer to the discovery of the X rays by Röntgen in 1895, the discovery of the radioactivity of uranium by Becquerel early in 1896, and the proof of the independent existence of the negative electron in 1897 by Sir J. J. Thomson, Wiechert and Kaufmann. In a sense these discoveries mark the beginning of a new epoch in physics, for they provided new and powerful methods for attacking the fundamental problems of physics, such as the nature of electricity and the constitution and relation of the atoms of the elements. While the rapid development of our knowledge in each of these new fields of enquiry has provided us with new and very valuable information on the nature of radiation and the interaction between radiation and matter, a new orientation of our views on this subject was given by the remarkable theory of quanta first put forward by Planck in 1900, although its full significance was not generally recognised for another decade. The application in 1913 by Bohr of the quantum theory to explain the origin of spectra and the arrangement of the electrons in the outer structure of the atom has proved of great significance to modern science. It has not only given us a general view of the motions and arrangements of the electrons in the outer atom, but has revolutionised our ideas of the origin of spectra and the interactions between radiation and matter.

While the development of each of these great discoveries has proceeded rapidly and has added much to our knowledge in its particular field, it is clear that these lines of enquiry are closely related, and have mutually interacted at all stages of the advance. This can be illustrated in a multitude of ways. For example, the production of a temporary conductivity in gases by X rays led to a clear understanding of the transport of electricity through gases by means of charged ions. This ionisation theory of gases was applied directly to the analogous conductivity of gases produced by the rays from radioactive substances and gave us a clear interpretation of the electrical method of comparing the activities of radioactive substances. The methods developed by Sir J. J. Thomson to measure the velocity and value of e/m of the electrons in a vacuum tube were directly applied by Kaufmann to study the variation with velocity of the mass of the swift β rays emitted by radium, and by Rutherford to determine the velocity and mass of the α particles ejected from radioactive substances.

The information gained from a close study of X rays was utilised to elucidate the nature of the very penetrating γ rays from radioactive bodies, while the application of the quantum theory gave us an actual measure of the wave-length of these high frequency radiations.

On the other hand, the study of the radioactive substances has given us a wealth of information on atomic structure which has proved of fundamental importance. It first brought out that the atoms of the heavier elements like uranium and thorium were not permanently stable, but were undergoing spontaneous transformation with the emission of characteristic radiations of new and powerful types. A close study of the successive transformations of uranium and thorium has disclosed the presence of a group of about thirty new elements which are characterised by their distinctive radioactive properties. The examination of the chemical properties of these radioactive elements, often existing in minute amounts, first brought to light the existence of isotopes, i.e. of elements of almost identical chemical and physical properties but differing in mass and in radioactive qualities. This has been followed by the proof that many of the ordinary elements also consist of a mixture of isotopes.

While radioactivity showed us that some of the atoms of the heavier elements were unstable and occasionally disintegrated with explosive violence, it at the same time gave us the most convincing evidence of the existence of atoms as definite units in the structure of matter and provided the first method for the detection and counting of the individual atoms of matter in the form of α particles.

The study of the transformation of the radioactive atoms brought to our attention a new type of sub-atomic change which was spontaneous and uncontrollable by any known physical or chemical agency, and was accompanied by an emission of energy per atom in the form of characteristic radiation, enormous compared with that involved in ordinary chemical reactions. The emission of the energetic

α particle, the swift β ray and the penetrating γ ray gave us new types of radiation to study, of an individual intensity that has so far not been equalled by laboratory methods. At the same time, the α particle provided us with a particle of such great swiftness and energy that it was able to pass freely through the structure of the atom. By studying the deflexion of individual α particles in their passage through a single atom, definite evidence was obtained of the nuclear structure of the atoms and of the magnitude of the nuclear charge and thus of the number of electrons in the outer atom.

In a similar way, the swift α particles have been used to bombard the atoms of the ordinary elements, resulting in some cases in the disintegration of the nucleus with the expulsion of a swift proton. While the property of radioactivity has been instrumental in giving us information of extraordinary value on the structure of the nuclei of atoms, it seems clear that a still closer study of the wonderful series of transformations in uranium and thorium cannot fail to yield further results of the greatest significance and importance, for as yet few even of the facts already obtained can be adequately interpreted.

It is now known that the fundamental radioactive processes have their origin, not in the electronic structure of the outer atom, but in the centre or nucleus of the atom. One of the most valuable methods for throwing light on the nature of these nuclear changes is a study of the radiations emitted during the transformation of an atom. In particular, investigations on the wave-length of the γ rays emitted by some radioactive bodies give us important information on the frequency of vibration within the nucleus of the atom, while the velocity of emission of the α and β particles gives us some indication of the magnitude of the electric fields close to the nucleus.

In this work, we shall deal mainly with the properties of the radiations emitted by radioactive elements. We shall only incidentally refer to the modes of analysis of the long series of successive changes in the primary radioactive elements uranium and thorium, except in so far as a knowledge of these processes is of importance in connection with the origin and nature of the radiations.

After a brief account of the theory of radioactive transformations, we shall deal in chapters II–VII with the nature and properties of the α particle and the varied effects which are observed in its passage through matter. In chapters VIII–XI we shall discuss the occasional large angle deflections suffered by an α particle in passing close to a nucleus and the information which has been obtained in this way on the nuclear structure of the atom and the nature and variation of the forces close to the nucleus. This will be followed by an account of experimental observations on the artificial transformation of certain light elements resulting from the close collisions of the α particle with the atomic nucleus. This has given us the first definite proof that the nuclei of some of the lighter elements can be transformed by the action of external agencies.

In chapters XII–XVI the nature and origin of the β and γ rays from radioactive matter and their absorption by matter are considered. Rapid advances in this field have been made in recent years and methods have been found to measure the wave-length of the stronger γ rays and to determine their relative intensity.

Chapter XVII gives an account of the present state of our knowledge of the mass and structure of the nuclei of the common elements and their relation to the radioactive elements. The energy changes involved in the formation and destruction of atomic nuclei are also considered.

In chapter XVIII some miscellaneous matters are discussed. Among these are the radioactivity of matter in general, with special reference to the type of radioactivity shown by potassium and rubidium; the counting of scintillations; and, very briefly, the methods of separation of radioactive elements and the preparation of sources.

§ 2. Discovery of Radioactivity. The discovery of the radioactivity of uranium by Professor Henri Becquerel in February 1896 was in a sense a direct consequence of the discovery of X rays by Röntgen a few months earlier. It is of interest to consider in some detail the sequence of events leading to this discovery which was afterwards to prove of such fundamental importance. The remarkable properties of the X rays had excited intense interest throughout the scientific world, but for some time the cause and nature of these rays were a matter of conjecture. It occurred to several that the origin of the rays might be connected with the brilliant phosphorescence of the glass of the X ray tube which appeared to accompany the emission of X rays. Following out this idea, several investigators tried whether substances like calcium sulphide, which phosphoresced under ordinary light, gave out penetrating radiations of the X ray type. After several negative experiments of this kind, it occurred to Becquerel to investigate a uranium salt, the double sulphate of uranium and potassium, which he had prepared fifteen years before and had shown to give a brilliant phosphorescence under the action of ultra-violet light. After exposure to light, the salt was wrapped in black paper and placed below a photographic plate with a small plate of silver between. After several hours' exposure a distinct photographic effect was observed, indicating the emission from the salt of a radiation of penetrating type. Subsequent experiments showed that the photographic action was quite independent of the phosphorescence and was shown equally by all the salts of uranium and the metal itself. In the light of later knowledge, it is clear that the photographic effect was due to the penetrating β rays emitted by uranium, for the easily absorbed α rays did not penetrate the black paper.

The photographic effect of the rays from uranium is feeble, and long exposures are required to produce a marked effect. As Becquerel first showed, the rays from uranium, like X rays, possess the property of discharging electrified bodies. This electric effect of the radiations is due to the volume ionisation of the gas by the radiations. Under suitable conditions, this electrical effect provides a sensitive quantitative method of studying the radiations from active bodies.

As a result of study of the penetrating power of the radiations emitted by uranium, Rutherford concluded that the rays could be divided into two types, one called the alpha (α) rays which were very easily absorbed but produced an intense ionisation, and a much more penetrating type called the beta (β) rays. When a still more penetrating type of radiation was later found to be emitted from radium, Villard named them the gamma (γ) rays.

In order to ascertain whether radioactivity was a general property of matter, most of the elements were examined for activity by Mme Curie, using the electrical method. Thorium, the next heaviest element to uranium, was found by Schmidt and Mme Curie to show an activity comparable with that of uranium. Apart from uranium and thorium and their numerous transformation elements, no other elements were found to be radioactive. Subsequently, in 1907, N. R. Campbell found that both potassium and rubidium showed a weak activity due to the emission of swift β particles. The activity, however, is very small, corresponding to only about 1/1000 of the β ray activity of an equal weight of uranium. On account of their weak activity, it is difficult to study the radioactive properties of these elements in detail. No evidence has been obtained of any successive transformations in these elements as in the case of uranium and thorium.

While the activity shown by these comparatively light elements, so far removed in atomic weight from uranium and thorium, is a matter of some surprise, the evidence so far obtained indicates that they are true radioactive elements, in which the atoms break up with the emission of swift electrons. The β rays from potassium are more penetrating than those from rubidium, but the number of β particles emitted per second is more numerous in the latter.

In 1898, Mme Curie compared by the electric method the activities of different uranium and thorium compounds and of a number of minerals containing uranium and thorium. In the case of uranium minerals, for example pitchblende from Bohemia, the activity observed was about four times that of uranium metal. This was a surprising result, for other observations had shown that the activity of uranium was an atomic property, i.e. it was proportional to the amount of uranium present and independent of its combinations with other inactive elements. It thus seemed probable that the high activity of the mineral must be due to the presence of small quantities of an unknown element or elements of activity greater than uranium or thorium. Relying on this hypothesis, Mme Curie made a systematic analysis of a uranium mineral to find evidence of a new radioactive element. This work was completely successful and soon led to the discovery of two new and very active substances called polonium and radium, the latter in the pure state having an activity several million times that of uranium. This important discovery was only possible by using the activity of the elements as a guide in their separation and concentration. By means of the electric method, the changes in activity due to the chemical treatment could be followed quantitatively. In this respect, the discovery of these bodies is quite analogous to the detection of rare elements by the methods of spectrum analysis.

Radium is normally separated with barium by treatment of the radioactive mineral. It can be completely separated from the barium by fractional crystallisation of the chloride, or, better still, the bromide. In this way, Mme Curie obtained preparations of pure radium salts, and determined the atomic weight of radium to be 226·5. A more recent determination by Hönigschmid, using about 1 gram of radium, gave a value 226. Radium metal has been isolated by Mme Curie by electrolysis of the fused salt. The metal is white in colour, but rapidly tarnishes on exposure to the air. It melts at about 700° C.

While radium is only one of a large number of radioactive elements which appear in the successive transformations of uranium and thorium, it possesses a special importance in radioactive work. Not only is it readily separated in the pure state, but it gives rise to a radioactive gas originally called the radium emanation but later named radon, which in turn produces an active deposit consisting of radium A, radium B and radium C. Radon and its active deposit have proved invaluable in providing convenient and intense sources of radiation for the study of the α, β and γ rays. Radium itself emits α rays, but in equilibrium with its rapidly changing products is a convenient and constant source of β and γ rays. The α rays are usually absorbed either in the radioactive material itself or in the envelope containing it.

The discovery of radium gave an impetus to the systematic chemical examination of uranium minerals' and this soon led to the detection of several new radioactive bodies. The most important of these are actinium, found by Debierne and Giesel, and an active form of lead called radiolead by Hoffmann and Strauss. It is now known that the active constituent in the latter is radium D, a radioactive isotope of lead which is ultimately transformed into polonium.

Subsequently Boltwood discovered in uranium minerals a radioactive element called ionium. This is an isotope of thorium and is thus separated with any thorium that may be present in the mineral. Ionium is of special interest as it is the direct parent of radium. A preparation of ionium, initially freed from radium, is found to grow radium at a rapid rate.

An interesting stage in the study of radioactive substances was the proof by Rutherford in 1900 that thorium emitted a radioactive "emanation" or gas which could be carried from point to point by a current of air. A similar property is shown by radium and actinium. Subsequent researches have shown that these radioactive gases are chemically inert and belong to the group of monatomic gases discovered in the atmosphere. These radioactive gases have proved of great interest and importance in radioactive work. They all give rise to an active deposit which can be produced and studied independently of the active matter which gives rise to the emanation.

At this stage, a number of radioactive substances had been found, some of which, like uranium and radium, appeared to show a constant activity and others, like the emanations and their active deposits, lost their activity in a few hours or days. This complicated mass of facts was reduced to order by the application of the transformation theory advanced by Rutherford and Soddy in 1903. On this theory, the atoms of the radio-elements, unlike the atoms of the ordinary elements, are not stable but undergo spontaneous disintegration accompanied by the expulsion of an α or β particle. After the disintegration, the resulting atom has physical and chemical properties entirely different from the parent atom. It may in turn be unstable and pass through a succession of transformations each of which is characterised by the emission of an α or β particle.

These processes may be illustrated by considering the changes in radium. At any moment, a number of the atoms, of atomic mass 226, become unstable and break up with explosive violence, expelling an α particle with characteristic speed. Since the α particle is a helium atom of mass 4, the resulting atom is lighter than before and becomes an atom of a new substance radon, of atomic mass 222, which is a radioactive inert gas. This in turn breaks up with the liberation of an α particle and is converted into an atom of a non-gaseous element radium A, and so on through a long series of successive transformations.

The activity of each of these successive products is not permanent, but decays with time according to a definite law. The activity I after a time t is given by $I/I_0 = e^{-\lambda t}$, where I_0 is the initial activity and λ a constant of transformation characteristic of the product. This law of decay may be expressed in another form. If N_0 be the number of atoms initially, the number N remaining after a time t is given by $N/N_0 = e^{-\lambda t}$, where λ is the same constant. In other words, the number of atoms which have escaped disintegration decreases exponentially with the time. Since $-\dfrac{dN}{dt} = \lambda N_0 e^{-\lambda t} = \lambda N$, the rate of disintegration is proportional to the number of atoms present and is equal to this number multiplied by the radioactive constant. This law of transformation applies universally to all radioactive products, but the constant λ is different for each product. The value of λ for any substance is a characteristic constant independent of all physical and chemical conditions. It is, for example, independent of temperature, of concentration and of the age of the atoms. If each atom has an equal chance of breaking up in a given interval, it is to be expected that the number disintegrating per second will be subject to the laws of chance and thus undergo fluctuations round the mean value. This important question is discussed in § 34.

Since $N/N_0 = e^{-\lambda t}$, the time T to reach half value, sometimes called the "period" of the substance, is given by $T = 0.693/\lambda$. The life of an atom, i.e. the time it can exist before transformation occurs, has theoretically all possible values from 0 to ∞. In practice, however, it is convenient to speak of the average life of a large number of atoms. This has a definite value which can be calculated. Suppose N_0 atoms are present initially. After a time t, the number transformed in a time dt is $\lambda N dt$ or $\lambda N_0 e^{-\lambda t} dt$. Each of these atoms has a life t, so that the average life of the whole number is given by $\int_0^\infty \lambda t e^{-\lambda t} dt = 1/\lambda$. The average life of an atom is consequently measured by the reciprocal of the radioactive constant.

To illustrate the processes of production and disappearance of a radio-element, we shall take the simple case of the production of radon by radium. The half value period of radium, 1600 years, is so long compared with the period of radon, 3·82 days, that we may suppose that radium is transformed at a constant rate during the relatively short time—a few days—under consideration. Suppose at first that the radium has been undisturbed for some time and that the amount of radon associated with it has reached an equilibrium value. Let P_0, Q_0 be the number of atoms of radium and radon in equilibrium and λ_1, λ_2 their constants of change. In equilibrium the number of atoms of radon formed per second from radium is equal to the number of radon atoms which break up per second, or $\lambda_1 P_0 = \lambda_2 Q_0$. Consequently $Q_0 = \dfrac{\lambda_1}{\lambda_2} P_0$, where Q_0 represents the number of atoms of radon in equilibrium with P_0 atoms of radium. Suppose now, either by solution or heating, the whole of the radon is removed. At the end of a time t, the number of its atoms remaining is $Q_0 e^{-\lambda_2 t}$, i.e. $Q_0(1 - e^{-\lambda_2 t})$ have disappeared. Since, however, the rate of the radioactive processes has not been changed by the mere removal of the radon, it is clear that to keep up the equilibrium quantity Q_0, $Q_0(1 - e^{-\lambda_2 t})$ new atoms of radon must have been formed by the radium in the time t. The amount of radon Q existing at a time t with the radium is thus given by $Q/Q_0 = 1 - e^{-\lambda_2 t}$. In Fig. 1, the decay curve of radon and the recovery curve are shown, where T is the period of radon and Q_0 is taken as unity. It is seen that the two curves are complementary to one another, the sum of the ordinates of the two curves being constant and equal to 1 in the figure. This argument applies to all transformations when the parent product has a period long compared with the succeeding product. The more formal proof is included in the general theory which we shall now consider.

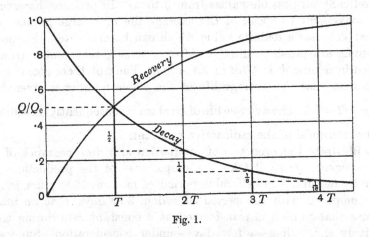

Fig. 1.

§ 3. Theory of successive transformations.

The radio-elements uranium, thorium and actinium undergo a long series of transformations, and it is often important to know the variation with time of the quantity or activity of a particular product under given initial conditions.

Suppose that P, Q, R, ... represent the number of atoms of the successive elements A, B, C, ... respectively at any time t. Let λ_1, λ_2, λ_3, ... be the radioactive constants of A, B, C, ... respectively. Each atom of A gives rise to one atom of B, one of B to one of C, and so on. The expelled α or β particles are non-radioactive and so do not enter directly into the calculation.

It is not difficult to deduce mathematically the number of atoms of A, B, C at any subsequent time if the initial values of P, Q and R are given. In practice, however, it is generally only necessary to consider three special cases of the theory which correspond, for example, to the changes in the active deposit, produced on a plate exposed to a constant amount of radon and then removed, (1) when the time of exposure is extremely short compared with the period of the changes, (2) when the time of exposure is so long that the amount of each of the products has reached a steady limiting value, and (3) for any time of exposure.

There is also another case of importance which is practically a converse of Case 2, viz. when the matter A is supplied at a constant rate from a primary source and the amounts of A, B, C are required at any subsequent time. The solution of this can, however, be deduced immediately from Case 2 without analysis.

CASE 1. *Suppose that the matter initially considered is all of one kind A. It is required to find the number of particles P, Q, R, S, ... of the matter A, B, C, D, ... respectively present after any time t.*

Then $P = P_0 e^{-\lambda_1 t}$, if P_0 is the number of particles of A initially present. Now dQ, the increase of the number of particles of the matter B per unit time, is the number supplied by the change of the matter A, less the number due to the change of B into C, thus

$$dP/dt = - \lambda_1 P \quad\dots\dots\dots\dots\dots\dots\dots\dots(1),$$
$$dQ/dt = \lambda_1 P - \lambda_2 Q \quad\dots\dots\dots\dots\dots\dots(2).$$

Similarly
$$dR/dt = \lambda_2 Q - \lambda_3 R \quad\dots\dots\dots\dots\dots\dots(3).$$

Substituting in (2) the value of P in terms of P_0,

$$dQ/dt = \lambda_1 P_0 e^{-\lambda_1 t} - \lambda_2 Q.$$

The solution of this equation is of the form

$$Q = P_0 (a e^{-\lambda_1 t} + b e^{-\lambda_2 t}) \quad\dots\dots\dots\dots\dots(4).$$

By substitution it is found that $a = \lambda_1/(\lambda_2 - \lambda_1)$.

Since $Q = 0$ when $t = 0$, $b = - \lambda_1/(\lambda_2 - \lambda_1)$.

Thus
$$Q = \frac{P_0 \lambda_1}{\lambda_2 - \lambda_1} (e^{-\lambda_1 t} - e^{-\lambda_2 t}) \dots\dots\dots\dots\dots\dots(5).$$

Substituting this value of Q in (3), it can readily be shown that

$$R = P_0 (a e^{-\lambda_1 t} + b e^{-\lambda_2 t} + c e^{-\lambda_3 t}) \quad\dots\dots\dots\dots(6),$$

where

$$a = \frac{\lambda_1 \lambda_2}{(\lambda_2 - \lambda_1)(\lambda_3 - \lambda_1)}, \qquad b = \frac{\lambda_1 \lambda_2}{(\lambda_1 - \lambda_2)(\lambda_3 - \lambda_2)},$$
$$c = \frac{\lambda_1 \lambda_2}{(\lambda_1 - \lambda_3)(\lambda_2 - \lambda_3)}.$$

Similarly it can be shown that

$$S = P_0 (a e^{-\lambda_1 t} + b e^{-\lambda_2 t} + c e^{-\lambda_3 t} + d e^{-\lambda_4 t}) \quad\dots\dots\dots(7),$$

where

$$a = \frac{\lambda_1 \lambda_2 \lambda_3}{(\lambda_2 - \lambda_1)(\lambda_3 - \lambda_1)(\lambda_4 - \lambda_1)}, \qquad b = \frac{\lambda_1 \lambda_2 \lambda_3}{(\lambda_1 - \lambda_2)(\lambda_3 - \lambda_2)(\lambda_4 - \lambda_2)},$$
$$c = \frac{\lambda_1 \lambda_2 \lambda_3}{(\lambda_1 - \lambda_3)(\lambda_2 - \lambda_3)(\lambda_4 - \lambda_3)}, \qquad d = \frac{\lambda_1 \lambda_2 \lambda_3}{(\lambda_1 - \lambda_4)(\lambda_2 - \lambda_4)(\lambda_3 - \lambda_4)}.$$

The method of solution of the general case of n products has been given in a symmetrical form by Bateman*. The amount of the nth product $N(t)$ at the time t is given by

$$N(t) = c_1 e^{-\lambda_1 t} + c_2 e^{-\lambda_2 t} + \ldots + c_n e^{-\lambda_n t} \quad \ldots\ldots\ldots\ldots(8),$$

where

$$c_1 = \frac{\lambda_1 \lambda_2 \ldots \lambda_{n-1} P_0}{(\lambda_2 - \lambda_1)(\lambda_3 - \lambda_1) \ldots (\lambda_n - \lambda_1)},$$

$$c_2 = \frac{\lambda_1 \lambda_2 \ldots \lambda_{n-1} P_0}{(\lambda_1 - \lambda_2)(\lambda_3 - \lambda_2) \ldots (\lambda_n - \lambda_2)},$$

$$\ldots\ldots\ldots\ldots\text{etc.}$$

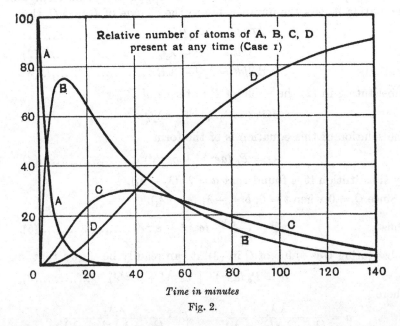

Fig. 2.

The variations in the values of P, Q, R, S with time t after removal of the source are shown graphically in Fig. 2, curves A, B, C and D respectively. The curves are drawn for the practical and important case of the first four products of the active deposit of radium, known as radium A, B, C and D. The matter is supposed to consist initially only of radium A. This corresponds to the case of a body exposed for a few seconds in the presence of radon. The values of $\lambda_1, \lambda_2, \lambda_3$ are taken as $3\cdot79 \times 10^{-3}$, $4\cdot31 \times 10^{-4}$, $5\cdot86 \times 10^{-4}$ (sec.)$^{-1}$ respec-

* Bateman, *Proc. Camb. Phil. Soc.* **15**, 423, 1910.

tively corresponding to the half value periods of A, B and C of 3·05, 26·8 and 19·7 minutes respectively. The half value period of radium D is about 25 years. Over the small interval under consideration in the figure, one may suppose that the atoms of radium D suffer no appreciable decrease by transformation.

The ordinates of the curves represent the relative number of atoms of the matter A, B, C and D existing at any time, and the value of P_0, the original number of atoms of the matter A deposited, is taken as 100. The amount of matter B is initially zero, and in this particular case passes through a maximum about 10 minutes later, and then diminishes with the time. In a similar way, the amount of C passes through a maximum about 35 minutes after removal. After an interval of several hours the amount of both B and C diminishes very approximately according to an exponential law with the time, falling to half value in both cases in 26·8 minutes.

Over the interval considered, the amount of D increases steadily with time, although very slowly at first. A maximum is reached when B and C have disappeared. Finally the amount of D would decrease exponentially with the time corresponding to a period of 25 years.

CASE 2. *A primary source supplies the matter A at a constant rate, and the process has continued so long that the amount of the products A, B, C, ... has reached a steady limiting value. The primary source is then suddenly removed. It is required to find the amounts of A, B, C, ... remaining at any subsequent time t.*

In this case, the number n_0 of particles of A, deposited per second from the source, is equal to the number of particles of A which change into B per second, and of B into C, and so on. This requires the relation

$$n_0 = \lambda_1 P_0 = \lambda_2 Q_0 = \lambda_3 R_0 \quad\dots\dots\dots\dots\dots(9),$$

where P_0, Q_0, R_0 are the maximum numbers of particles of the matter A, B and C when a steady state is reached.

The values of Q, R and S at any time t after removal of the source are given by equations of the same form as (4), (6) and (7) for a short exposure. Remembering the condition that initially

$$P = P_0 = n_0/\lambda_1,$$
$$Q = Q_0 = n_0/\lambda_2,$$
$$R = R_0 = n_0/\lambda_3,$$

it can readily be shown that

$$P = \frac{n_0}{\lambda_1} e^{-\lambda_1 t} \quad \dots\dots\dots\dots\dots\dots\dots\dots\dots\dots(10),$$

$$Q = \frac{n_0}{\lambda_1 - \lambda_2} \left(\frac{\lambda_1}{\lambda_2} e^{-\lambda_2 t} - e^{-\lambda_1 t} \right) \quad \dots\dots\dots\dots\dots(11),$$

$$R = n_0 \left(a e^{-\lambda_1 t} + b e^{-\lambda_2 t} + c e^{-\lambda_3 t} \right) \quad \dots\dots\dots\dots(12),$$

where

$$a = \frac{\lambda_2}{(\lambda_2 - \lambda_1)(\lambda_3 - \lambda_1)}, \qquad b = \frac{\lambda_1}{(\lambda_1 - \lambda_2)(\lambda_3 - \lambda_2)},$$

$$c = \frac{\lambda_1 \lambda_2}{\lambda_3 (\lambda_1 - \lambda_3)(\lambda_2 - \lambda_3)}.$$

Similarly for four changes it can be shown that

$$S = n_0 \left(a e^{-\lambda_1 t} + b e^{-\lambda_2 t} + c e^{-\lambda_3 t} + d e^{-\lambda_4 t} \right) \quad \dots\dots\dots(13),$$

where

$$a = \frac{\lambda_2 \lambda_3}{(\lambda_2 - \lambda_1)(\lambda_3 - \lambda_1)(\lambda_4 - \lambda_1)}, \qquad b = \frac{\lambda_1 \lambda_3}{(\lambda_1 - \lambda_2)(\lambda_3 - \lambda_2)(\lambda_4 - \lambda_2)},$$

$$c = \frac{\lambda_1 \lambda_2}{(\lambda_1 - \lambda_3)(\lambda_2 - \lambda_3)(\lambda_4 - \lambda_3)}, \qquad d = \frac{\lambda_1 \lambda_2 \lambda_3}{\lambda_4 (\lambda_1 - \lambda_4)(\lambda_2 - \lambda_4)(\lambda_3 - \lambda_4)}.$$

Bateman has pointed out that the solutions for Case 2 can be derived from Case 1. This is obvious when it is remembered that the amount of C, for example, remaining after a definite interval t is made up of (1) supply from A through B, (2) supply from B, (3) part of C remaining. Remembering that the amounts of A, B, C initially present are n_0/λ_1, n_0/λ_2, n_0/λ_3 respectively, the solution can be written down with the aid of equations (5) and (6).

The relative numbers of atoms of A, B, C existing at any time are shown graphically in Fig. 3, curves A, B, C respectively. The number of atoms Q_0 is taken as 100 for comparison, and the values of λ_1, λ_2, λ_3 are taken corresponding to the 3·05, 26·8, and 19·7 minute changes in the active deposit of radium. It should be pointed out that the curves show the variation of P, Q and R on removal of a body which has been exposed for a long interval to a *constant* amount of emanation. The curves are slightly different for the practical case where the body is exposed in a *decaying* source of emanation.

A comparison with Fig. 2 for a short exposure brings out very clearly the variation in the relative amounts of A, B, C corresponding to Cases 1 and 2. In Case 2 the amount of C decreases at first very slowly. This is a result of the fact that the supply of C due to the breaking up of B at first nearly compensates for the breaking up of C. The values of Q and R after several hours decrease exponentially, falling to half value in 26·8 minutes.

A consideration of the formulae for Cases 1 and 2 brings out the interesting point that the amount of C ultimately decreases according to an exponential law with the period of 26·8 minutes, which is

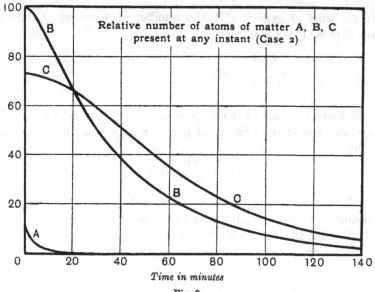

Relative number of atoms of matter A, B, C present at any instant (Case 2)

Time in minutes

Fig. 3.

characteristic of radium B and not of radium C. This is an expression of a general result that the product of longest period ultimately governs the decay curve in all cases.

CASE 3. *Suppose that a primary source has supplied the matter A at a constant rate for any time T and is then suddenly removed. Required the amounts of A, B, C at any subsequent time.*

Suppose that n_0 particles of the matter A are deposited each second. After a time of exposure T, the number of particles P_T of the matter A present is given by

$$P_T = n_0 \int_0^T e^{-\lambda_1 t}\, dt = \frac{n_0}{\lambda_1}\left(1 - e^{-\lambda_1 T}\right).$$

At any time t, after removal of the source, the number of particles P of the matter A is given by

$$P = P_T e^{-\lambda_1 t} = \frac{n_0}{\lambda_1}\left(1 - e^{-\lambda_1 T}\right) e^{-\lambda_1 t} \quad \ldots\ldots\ldots\ldots(14).$$

Consider the number of particles $n_0 dt$ of the matter A produced during the interval dt. At any later time t, the number of particles dQ of the matter B, which result from the change in A, is given (see equation 5) by

$$dQ = \frac{n_0 \lambda_1}{\lambda_1 - \lambda_2} (e^{-\lambda_2 t} - e^{-\lambda_1 t}) \, dt = n_0 f(t) \, dt \quad \ldots\ldots(15).$$

After a time of exposure T, the number of particles Q_T of the matter B present is given by

$$Q_T = n_0 \left[f(T) dt + f(T - dt) \, dt + \ldots\ldots + f(0) dt \right]$$

$$= n_0 \int_0^T f(t) \, dt.$$

If the body is removed from the emanation after an exposure T, at any later time t the number of particles of B is in the same way given by

$$Q = n_0 \int_t^{T+t} f(t) \, dt.$$

It will be noted that the method of deduction of Q_T and Q is independent of the particular form of the function $f(t)$.

Substituting the particular value of $f(t)$ given in equation (15) and integrating, it can readily be deduced that

$$\frac{Q}{Q_T} = \frac{ae^{-\lambda_2 t} - be^{-\lambda_1 t}}{a - b} \quad \ldots\ldots\ldots\ldots\ldots\ldots(16),$$

where
$$a = \frac{1 - e^{-\lambda_2 T}}{\lambda_2}, \qquad b = \frac{1 - e^{-\lambda_1 T}}{\lambda_1}.$$

In a similar way, the number of particles R of the matter C present at any time can be deduced by substitution of the value of $f(t)$ in equation (6).

$$\frac{R}{R_T} = \frac{ae^{-\lambda_1 t} + be^{-\lambda_2 t} + ce^{-\lambda_3 t}}{a + b + c} \quad \ldots\ldots\ldots\ldots(17),$$

where
$$a = \frac{\lambda_2}{(\lambda_2 - \lambda_1)(\lambda_3 - \lambda_1)} (1 - e^{-\lambda_1 T}),$$

$$b = \frac{\lambda_1}{(\lambda_1 - \lambda_2)(\lambda_3 - \lambda_2)} (1 - e^{-\lambda_2 T}),$$

$$c = \frac{\lambda_1 \lambda_2}{\lambda_3 (\lambda_1 - \lambda_3)(\lambda_2 - \lambda_3)} (1 - e^{-\lambda_3 T}).$$

In a similar way the amount of any product may be written down.

CASE 4. *The matter A is supplied at a constant rate from a primary source. Required to find the number of particles of A, B, C at any subsequent time t, when initially A, B, C are absent.*

The solution can be simply obtained in the following way. Suppose that the conditions of Case 2 are fulfilled. The products A, B, C are in radioactive equilibrium and let P_0, Q_0, R_0 be the number of particles of each present. Suppose the source is removed. The values of P, Q, R at any subsequent time are given by equations (10), (11) and (12) respectively. Now suppose the source, which has been removed, still continues to supply A at the same constant rate and let P_1, Q_1, R_1 be the number of particles of A, B, C again present with the source at any subsequent time. Now we have seen that the rate of change of any individual product, considered by itself, is independent of conditions and is the same whether the matter is mixed with the parent substance or removed from it. Since the values of P_0, Q_0, R_0 represent a steady state where the rate of supply of each kind of matter is equal to its rate of change, the sum of the number of particles A, B, C present at any time with the source, and in the matter from which it was removed, must at all times be equal to P_0, Q_0, R_0, that is

$$P_1 + P = P_0,$$
$$Q_1 + Q = Q_0,$$
$$R_1 + R = R_0.$$

This must obviously be the case, for otherwise there would be a destruction or creation of matter by the mere process of separation of the source from its products; but, by hypothesis, neither the rate of supply from the source, nor the law of change of the products, has been in any way altered by removal.

Substituting the values of P, Q, R from equations (10), (11), and (12), we obtain

$$\frac{P_1}{P_0} = 1 - e^{-\lambda_1 t} \quad\dots\dots\dots\dots\dots\dots\dots\dots\dots\dots(18),$$

$$\frac{Q_1}{Q_0} = 1 - (\lambda_1 e^{-\lambda_2 t} - \lambda_2 e^{-\lambda_1 t})/(\lambda_1 - \lambda_2) \quad\dots\dots\dots\dots(19),$$

$$\frac{R_1}{R_0} = 1 - \lambda_3 (ae^{-\lambda_1 t} + be^{-\lambda_2 t} + ce^{-\lambda_3 t}) \quad\dots\dots\dots\dots(20),$$

$$\dots\dots\dots\dots\dots\dots\text{etc.},$$

where a, b, and c have the values given after equation (12). The curves representing the increase of P, Q, R are thus, in all cases, complementary to the curves shown in Fig. 3. The sum of the ordinates of thé two curves of rise and decay at any time is equal to 100.

§ 4. Secular and transient equilibria. The theory of Cases 2 and 4 has been worked out on the assumption that there is a permanent equilibrium between the successive products of transformation. This is impossible to realise completely in practice, since the amount of every radioactive substance is always decreasing with time. No sensible error, however, is introduced when the *primary* source is transformed so slowly that there is no appreciable change in its amount in an interval of time required for the later products to attain approximate equilibrium with the primary source. This condition is very nearly fulfilled, for example, in the case of radium and its product radon, where the period of the former is 1600 years, and of the latter 3·82 days. The latter approaches its equilibrium value very closely after the radon has been supplied continuously from the radium for an interval of 2 months. During this time, the fraction of the radium transformed is only about 7/100,000, so that for the interval under consideration it may be regarded as a constant source without sensible error. It is convenient to apply the term "secular" equilibrium to this and similar cases.

Consider next the important case of radon and its products, radium A, B and C. A stage of equilibrium between radon and its products is reached after the radon has been stored about 5 hours, and the amount of each of the products finally decays exponentially with the period of radon. This is a case of "transient" equilibrium, for the amounts of the products are changing comparatively rapidly. The amount of radium A, B or C at any subsequent time is always appreciably greater than the amount for secular equilibrium when the supply of radon is kept constant.

Consider, for example, the case of radium C. Using the notation of § 3, the number of atoms S of radium C at any time t is given by

$$S = ae^{-\lambda_1 t} + be^{-\lambda_2 t} + ce^{-\lambda_3 t} + de^{-\lambda_4 t},$$

where λ_1, λ_2, λ_3, λ_4 are the constants of transformation of radon, radium A, B and C respectively. Since the value of λ_1 is much smaller than λ_2, λ_3 or λ_4, by making t very large only the first term becomes important, i.e. when t is large

$$S = \frac{\lambda_1 \lambda_2 \lambda_3 P_0}{(\lambda_2 - \lambda_1)(\lambda_3 - \lambda_1)(\lambda_4 - \lambda_1)} e^{-\lambda_1 t} \dots\dots\dots\dots(21).$$

For the instant t, the amount S_0 of radium C which would be in secular equilibrium with the radon is given by

$$\lambda_1 P_0 e^{-\lambda_1 t} = \lambda_4 S_0.$$

Consequently $\dfrac{S}{S_0} = \dfrac{\lambda_2 \lambda_3 \lambda_4}{(\lambda_2 - \lambda_1)(\lambda_3 - \lambda_1)(\lambda_4 - \lambda_1)} = 1 \cdot 0089$

when the values of λ are substituted in the equation.

This shows that the amount of radium C present is $0 \cdot 89$ per cent. *greater* than corresponds to secular equilibrium. For example, consider a certain quantity of radon existing alone by itself and an equal quantity of radon associated with the amount of radium with which it is in equilibrium. The amount of radium C in transient equilibrium with the radon in the first case is $0 \cdot 89$ per cent. greater than in the latter. Now the γ radiation (after passing through 2 cm. of lead) from radium in equilibrium or from a tube filled with radon arises mainly from radium C. If the γ radiation from the tube containing radon is compared with that due to a standard radium preparation through 2 cm. of lead, the amount of radon actually present in the tube at the time of observation is obviously $0 \cdot 89$ per cent. *less* than that deduced by direct measurements of the γ ray effects.

In a similar way, it can be shown that the amount of radium A and radium B present are $0 \cdot 054$ and $0 \cdot 54$ per cent. greater respectively than the true equilibrium amount.

It is clear from these considerations that the decay curve of the active deposit after a long exposure to a source of radon stored by itself does not follow exactly the theory given in § 3, Case 2, for the relative ratios of radium A, B and C above the true equilibrium are as $1 \cdot 00054 : 1 \cdot 0054 : 1 \cdot 0089$. It is obvious, however, that the differences from the theoretical curves given in Fig. 3 will be small.

§ 5. **Applications to some practical cases.** We shall now consider a few practical examples to illustrate the application of the theory to follow changes in activity under given initial conditions, paying special attention to some cases which are of importance in obtaining sources for experiment on the α, β and γ rays.

First consider the changes of activity due to a body which has been exposed to a constant supply of radon for a sufficient time to reach equilibrium. On removal, the active matter consists of radium A, radium B and radium C in equilibrium, so that the same number of atoms of each break up per second. The succession of changes in the active deposit is shown below. The periods and radiations of the products are added.

$$\text{Radium A} \xrightarrow{\alpha} \text{Radium B} \xrightarrow{\beta + \gamma} \text{Radium C} \xrightarrow{\alpha + \beta + \gamma}$$
$$\text{3·05 min.} \qquad \text{26·8 min.} \qquad \text{19·7 min.}$$

The variation with time of the relative number of atoms of A, B and C have been shown in Fig. 3. Radium A emits α rays, radium B only β and γ rays, while, in consequence of the complex changes occurring in radium C, this product may be supposed for this purpose to emit α, β and γ rays.

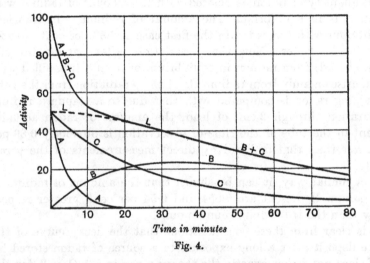

Fig. 4.

The number of atoms P, Q, R of radium A, B and C with constants $\lambda_1, \lambda_2, \lambda_3$ respectively have been given in equations (10), (11) and (12). The activity measured by the number of α particles emitted per second is equal to $\lambda_1 P + \lambda_3 R$ and can thus be calculated. Obviously, at the instant of removal, the same number of α particles is emitted per second from radium A and from radium C. Taking the total number initially to be 100, the decay curves are shown in Fig. 4. The curve AA represents the decay of activity due to radium A alone, $A + B + C$ due to radium A and radium C together, while the curve $LL . B + C$ which meets the axis of ordinates at 50, gives the α ray activity due to radium C. The activity due to radium C alone decays slowly at first, reaches half value in 55 minutes and ultimately decays exponentially according to the period of radium B, viz. half value in 26·8 minutes. This is an illustration of the general fact that the product of longest period always finally controls the rate of decay of the bodies that follow.

The curve $B + C$ is obviously made up of two components, one due to an exponential decay of the radium C present at the moment of removal represented by the curve CC, and the other to the fresh production of radium C, arising from the subsequent transformation of radium B into radium C represented by the curve BB. The results would be very different if we measured the change of activity by means of the β and γ rays. Assuming one β ray is emitted per atom of radium B and of radium C, the number of β rays per second is $\lambda_2 Q + \lambda_3 R$. Since, however, the β rays from radium B and radium C vary very widely in penetrating power, the activity measured in an electroscope would be proportional to $\lambda_2 Q + K\lambda_3 R$, where Q and R are given by equations (11) and (12) and K is a constant representing the relative ionisation effect in the electroscope of the average β particle emitted by radium C compared with that emitted by radium B.

A similar argument applies if the decay is measured by γ rays. The γ rays emitted by radium B are on the average much less intense and penetrating than the γ rays from radium C. Moseley and Makower showed that if the activity is measured through a lead sheet 2 cm. thick, practically all the γ rays from radium B are stopped and the γ ray activity falls off according to the amount of radium C shown by the curve $B + C$ in Fig. 4. This is an important result, for the γ ray effect after passing through 2 cm. of lead affords a direct measure of the amount of radium C present. Since the penetrating γ rays from radium in equilibrium are due to its product radium C, the activity of the radium C source can thus be expressed in terms of the γ ray activity shown by a given weight of radium element in equilibrium.

Tables of the decay of radium A, radium B, and radium C after a long exposure to a source of radon have been given by Meyer and Schweidler in their book *Radioaktivität*, p. 437 (2nd Edit., Teubner, Berlin).

If the radium C is at any time separated from the active deposit, for example by dipping a nickel plate in a solution of the deposit, the activity, whether measured by the α, β or γ rays, decays exponentially with a period of 19·7 minutes.

To illustrate the variation of activity with time, we shall consider also the decay of the active deposit due to a very short exposure—

a few seconds—to radon. In this case, the α ray activity at the moment of removal is due to radium A and subsequently to radium A and radium C together. Curve AA, Fig. 5, represents the activity due to A which decays exponentially with a period of 3·05 minutes. In order to show clearly in the figure the relatively small activity due to C, the activity due to A is plotted after an interval of 6 minutes when the activity has been reduced to 25 per cent. of its maximum value. The activity due to C is proportional to $\lambda_3 R$ (equation (6)),

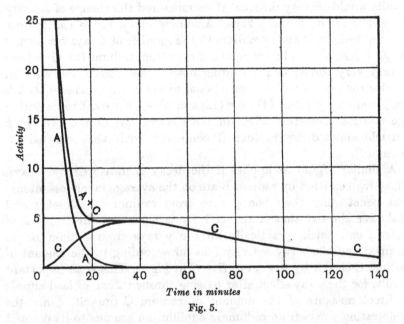

Fig. 5.

and in order to represent the activity due to C on the same scale as A it is necessary to reduce the scale of the ordinates of curve CC in Fig. 2 in the ratio λ_3/λ_1. The activity due to C is thus represented by curve CC, Fig. 5, and the total α activity by curve $A + C$ whose ordinates are the sum of the ordinates of curves A and C. The activity due to C passes through a maximum after about 35 minutes.

If the activity were measured by the γ rays through 2 cm. of lead, it would follow the curve CC.

There is another interesting case which has proved of importance in experiments to test whether α rays in traversing matter can excite penetrating γ rays (§ 111). A quantity of radon is suddenly introduced into a closed tube. It is required to find the variation in

the amount of radium C for the first few minutes after the introduction of the radon. Obviously the amount R_1 of radium C, after a
time t, is given by equation (20), where R_0 is the ultimate equilibrium
value. The rise of activity during the first 10 minutes is shown by
the curve E in Fig. 6. This corresponds to the rise of γ ray activity
measured through 2 cm. of lead. It is seen that the amount of
radium C after 4 minutes is less than 1/500 of the ultimate maximum.
The curve F gives the γ ray effect due to both radium B and C

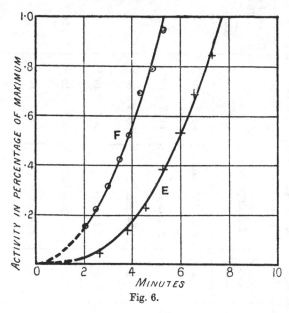

Fig. 6.

together under conditions when the absorbing screens corresponded
to 3 mm. of lead. The effect of the γ rays from radium B is most
markedly shown in experiments of this kind.

§ 6. Radioactive products. The elements produced by the successive transformations of the three radioactive elements, uranium,
thorium, and actinium, are given in the following table. In addition
to the products in the main line of descent, there are several side
or branch products produced in relatively small amount to which
reference will be made later (§§ 7 and 8). In a number of cases,
individual products have been separated by direct chemical or electrochemical methods; in other cases by the use of the powerful method
of radioactive recoil. In the case of uranium, the existence of an

TABLE OF ELEMENTS

It will be observed that in the last column the end products of the lines of a branch are indicated separately. While on the scheme of changes outlined the resulting element from each branch has the same atomic number, there is no definite evidence to decide whether the nuclei have identical structures. Our information on this interesting question is very uncertain. The difficulties involved are considered later in §§ 7 and 8. For three elements—U II, Ac C', and Th C'—the periods of transformation have not been measured but have been estimated from the ranges of the α particles, using the Geiger-Nuttall relation (§ 35). For radium C, the α rays have not been observed but are inferred from the intensity of the recoil of radium C''.

Element	Atomic weight	Atomic number	Type of disin-tegration	T (half period)	λ sec.$^{-1}$	Range of α 15° C. 760 mm. air (cm.)	Sequence
Uranium I U I	238·18	92	α	$4\cdot5 \times 10^{9}$ years	$4\cdot9 \times 10^{-18}$	2·70	
Uranium X₁ U X₁	(234)	90	β	24·5 days	$3\cdot275 \times 10^{-7}$	—	
Uranium X₂ U X₂	(234)	91	β	1·14 min.	$1\cdot013 \times 10^{-2}$	—	
Uranium Z U Z	(234)	91	β	6·7 hours	$2\cdot87 \times 10^{-5}$	—	
Uranium II U II	(234)	92	α	ca. 10^{6} years(?	ca. 2×10^{-14}	3·28	
Uranium Y U Y	(?)	90	β	24·6 hours	$7\cdot83 \times 10^{-6}$	—	
Ionium Io	(230)	90	—	—	—	—	
Protactinium Pa	(231)	91	—	—	—	—	

Ionium Io	(230)	90	α	$7 \cdot 6 \times 10^4$ years	$2 \cdot 9 \times 10^{-13}$	3·194
Radium Ra	225·97	88	α	1600 years	$1 \cdot 373 \times 10^{-11}$	3·389
Radon Rn	(222)	86	α	3·825 days	$2 \cdot 097 \times 10^{-6}$	4·122
Radium A Ra A	(218)	84	α	3·05 min.	$3 \cdot 79 \times 10^{-3}$	4·722
Radium B Ra B	(214)	82	β	26·8 min.	$4 \cdot 31 \times 10^{-4}$	—
†Radium C Ra C	(214)	83	αβ	19·7 min.	$5 \cdot 86 \times 10^{-4}$	(?)
Radium C' Ra C'	(214)	84	α	ca. 10^{-6} sec.	ca. 10^6	6·971
Radium C'' Ra C''	(210)	81	β	1·32 min.	$8 \cdot 75 \times 10^{-3}$	—
Radium D Ra D	(210)	82	β	22 years	$1 \cdot 00 \times 10^{-9}$	—
Radium E Ra E	(210)	83	β	5·0 days	$1 \cdot 60 \times 10^{-6}$	—
Radium F Ra F	(210)	84	α	136·3 days	$5 \cdot 886 \times 10^{-8}$	3·925
Radium G Ra G	206·05	82	Stable	—	—	—

† The α rays from radium C have recently been detected and found to be complex. The main group of α particles has a range about 4·14 cm. (Rutherford, F. A. B. Ward and C. E. Wynn Williams, *Proc. Roy. Soc.* A, 129, 211, 1930.)

Element	Atomic weight	Atomic number	Type of disintegration	T (half period)	λ sec.$^{-1}$	Range of α 15° C. 760 mm. air (cm.)	Sequence
Protactinium Pa	(231)	91	α	$1 \cdot 25 \times 10^{4}$ years	$1 \cdot 80 \times 10^{-12}$	3·673	
Actinium Ac	(227)	89	β	13·4 years	$1 \cdot 64 \times 10^{-9}$	—	
Radioactinium Rd Ac	(227)	90	α	18·9 days	$4 \cdot 24 \times 10^{-7}$	4·676	
Actinium X Ac X	(223)	88	α	11·2 days	$7 \cdot 16 \times 10^{-7}$	4·369	
Actinon An	(219)	86	α	3·92 sec.	0·177	5·789	
Actinium A Ac A	(215)	84	α	$2 \cdot 0 \times 10^{-3}$ sec.	$3 \cdot 5 \times 10^{2}$	6·584	
Actinium B Ac B	(211)	82	β	36·0 min.	$3 \cdot 21 \times 10^{-4}$	—	
Actinium C Ac C	(211)	83	$\alpha\beta$	2·16 min.	$5 \cdot 35 \times 10^{-3}$	5·511	
Actinium C' Ac C'	(211)	84	α	ca. 5×10^{-3} sec.(?)	ca. $1 \cdot 4 \times 10^{2}$	6·5	
Actinium C'' Ac C''	(207)	81	β	4·76 min.	$2 \cdot 43 \times 10^{-3}$	—	
Actinium D Ac D	(207)	82	Stable	—	—	—	

Thorium Th	232·12	90	α	$1\cdot65 \times 10^{10}$ years	$1\cdot33 \times 10^{-18}$	2·90
Mesothorium 1 Ms Th 1	(228)	88	β	6·7 years	$3\cdot28 \times 10^{-9}$	—
Mesothorium 2 Ms Th 2	(228)	89	β	6·13 hours	$3\cdot14 \times 10^{-5}$	—
Radiothorium Rd Th	(228)	90	α	1·90 years	$1\cdot16 \times 10^{-8}$	4·019
Thorium X Th X	(224)	88	α	3·64 days	$2\cdot20 \times 10^{-6}$	4·354
Thoron Tn	(220)	86	α	54·5 sec.	$1\cdot27 \times 10^{-2}$	5·063
Thorium A Th A	(216)	84	α	0·145 sec.	4·78	5·683
Thorium B Th B	(212)	82	β	10·6 hours	$1\cdot82 \times 10^{-5}$	—
Thorium C Th C	(212)	83	αβ	60·5 min.	$1\cdot91 \times 10^{-4}$	4·787
Thorium C' Th C'	(212)	84	α	ca. 10^{-11} sec. (?)	10^{11}	8·617
Thorium C'' Th C''	(208)	81	β	3·20 min.	$3\cdot61 \times 10^{-3}$	—
Thorium D Th D	207·77	82	Stable	—	—	—

element called uranium II has been inferred from the observed complexity of the α radiation. In successive columns are given the atomic weights and atomic numbers of the elements, the nature of the radiation emitted, the time period T required for the element to be half transformed, the constant of transformation λ, and the range of the α rays measured in air at 15° C. and 760 mm.

The system of nomenclature is not ideal. In some cases intermediate products have been discovered after a system of nomenclature has been accepted, and have had to be named to indicate as far as possible their positions in the series. As the radioactive emanations are now known to be isotopic elements, the names radon, thoron and actinon have been suggested as distinctive and indicative of the similarity of these products. The successive products following the emanations are in all cases denoted by the letters A, B, C, etc.

It will be seen that the value of T, which is a measure of the relative stability of atoms, varies over an enormous range from $1·65 \times 10^{10}$ years (thorium) to $0·002$ second (actinium A) and about 10^{-6} second for radium C' (§ 7).

The atomic weights of uranium, radium, uranium lead, thorium, thorium lead have been determined directly by ordinary chemical methods and their atomic numbers by X ray methods. The atomic weights and atomic numbers of the others are deduced on the assumption that the expulsion of an α particle (helium nucleus) of charge two units and mass 4 lowers the atomic number of the succeeding element by two units and the atomic weight by four. The expulsion of a β particle raises the atomic number by one unit, but it is not supposed to influence the atomic weight to a detectable degree. The atomic weight of any member of the actinium series has not been directly measured, but recently strong evidence has been obtained that the end-product actinium lead has an atomic weight 207 (*vide* § 8 a) and from this the atomic weights of all the members of the series have been deduced.

§ 7. **Branch products.** In the great majority of cases each of the radioactive elements breaks up in a definite way, giving rise to one α or β particle and to one atom of the new product. Undoubted evidence, however, has been obtained that in a few cases the atoms break up in two or more distinct ways, giving rise to two or more products characterised by different radioactive properties.

It will be seen that a branching of the uranium series was early demanded to account for the origin of actinium. The most striking cases of branching occur in the "C" products of radium, thorium and actinium, each of which breaks up in two or more distinct ways. In the case of radium C, a new substance called radium C″ is obtained by recoil from a nickel plate coated with radium C. This product emits only β rays and has a period of 1·4 minutes. Fajans estimated that the amount of the product is only 1/3000 of that of radium C. To account for these results the following scheme of transformation

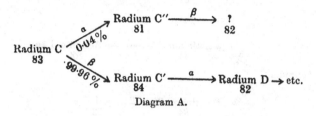

Diagram A.

has been proposed. The periods of transformation and atomic numbers are shown in Diagram A, where in the main branch a β particle is first expelled, giving rise to radium C′, which emits an α particle. The reverse process is assumed to take place in the other branch. Radium C′, which emits a swift α particle, has an exceedingly short period of transformation, which has been measured approximately by Jacobsen, and found to be about 10^{-6} second. It is uncertain whether the radium C″ branch ends after the expulsion of a β particle. The resulting product must be an isotope of lead, like radium D in the main branch. The α particles, expelled directly from radium C, have not been detected in the presence of the main group of longer range particles from radium C′. Their presence has been inferred from the vigour of the recoil of radium C″. From the relation between the range of an α particle and the life of a product (§ 35), the range in air of this small group of α particles is estimated to be about 4 cm.

In the case of thorium C, two sets of α particles are observed, two-thirds of the total number having a range of 8·6 cm. and the remainder 4·8 cm. Here, as in radium C, Marsden supposed that the main series goes by a β ray change to the C′ product. The scheme of changes is seen in Diagram B.

A similar scheme of dual transformation, Diagram C, is assumed to account for the emission of two groups of α particles by actinium C. 99·7 per cent. of the C atoms emit α particles of range 5·5 cm., and the remaining 0·3 per cent. goes to C′, which gives particles of range 6·4 cm. It is of interest to note that, while in both radium C and thorium C the β ray branch predominates, in actinium C the reverse holds and most of the atoms follow the α ray branch.

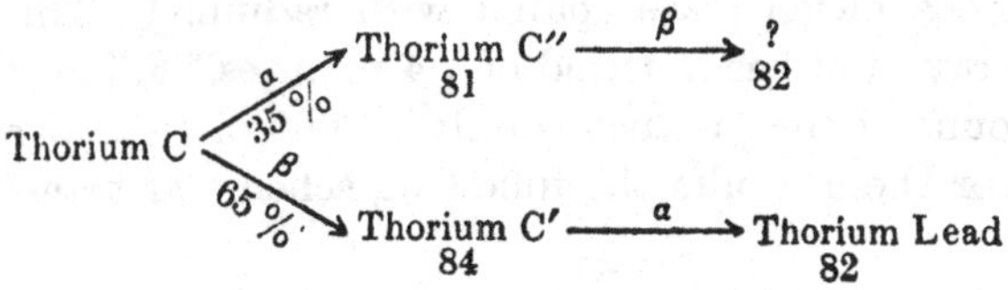

Thorium C <α 35%> → Thorium C″ 81 —β→ ? 82
          <β 65%> → Thorium C′ 84 —α→ Thorium Lead 82

Diagram B.

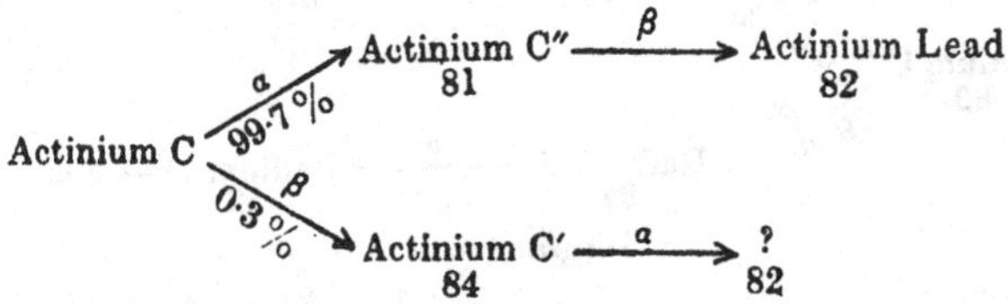

Actinium C <α 99.7%> → Actinium C″ 81 —β→ Actinium Lead 82
           <β 0.3%> → Actinium C′ 84 —α→ ? 82

Diagram C.

Subsequent research has shown that the modes of transformation of the "C" bodies, including the dual transformations, are even more complicated than was at first supposed. Radium C in addition emits one group of α particles of range about 9·1 cm., comprising about 1/30,000 of the main group, and possibly a still smaller group of range 11·3 cm. Similarly thorium C emits a group, about 1/10,000 of the total, of range 11·7 cm., and another smaller group of range 9·9 cm.* A more detailed account of our knowledge of these groups of "long range" particles will be given in § 20, together with a discussion of their origin. There seems to be no doubt that these long-range particles are true α particles of mass 4 which arise from the transformation of the radioactive atoms. It was at one time thought possible that these particles might arise from the artificial disintegration of the nitrogen or oxygen atoms in the path of the α particles, but careful observation by Rutherford and Chadwick negatived this supposition. It seems probable that the emission of these groups of long-range particles is connected with different modes of disintegration of the "C" bodies.

The emission of these groups of swift α particles is of great interest not only in giving us evidence on the intensity of the atomic explosions in certain atoms, but in indicating that all atoms do not

* Recent experiments have shown that the α rays from Thorium C itself are remarkably complex. See footnote, p. 47.

break up in identical ways. It will be seen (§ 65) that the presence
of these groups of long-range particles has to be taken into account
in experiments on the artificial disintegration of the elements when
radium C and thorium C are used as sources of α rays.

§ 8 a. Branch products of uranium series and origin of actinium.

It seems clear that the main line of transformation of the uranium
series is given by the scheme below:

$$\underset{92}{\text{U I}} \xrightarrow{\alpha} \underset{90}{\text{U X}_1} \xrightarrow{\beta} \underset{91}{\text{U X}_2} \xrightarrow{\beta} \underset{92}{\text{U II}} \xrightarrow{\alpha} \underset{90}{\text{Io}} \xrightarrow{\alpha} \underset{88}{\text{Ra}} \xrightarrow{\alpha} \ldots$$

In addition to these, however, two other products whose origin is
uncertain have been separated from uranium, called uranium Y
and uranium Z. In 1911, Antonoff observed a new β ray product
which is half transformed in 24·6 hours, which he named uranium Y
(U Y). It seems clear that this product which emits soft β rays is
of atomic number 90, isotopic with uranium X_1 and thorium and exists
in only about 3 per cent. of the equilibrium amount to be expected
if it were in the direct line of descent. Either therefore uranium Y
is a branch product at some point of the uranium series, or arises
from some unknown radioactive isotope of uranium. It has been
suggested that uranium Y is a branch product of uranium ·I or II
and is the head of the actinium series. The amount of uranium Y
is of the right order of magnitude to fit in with this view.

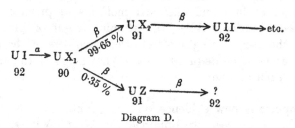

Diagram D.

Uranium Z, discovered by Hahn in 1921, is a product of very
weak activity which emits easily absorbed β rays and is transformed
with a period of 6·7 hours. It is estimated that the amount of this
product is only about 0·35 per cent. of that to be expected if it were
in the main line of descent in the uranium series. In chemical properties, it behaves like an element of number 91 and is thus isotopic
with protactinium. The proportion U Z to U X_1 is always constant
in old U X_1 preparations. The exact origin of U Z is uncertain. It

may be either a branch product of UX_1, in which case the scheme of dual transformation is shown in Diagram D, or it may be one of the products of an unknown isotope of uranium of weak activity.

The origin of actinium and of its parent protactinium has always been a question of great interest and importance and it is still unsettled. Boltwood early showed that in uranium minerals the quantity of actinium was always proportional to the amount of uranium, indicating that a genetic relation existed between them. On the other hand, the activity of actinium with its whole series of α ray products in a uranium mineral is much less than that shown by a single α ray product of the radium series. The discovery of the new product uranium Y led to the suggestion that it might prove to be the head of the actinium series, for it was present in about the required amount. On this view, the more recently discovered product protactinium is the missing link between uranium Y and actinium. On the other hand, it has also been suggested that actinium has its ultimate origin not in a branch product of the uranium series but in an isotope of uranium existing in small relative amount. It is, however, very difficult to reach a definite decision on these points with the data at present available. Much light may be thrown on this question if it proves possible to purify either actinium or protactinium in sufficient quantity to determine its atomic weight with accuracy. This seems likely to be done for protactinium in the near future, for it exists in uranium minerals in quantity comparable with that of radium and chemical methods of purification have already been devised. When the atomic weight of protactinium is known, the atomic weight of all other members of the actinium series can at once be deduced from a knowledge of the radiations emitted by each product.

§ 8 b. Important new evidence on the problem of the origin of actinium has been recently obtained by Aston (*Nature*, March 2, 1929, p. 313), by examining with a mass-spectrograph the isotopic constitution of lead obtained from a uranium mineral. A sample of lead, separated by Mr Piggot of Washington from the radioactive mineral Norwegian bröggerite, was compared with the spectrum of ordinary lead. The mass spectrum of the lead from the mineral showed a strong line 206, a faint line 207, and a still fainter line 208. The line 206 is no doubt due to uranium-lead, but the line 207 cannot be due to the presence of ordinary lead as an impurity or to thorium-lead, for in the mass spectrum of ordinary lead the line 208 is about twice as strong as 207. Aston concludes that the line 207 must be in part due to the end-product of the actinium series, actinium-lead. This, if correct, fixes the atomic weight of all the members of the actinium series. Since six α particles are expelled in the transformation of protactinium into actinium lead, the atomic weight of protactinium should on this view be 231. The atomic weights of the actinium series, given in the table of the radioactive elements, have been calculated on this basis. The evidence indicates that protactinium must arise from an isotope of uranium. The simplest assumption to make is that this isotope, number 92, which will be termed actino-uranium, has a mass 235, and is transformed by the emission of an α particle into a β ray product of number 90 which gives rise to protactinium. The β ray product is probably to be identified with uranium Y.

It was not found possible to measure the relative intensity of the isotopes with accuracy. Aston estimates that if the line 206 be taken as 100, the intensity of 207 is 10·7 ± 3 and of 208 is 4·5 ± 2. In a following letter (*Nature, loc. cit.*) Rutherford estimates that if the intensity of the line 207 is taken as 7 per cent. of the line 206, the half period of transformation of the new isotope is $4·2 \times 10^8$ years and that the actino-uranium is present in only about 0·28 per cent. of the main uranium isotope—a quantity too small to have much influence on the atomic weight of uranium as ordinarily measured. In order to draw definite conclusions, more accurate data are required on the relative amounts of the isotopes of lead in old uranium minerals. A determination of the atomic weight of protactinium will afford a crucial test of the general correctness of this new method of attack.

§ 9. The radio–elements and the Periodic Law. The great majority of the radio-elements exist in such minute quantity that it is impossible to obtain a weighable amount for an ordinary chemical analysis. In the case of radium, however, a sufficient quantity has been separated in a pure state to examine its chemical properties in detail, and to determine its atomic weight and spectrum. Notwithstanding the minute amount of the rapidly decaying elements, it was found possible in a number of cases to compare their chemical properties with those of the ordinary elements, using the activity of the radio-element as a method of quantitative analysis. Certain striking facts soon emerged. McCoy and Ross found that it was impossible to separate by chemical methods radiothorium from thorium. Later Soddy and Marckwald found in a similar way that radium and mesothorium were inseparable, while as the result of extensive investigations it has been found impossible to isolate radium D from the lead with which it appears in the analysis of a uranium mineral.

These results indicated that, in certain cases, the chemical properties of two elements were so closely allied that no chemical method was effective in producing the least degree of separation. Soddy concluded that such inseparable elements must occupy the same position in the periodic table of the elements and gave them the name of "isotopes." It will be seen that a number of such isotopes occur among the radio-elements. From these results, it seemed probable that some of the ordinary elements consisted of a mixture of isotopes of different atomic weights and this conclusion has been completely established in recent years by the researches of Aston. We now know that a number of elements consist of a mixture of two or more isotopes which differ in mass but have nearly identical chemical and physical properties.

It was early pointed out by Soddy that there appeared to be a connection between the radiation emitted from a radio-element and the chemical properties of the succeeding element. In a number of cases, the expulsion of an α particle causes a change in the position of the radio-element in the Periodic Table by two places in the direction of diminishing mass. For example, the expulsion of an α particle from radium, which belongs to group II of the Periodic Table, gives rise to radon, a radioactive inert gas belonging to the group 0. It was not, however, until the main sequence of transformations had been unravelled, and the chemical and electrochemical behaviour of many of the elements had been investigated by Fleck and Hevesy, that a relation of extraordinary simplicity was seen to connect the chemical properties of an element with the radiation emitted from the parent product. This law, known as the *displacement law*, was enunciated almost simultaneously by A. S. Russell and in more detail by Soddy and Fajans. The expulsion of an α particle causes the element to shift its position in the Periodic Table by two places in the direction of diminishing mass. The expulsion of a β particle causes a shift of one place in the opposite direction. This simple rule has been found of universal application to all the series of radioactive elements.

This generalisation can be looked at from another point of view based on the conception of the nuclear atom. The radioactive changes occur in the nucleus of the atom which carries a charge represented by its atomic number. The loss of an α particle, which carries two positive units of charge, from the nucleus, lowers its nuclear charge, i.e. its atomic number, by two units and its mass by four units. The expulsion of a β particle which carries a unit negative charge has the effect of raising the nuclear positive charge by one unit but does not sensibly alter the mass. For example, U I of atomic mass 238 and atomic number 92 by loss of an α particle changes into U X_1 of mass 234 and number 90. By loss of a β particle this changes into U X_2 of mass 234 and number 91, and again by loss of a β particle into U II of mass 234 and number 92. Expressed in another way, an α ray change followed by two β ray changes results in an element of the same number and nuclear charge as the first. It is seen that U I and U II, which have the same nuclear charges but different masses, are isotopic elements. This conclusion is confirmed by our inability to separate these two elements by chemical methods.

The displacement law as well as the change of atomic number with the radiation is clearly illustrated on page 36. An arrow in the downward direction represents an α ray transformation and an arrow sloping upwards a β ray transformation. The atomic numbers 81, 82, 83 are represented by the ordinary non-radioactive elements Tl, Pb, Bi respectively. All elements in the same horizontal column are isotopic and are thus inseparable chemically, although they may differ markedly in their radioactive properties. This is not unexpected, since the ordinary chemical and physical properties of an atom apart from its mass are determined by the arrangement of the outer electrons which are controlled by the charge on the nucleus. The instability of a nucleus which determines its radioactive behaviour is, however, dependent on the structure of the nucleus which may be very different for atoms of the same nuclear charge but different masses.

It is seen that there are many isotopic elements corresponding to atomic numbers from 82 to 92 inclusive, with the exception of the numbers 85, 87, which are not represented by any known elements. Each of these groups of isotopic elements should show identical physical and chemical properties apart from mass. Even the spectrum should be nearly identical with the exception of a minute change of frequency of some lines due to the difference of mass of the nucleus.

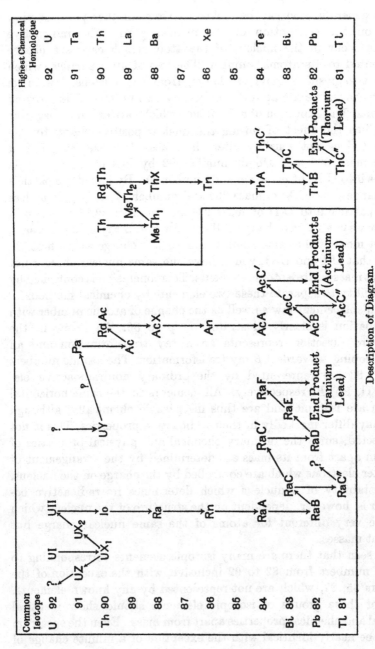

Description of Diagram.

Change of atomic number in the radioactive transformations.
Elements in the same horizontal column are isotopic.

When the types of transformation are known, we can at once write down not only the atomic number of the element but also its atomic weight. For example, uranium of atomic weight 238, after loss of three α particles of total mass 12, changes into radium of known atomic weight 226. After loss of three α particles radium changes into radium B, an isotope of lead of atomic weight 214, and so on.

In the scheme of radioactive changes outlined above, the end products of the uranium, thorium, and actinium series should be of atomic number 82 and thus be isotopes of ordinary lead. Since uranium lead arises from radium 226 which has lost five α particles, the atomic weight of uranium lead should be 206. Similarly the atomic weight of thorium lead should be 208. If these views be correct, the lead formed in old primary uranium minerals should have a lower atomic weight (206) than ordinary lead (207), while thorium lead in minerals should be higher, viz. 208. These conclusions have been completely confirmed by direct analysis of uranium and thorium lead by Richards, Hönigschmid, Soddy and others. It is of interest to note that this is the only occasion where any isotope of a complex element has been found isolated in nature. No difference in average atomic weight has been detected for any of the ordinary elements, although the material may be obtained from widely different sources.

We have seen (§ 8 a) that an examination of the isotopes of lead from a uranium mineral indicates that actinium lead has an atomic weight 207 and exists in small quantity compared with ordinary uranium lead (206). Aston has also determined the isotopic constitution of ordinary lead, and has found the following isotopes: 206 (4), 207 (3), 208 (7), and probably 209 and others in small proportion. The numbers in brackets give the approximate relative intensity of the photographic bands. It is clear from these results that, while the atomic masses 206 and 208 are predominant, the isotope (207) is relatively much too intense to be ascribed entirely to actinium lead. It is unlikely therefore that ordinary lead in nature is derived entirely from the transformation of uranium, thorium and actinium.

It will be seen, from the brief account given above, that the properties of the radio-elements are connected in an exceedingly simple way with the nature of the radiations emitted during their transformation. This simplicity is connected with the fact that the α and β particles originate in the nucleus of the atom. The displacement rule in its general form affords very convincing evidence of the correctness of the modern conception of the nuclear constitution of the atom.

CHAPTER II

THE α RAYS

§ 10a. Comparison of the radiations. All the radioactive substances possess in common the power of acting on a photographic plate and of ionising the gas in their immediate neighbourhood. The intensity of the radiations may be compared by means of their photographic or electrical action, and in the case of the strongly radioactive substances by the luminosity excited in a phosphorescent screen.

Two general methods have been used to distinguish the types of radiation given out by a radioactive matter depending upon

(1) a comparison of the relative absorption of the rays by solids and gases,

(2) observations on the direction and magnitude of the deflection of the rays when exposed to the action of strong magnetic and electric fields.

Examined in this way, it has been found that there are three distinct primary types of radiation emitted from radioactive bodies which for brevity and convenience have been termed the α (alpha), β (beta), γ (gamma) rays.

(1) *The α rays*, which are very readily absorbed by thin metal foil or by a few centimetres of air, consist of a stream of positively charged atoms of helium, initially projected from the radioactive matter with high velocity, which varies for different substances between $1 \cdot 4 \times 10^9$ and $2 \cdot 2 \times 10^9$ cm./sec. Normally the α particle at the moment of its expulsion carries two positive charges and is to be identified with the nucleus of the helium atom.

(2) *The β rays*, which are on an average far more penetrating in character than the α rays, consist of a stream of electrons projected from the active matter with a wide range of velocity and energy. The average velocity of the β particle is in general about ten times greater than that of the α particle, and in some substances β rays are liberated with a velocity closely approaching that of light and with an individual energy of the same order of magnitude as the α particle. They are far more readily deflected by magnetic fields than the α particle and are identical in type with the cathode rays produced in a vacuum tube.

(3) *The γ rays* are extremely penetrating and are not deviated by either a magnetic or electric field. They constitute a type of electromagnetic radiation like X rays, but in general have a much higher frequency of vibration and are far more penetrating than the X rays produced in an ordinary discharge tube. The γ rays from a radioactive product give a line spectrum corresponding to a number of discrete frequencies in the emitted radiation.

The types of radiation emitted from radioactive matter thus present a very close analogy with the kinds of radiation produced by the electric discharge through a gas at low pressure. The α rays are analogous to the "canal" or "positive" rays which consist of positively charged atoms or molecules of matter moving with high speed. The β rays are identical with the cathode rays, while the γ rays correspond to X rays. It should be pointed out, however, that while the "positive" rays in the discharge tube are produced by the action of the electric field or other agencies in removing one or more of the outer electrons from the atoms or molecules, the α rays, which are simply helium nuclei, are spontaneously expelled from the nucleus of the atom. In a similar way, much of the β and γ radiation observed has its origin in the radioactive nucleus and not in the outer electronic distribution of the atom.

Of the three types of rays, the α rays produce most of the ionisation in the gas and the γ rays the least. For example, using a thin layer of a radium compound spread on the lower of two parallel plates 5 cm. apart, the amount of ionisation due to the α, β, and γ rays is of the relative order 10,000, 100 and 1. These numbers are only rough approximations and the differences become less marked as the thickness of the radioactive layer is increased. Since each type of rays from radioactive substances is usually complex and consists of radiations which are absorbed to an unequal extent, it is difficult to give an accurate comparison of the relative penetrating power of the three types of rays. As a rough working rule, it may be taken that the β rays are about 100 times as penetrating as the α rays, and the γ rays from 10 to 100 times as penetrating as the β rays.

It is often convenient to know what thickness of matter is sufficient to absorb a specific type of radiation. A thickness of ·006 cm. of aluminium or mica or a sheet of ordinary writing-paper is sufficient to absorb completely all the α rays. With such a screen over the active material, the external effects are due only to the β and γ rays, which pass through with a very slight absorption. Most of the β rays are absorbed in 5 mm. of aluminium or 1 mm. of lead. The radiation passing through such screens consists very largely of the γ rays. As a rough working rule, it may be taken that a thickness of matter required to absorb β or γ rays is inversely proportional to the density of the substance, i.e. the absorption is proportional to the density. This rule holds approximately for light substances, but, in heavy substances like mercury and lead, the γ radiations are more readily absorbed than the density rule would lead us to expect.

It is difficult to make quantitative or even qualitative measurements of the relative intensity of the three types of rays from active substances. The three general methods employed depend upon the action of the rays in ionising the gas, in acting on a photographic plate, and in causing phosphorescent or fluorescent effects in certain substances. In each of these methods the fraction of the rays which is absorbed and transformed into another form of energy is different for each type of ray. Even when one specific kind of ray is under observation, comparative measurements are rendered difficult by the complexity of that type of ray. For example, the β rays from radium consist of negatively charged particles projected with a wide range of velocity, and, in consequence, they are absorbed in different amounts in passing through a definite thickness of matter. In each case, only a fraction of the energy absorbed is transformed into the particular type of energy, whether ionic, chemical, or luminous, which serves as a means of measurement.

§ 10 b. The α rays. After the discovery of radioactivity, attention was at first directed to the β rays on account of their great penetrating power and their marked action on a photographic plate, and in producing phosphorescence in certain substances. Towards the end of 1899, at an early stage in the history of radioactivity, the β rays were found to be easily deflected by a magnetic field, and it was correctly concluded that the β rays consisted of a stream of negatively charged electrons projected at high speeds from the radioactive matter. Some of the particles were much more easily bent than others, indicating that the β rays from a radium preparation were expelled over a wide range of speeds, the velocity of the swifter particles approaching that of light. This complexity of the β rays from radium was later used by Kaufmann and Bucherer to show that the mass of the electron was not constant but increased rapidly as the velocity of the electron became comparable with the velocity of light.

In comparison with the β rays, the α rays were at first little studied, for a magnetic field which easily bent the β rays had no noticeable effect on the α rays. Certain peculiarities in the absorption of the α rays by matter suggested that they might consist of massive projected particles. If the particles carried an electric charge, they should be deflected in passing through a strong magnetic or electric field. This was difficult to test until active preparations of radium were available so that narrow but intense beams of radiation could be used. Using an ionisation method to detect the rays, Rutherford showed in 1903 that the α rays from radium were deflected both by a magnetic and electric field, as if they were positively charged particles projected with great velocity. If particles each of mass M and charge E, moving with a velocity u, are acted on by a magnetic field, the value Mu/E can be measured. By observing the deflection in a strong electric field, the value Mu^2/E is determined. From these two observations, the value E/M of the particles and the velocity u can be deduced. In this way it was estimated that the α particles had a velocity $u = 2 \cdot 5 \times 10^9$ cm./sec. and a value $E/M = 6000$ electromagnetic units. This result was soon confirmed by Des Coudres with the photographic method, using pure radium bromide as a source of rays. The value $E/M = 6400$ was obtained. Later Mackenzie found for radium rays a value of $E/M = 4600$ and Huff for polonium rays a value 4300.

These results showed that the α particles were of atomic mass comparable with that of the hydrogen atom. The value e/m for the hydrogen atom is 9643, indicating that if the α particle carried the same charge its mass should be about twice that of the hydrogen atom.

In these experiments, the radioactive sources emitted α particles over a wide range of velocity, so that only an approximate estimate could be made of the average velocity and the value E/M. In the meantime, it became of great importance to fix the value E/M of the α particle with precision. It had been suggested that helium, which is always found associated with radioactive minerals, might be a product of the transformation of radioactive bodies, and the value of E/M observed for the α particle was of about the magnitude to be expected if it were a helium atom carrying two positive charges, for which E/M should be 4826. It will be seen that the identity of the α particle with the helium nucleus was finally established by several methods, which will be referred to later, but at this point we shall pass on to the development of experimental methods to determine the value of E/M with the necessary precision to settle this very important question. Greater accuracy was obtained by making use of homogeneous sources of α rays. The most convenient source for this purpose is a platinum wire which has been exposed for several hours to a strong source of radon. It becomes coated with an invisible active deposit containing the products radium A, radium B, radium C. When the wire is removed, the radium A rapidly dies away and the α ray activity is then due to the product radium C, for radium B emits only β rays. The α rays from such a radium C source are nearly homogeneous, having a range in air of about 7 cm.* The magnetic deflection of the α rays from such a source in a vacuum is most conveniently determined by the photographic method in an

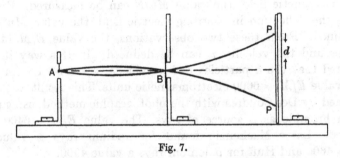

Fig. 7.

apparatus shown in Fig. 7. The rays from the active wire A pass through a narrow slit B parallel to the wire, and fall on the photographic plate PP. A carrier on which is mounted the source, slit, and plate is placed in an exhausted vessel between the pole pieces of a large electromagnet so that the magnetic field is uniform between A and P and parallel to the slit. The electromagnet is excited by a constant current, which is reversed at intervals. For a distance between source and photographic plate of about 10 cm., two well-defined bands are observed on the plates if the strength of the source corresponds in γ ray activity to several milligrams of radium. If ρ is the radius of curvature of the circular arc described by the rays in a field of strength H, then $H\rho = Mu/E$, where u is the velocity of the rays. If $2d$ = distance between centres of two bands obtained

* The α rays of range 7 cm. originate not from radium C itself but from the subsequent branch product radium C′ which has a very short average life. As radium C′ always accompanies radium C and is in equilibrium with it, it is found convenient in practice to use the term radium C rather than radium C + C′ or radium C′.

by reversal of the field, a = distance between source and slit, b = distance between slit and plate, then, if d is small,

$$2\rho d = b\,(a+b).$$

If d is comparable with a or b, the correct value of ρ is given by

$$\rho^2 = \frac{1}{4d^2}\,(b^2 + d^2)\,(\overline{a+b}^2 + d^2).$$

In this way, Rutherford and Robinson* made an accurate determination of the deflection of the α rays from radium C and found a value $Mu/E = 3\cdot983 \times 10^5$, where E is measured in electromagnetic units. Special precautions were taken to determine the magnetic field and its variations over the path of the rays. For $a = 6\cdot540$ cm., $b = 6\cdot618$ cm., the deflection $2d = 13\cdot63$ mm. for an effective field of 6236 gauss.

Another determination of this important constant has been made recently by Briggs†, who found a concordant value of $H\rho = 3\cdot993 \times 10^5$. Special precautions were taken to ensure the constancy of the field during the experiment, and to measure its strength; the result is believed to be correct to 1 in 1000.

It seems clear that the majority of the particles are projected from radium C with identical speed. This can best be tested by using a very thin pencil of the α rays and obtaining a deflection of several centimetres in a magnetic field, and comparing with a microphotometer the width of the deflected and undeflected pencil on a photographic plate. In this way, Briggs showed that the velocity of the main group of α particles was constant within 1 in 1000. I. Curie‡ carefully examined the α rays of polonium and found them homogeneous within the limit of measurement 1 in 1000. It thus seems probable that α particles from all products are homogeneous in velocity.

A determination of the electric deflection of the α rays from radium C was made by Rutherford§ by passing the rays in a vacuum between two parallel plates kept at a constant difference of potential. The deflection was small under the experimental conditions and difficult to measure with accuracy. A value $E/M = 5070$ was

* Rutherford and Robinson, *Phil. Mag.* **28**, 552, 1914.
† Briggs, *Proc. Roy. Soc.* A, **118**, 549, 1928.
‡ I. Curie, *Ann. de physique*, **3**, 299, 1925.
§ Rutherford, *Phil. Mag.* **12**, 348, 1906; also Rutherford and Hahn, *Phil. Mag.* **12**, 371, 1906.

obtained. About the same value of E/M was found for the α particles from a number of other radioactive products.

In 1914, Rutherford and Robinson* made a new series of measurements of the electric deflection of the α rays. For precise measurements, it was necessary to obtain a total deflection of the α rays of the order of about 1 cm., under conditions when all the quantities involved could be measured with precision. With the steady voltage available, about 3000 volts, this could only be obtained by observing the deflection of the rays at a distance of about 1 metre from the source. Under these conditions, the photographic effect due to a very active wire coated with radium C would be too small for detection. To overcome this difficulty the source finally employed was a thin-walled α ray tube filled with a

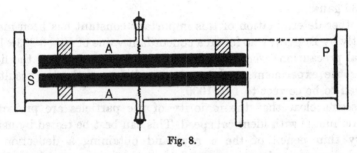

Fig. 8.

large quantity of radon. The apparatus employed for the electrostatic deflection is shown in Fig. 8. The electric field was applied between two silvered strips of plate glass AA, each 35 cm. long, 2·5 cm. wide and 1 cm. thick, which were held 4 mm. apart by ebonite stops. The α ray tube S was supported centrally near one end of the glass plates. At the other end of the plates was a mica slit 1/6 mm. wide. The apparatus was placed in a large glass tube, closed at the ends by ground glass plates. The photographic plate P, wrapped in aluminium leaf to protect it from light and from the glow of the source, was fixed to the glass end plate 50 cm. from the slit and 85 cm. from the source. Electrical contacts to the plates were made by metal springs passing through ground joints in the side of the tube. The path of the rays from the slit to the photographic plate was shielded from the stray electric field by a long metal cylinder closed at the end next the slit except for a narrow opening for the passage of the rays.

* Rutherford and Robinson, *Phil. Mag.* **28**, 552, 1914.

PLATE I

Fig. 1. Electrostatic deflection of α particles from an α ray tube filled with radon.

Fig. 2. Magnetic deflection of α particles from the same tube.

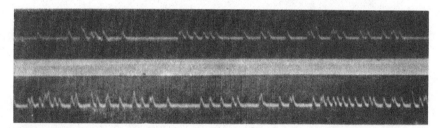

Fig. 3. Photographic registration of α particles with Rutherford-Geiger counter and string electrometer. The upper record corresponds to the entrance of particles at a rate of 600 per minute into the counter, the lower to a rate of 900 per minute.

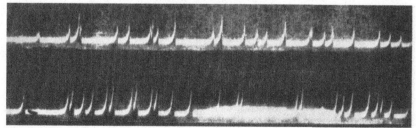

Fig. 4. The upper record shows the registration of β particles with the Geiger point counter, the lower the registration of α particles.

After the source and the photographic plate had been fixed in position, the vessel was rapidly exhausted and a constant potential difference of about 2000 volts was maintained between the plates by a battery of small cells. The exposure was usually about 6 hours for a tube containing 150 millicuries* of radon. An enlarged photograph showing the electrostatic deflection of the α rays is shown in Plate I, fig. 1. The central band is the result of the direct action of the rays before the application of the field, while the three bands on each side show the traces produced on reversal of the field by the α rays from radium C, radium A and radon respectively. The inner bands are due to the α rays from radium C, which have the highest velocity. It will be seen that the inside edges of the bands are very sharply defined. This peculiarity, which was of material assistance in making accurate measurements, can be readily explained by taking into account the fact that the rays which make up the band travel through different thicknesses of glass in escaping from the radon tube. Thus rays emitted at a certain angle with the central line passing through the source and slit suffer a minimum deflection. There is consequently a concentration of the radiation at the inner edge of the band due to this lens-like action of the radon tube.

It will be seen that the outer edges of the band are not sharp. This is not due to any lack of homogeneity of the rays on emission from the radioactive matter, but results from the fact that the rays escaping from near the edge of the tube pass through a greater thickness of glass than those escaping near the centre and are thus reduced in velocity. This broadening of the bands is clearly seen in the original photographs and is especially marked for the radon and radium A bands owing to the lower velocity of the α particles.

In order to avoid any correction for the reduction of velocity of the α rays in passing through the glass walls of the α ray tube, the magnetic deflection of the α rays was measured for the *same* source. Special care was taken to determine with accuracy the intensity of the magnetic field, and its variation along the path of the rays. The photograph of the magnetic deflection is shown in Plate I, fig. 2. It will be noticed that the separation of the bands is less marked than for

---

* The amount of radon in equilibrium with one gram of radium element is called a "curie" in honour of the discoverers of radium. A millicurie (1/1000 curie) is the amount of radon in equilibrium with one milligram of radium.

the electrostatic deflection, a result to be expected since the magnetic deflection is proportional to $1/u$, while the electrostatic deflection is proportional to $1/u^2$, where u is the velocity.

From the combined observations of the deflection in the magnetic and electrostatic fields, the values of E/M and u for each group of rays can be deduced. The results obtained from two distinct experiments are shown in the following table.

Exp. 1. Voltage 1435. Stopping power of glass tube 2·00 cm. of air.
Exp. 2. Voltage 1958. Stopping power of glass tube 2·00 cm. of air.

Source of α rays	Exp.	Mu/E	Mu^2/E	E/M
Radium C	1	$3 \cdot 555 \times 10^5$	$6 \cdot 083 \times 10^{14}$	4813
	2	$3 \cdot 555 \times 10^5$	$6 \cdot 110 \times 10^{14}$	4826
Radium A	1	$2 \cdot 941 \times 10^5$	$4 \cdot 174 \times 10^{14}$	4824
	2	$2 \cdot 941 \times 10^5$	$4 \cdot 185 \times 10^{14}$	4837
Radon	1	$2 \cdot 717 \times 10^5$	$3 \cdot 560 \times 10^{14}$	4822
	2	$2 \cdot 717 \times 10^5$	$3 \cdot 563 \times 10^{14}$	4826

The values of E/M for radium C, radium A and radon are seen to be in close accord. The final results for radium C, which can be determined with most accuracy, gave a mean value $E/M = 4820$, while the calculated value is $E/M = 4826$ for the doubly charged helium atom of radium C [*]. Within the probable error of experiment, about 1 in 400, it is seen that the value of E/M for the α particles from the three products agrees with the value for the helium atom. We may thus conclude with certainty that the α particles are helium nuclei carrying two unit charges expelled at high speed from the disintegrating atoms.

All the evidence, both direct and indirect, shows that the α rays from all radioactive substances are identical in mass but vary only in the velocity of their emission. As far as observation has gone, the α particles from each product are initially expelled with an identical velocity characteristic for each product. It will be seen (§ 22) that the α particles diminish in velocity in traversing matter. Consequently the α rays from a thick layer of radioactive material have all velocities between zero and the maximum. It is only by using thin films of radioactive matter that homogeneous pencils of α rays can be obtained.

* See Briggs, *Proc. Roy. Soc.* A, **118**, 549, 1928.

§ 10 c. **Velocity of emission of rays.** It has been shown that for the α particles from radium C, the magnetic deflection gave a value $Mu/E = 3.985 \times 10^5$.

Taking $E/M = 4826,$
$$u = 1.922 \times 10^9 \text{ cm./sec.},$$
or about 1/16 of the velocity of light.

We have seen that a recent accurate determination by Briggs (*loc. cit.*) gives

$$H\rho = Mu/E = 3.993 \times 10^5$$

for the α particles from radium C. Taking Aston's value 4·00216 as the mass of the helium atom, the mass of the doubly charged helium atom or helium nucleus is 4·00107 at slow speeds. Taking the Faraday as 9649·4 and making the relativity correction for the mass of the α particle, the value **E/M = 4813** for the fast α particle under the conditions of measurement. This gives a value $u = 1.922 \times 10^9$ cm./sec., agreeing with that deduced above on less accurate data.

In addition to the main group of α particles of range 7 cm. in air, radium C also emits a small number of α particles of range 9·3 cm. These constitute such a minute fraction of the total number of α particles that their photographic effect is too small to give a detectable effect under ordinary experimental conditions.

While the experimental evidence strongly supports the view that the great majority of the α particles from radium C are initially expelled with identical velocity and charge*, the radiation obtained from active wires in an experiment such as is shown in Fig. 7 is not completely homogeneous. Some of the α particles are able to escape from the edge of the wire source by penetrating the material of the wire. Superimposed on the main homogeneous beam, there is in

---

* S. Rosenblum (*C.R.* **188**, 1401, 1929) has examined the fine structure of the α rays from thorium C (range 4·8 cm.) by the focussing method so much employed in the examination of β ray spectra. For this purpose, the large electromagnet of the Academy of Sciences, constructed by Cotton, was used and the rays were bent in a circle of diameter about 25 cm., using a field of about 36,000 gauss. In this way, he was able to show that three faint groups of α rays accompanied the main group of which the velocities were 1·003, ·975, ·961, in terms of the velocity of the main group. The discovery of these additional groups illustrates the great complexity in the emission of α rays from thorium C. In a subsequent paper, Rosenblum (*C.R.* **188**, 1549, 1929) finds no certain evidence of additional groups of α particles from radium A, polonium, radium C', and thorium C'.

This new method of attack on the problem of the homogeneity of the α-rays is of much interest and importance and promises to give us valuable data on this question.

consequence a distribution of α particles having all velocities between zero and the maximum. Usually the number of these is only a few per cent. of the total. They are not easily noted in photographic experiments as they are bent away from the main band by the magnetic field. A certain small complexity in the beam is also introduced by the scattering of the α particles by the material of the source and at the edge of the slits. The radiation in a high vacuum is also complicated by another factor. It will be seen in § 24 that a small fraction of the α particles, even from an uncovered source, escape with a single charge. This is probably due to the capture of an electron by some of the α particles in their escape from the active material.

§ 11. Nature of the α particle. We have seen that the value of E/M of the α particle agrees closely with the value to be expected if the α particle is a helium nucleus. Later it will be shown, by direct measurement of the charge carried by a known number of α particles, that each α particle carries twice the fundamental unit of charge. The evidence of the identity of the α particle with the helium nucleus is thus very strong. On this view, we can offer a simple explanation of the production of helium by radium first observed by Ramsay and Soddy. The α particles are absorbed in either the radioactive material or the walls of the containing vessel, and after losing their charge become neutral atoms of helium. This helium can be liberated by the action of heat or by dissolving the radium.

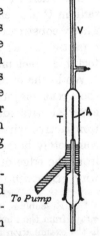

Fig. 9.

This conclusion was confirmed by a direct experiment of Rutherford and Royds*, who obtained spectroscopic evidence that the accumulated α particles, quite independently of the matter from which they are expelled, give rise to helium. The experimental method will be clearly seen from Fig. 9. A large quantity of purified radon was introduced into the thin-walled glass tube A. This was sufficiently thin to allow most of the α particles to be shot through the walls. The tube A was surrounded by a glass tube T to which a small spectrum tube V was attached. The tube T was completely exhausted and by means of a side tube any gases

* Rutherford and Royds, *Phil. Mag.* **17**, 281, 1909.

collecting in T could be compressed into the spectrum tube and their nature determined spectroscopically. Two days after the introduction of the radon into A the yellow line of helium was clearly seen in the spectrum tube, and after six days the whole spectrum was visible. Control experiments showed that helium gas could not penetrate through the thin walls of A. It is thus clear that the α particles gave rise to the helium observed.

Most of the α particles from the tube A are fired into the walls of the outer tube and lose their charge. Some of the atoms gradually diffuse out into the exhausted space. The presence of helium in the spectrum tube could be detected earlier if a thin cylinder of lead were placed over the tube A, since the helium diffuses out from lead more rapidly than from glass. If a thin sheet of lead is placed over an α ray tube filled with radon in the open air, and left for several hours to be bombarded by α particles, helium is always detected in the lead. This was tested by melting the lead *in vacuo* and examining the residual gases.

It is thus clear that the helium, which is produced by all radio-active substances emitting α particles, arises from the accumulated α particles. This has been confirmed by direct measurement (§ 33).

§ 12. **Detection of a single α particle.** The α particle expelled from radioactive substances is the most energetic atomic projectile known to science. The kinetic energy of the α particle expelled from a source of radium C is $1·2 \times 10^{-5}$ erg, corresponding to the energy acquired by an electron in passing between two points in a vacuum differing in potential by 7·66 million volts. Its energy is consequently much greater than can be acquired by an electron or charged atom in a discharge tube in the laboratory, where the difference of potential applied is seldom more than 100,000 volts.

The α particle is thus differentiated from the ordinary atom of matter by its great energy of motion, and it has been found possible to detect and even to count single α particles by a variety of methods. The different methods of detection no doubt depend in an ultimate analysis on the property of the α particle of producing ions in its passage through matter whether solid, liquid or gaseous. In some cases, the power of the α particle of dissociating complex molecules may also be important. These different methods may be conveniently discussed under the headings (1) electrical, (2) optical or scintillation, (3) photographic, (4) expansion method.

§ 12 a. **Electrical method**. It is known that an α particle from radium C in its path in air produces $2 \cdot 2 \times 10^5$ pairs of ions each carrying a charge $4 \cdot 77 \times 10^{-10}$ E.S.U. This corresponds to a transfer of about 10^{-4} E.S.U. of electricity in an electric field. Consequently if this quantity were communicated to an electrometer circuit of capacity, for example, 50 cm., it would alter its potential by 6×10^{-4} volts. Using an electrometer of the Compton type, which gives a deflection of 10,000 divisions per volt, the entrance of an α particle into the detecting vessel would cause a throw of the electrometer needle of six divisions. No doubt, if suitable precautions were taken, it would be quite possible to detect a single α particle in this way by its direct ionisation effect *. On account of the slow motion of the needle, such a method is unsuited to count α particles at a rapid rate.

In order to increase the electrical effect of a single α particle, Rutherford and Geiger† in 1908 used the method of ionisation by collision to magnify automatically the small ionisation produced by an α particle. If a strong electric field acts on a gas at low pressure, the number of ions is greatly increased owing to the production of new ions by collision. If the voltage applied is near the sparking potential, the ionisation current may in this way be increased several thousand times.

In the experiments of Rutherford and Geiger it was arranged that the α particles were fired through a gas at low pressure exposed to an electric field somewhat below the sparking value. The small current through the gas due to the entrance of an α particle into the detecting vessel was magnified by this method sufficiently to give a marked deflection to the needle of an electrometer of moderate sensibility.

The experimental arrangement is shown in Fig. 10. The detecting chamber A consisted of a brass vessel about 25 cm. in length, 1·77 cm. in diameter, with a central insulated wire passing through ebonite plugs at the end. This wire was connected with the electrometer while the outside tube was connected with the negative terminal of a battery of accumulators, the other pole being earthed. In the ebonite plug C was fixed a short glass tube D, in the end of which was a circular hole 1·5 mm. diameter covered with a thin sheet of

---

* This has been done by Hoffmann using a very sensitive binant electrometer. With the same instrument, Ziegert (*Zeit. f. Phys.* **46**, 668, 1928) has measured the total ionisation due to a single α particle.

† Rutherford and Geiger, *Proc. Roy. Soc.* A, **81**, 141, 1908.

mica corresponding in stopping power for the α particle to about 5 mm. of air. A long glass tube with a wide stopcock F served as a continuation of the tube D. The detecting tube was exhausted to a pressure of a few centimetres of mercury and the applied voltage was adjusted until the current due to a source of γ rays was magnified several thousand times by collision. The active matter in the form of a thin film of small area was placed in the long tube E, which was then pumped out. When the stopcock was opened, a small fraction of the α particles from the source entered the aperture in the detecting vessel. If the strength and distance of the source were adjusted so that about three to five α particles entered per minute, the entrance of each α particle was indicated by a ballistic throw of the electrometer

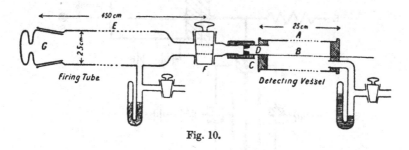

Fig. 10.

needle. This experiment has a certain historical interest, for it was the first time that it was shown to be practicable to detect a single atom of matter in swift motion by its electrical effect. If the electrometer circuit were connected to earth through a high resistance of the order of 100 megohms, the charge was able to leak away completely in the interval between the entrance of successive particles. In this way it was found possible to count the number of α particles entering the detecting vessel in a given time and thus estimate the total number emitted by a known quantity of radium (§ 13).

The rapidity of counting was greatly increased by using a string electrometer instead of the ordinary quadrant electrometer*. By suitably adjusting the tension of the string and recording its movements photographically on a moving film, it was found possible to record distinctly the entrance of each α particle for an average rate of 1000 per minute. The detecting vessel in this case consisted of

* Geiger and Rutherford, *Phil. Mag.* 24, 618, 1912.

a central spherical electrode in a concentric hemisphere, with a small aperture covered with mica for the entrance of α particles. An example of the record so obtained is shown in Plate I, fig. 3. The movement from the horizontal line records the entrance of each α particle. The string, after deflection, was brought back rapidly to its zero position by the aid of a suitable leak to earth in the electro-meter circuit. It will be seen that the α particles enter the detecting vessel at irregular intervals. Detailed investigations by this and other methods show that the number of the α particles entering the detecting vessel in a given time is subject to fluctuation obeying the ordinary laws of probability (§ 34).

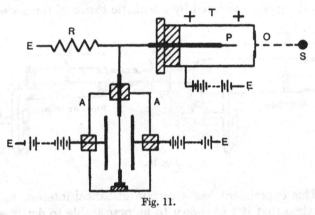

Fig. 11.

The type of detecting vessel shown in Fig. 10 gives a large effect for an α particle but a very slight effect when a β particle enters the vessel. Geiger* subsequently developed a variant of this method which gives much larger magnifications and is equally effective for detecting a β particle as well as an α particle. The detecting vessel and connections are shown in Fig. 11. A rod, ending in a fine, ground point P, is insulated inside a short brass tube T, which contains a small opening O, opposite the point, through which the α and β particles enter. For work at reduced pressures, the opening is hermetically closed with a very thin sheet of mica. The needle is connected with the string electrometer A, with a suitable high re-sistance R, in parallel. A high resistance of convenient form can be made by filling a capillary tube with a mixture of alcohol and xylol. A variable high resistance is readily provided by the use of a ther-

* Geiger, *Verh. d. D. Phys. Ges.* **15**, 534, 1913; *Phys. Zeit.* **14**, 1129, 1913.

mionic valve. The plate and grid are connected with the needle of
the counter and one end of the filament is earthed. The effective
resistance of this combination is controlled by the temperature of
the filament. If the detector contains air at ordinary pressure, a
voltage of from 1000 to 1500 is required, the outer tube being positive.
The detector works well at reduced pressure, when a few hundred
volts suffices. An example of a photographic record of α and β rays
obtained by Geiger is shown in Plate I, fig. 4.

The action of the counter has been the subject of numerous investigations. When the voltage is raised nearly to the discharging
point, the small ionisation near the point due to the entrance of an
α or β particle causes a momentary strong discharge from the point.
In the case of the point-detector, the local ionisation exerts a trigger
action and causes a momentary discharge, the magnitude of which
is largely independent of the initial ionisation. For this reason, the
weak ionisation of the gas due to a β particle produces an effect of
about the same magnitude as the intense ionisation due to an α
particle. If the point is examined under a microscope, a sharp flash
is seen at the point corresponding to the entrance of each α particle.
In this respect the point-detector has an entirely different action
from the detector shown in Fig. 10, where the momentary current
due to collision of ions is proportional to the initial ionisation. For
this reason, if α particles alone are to be counted in the presence of
β rays, the detector of the type shown in Fig. 10 is preferable to the
point type.

The effect of the capacity and resistance used with the counter,
as well as the effect of pressure, has been examined by several observers, including Geiger, Appleton, Emeléus *, Taylor † and others. It
has been shown that the magnitude and duration of the momentary
discharge depends on the capacity. Emeléus has given an interesting
account of the various factors involved in the working of the counter.
Since the momentary current in the counter may in some cases be
as high as 5×10^{-5} amps, it is not difficult to magnify the current
by thermionic valves so as to work a relay or a telephone. This has
been done by Kovarik, Mme Curie and Greinacher. In this way the
entrance of an α or β particle into the detector may be automatically
recorded or made visible by the lighting of a lamp. By magnifying
the actual small ionisation current due to an α particle by means of

* For literature of this subject see Emeléus, *Proc. Camb. Phil. Soc.* 22, 676, 1925.
† J. Taylor, *Proc. Camb. Phil. Soc.* 24, 251, 1928.

a series of amplifiers, Greinacher* has recorded photographically the entrance into the detecting vessel of swift hydrogen particles as well as α particles. Such a method does not record β particles and is thus very serviceable in counting α or swift H particles even when a number of β particles may be present. It will be seen (§ 13) that this method has been applied to count accurately the number of α particles expelled per second per gram of radium. Geiger and Klemperer† have recently found that if the needle of the detector ends in a small sphere, and the outer vessel is charged negatively over a limited range of voltage, the detector counts α particles and not β particles, but above a critical voltage it counts α and β particles indiscriminately. In other words, in the first case the current from the detector is proportional to the ionisation, and in the second it acts like the ordinary β ray counter. By an adaptation of the method originally used by Rutherford and Geiger to count α particles, Geiger finds that β particles can be counted over a large surface area. The β particles from the air and walls of an ordinary room give a large number of throws per minute and, by careful shielding, the β rays arising from the penetrating radiation in the atmosphere may be counted.

§ 12 b. **Scintillation method.** Sir William Crookes and also Elster and Geitel discovered in 1903 a property of the α rays that has subsequently proved of great importance in radioactive measurement. If a screen coated with the small crystals of phosphorescent zinc sulphide is exposed to α rays, a brilliant luminosity is observed. On viewing the surface of the screen with a magnifying glass, the light from the screen is seen not to be distributed uniformly but to consist of a number of scintillating points of light scattered over the surface and of short duration. Crookes devised a simple apparatus called a "spinthariscope" to show the scintillations. A small point coated with a trace of radium is placed several millimetres away from a zinc sulphide screen which is fixed at one end of a short tube and viewed through a lens at the other end. In a dark room the surface of the screen is seen as a dark background dotted with brilliant points of light which come and go with great rapidity. This beautiful experiment brings vividly before the observer the idea that the radium is shooting out a stream of projectiles each of which

---

* Greinacher, *Zeit. f. Phys.* **23**, 361, 1924; **36**, 364, 1926; **44**, 319, 1927.
† Geiger and Klemperer, *Zeit. f. Phys.* **49**, 753, 1928.

causes a flash of light on striking the screen. The correctness of this impression has been verified by direct experiment, for each particle which falls on the crystals of zinc sulphide has been found to give a scintillation. A similar effect is observed in certain kinds of diamonds, but the scintillation is much weaker than in zinc sulphide.

Regener* was the first to devise methods of counting scintillations in order to determine the number of α particles incident on the screen. It has been found best for this purpose to use a microscope of magnification about 30 with an objective of large numerical aperture to increase the apparent brightness of the scintillations. The experiments should be made in a dark room when the eye is well rested, and the surface of the zinc sulphide should be faintly lighted to keep the eye focussed upon it.

The scintillation method of counting α particles has been widely used in many experiments and has proved a method of very great delicacy and power. By its aid we are able to count directly the number of α particles falling on the screen in a given time, and also to count the high speed hydrogen nuclei which are set in motion when an α particle collides with a hydrogen atom or are liberated during the disintegration of certain atoms.

We have seen that such swift ionising particles can also be detected and counted by the electric method under certain conditions, but it is in general found impracticable to count α particles by the electric method in the presence of a strong radiation due to β and γ rays. The zinc sulphide screen has the great advantage that it is comparatively insensitive to β and γ rays and can be employed to count α particles even when exposed to such a strong γ radiation that the electric method is quite inapplicable. For this reason the scintillation method has proved invaluable in experiments to determine the laws of the scattering of α particles by atomic nuclei and in investigating the artificial disintegration of atoms due to violent nuclear collisions. On account of its importance as a powerful and extraordinarily delicate method of research, a description is given in §§ 126 a, 126 b of our knowledge of the origin, duration and efficiency of these scintillations and the limits of their detection. A discussion is also given of the arrangements suitable for counting scintillations under various conditions and of the methods for testing the efficiency of counters.

* Regener, *Verh. d. D. Phys. Ges.* **19**, 78 and 351, 1908.

§ 12 c. Photographic action of the α rays. The α, β and γ rays all produce a marked action on the photographic plate but, owing to the difference in the absorption of these rays by a photographic film, it is difficult to make any quantitative comparison of their effects. A study of the photographic action of a pencil of α particles has, however, thrown much light on the processes involved. Kinoshita* in 1910 made a detailed study of the action of the homogeneous α rays from radium C on the photographic plate. He found that the photographic action was nearly constant along the range of the α particle but fell off rapidly near the end of its range. In this respect, the photographic action differs from the ionisation effect of the α particles which passes through a well-marked maximum near the end of the range. It is believed that a small number or even a single pair of ions formed in a grain renders it developable. The ionisation along the track of an α particle is so dense that every grain through which it passes must be ionised several times and so be equally developable independently of the density of the ions. By counting with a microscope, the number of blackened grains in a thin film after development was found equal to the number of α particles incident on the film for small photographic densities of the film. This is an important result, for it shows that each α particle produces a detectable photographic effect. Under suitable conditions this method can be used to count α particles.

The effect of a single α particle is more closely brought out by allowing α particles to fall nearly tangentially to the surface of the photographic film, when the track of each α particle is rendered clearly visible by the succession of grains produced along the track. A simple way of showing such tracks is to infect a photographic plate by touching it with a needle point containing a trace of active matter and allowing the activity to decay *in situ* before development. Under a high magnification, the radiating tracks due to the individual α particles are clearly visible. An example of such photographs is shown in Plate II, fig. 1, where two centres are visible. Photographs of this kind have been obtained by Reinganum, Michl, Makower, Walmsley, Kinoshita, Ikeuti, Sahni and others. Owing to the scattering of the air, some of the α particles may enter the film at some distance from the active point. In consequence, the tracks are not all of equal length. To avoid this effect Mühlestein took photographs with the film covered with mercury. Under such

* Kinoshita, *Proc. Roy. Soc.* A, **83**, 432, 1910.

PLATE II

Fig. 1. Tracks of α particles from infected
points on a photographic plate.

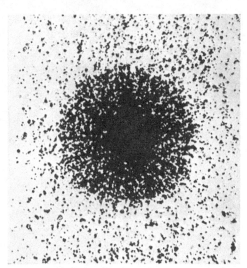

Fig. 2. Similar to Fig. 1, except that the plate
was covered with a layer of mercury.

Fig. 3. Tracks of α particles from thorium (C + C') in a
Wilson chamber, showing the two ranges.

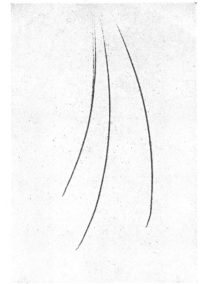

Fig. 4. Curved tracks of α particles
in a magnetic field of 43,000 gauss.

conditions the α particles show a definite range in the film and form a well-developed halo of radius depending on the velocity of the α particles (§ 38); an example of such a halo obtained by Harrington* is shown in Plate II, fig. 2. The very thin films of Schumann plates are especially serviceable in studying α particles of small velocity or other radiations of very small penetrating power. In such thin films the effect of β and γ rays which may be present is relatively very small compared with that of the α particle.

§ 12 d. Expansion method. The expansion method of detecting single α and β particles depends on the discovery by C. T. R. Wilson that the ions produced in gases by these radiations act as nuclei of condensation of water vapour when a certain supersaturation is reached. This supersaturation is produced by the sudden expansion of dust free air saturated with water vapour. If V_1 be the initial volume of the air, V_2 the volume after expansion, the negative ions act as centres of condensation when $V_2/V_1 > 1\cdot25$ and the positive ions when $V_2/V_1 > 1\cdot31$. For $V_2/V_1 = 1\cdot38$, a dense cloud appears even in the absence of all ions. Over a certain range, therefore, each ion becomes the nucleus of a visible drop of water which can be photographed under suitable illumination.

A great advance was made in 1912 when Wilson† found by an application of this method that the tracks of individual α and β particles could be clearly seen and photographed. By a timing arrangement instantaneous photographs were taken of the tracks of α particles at short intervals after the expansion. The beautiful photographs obtained by this method bring out in a striking and concrete way the path of these particles through the gas. Owing to the density of the ionisation, the path of the α particle shows as a continuous line of water drops. A swift β particle, on the other hand, gives so much smaller ionisation that the individual ions formed along its track can be counted (v. Plates IV and VIII). The photographs of the tracks of α particles (see Plate III) show clearly that they travel in nearly a straight line through the gas. A marked deviation of the α particle is occasionally observed, most frequently near the end of the track (cf. Plate III, fig. 2). This is due to the deflection of the α particle by collisions with the nuclei of atoms in their path.

* Harrington, *Phil. Mag.* 6, 685, 1928.
† C. T. R. Wilson, *Proc. Roy. Soc.* A, 85, 285, 1911; 87, 277, 1912; 104, 1, 1923.

By employing two cameras, stereoscopic pictures can be obtained which show the tracks of ionising particles in space and allow us to study in great detail the effects produced by a single α or β particle or by a weak beam of X rays or γ rays in traversing a gas. The method has the great advantage that the individual action of a single α or β particle can be studied at leisure on the photographs. In the hands of Wilson and others, it has proved a most powerful method of throwing light on all the complicated phenomena which occur in the passage of an ionising radiation through a gas.

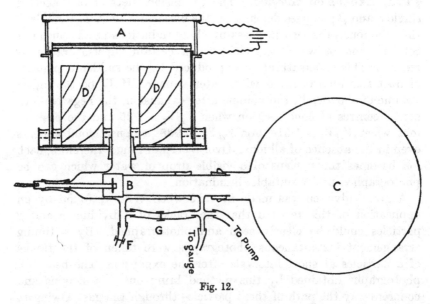

Fig. 12.

The apparatus used by Wilson is shown in Fig. 12. The closed chamber is cylindrical in shape, 16·5 cm. in diameter and 3·4 cm. high. The expansion is effected by opening the valve B in communication with the vacuum chamber C. The floor of the expansion chamber drops suddenly until brought to a sudden stop by striking an india-rubber base plate. The wooden cylinders D were added merely to reduce the volume of air required to be displaced during an expansion. The valve B is released by a falling pendulum which also times a Leyden jar discharge through mercury vapour to give a momentary strong illumination for the photographs. When necessary an electric field can be applied between the ceiling and floor of the expansion chamber to remove ions present before the expansion.

PLATE III

Fig. 1. Complete track of α particle ejected from atom of radon in air.

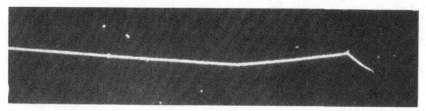

Fig. 2. End portion of track of α particle, showing two nuclear deflections (in air).

Fig. 3. Track of α particle in hydrogen, showing trails of δ rays.

Fig. 4. Track of α particle in hydrogen, showing effect of
magnetic field on δ rays.

By accurate timing, very sharp and clear stereoscopic photographs can be obtained with this apparatus. In a study, however, of rare phenomena, like the single scattering of α particles and the disintegration of atomic nuclei by bombardment with α particles, it is often necessary to photograph a very great number of α particles, of the order of one million. For this purpose an automatic type of apparatus is required which will allow a large number of expansions at short intervals. For this purpose, Shimizu* devised a simple form of expansion vessel which allows expansions to be repeated rapidly.

The working of this apparatus will be clearly seen from Fig. 13. By means of a motor, the tightly fitting piston of the expansion

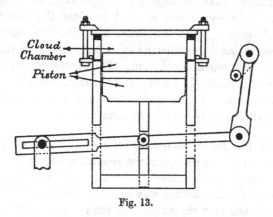

Fig. 13.

chamber is given a reciprocating motion of appropriate amount. Although the expansion is not sudden, as in Wilson's apparatus, it yet shows clear α ray tracks if the latter are allowed to enter the cloud chamber just at the end of the stroke of the piston. By arrangement of an electrical field to clear the cloud chamber between expansions, it is possible to make an effective expansion every few seconds. Shimizu developed a method of taking simultaneous photographs of the α ray tracks in two directions perpendicular to each other on a moving film, so that the orientation of the tracks in space could be determined. This general method has been used by Shimizu, Blackett†, Harkins and Ryan‡, Auger and Perrin§ and

* Shimizu, *Proc. Roy. Soc.* A, **99**, 425, 432, 1921.
† Blackett, *Proc. Roy. Soc.* A, **102**, 294, 1922; **103**, 62, 1923.
‡ Harkins and Ryan, *Phys. Rev.* **21**, 375, 1923; **23**, 308, 1924.
§ Auger and Perrin, *C.R.* **175**, 340, 1922.

others to study rare occurrences in the life of an α particle. Blackett, in particular, has used it to study in detail the nuclear scattering of α rays in different gases and to obtain visual evidence of the collisions in nitrogen which lead to the expulsion of a proton from the nitrogen nucleus. In order to obtain sharp and clear tracks, he has found it best to discard the reciprocating motion and to arrange by the rotation of a cam that a sudden expansion occurs at the right moment. It is necessary to time carefully the opening of a shutter admitting the α particles to the apparatus, the expansion and opening of the shutter of the film camera and the illumination. In practice the apparatus works automatically, being driven by a motor, and it is arranged that, in alternate expansions, no α particles are admitted, but the expansion is used to clear the cloud chamber. Examples of photographs obtained by Blackett are given in Plate IX illustrating the collisions of α particles with nuclei of the lighter gases. Photographs showing the disintegration of nitrogen nuclei are given in Plate X. Each photograph contains two pictures of the α particles viewed in perpendicular directions so that the orientation of the tracks in space can be determined.

§ 13. **Number of α particles from radium.** The number of α particles Q emitted from radium per gram per second is an important physical constant which it is desirable to know with precision. The first determination of this number was made in 1908 by Rutherford and Geiger*, using the electric method of counting α particles shown in Fig. 10. A thin film of the active deposit of radium was used as a source of α rays. After radium A has decayed, the α particles come entirely from radium C. The activity of this source was compared by the γ ray method in terms of radium in equilibrium with its products. They found a value $Q = 3\cdot4 \times 10^{10}$ in terms of the laboratory standard. This was subsequently calibrated in terms of the International Standard, giving a value $Q = 3\cdot57 \times 10^{10}$.

This gives the number of α particles from radium itself, freed from its subsequent products. When radium is in equilibrium with its products, the same number of particles are emitted per second by each of its α ray products, viz. radon, radium A and radium C. Consequently the number of α particles from 1 gram of radium in equilibrium with the short-lived products is $4Q$. The value of Q so found agrees fairly well with a number of measurements which involve

* Rutherford and Geiger, *Proc. Roy. Soc.* A, **81**, 141, 1908.

Q, viz. the production of helium by radium, the volume of radon, and the heating effect due to the α particles.

Subsequent determinations of Q have been made by Hess and Lawson* in Vienna in 1918, and by Geiger and Werner† in Berlin in 1924. Hess and Lawson used radium C as a source of α rays and counted the α particles by the electrical method. The detecting chamber was hemispherical in shape, containing a spherical electrode opposite the aperture where the α particles entered. In order to reduce the effect of the β and γ rays compared with the α rays, they found it best to use a mixture of 54 per cent. CO_2 and 46 per cent. air at about 40 mm. pressure in the detecting chamber. The string electrometer available had too slow a period for rapid photographic registration of the α particles, so the final results were obtained by eye observation of the throws of the string of the electrometer. As the result of a large number of fairly concordant observations, they found a value $Q = 3 \cdot 72 \times 10^{10}$.

Geiger and Werner made an investigation of the efficiency of the point-counter and found that it could not be relied upon completely for accurate work, since the number of α particles registered was never constant but increased slowly with the voltage applied to the counter. After a very careful investigation they concluded that Q was not greater than $3 \cdot 5 \times 10^{10}$. They then turned their attention to the scintillation method of counting α particles. The counting screen was completely covered with zinc sulphide crystals and viewed by two microscopes. The scintillations were counted simultaneously by two observers and independently recorded side by side on the same moving tape. From the number of scintillations counted by both observers and the number counted by one and not the other, they were able, by an ingenious treatment of the observations, to obtain the recording error of each observer and the true number of scintillations. A description of this method of testing the efficiency of counters is given in § 126 b. The source of α rays consisted of radon and its products contained in a small conical tube, one end of which was covered with a thin sheet of mica. A large number of scintillations were counted, and they concluded that $Q = 3 \cdot 40 \times 10^{10}$. This is a value 4 per cent. lower than Rutherford and Geiger's original number and 10 per cent. lower than the value found by

* Hess and Lawson, *Wien. Ber.* **127**, 405, 1918.

† Geiger and Werner, *Zeit. f. Phys.* **21**, 187, 1924; Geiger, *Verh. d. D. Phys. Ges.* **5**, 12, 1924.

Hess and Lawson. This is a large discrepancy, and the possible sources of error in the various methods have been a subject of some discussion. There is no doubt that a direct determination of Q is beset with considerable difficulties, partly connected with the source of α rays and partly with the efficiency of the electrical and scintillation methods of counting α particles. While there can be no doubt that the measurements of Geiger and Werner were carried out with great care and accuracy, there is some uncertainty as to the reliability of their source of α rays. It is possible that some of the emanation was absorbed in the material of the tube so that the full number of radioactive atoms, as measured by the γ rays, was not effective in the counting of the α particles.

In view of these discrepancies, new measurements of Q have been made by measuring the positive charge carried by the α particles from radium C, assuming the charge on the α particle is $2e$, where e is Millikan's precision value $4 \cdot 774 \times 10^{-10}$ E.S.U. The general method for measuring the charge carried by the α particles is described in the next section. H. Jedrzejowski* found in this way $Q = 3 \cdot 50 \times 10^{10}$, while Braddick and Cave†, as a result of a series of careful measurements, found $Q = 3 \cdot 68 \times 10^{10}$. In both experiments, special precautions were taken in the preparation of a suitable source of radium C of small area and in the removal of the β rays by a strong magnetic field. The collecting plate was of just sufficient thickness to absorb completely the α particles. The effect of the γ rays on the measurement of the charge was negligible in the experiments of Cave and Braddick. It is thus seen that these measurements by the same method also differ considerably. In order to throw further light on this question, experiments have been made by Ward, Wynn Williams and Cave‡ in another way. Using the Greinacher method, the actual ionisation current produced by the entering α particle is magnified by valves to give a deflection of the string of an Einthoven galvanometer which is recorded photographically on a moving film. The results obtained by this method, which should be free from serious errors, give a value $Q = 3 \cdot 66 \times 10^{10}$, in fair agreement with the value of Lawson and Hess, $3 \cdot 72 \times 10^{10}$, and Cave and Braddick, $3 \cdot 68 \times 10^{10}$, and also with the value deduced from the heating effect of the α rays.

When we consider the care taken in the measurements, it is

---

* Jedrzejowski, *C.R.* **184**, 1551, 1927.

† Braddick and Cave, *Proc. Roy. Soc.* A, **121**, 367, 1928.

‡ Ward, Wynn Williams and Cave, *Proc. Roy. Soc.* A, **125**, 713, 1929.

difficult to account for the divergence between the various values of Q, and particularly the low values in some determinations. One of the main difficulties in these experiments lies in the choice and preparation of a suitable source of α rays which is free from all uncertainty and can be measured with precision.

Considering the evidence as a whole, it is probable that the value of Q lies between $3\cdot65 \times 10^{10}$ and $3\cdot70 \times 10^{10}$, and for purposes of calculation a value $Q = 3\cdot70 \times 10^{10}$ will be used in all subsequent calculations. It is of interest to consider estimations of Q which can be obtained from a variety of indirect data.

The volume of helium produced by 1 gram of radium in equilibrium with its products has been carefully determined by Boltwood and Rutherford* and found to be 156 c.mm. per year in terms of the International Radium Standard. Since each α particle on losing its charge becomes an atom of helium, we can at once deduce the value of Q by using Millikan's data that the number of atoms in 1 c.c. of helium is $2\cdot705 \times 10^{19}$. The value so found is $Q = 3\cdot50 \times 10^{10}$.

Method	Investigator	Date	Q
Electric counting	Rutherford and Geiger	1907	$3\cdot57 \times 10^{10}$
	Hess and Lawson	1918	$3\cdot72 \times 10^{10}$
	Geiger and Werner	1924	$3\cdot5 \ \times 10^{10}$
	Ward, Williams, and Cave	1929	$3\cdot66 \times 10^{10}$
Scintillation	Geiger and Werner	1924	$3\cdot40 \times 10^{10}$
Total charge	Jedrzejowski	1927	$3\cdot50 \times 10^{10}$
	Braddick and Cave	1928	$3\cdot68 \times 10^{10}$
Production helium	Boltwood and Rutherford	1911	$3\cdot50 \times 10^{10}$
	Dewar	1910	$3\cdot69 \times 10^{10}$
Life of radium	Gleditsch	1919	$3\cdot55 \times 10^{10}$
Heating effect	Rutherford and Robinson	1912	$3\cdot72 \times 10^{10}$
	Watson and Henderson	1928	$3\cdot70 \times 10^{10}$
Volume of radon	Wertenstein	1928	$3\cdot7 \ \times 10^{10}$

Rutherford and Robinson† found that the heating effect of the α rays of radon in equilibrium with its products was 97·4 gram calories per hour after correcting for the heating effect of the recoil atoms and of the β and γ rays. Assuming that the heating effect is a measure of the energy of the expelled α particles (using Geiger's data for the ranges), we find $Q = 3\cdot72 \times 10^{10}$. A concordant value, $Q = 3\cdot70 \times 10^{10}$, was obtained by Watson and Henderson in their recent measurement of the heating effect of radium products.

* Boltwood and Rutherford, *Wien. Ber.* 120, 313, 1911; *Phil. Mag.* 22, 586, 1911.
† Rutherford and Robinson, *Phil. Mag.* 25, 312, 1913; *Wien. Ber.* 121, 1491, 1912.

E. Gleditsch found by direct measurement that the transformation constant of radium is $4\cdot18 \times 10^{-4}$ (year)$^{-1}$. Since one α particle is expelled from each disintegrating atom of radium, it can at once be calculated from atomic data that $Q = 3\cdot55 \times 10^{10}$. In some recent experiments, Wertenstein* measured the volume of radon in equilibrium with 1 gram of radium, and found it to be $0\cdot7$ c.mm. From this result the value Q can be calculated to be $3\cdot7 \times 10^{10}$. For convenience we include in the above table the estimates of Q made in these various ways.

The values obtained vary between $Q = 3\cdot4 \times 10^{10}$ and $Q = 3\cdot72 \times 10^{10}$. Considering the difficulty of accurate measurement by some of the indirect methods, there is a fair accord. We have seen that the true value is probably close to $Q = 3\cdot70 \times 10^{10}$.

§ 14. The charge carried by the α particle. By measuring the total charge carried by a counted number of α particles, the charge E carried by each particle can be found. Certain difficulties were encountered at first in measurements of the charge on account of the vigorous emission of electrons (δ rays) by surfaces on which the α particles fall (see § 29). The charge was measured by Rutherford and Geiger† by the method shown in Fig. 14. The source of radiation was radium C deposited on a glass vessel R. The α rays, after passing through a thin aluminium foil which served to stop recoil atoms, fell on a plate AC, the lower surface of which was covered with thin aluminium foil. The fraction of the α particles falling on AC was defined by the diaphragm B. The whole apparatus was placed in a strong magnetic field perpendicular to the stream of particles, in order to bend away the β rays from the source and

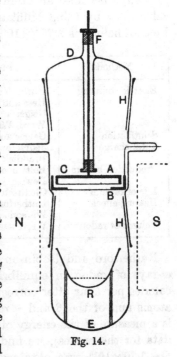

Fig. 14.

to reduce to a minimum the effect of the δ rays. The charge communicated to AC in a good vacuum was measured by an electrometer.

* Wertenstein, *Phil. Mag.* **6**, 17, 1928.

† Rutherford and Geiger, *Proc. Roy. Soc.* A, **81**, 141, 1908.

The strong magnetic field also deflects the electrons liberated by the α rays so that they return to the plate from which they started. It will be seen from § 29 that the maximum speed of the liberated electron is twice the velocity of the α particle. In order to prevent the escape of the electrons emitted obliquely from the surface, the collecting plate was surrounded by a shallow copper cup. In this way the application of a strong magnetic field is entirely successful in preventing the escape of the electrons, so that the charge carried by the α particles themselves can be measured. By this method Rutherford and Geiger found that the charge carried by the α rays from radium C in equilibrium with 1 gram of radium was 33·2 E.S.U. per second on the International Standard. Taking the number of α particles from 1 gram of radium per second as $3·57 \times 10^{10}$, the charge on the α particle was found to be $9·3 \times 10^{-10}$ E.S.U. The general evidence at that time indicated that the α particle carried two charges, so that the value of the unit charge e was $4·65 \times 10^{-10}$ E.S.U. This may be compared with the precision measurement made later by Millikan by other methods, viz. $e = 4·774 \times 10^{-10}$ E.S.U. At that time, 1908, this radioactive method provided the most reliable estimate of e, and for a considerable time it was widely used in radioactive and other calculations. If we assume the precision value of e, we can conversely deduce the number of α particles emitted per second per gram of radium, viz. $3·48 \times 10^{10}$.

Further measurements of the charge carried by the α particle were made by Duane and Danysz* in 1912 by a similar method. A careful investigation was made of the effect on the measurements of varying the magnetic field and of applying an electric field in two directions. As source, a very thin α ray bulb containing radon was used. They found the charge carried per second by the α rays from 1 gram of radium itself was 30·3 E.S.U. in terms of their standard, a value somewhat smaller than that found by Rutherford and Geiger. It was, however, difficult in their experiments to make certain that all the α rays from the active matter escaped through the walls of the bulb.

We have discussed in the last section some new determinations of the total charge by Jedrzejowski, Braddick and Cave, who find values 33·4 and 35·1 E.S.U. per second respectively.

It is of interest to record an early measurement of the charge, made by Regener†. The α particles from a source of polonium were

* Duane and Danysz, *Le Radium*, **9**, 417, 1912.
† Regener, *Ber. Preuss. Akad. d. Wiss.* **38**, 948, 1909.

counted by observing the scintillations produced on a small diamond, while the charge carried by the α rays was measured. He found the charge on each particle to be 9.58×10^{-10}, giving a value of $e = 4.79 \times 10^{-10}$, in good accord with Millikan's later value. In his experiments, however, the polonium source was uncovered, so that the charge, if any, of the recoil atoms would be measured with the α particles. It may be that most of the recoil atoms, as Wertenstein observed in the case of recoil from radium C, were uncharged, so that the result was not affected.

PLATE IV

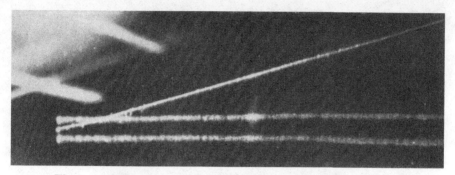

Fig. 1. α tracks from actinium emanation and actinium A atoms.

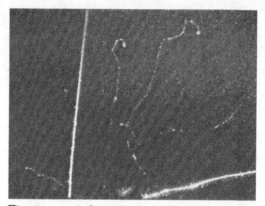

Fig. 2. α and β tracks (in hydrogen), showing
the difference in ionizing power of the particles.

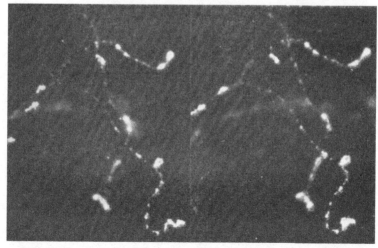

Fig. 3. Stereoscopic pair showing the collision of a slow β particle
with an electron.

CHAPTER III

ABSORPTION OF THE α RAYS

§ 15. Absorption of α rays. Historical. The absorption of α rays by matter was first investigated by Rutherford in 1899 by the electric method. A layer of radioactive material was spread on a plate and the variation of the saturation current between this plate and another placed parallel to it was examined by an electrometer when successive screens of absorbing matter were placed over the active matter. The current was found at first to decrease approximately according to an exponential law with the thickness of matter traversed, but ultimately fell off more slowly. Experiments of this kind were first made with uranium, and led to the divisions of the radiation into two types, called the α and β rays. Observations by similar methods showed also that the α rays were rapidly absorbed in air and other gases. Many experiments of this kind were made in the early days of radioactivity by Owens, Rutherford and Miss Brooks, Meyer and Schweidler and others.

When more active materials were available, it was possible to obtain easily measurable effects with a definite pencil of α rays. Mme Curie examined the absorption of the α rays from a thin film of polonium by a different method. The rays from polonium passed through a circular opening in a metal plate covered with a wire gauze or a thin metal foil, and the ionisation of the rays, after passing through the hole, was measured between this plate and a plate placed 3 cm. above it. No current was observed when the polonium was 4 cm. below the hole, but as this distance was diminished beyond a certain value, the current rose rapidly so that a small change in the distance caused a large alteration in the current. This effect was entirely different from that to be expected if the α rays were absorbed by matter according to an exponential law, but indicated that the ionisation of the α rays from polonium ceased suddenly after traversing a certain thickness of air or other absorber.

In 1903 Rutherford had shown that the α rays consisted of positively charged atoms projected with high velocity. W. H. Bragg early in 1904 suggested that the peculiarities observed in the absorption of α rays could be explained by supposing that the massive

α particle of high velocity travelled in nearly a straight line through the gas, expending its energy in ionisation until its velocity fell to a certain value. The α particle should on this view have a definite range of travel in the gas depending on its initial velocity. Experiments were at once begun to test this hypothesis by examining the ionisation along the path of a pencil of α rays emitted by a thin layer of radium. This brought to light the characteristic ranges of the α particles arising from the individual active products radium, radon, radium A, radium C present when the radium is in equilibrium. This led to experiments on the ranges of α rays in different gases and metals, and a study of the laws of absorption of the α particle.

The essential correctness of these views of the nature of the absorption of the α particle was confirmed by a study of the deflection of a pencil of α rays in a magnetic field (Rutherford and Geiger). The α particles were shown to be ejected initially at the same speed from a thin film of active matter like radium C and to retain their homogeneity to a large extent in traversing matter. The law of retardation of the α particle was carefully examined. Subsequent researches by a large number of investigators have been concerned with a detailed study of the phenomena occurring in the passage of α particles through matter in the gaseous, solid and liquid state, such as the ionisation and range of the α particles in different materials, the laws of the scattering and straggling of the particles and, more recently, the variation of charge during the flight of the α particle. These investigations, both experimental and theoretical, have thrown much light on this most interesting and important problem, and we have now a fairly clear idea of the events that may happen in the short life of the α particle after it has been liberated from a radioactive atom. The information obtained from the various directions of investigation will now be considered in detail. In this chapter, we shall discuss the variation of the ionisation along the path of the α particle in different gases, methods of measurement of the range of α particles and the relative stopping power of the α particle by different materials.

§ 16. Ionisation curve for α particles. We have already seen that Bragg first observed that the α rays from a thin layer of one kind of radioactive matter had a definite range of travel in air, and that the ionisation produced by the rays fell off abruptly at the end of their path. This "range" of the α particle depends on its velocity and the nature of the absorbing material, whether solid, liquid or gaseous. The variation of the ionisation along the track of the α particle in a gas can be best investigated by the electric method,

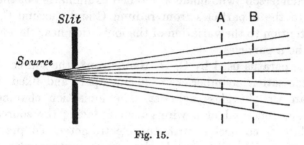

Fig. 15.

using a narrow pencil of homogeneous rays. Suppose the saturation ionisation current produced by a narrow pencil of α rays between parallel gauzes A and B, Fig. 15, is measured at different distances from the source S. The ionisation current between A and B, which may be close together, is a measure of the number of pairs of ions

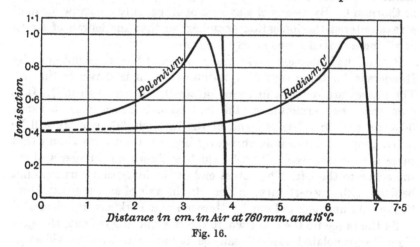

Fig. 16.

produced in the gas between the two plates. The current is found to increase with distance from the source, pass through a maximum near the end of the range and then fall off rapidly to zero. The curves connecting the amount of ionisation at each point of the range of the α particles from radium C′ and polonium obtained by I. Curie are shown in Fig. 16. The ionisation curve can be obtained either by

altering the distance between the source and AB, keeping the pencil
of α rays constant, or by keeping the source and AB fixed at a convenient distance apart in a closed vessel and varying the pressure
of the gas. This latter method was adopted by Geiger, Lawson and
later by Henderson, who made a detailed examination of the ionisation due to the α particles from radium C and thorium C, paying
special attention to the variation of the ionisation near the end of the
range of the α particles.

In the apparatus used by G. H. Henderson[*], the source of α rays
and the ionisation chamber were 38 cm. apart and fixed near the
ends of an airtight glass cylinder. The ionisation chamber, only
1 mm. deep, consisted of a wire gauze B facing the source and a
copper plate C connected with the electrometer. To prevent the
collection of ions formed outside the gauze, another wire gauze A
connected with earth was placed 1 mm. in front of B, which was kept
at a potential sufficient to ensure a saturation current. The pressure
of the air in the chamber could be adjusted and accurately measured.
The source consisted of a polished disc of brass or nickel 1 cm.
diameter, which could be activated by a deposit of radium B and C
or thorium C. By means of a system of stops, only a narrow beam of
α rays entered the ionisation chamber, so that the length of travel
of all the α particles was nearly the same.

Part of the ionisation curve near the end of the range obtained by
Henderson for the α rays from radium C in air is shown in Fig. 17.
The ranges are reduced in terms of air at 760 mm. and 0° C. It is
seen that the maximum of the ionisation is reached for a range
6·13 cm. and the observed end of the range is 6·70 cm. It will be
noted that the curve falls off nearly linearly from C to B and that
there is a pronounced tail (BA) which is shown on a larger scale in
the curve to the left. The actual end of the ionisation curve cannot
be fixed with any certainty, as it meets the axis of abscissae asympto-
tically. As first pointed out by Marsden and Perkins[†], it is desirable
to fix the range by the point where the straight line BC cuts the axis.
This "extrapolated range" can be determined with considerable
accuracy, and should be used to fix the range of all groups of α
particles. The maximum range, which depends on the refinements
of measurements, is too indefinite and uncertain for this purpose.
Henderson found by this method that the extrapolated range of

* Henderson, Phil. Mag. 42, 538, 1921.
† Marsden and Perkins, Phil. Mag. 27, 690, 1914.

radium C particles in air is 6·592 cm. at 0° C. and 760 mm. There is thus a difference of about 1·1 mm. between this range so defined and the maximum range detected under his experimental conditions.

Exactly similar end curves were obtained for the α rays from thorium C and the slope of the straight portion was identical. The

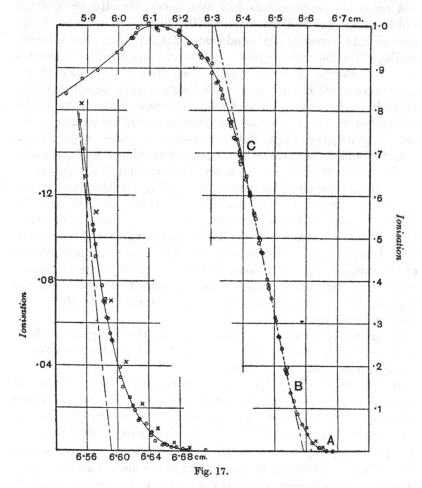

Fig. 17.

range of one group of rays was found to be 4·529 cm. and the other (ThC′) 8·167 cm. at 760 mm. and 0° C.

Concordant values for these ranges have been obtained by Geiger*, viz. 6·608 cm. for radium C, 4·538 cm. for thorium C, and 8·168 cm.

* Geiger, *Zeit. f. Phys.* **8**, 45, 1921.

for thorium C'. I. Curie and Behounek* found a range 6·60 cm.
for radium C. When once these extrapolated ranges are fixed with
precision for such convenient sources of α rays, it is possible by
comparison to fix the ranges of α particles from other substances
(§ 19).

A series of measurements has been made recently by I. Curie
and Behounek to determine the ionisation curve with accuracy under
definite and reproducible conditions. They discuss the methods
employed for this purpose and in particular the possible errors due
to the scattering of the α rays, arising from the position of the
diaphragm which is used to obtain a nearly parallel pencil of α rays.
In their experiments the diaphragm is placed close to the source
and moves with it. The ionisation chamber consists of parallel plates
provided with guard rings. Both the distance of source and diaphragm
from the ionisation chamber and the pressure of gas can be varied.

The curves obtained for the α rays from radium C' and polonium,
at 15° C. and 760 mm., have been given in Fig. 16. They confirm the
straight end part of the curve and the tail of the curve as found by
Henderson but differ from him somewhat in the shape of the maximum and of the earlier part of the curve. They find that their curve
falls above that of Henderson, the relative ionisation at the beginning
of the range being 0·41 instead of 0·31. In a later careful investigation I. Curie and F. Joliot† find the ionisation curve near the
beginning of the range to lie between the curves found by Henderson
and Curie and Behounek.

It should be pointed out that the long range α particles from
radium C and thorium C are much too few in number to influence
the results of these experiments.

It will be shown (§ 22) that the velocity V of α particles of
range R is given very approximately by the relation $V^3 = aR$, where
a is a constant. If it be supposed that the ionisation produced by
an α particle in a short distance dR is a measure of the energy dE
lost by the α particle, then $dE/dR \propto R^{-\frac{1}{3}} \propto 1/V$, or, in other words,
the ionisation produced by an α particle should be inversely pro-
portional to its velocity. This deduction is in fair accord with observations of the ionisation in air for α particles of ranges between
8·6 and 2 cm., but is probably only a rough approximation to a much
more complicated law of variation (see §§ 22, 22 a).

* I. Curie and F. Behounek, *Journ. de Phys.* **7**, 125, 1926.
† I. Curie and F. Joliot, *C.R.* **187**, No. 1, p. 43, 1928.

Geiger early suggested that the general shape of the ionisation curve could be explained by supposing it to be the average effect of a number of α particles of different ranges. He supposed that the ionisation curve of a single α particle varied nearly inversely as the velocity until close to the end of the range and then fell off sharply to zero. At that time, however, he relied on the straggling curve obtained by the scintillation method (§ 23 a), but this, as we now know, greatly exaggerated the true straggling.

It is obvious that the shape of the ionisation curve near the end of the range is closely connected with the straggling of the rays, i.e. to the inequalities in the range of α particles all expelled initially with the same speed. A full discussion of this effect is given in § 23. I. Curie determined the straggling directly by measuring the lengths of the tracks formed in an expansion chamber by α particles from polonium. Knowing the actual distribution of particles as regards range, and the ionisation curve, it should be possible to deduce the true ionisation curve of an α particle or group of particles which all have the same range. This interesting question has been attacked by I. Curie from data obtained on the α rays of polonium. By taking the ionisation per unit path to be proportional to the loss of energy of the α particle, it was found possible by graphical analysis, using certain simplifying assumptions, to deduce the ionisation curve for an average α particle. It was concluded that the true ionisation curve of the average particle is very similar in shape to the ordinary ionisation curve, i.e. the ionisation passes through a maximum about 4·5 mm. from the end of the range in air and then falls off, at first gradually and then more rapidly near the end of the range. It is, however, difficult to determine the exact shape of the curve by such methods.

Another interesting method of attack on this question has been used recently by Feather and Nimmo*. Observations of the density of photographs of the tracks of the α particles obtained by the expansion method show clearly that the density falls off rapidly in the last few millimetres of the range in air. Measurements of the density of a number of tracks in air, hydrogen and helium were carefully made by a micro-photometer. Averaging the results for 28 tracks, the density was found to be a maximum at a distance 3·0 mm. from the end of the track reduced to air at 15° C. and 760 mm. The corresponding distances in air for hydrogen and helium, corrected for the known differences in stopping power near the end of the range, were

* Feather and Nimmo, *Proc. Camb. Phil. Soc.* **24**, 139, 1928.

2·25 mm. for hydrogen and 2·55 mm. for helium. The forms of the curves for air and helium showed marked differences. The evidence as a whole indicates that the density of a uniformly illuminated track, corrected for the characteristic curve of blackening of the plate, is a measure of the number of droplets formed and of the ionisation of the gas at that point. It will be observed that the maximum of the ionisation found by this method varies for the different gases and is considerably nearer the end of the range, viz. 3·0 mm., than that deduced by I. Curie, viz. 4·5 mm. It is suggested that the smaller distance for H_2 and He may be connected with the smaller probability of capture of electrons by the α particles in these gases.

From the measurements of Briggs (§ 22), the velocity of an α particle of range 3·0 mm. is $0·22 V_0$ or $4·2 \times 10^8$ cm./sec., corresponding in velocity to a 50 volt electron. On the other hand, the ionisation per cm. for an electron passes through a maximum for a speed corresponding to about 200 volts*.

It is of interest also to compare the rate of loss of energy of the α particle with the ionisation produced at each element of its path. From the data given in § 22 for the retardation of the α particle in air, Briggs compared the variation of dT/dx, where T is the energy of the α particle from radium C (Curve 1, Fig. 18), with the ionisation curves obtained by Henderson (Curve 2) and Curie and Behounek (Curve 3). The results are shown in Fig. 18, where the value of the maximum ionisation in each case is made the same. It is seen that the shapes of the curves are very similar and pass through a maximum at about the same range. A similar result has been found by I. Curie and F. Joliot (loc. cit.), who re-determined the ionisation curve of the α particles from radium C. They conclude that for equal expenditure of energy, the ionisation produced is about 2 per cent. greater for the first 3 cm. of the range than for the latter part of the range. This excellent agreement between expenditure of energy and ionisation is somewhat surprising since it is known (§ 18) that, with the exception of the rare gases, only about 50 per cent. of the energy of the α particle can be accounted for as ionisation. It thus seems that the fraction of the energy that appears as ionisation in air is sensibly constant and independent of the velocity of the α particle. Since, as will be seen, the ionisation curves in different gases are not always identical in shape, this relation between energy and ionisation cannot be generally true.

* Lehmann, Proc. Roy. Soc. A, 115, 624, 1927.

While it is difficult to fix the true ionisation curve for a single average particle with precision, the various methods give results in fair accord. When we consider the rapid interchange of charges of the α particle near the end of its range and the fact that singly charged and neutral α particles predominate (*vide* § 24), it is obviously not easy to obtain any simple theory of the variation of the ionisation with velocity for such complicated processes. It is necessary to take into account not only the velocity of the α particle and its charge, but also the motion of the electron in the atoms of the gas.

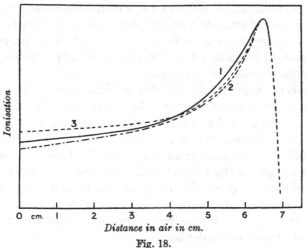

Fig. 18.

Henderson (*loc. cit.*) has shown that the pronounced tail of the ionisation curve (*vide* Fig. 17) is to be expected from the theory of straggling given by Bohr (§ 23). There is always a certain probability that a small fraction of the α particles will lose much less energy by collisions than the average particle and thus travel farther. The presence of a very small fraction of such favoured particles can be detected several millimetres beyond the "extrapolated range" when the tracks of an intense beam of α particles in an expansion chamber are photographed. Henderson has estimated the effect of this straggling on the ionisation curve, and the calculated values are shown by crosses in Fig. 17 (left) for the end of the curve.

The shape of the ionisation curve beyond the maximum, including the straight portion and the tail of the curve, can thus be accounted for in a general way by the theory of straggling and variation of the ionisation of the α particle near the end of its path. Gibson and Eyring* have pointed out that the straggling in air and other gases can be simply estimated on certain assumptions by determining the slope of the ionisation curves near the end of the range and with results in good accord with other observations.

We have so far considered the ionisation curve in air only. It was early noted by T. S. Taylor† that the maximum of the ionisation curve in hydrogen was much more pronounced than in air. In subsequent experiments, the ionisation curves for the α rays from polonium were compared for hydrogen, helium, and air. F. Hauer‡ determined the relative ionisation at different parts of the range for air, O_2, CO_2 and He and found small but evident differences. He concluded that the loss of energy of the α particles was not accurately proportional to the ionisation but was a function of the velocity of the α particle and the molecular weight of the gas.

Some recent experiments on this subject have been made by G. Gibson and E. Gardiner§, using the α rays from polonium for the purpose. Allowing for differences in stopping power, they found the shape of the ionisation curves very similar in helium, nitrogen and air, but differing markedly from the curves for hydrogen and neon. They also compared the relative ionisation and stopping powers in a number of complex gases. For example, Gibson and Eyring (loc. cit.) found the ionisation in nitrous oxide, nitric oxide and air to be proportional at all points of the range, but the ionisation in methyl iodide showed systematic deviations.

We have seen that the fraction of the absorbed energy of the α particle spent on ionisation in air appears to be nearly constant at all points of the path of the α particle. From the above results we should expect a similar result to hold for helium and nitric and nitrous oxides. If this be the case, the fraction cannot be constant in the case of hydrogen but must depend on the velocity of the α particle.

In general, the results so far obtained indicate that the ionisation is not in all cases proportional to the absorption of energy from the α particle but depends on the velocity of the α particle and on the nature of the gas. Such a result is not unexpected since, in the case

        * Gibson and Eyring, *Phys. Rev.* **30**, 553, 1927.
        † T. S. Taylor, *Phil. Mag.* **21**, 571, 1911; **26**, 402, 1913.
        ‡ Hauer, *Wien. Ber.* **131**, 583, 1922.
        § Gibson and Gardiner, *Phys. Rev.* **30**, 543, 1927.

of diatomic and polyatomic gases, it is known that only about half of the energy absorbed can be accounted for by ionisation. It is rather a matter of surprise that there should be such a close relative agreement between ionisation and absorption of energy for a number of gases over a wide range of velocity of the α particle. A further discussion of this general question will be found in §§ 18 and 28.

§ 17. **Number of ions produced by an α particle.** It is of considerable importance to know accurately the total number of ions produced in gases by the complete absorption of an α particle of known energy, as this gives us information about the average energy required to produce a pair of ions under definite conditions. The first accurate determination was made by Geiger* in 1909, in which the ionisation produced by a definite pencil of α rays from radium C was measured in hydrogen. He found the total number to be $2 \cdot 37 \times 10^5$, but this depended on data which are now obsolete. The calculation was based on the assumption that the ionic charge was $4 \cdot 65 \times 10^{-10}$, while the accepted value is now $4 \cdot 77 \times 10^{-10}$ E.S.U. The number Q of α particles expelled from 1 gram of radium per second was taken as $3 \cdot 57 \times 10^{10}$. Taking $Q = 3 \cdot 7 \times 10^{10}$ (§ 13), the number of ions reduces to $2 \cdot 2 \times 10^5$. Another determination has been made by Fonovits-Smereker† in 1922, where the ionisation current in air due to α particles from radium C was measured by a galvanometer. She found the number of ions produced by an α particle from radium C to be $(2 \cdot 20 \pm 0 \cdot 02) \times 10^5$. This is seen to be in good accord with Geiger's value.

New determinations of this quantity have been made by Bracelin‡ and by I. Curie and F. Joliot§, who both find a value $2 \cdot 2 \times 10^5$ for $Q = 3 \cdot 70 \times 10^{10}$. Assuming this value, the average energy to produce a pair of ions can at once be deduced. The energy of an α particle from radium C, using the latest data (§ 19), is found to be $7 \cdot 68 \times 10^6$ electron-volts. The energy required to produce a pair of ions in air is consequently about 35 volts.

If we take a pencil of α particles from radium C, for example, the ionisation per mm. of path in air at 15° C. and 760 mm. for the average α particle can be deduced from the ionisation curve shown

* Geiger, *Proc. Roy. Soc.* A, **82**, 486, 1909.
† Fonovits-Smereker, *Wien. Ber.* **131**, 355, 1922.
‡ Bracelin (unpublished).
§ I. Curie and F. Joliot, *C.R.* **187**, 43, 1928.

in Fig. 16. The total number of ions produced by absorption of the average α particle is taken as $2 \cdot 20 \times 10^5$. The range of the α particle is 6·97 cm.

Extrapolated range of α particle at 15° C. and 760 mm. (cm.)	No. of ions per mm. of path in air at 15° C. and 760 mm.
7·0	2440
6·0	2480
5·0	2540
4·0	2680
3·0	2880
2·0	3440
1·5	3960
1·0	4800
0·47	6000
0·21	4500

The results are only approximate and apply to a particle of average range. As we have seen (§ 16), the ionisation in air of a single track is a maximum about 3 mm. from the end of its range.

It is known that a considerable part of the ionisation produced by an α particle is secondary in character, due to the ionisation produced by the swift electrons (δ rays) liberated by the passage of the α particle through matter (§§ 28 and 29). The average energy required to produce an ion with α rays should consequently be about the same as that observed for electrons of moderate speeds corresponding to energies of a few hundred volts. It is of interest to note that Lehmann* finds the average energy to produce an ion in hydrogen is 37·5 volts and in helium 33 volts for 400 volt electrons. The value found for air is higher, about 45 volts (v. p. 499). There is consequently a rough agreement between the average ionisation energies observed for α particles and slow electrons.

Taking Geiger's relation $V^3 = aR$ connecting the velocity and range of α particles, $V^2 \propto R^{2/3}$, and assuming that the total ionisation due to an α particle is proportional to the energy of the α particle, the total ionisation of a particle of range R should be proportional to $R^{2/3}$. This simple relation between range and total ionisation has been found to be in fair accord with observation. It has proved very serviceable in estimating the total α ray ionisation to be expected from a thin film containing a series of products in equilibrium.

* Lehmann, *Proc. Roy. Soc.* A, **115**, 624, 1927.

Measurements of this kind have been made by Boltwood*, McCoy and Leman†. For example, it can be calculated that the ionisation due to radium in equilibrium with its products, radon, radium A and radium C, should give 4·09 times the ionisation due to the radium alone. The experimental value was 4·11. Similarly concordant results have been observed among the other series of radioactive elements.

Using a very sensitive electrometer, H. Ziegert‡ has detected individual α particles and measured directly the total ionisation in air due to each. He found the total ionisation due to the α particle from U I to be $1·16 \times 10^5$, from U II $1·29 \times 10^5$ and Ra $1·36 \times 10^5$. These values are in fair accord with the values calculated from the ranges.

§ 18. Ionisation in different gases. It is of great interest to compare the relative ionisation produced in different gases corresponding to definite absorption of energy from a beam of α particles, as this gives valuable information on the relative energy required to produce ions in different gases. This question was early examined by a number of investigators including Rutherford, W. H. Bragg, Laby, Kleeman and more recently by Hess and Hornyak, and Gurney. One of the chief difficulties in such comparisons is to ensure complete saturation in the ionised gas. This is particularly marked in complex gases and vapours of high molecular weight on account of the great density of the ionisation along the track of the α particle and the slow mobility of the heavy ions. Some of the discrepancies in earlier observations are no doubt to be ascribed to this cause.

It is to be anticipated that the average energy required to produce an ion would depend markedly upon the ionisation potential of the atom. It is of special interest to compare the relative ionisation in the monatomic gases, for which the ionisation potentials are known, with the ionisation in the simpler diatomic gases, where energy may be absorbed in dissociating the molecules as well as in ordinary ionisation and excitation.

A careful comparison of the ionisation in the monatomic gases for corresponding parts of the range of the α particles has been made

* Boltwood, *Amer. Journ. Sci.* 25, 269, 1908.

† McCoy and Leman, *Phys. Rev.* 6, 184, 1915.

‡ H. Ziegert, *Zeit. f. Phys.* 46, 668, 1928.

by Gurney*. A thin film of polonium served as a convenient source of α rays. A pencil of the rays passed through a mica window and was completely absorbed in an upper ionisation chamber to which an electroscope was attached. The average range of the rays entering the chamber could be varied by adjusting the pressure of gas between the source and the mica window. The results obtained are illustrated in Fig. 19, where the ionisations for different gases are compared with that for air taken as unity for α particles of ranges between 3 and

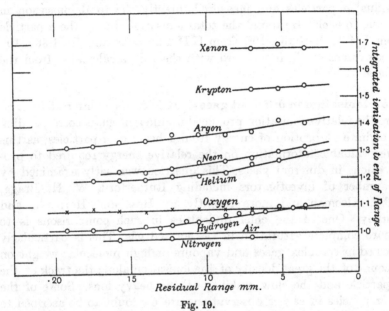

Fig. 19.

20 mm. of air. The ionisation shown in the curve represents the total ionisation due to absorption of the α particles of given range. It is seen that the relative ionisation depends not only on the gas but also on the average velocity of the α particles.

The curves for the rare gases and also hydrogen show a similar slope indicating a marked increase of the ionisation compared with air near the end of the range.

The relative ionisations by α particles of residual range 7 mm. are shown in the table.

Air	1·00	Nitrogen	0·98	Argon	1·38
Hydrogen	1·07	Oxygen	1·08	Krypton	1·53
Helium	1·26	Neon	1·28	Xenon	1·68

* Gurney, *Proc. Roy. Soc.* A, **107**, 332, 1925.

It is of interest to compare the relative ionisation with the atomic number and lowest ionisation potentials of these gases.

Gas	Atomic number	Energy spent per ion pair (volts)	Ionisation potential (volts)	Energy difference (volts)
Hydrogen	1	33·0	16·5	16·5
Helium	2	27·8	24·6	3·2
Nitrogen	7	35·0	17·0	18·0
Oxygen	8	32·3	15·5	16·8
Neon	10	27·4	21·5	5·9
Argon	18	25·4	15·3	10·1
Krypton	36	22·8	?13·0	? 9·8
Xenon	54	20·8	?11·0	? 9·8

The exact values of the ionisation potential of krypton and xenon are not known with certainty but have been estimated. The third column gives the energy expressed in electron-volts required to produce a pair of ions, assuming that the average energy required to produce a pair of ions in air corresponds to 35 volts (§ 17) and not 33 volts as taken by Gurney.

It is at once seen that for the diatomic gases examined, viz. H_2, N_2, O_2, the difference between the energy spent and the minimum energy required to ionise the atom shown in the last column is very marked. In fact only about half the energy spent is required to ionise the atom. This would indicate that a considerable part of the energy of the α particle is used up in processes which do not involve ionisation, i.e. in excitation or dissociation of the molecules. On the other hand, the differences are comparatively small for helium and neon, indicating that only a small part of the energy of the α particles goes in excitation in these gases. The difference is smallest in the case of helium. This is especially noteworthy when we recall that Wilkins[*] found marked evidence of double ionisation in helium amounting, near the end of the range, to about 10 per cent. Since it requires 1·6 times more energy to eject both electrons from a helium atom than to liberate two separately, the average ionisation potential to be taken for helium should be about 27 volts. This would indicate that very little, if any, energy is lost in helium by excitation. It will be seen from the theoretical discussion in § 28, that practically all the available energy appears as ionisation in the case of helium.

[*] T. R. Wilkins, *Phys. Rev.* **19**, 210, 1922; see also Millikan, Gottschalk and Kelly, *Phys. Rev.* **15**, 157, 1920.

The total ionisation relative to air in a large number of gases and vapours has been determined by various investigators using different methods. As already pointed out, there are in some cases considerable discrepancies between the values due in part to the difficulty of obtaining complete saturation. For example, values for oxygen have been found varying between 1·03 and 1·13 compared with air. The most probable value is about 1·08, found by Gurney. The relative values for the rare gases and also H_2, N_2, O_2 for different maximum ranges of the α particles can be read off from Fig. 19 due to Gurney. The following table, compiled by Kovarik and McKeehan*, gives the relative total ionisation observed, mean values being taken. Air 100 is taken as the standard.

Gas	Mean relative total ionisation	Gas	Mean relative total ionisation
H_2	99	C_2H_4O	105
N_2	96	HCl, HBr, HI	129
CO	101	CH_4	117
NH_3	90	CH_4O	122
CS_2	137	C_2H_2	126
SO_2	103	CH_3I	133
C_2H_4	122	C_2H_5I	128
C_2H_6	130	$CHCl_3$	129
C_5H_{12}	135	C_2H_5Cl	129
C_2H_6O	123	CCl_4	132
$C_4H_{10}O$	132	CH_3Br	132
C_6H_6	129		

It will be seen that the total ionisation does not vary much with the molecular weight or constitution of the heavier molecules. The ionisation potentials of these gases are not known with sufficient precision to make comparisons such as have been done for the rare gases and simple diatomic gases.

§ 19. **Range of the α particles in gases.** The range of the α particles can best be determined by the ionisation method first employed by Bragg. In some cases the scintillation method has proved useful, but on account of the difficulty of counting the weak scintillations produced by α particles near the end of their range, this method is troublesome and tends to give a slightly lower value of the range than the electric method. With strong sources of α radiation, however, scintillations give a rapid method of estimating the stopping power of absorbing screens when great accuracy is not desired.

* *Radioactivity Bull. Nat. Res. Council, U.S.A.* no. 51, p. 72, 1929.

A large number of experiments have been made to determine the range of the α particles expelled by the various radioactive substances. The α particles are expelled from each element at a characteristic speed and the total range of the α particle in air or other gases allows us to deduce the initial velocity of expulsion of the α particle.

As an example of a simple method for determining the ranges of α particles where only weak sources of radiation are available, we may refer to the experiments of Geiger and Nuttall *. The general arrangement is shown in Fig. 20. A thin layer of the active material is placed in the centre of a bulb on the small metal disc B which is connected with an electrometer through the rod H. The bulb was spherical and of 7·95 cm. radius and silvered on the inside. A high voltage was applied to the bulb and the saturation current measured for different pressures of air. For low pressures, the current is found to vary nearly proportionally with the pressure; but when the pressure reaches a value such that the α particles are completely absorbed in the gas, the current reaches a maximum value and there is no further change with increase of pressure. In order to obtain the best results the active matter on the disc should be in

Fig. 20.

the form of a thin film of a few millimetres diameter. Illustrations of the results obtained are shown in Fig. 21, where the ordinates represent the ionisation current and the abscissae the pressure of air in the bulb. It is seen that there is a sudden break in each curve, at a definite pressure corresponding to the point where the α particles just reach the surface of the bulb. Since the range of the α particle is inversely proportional to the pressure, the range in air at atmospheric pressure can at once be deduced.

By an adaptation of a method employed by Geiger and Nuttall in 1912, a very complete and accurate determination of the ranges of α particles from the radioactive elements was made by Geiger † in 1921. The general arrangement of the apparatus is shown in Fig. 22. The upper and lower vessels Z_1 and Z_2 were separated by a brass plate AB in which 500 holes were bored, each 15 mm. long and 3 mm.

---

* Geiger and Nuttall, *Phil. Mag.* 22, 613. 1911; 23, 445, 1912.
† Geiger, *Zeit. f. Phys.* 8, 45, 1921.

diameter. This served to produce a pencil of nearly parallel rays entering Z_1. A thin sheet of mica was waxed over the openings so that the vessels Z_1 and Z_2 could be separately exhausted. If the source emitted radioactive gases, these were prevented from entering the chamber by waxing a thin mica plate over the source. During an experiment, the pressure of air in Z_1 was adjusted between 3 and 6 mm. and the ionisation in Z_1 measured by an electroscope for

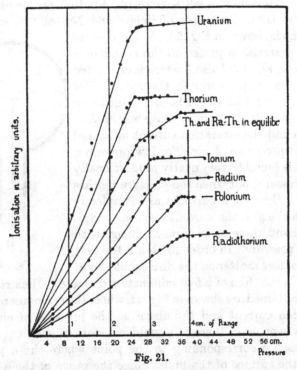

Fig. 21.

different pressures of air in the lower chamber. In this way, the end of the Bragg curve could be traced and the "extrapolated" range determined for each set of α rays by the method described in § 16. The apparatus was calibrated by using radium C as a source of α rays for which the extrapolated range was found in a separate experiment. The extrapolated range found by Geiger was 6·971 cm. at 15° C., a value in good accord with the measurements of Henderson, 6·953 cm., and I. Curie, 6·96 cm., at the same temperature.

All the ranges given in the table on p. 86 were carefully redetermined by Geiger, except the ranges of α particles from U I and U II and thorium. The ranges for the uranium products were deduced from the original experiments of Geiger and Nuttall, by comparison with the range of the α rays from ionium. These ranges have also been measured by Gudden (see § 38), by observations of the diameter of the rings in the haloes of radioactive origin found in fluorite. The two haloes due to the two sets of α particles from uranium were distinctly seen, and the ranges estimated in terms of air at 15° C. were 2·83 cm. and 2·91 cm., values somewhat smaller than those found by Geiger.

Recently the ranges for U I and U II have been measured by Laurence* by a new method. A thin film of uranium was placed on the cylindrical surface of an expansion vessel of the Shimizu type (§ 12 d). A large number of α ray tracks were photographed and the ranges were measured. In terms of air at 15° C. and 760 mm., the range of the α particles for U I was found to be 2·73 cm., for U II 3·28 cm. These ranges, which are believed to be correct within 1 per cent., are distinctly greater than those found by the two other methods.

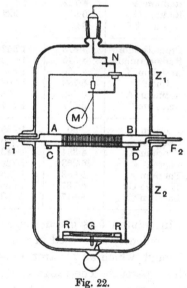

Fig. 22.

The range of the α rays from thorium is uncertain, as it is difficult to free it from radiothorium. The probable error of each determination is given in the original paper of Geiger. The range for protactinium is not known with certainty.

The results obtained by Geiger for the ranges of the α particles in air at 760 mm. and 15° C. are given in column 2 of the following table. In column 3 are given the velocities calculated on Geiger's rule; in column 4 the velocities estimated by Briggs† from the direct determination of the velocities of different groups of particles of known range, viz. radium C′, radium F, thorium C′ and thorium C. The values given by Briggs for the velocities and energies have been

* Laurence, Trans. Nova Scotia Inst. Sec. 17, pt 1, p. 103, 1927; Phil. Mag. 5, 1027, 1928.

† Briggs, Proc. Roy. Soc. A, 118, 549, 1928.

slightly changed to fit in with the latest value of E/M for the helium atom given in § 10 b.

Substance	Range in cm. at 15° C.	Velocity in cm. per sec.		Energy in electron volts
		$V \propto \sqrt[3]{R}$	From velocity curve for radium C′	
Uranium I	2·73	1·401 × 10⁹	1·391 × 10⁹	4·049 × 10⁶
Uranium II	3·28	1·495 × 10⁹	1·492 × 10⁹	4·626 × 10⁶
Ionium	3·194	1·482 × 10⁹	1·479 × 10⁹	4·545 × 10⁶
Radium	3·389	1·511 × 10⁹	1·511 × 10⁹	4·744 × 10⁶
Radon	4·122	1·613 × 10⁹	1·618 × 10⁹	5·441 × 10⁶
Radium A	4·722	1·688 × 10⁹	1·695 × 10⁹	5·972 × 10⁶
Radium C′	6·971	1·922 × 10⁹	1·922 × 10⁹	7·683 × 10⁶
Radium F*	3·925	1·587 × 10⁹	1·590 × 10⁹	5·253 × 10⁶
Protactinium	3·673	1·552 × 10⁹	1·553 × 10⁹	5·013 × 10⁶
Radioactinium	4·676	1·683 × 10⁹	1·688 × 10⁹	5·923 × 10⁶
Actinium X	4·369	1·645 × 10⁹	1·650 × 10⁹	5·659 × 10⁶
Actinon	5·789	1·807 × 10⁹	1·810 × 10⁹	6·819 × 10⁶
Actinium A	6·584	1·886 × 10⁹	1·887 × 10⁹	7·405 × 10⁶
Actinium C	5·511	1·777 × 10⁹	1·783 × 10⁹	6·610 × 10⁶
Thorium	2·90	1·435 × 10⁹	1·437 × 10⁹	4·231 × 10⁶
Radiothorium	4·019	1·600 × 10⁹	1·604 × 10⁹	5·347 × 10⁶
Thorium X	4·354	1·643 × 10⁹	1·648 × 10⁹	5·645 × 10⁶
Thoron	5·063	1·728 × 10⁹	1·733 × 10⁹	6·244 × 10⁶
Thorium A	5·683	1·796 × 10⁹	1·800 × 10⁹	6·737 × 10⁶
Thorium C†	4·787	1·696 × 10⁹	1·701 × 10⁹	6·015 × 10⁶
Thorium C′	8·617	2·063 × 10⁹	2·052 × 10⁹	8·761 × 10⁶

In column 5 is given the energy of the α particle; this is most conveniently expressed in volts, i.e. the potential difference in volts through which an electron must fall freely to acquire the energy of the α particle. The range of an α particle in a gas depends only on the number of atoms which it encounters in its path, and thus varies inversely as the pressure of the gas and directly as its absolute temperature. It is often convenient to express the ranges of α particles in air under normal laboratory conditions. For this reason the ranges are given for normal pressure at 15° C.

The number of pairs of ions produced by complete absorption of an α particle in air is approximately proportional to $R^{2/3}$, where R

* G. I. Harper and E. Salaman (*Proc. Roy. Soc.* A, **127**, 175, 1930) find the range of the α rays from radium F to be 3·87 cm. in air at 15° C. This is lower than Geiger's value but agrees with the observations by I. Curie (*Annales de physique*, **3**, 299, 1925). The ranges in hydrogen and argon have also been measured.

† Rosenblum (*C.R.* **190**, 1124, 1930) has measured the energies of the different groups of α particles emitted during the disintegration of thorium C (v. p. 47).

is its range and can be at once calculated since it is known that the total number of ions produced by an α particle from radium C' is $2 \cdot 20 \times 10^5$ (§ 7).

An approximate relation between the velocities of the α particles from different products and their period of transformation is discussed in § 35.

§ 20. **The long-range α particles.** In addition to the groups of α particles which have been assigned to a definite product in the radioactive series, there have been found associated with the complex bodies radium C and thorium C a few particles of abnormally long range. The first observation of these long-range particles was made by Rutherford and Wood* in 1916. When absorbing screens sufficient to stop completely the α particles of 8·6 cm. range were placed between a thorium C source and a zinc sulphide screen they still observed a small number of scintillations, which they attributed to an unknown group of long-range α particles. The range of these particles was found to be about 11·3 cm. in air, and their number was about 100 for every 10^6 α particles emitted from the source. They obtained indications that roughly one-third of these particles had a range of about 10 cm., but the evidence was not definite.

Later Rutherford† observed, during the investigations which led to the discovery of the artificial disintegration of nitrogen, that the α particles of radium C were accompanied, in their passage through nitrogen and oxygen, by a small number of particles which had a range in air of about 9 cm. On account of their small number—about 28 in 10^6 of the normal α particles—it has been difficult to fix with certainty the origin and nature of this group of long-range particles. For various reasons Rutherford at first concluded that these particles were not emitted by the source of radium C but arose in the volume of gas exposed to the collisions of the α particles, and suggested that they might prove to be atoms of nitrogen and oxygen set in rapid motion by close collision with an α particle. In later experiments‡ he examined the deflection of these particles in a magnetic field and concluded from his results that the particles probably had a mass of 3 and carried a double positive charge, and that they were indeed isotopes of the ordinary α particles. Subsequently, however, similar measurements§ of the magnetic deflection of the more numerous long-range particles from thorium C showed that

* Rutherford and Wood, *Phil. Mag.* **31**, 379, 1916.

† Rutherford, *Phil. Mag.* **37**, 537, 1919.

‡ Rutherford, *Proc. Roy. Soc.* A, **97**, 374, 1920.

§ Rutherford, *Phil. Mag.* **41**, 570, 1921.

these particles had a mass of 4, that is, were ordinary α particles. This result threw grave doubt on the former conclusions about the nature of the particles from radium C, and further investigation showed that these also were α particles, and, moreover, that most probably they were emitted by the source of radium C and were not produced by collision in the gases traversed by the α particles.

A detailed search for the presence of long-range particles from radium C, thorium C, actinium C, and polonium was made by Bates and Rogers* in 1924. They reported the presence of new groups of long-range particles from all these products, but the existence of most of these was not substantiated by the later investigations of Yamada†, Philipp‡, Mercier§ and others. While there is no certain evidence that actinium C or polonium emit any particles with ranges greater than those of the known α particles, it is now generally accepted that both radium C and thorium C give two groups of long-range particles. The earlier suggestion of Rutherford and Wood that the long-range particles from thorium C consist of two sets was definitely proved by Meitner and Freitag ||, who obtained tracks of the particles in a Wilson expansion chamber, and by Philipp, by the scintillation method. The ranges of the groups were given as about 9·5 cm. and 11·5 cm., the latter group being the more numerous. The second group of particles from radium C was found by Bates and Rogers, and confirmed by Rutherford and Chadwick ¶. In both cases the scintillation method was used. The particles have a range in air of about 11·2 cm., and are fewer in number than those of 9 cm. range.

The origin and nature of these long-range particles was a matter of some importance, for it seemed not improbable that they might be the result of the artificial disintegration of matter exposed to the bombardment of the ordinary α particles. A series of experiments to decide this question in the case of the particles from radium C was carried out by Rutherford and Chadwick. Attention was directed mainly to the particles of 9 cm. range, for the number of the 11 cm. particles was found to be too small to permit a detailed investigation.

In the first place, the number and range of the particles was

* Bates and Rogers, *Proc. Roy. Soc.* A, **105**, 97, 1924.
† Yamada, *C.R.* **180**, 436 and 1591, 1925; **181**, 176, 1925 and more.
‡ Philipp, *Naturwiss.* **12**, 511, 1924.
§ Mercier, *C.R.* **183**, 962, 1926.
|| Meitner and Freitag, *Zeit. f. Phys.* **37**, 481, 1916.
¶ Rutherford and Chadwick, *Phil. Mag.* **48**, 509, 1924.

examined when the α rays of radium C were passed through different materials, in particular through the gases helium, oxygen, carbon dioxide, and xenon. The long-range particles appeared in equal amount and with the same energy in all these cases. As it is improbable that different atoms, and especially the helium atom, would be disintegrated in exactly the same way and to the same extent, these results suggested strongly that the particles are emitted by the source.

This view was confirmed in a simple way by a shadow method, the principle of which will be clear from the diagram of Fig. 23. When the pressure of the gas, in this experiment carbon dioxide, was reduced to allow the main α rays from the source S to reach the zinc sulphide screen Z, a sharp shadow of the edge B was formed on the screen at P. The position of the observing microscope was adjusted so that a cross-wire in the eye-piece coincided with the edge

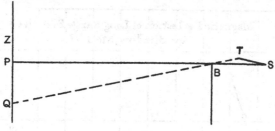

Fig. 23.

of the shadow. The pressure of the gas was then raised to cut off the main α rays, and the scintillations due to the long-range particles were observed. It was found that no scintillations occurred below the edge of the shadow. This showed either that the long-range particles are emitted from the source, or, if they arise in the gas, that their direction of emission cannot make an angle of more than a few degrees with the direction of the colliding α particle. For example, if the long-range particles were produced in the gas and if their direction of emission made even a small angle with the direction, say ST, of the impinging α particle, then scintillations would be observed on the screen below P, say at a point Q.

Taken in conjunction with the previous evidence, this experiment gave a simple and direct proof that the long-range particles are emitted by the source. There still remained the possibility that the particles were produced in the material on which the radium C was deposited. As, however, no difference in the number or range of the particles could be found when the active matter was deposited on platinum, copper, brass, or nickel, it was concluded that the particles are emitted directly by radium C.

The magnetic deflection of the particles of 9 cm. range was measured by two methods, of which the first will be described. An arrangement similar to that of Fig. 23 was used, in which the particles passed through an atmosphere of carbon dioxide and threw a shadow of a graphite edge on a zinc sulphide screen. If a strong magnetic field were applied at right angles to the path of the rays and the plane of the paper, the particles would be deflected and the shadow of the edge would move downwards, say. On reversing the direction of the field, the shadow will move upwards. The distance between the two positions of the shadow gives a measure of the value of mv/e for the particles.

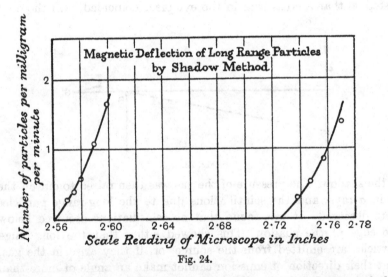

Fig. 24.

The apparatus was calibrated by observing the deflection of the shadow cast by the α rays of range 8·6 cm. emitted from a source of thorium (B + C). The number of scintillations was so great as to give a definite edge to the beam of rays and the distance between the two positions could be measured accurately. The number of long-range particles was not sufficient to show at once the edge of the shadow, and for these the position of the edge was found by counting the number of scintillations which appeared as the microscope was moved into the beam of particles. An example of such a measurement is shown in Fig. 24. The ordinates give the numbers of particles counted for the different positions of the microscope given by the abscissae. The point of intersection of the curve with the base line gives the position of the edge of the shadow. The distance between the two positions of the edge found in this experiment was 0·165 inch, and in another similar experiment 0·166 inch. In the calibration experiment with α particles of 8·6 cm. range the distance between the two positions was 0·173 inch. The measurements showed, therefore, that the long-range particles have very closely the same value of mv/e as α particles of the same range.

The experiments outlined above led, therefore, to the conclusion that the long-range particles observed with sources of radium C are α particles arising from the disintegration. While the evidence in the case of the particles of 9 cm. range is very convincing, that for the 11 cm. particles is, on account of the difficulties due to their small number, less conclusive. The long-range particles of thorium C have not been investigated in such detail, but there is no reason to doubt that these also are α particles resulting from new types of disintegration of thorium C.

§ 20 a. Numbers and ranges of the particles. The data which have been obtained by different experimenters of the numbers of the long-range particles relative to the number of particles emitted in the main type of disintegration are not in good agreement. The most reliable estimates are probably those of Rutherford and Chadwick for the particles from radium C, and of Philipp for the particles from thorium C. In both cases the observations were made by the scintillation method. The results of Philipp agree well with those of Meitner and Freitag deduced from photographs of tracks obtained in the expansion chamber.

The following table gives the relative proportions of the long-range particles and the normal particles.

Radium C.	28 of ca. 9 cm. range
	5 of ca. 11 cm. range
	10^6 of normal particles of 7 cm. range.
Thorium C.	65 of ca. 9·5 cm. range
	180 of ca. 11·5 cm. range
	10^6 of normal particles of 8·6 cm. range.

The ranges of the particles have been measured recently with some accuracy by Nimmo and Feather*. Tracks of the particles were obtained in an expansion chamber operated under well-defined conditions and photographed. The mean range of the particles was obtained from a statistical study of the lengths of the tracks and

* Nimmo and Feather, Proc. Roy. Soc. A, 122, 668, 1929.

a comparison with the lengths of the tracks of the normal α particles taken under the same conditions of experiment. The results obtained with sources of thorium C are shown in Fig. 25, in which the ranges of the tracks of about 550 long-range particles are recorded. The ranges are given in centimetres of standard air, i.e. at 15° C. and 760 mm. pressure, on the assumption that the mean range of the

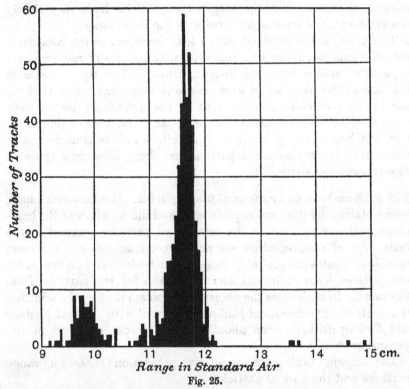

Range in Standard Air

Fig. 25.

normal α particle of thorium C′ is 8·54 cm. It will be seen that, in confirmation of previous experiments, the long-range particles from thorium C fall into two groups. The mean ranges, on the above assumption, are 9·82 cm. and 11·62 cm. in standard air. The corresponding ranges extrapolated from an ionisation curve were taken to be 9·90 cm. and 11·70 cm.

The number of particles in the group of longer range was about five times the number in the shorter group, as compared with a ratio of three to one found both by Meitner and Freitag and by Philipp. A few tracks were observed which had ranges greater than 12·5 cm.

and which appeared to be due to α particles; their number was however too small to allow any definite conclusion to be drawn.

The results obtained by Nimmo and Feather when sources of radium C were used are shown in Fig. 26. It is seen that the results do not permit such a clear interpretation as in the case of thorium C. There appears a well-defined group of particles with a mean range of 9·08 cm. (in air at 15° C. and 760 mm. pressure), but in addition there is a fairly general distribution of the particles with ranges between about 7·5 cm. and 12 cm. The main features of this distribution can

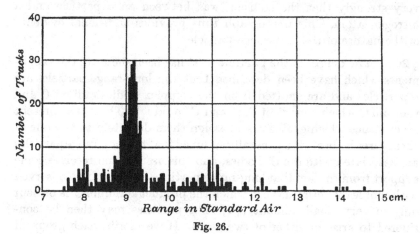

Fig. 26.

be explained by assuming the presence of two other groups with mean ranges of 8·0 cm. and 11·0 cm., and possibly a further group of mean range of about 10 cm. Such an explanation is perhaps preferable to a suggestion of a continuous spectrum of α particles.

Similar experiments on the long range α particles from radium C have been carried out by Philipp and Donat*, using the expansion method. There is in general a fair agreement with the results of Nimmo and Feather, both as regards the relative number and ranges of the main groups of α particles.

While the conclusions deduced from the experiments differ markedly from those from the scintillation experiments previously quoted, it is easy to see that a distribution of particles as shown in Fig. 26 would produce an effect in the latter experiments similar to that of a group of particles of 11 cm. range. The number of tracks of long-range particles was too small to give more than tentative

* Philipp and Donat, *Zeit. f. Phys.* **52**, 759, 1929.

values for the ranges of the less prominent groups, or to give a reliable estimate of the relative intensities of the groups.

Some of the photographs of long-range tracks obtained by Nimmo and Feather are shown in Plate V, figs. 2, 3 and 4; fig. 4, taken with thorium C, shows on the left a track of range 11·47 cm. and on the right a long-range track which has experienced a close collision. It is clear from the sharpness of the track that the plane of the collision and the object plane of the camera nearly coincide, and consideration of the angles of the fork and the length of the short spur indicates very strongly that the collision was between an α particle and a nitrogen atom. This photograph thus provides additional evidence of the nature of the long-range particles.

§ 20 b. The origin of the particles. It has been shown in the experiments which have been described that the long-range particles are α particles and are emitted from the complex bodies radium C and thorium C which consist of C, C′ and C″ products. There is at present no evidence which enables us to assign them definitely to any one of these three bodies*. It is generally assumed that the long-range particles are associated with the C′ bodies, and this assumption receives some support from the fact that long-range particles have not been observed with sources of actinium C, in which the product actinium C′ is present only in very small amount (§ 7). The particles may then be considered to arise in either of two ways. If we ascribe each group of particles of definite range to a separate radioactive product, we must suppose that there are three or perhaps four isotopes of radium C′, and two of thorium C′, which have been formed by extra β ray branches at the parent bodies radium C and thorium C. If the Geiger-Nuttall relation holds for these new bodies, their periods of decay must be exceedingly short. On the other hand, we may suppose that the long-range particles do not exist in the nucleus with their abnormally high energies, but that they are α particles of the normal groups which have received in some way, just before their ejection, an extra supply of energy. This assumption is perhaps the simplest which can be made at present. While there may be some difficulty in providing a source for the extra energy required, which is of the order of one to two million electron volts, it may be noted that some of the γ rays emitted by radioactive nuclei have energies of almost the same amount.

---

* A recent experiment (E. Stahel, *Zeit. f. Phys.* **60**, 595, 1930) shows that the long-range particles of thorium are not emitted by the product thorium C″.

PLATE V

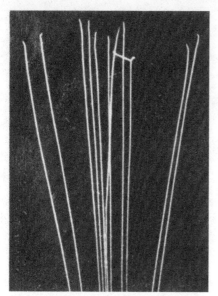

Fig. 1. Normal particles (8·6 cm.) from thorium C′ showing straggling of ranges. (The last 3 cm. of range.)

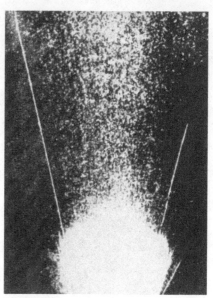

Fig. 2. Long-range particles from radium C. Ranges 11·04 cm., 8·93 cm.

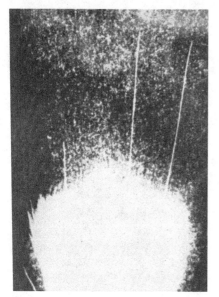

Fig. 3. Long-range particles from thorium C. Ranges 9·60 cm., 11·55 cm., 11·52 cm.

Fig. 4. Long-range particles from thorium C. Range of particle on left, 11·47 cm. Particle on right shows a collision with N atom.

In this connection it is of interest to compare the groups of long-range particles emitted from the two bodies. In each case the most abundant group is accompanied by a group of particles of lower velocity containing about one-fifth as many particles. Further, the energies of the normal particles of 7 cm. range and of the 8·1 cm. and 9·2 cm. particles from radium C are in the ratios 1 : 1·09 : 1·18, while the ratios for the energies of the normal particles of 8·6 cm. range and the two groups of long-range particles from thorium C are 1 : 1·09 : 1·20. The two groups of long-range particles from thorium C may then correspond to the 8·1 cm. and 9·2 cm. particles from radium C. There is, however, no group from thorium C to correspond to the 11 cm. particles of radium C, and the analogy cannot be pursued further. Our knowledge of these groups of long-range particles has been discussed in some detail, not only because of their intrinsic importance in throwing light on the varied modes of disintegration of a radioactive nucleus, but also on account of the evidence they give us on the maximum energy that can be released in a transformation. In addition, it has been of great importance to fix definitely the nature and origin of these particles and to be certain that they were not of secondary origin but were helium nuclei expelled from the radioactive nucleus. The energy of the swiftest α particle from thorium C, which has a range of 11·5 cm., corresponds to about 10·6 million volts, and this represents the highest emission of energy observed in any radioactive transformation.

§ 21. **Stopping power of substances.** When a sheet of matter is placed in the path of a pencil of α rays in air, the emergent range is reduced by a definite amount and the stopping power of the absorbing screen may be expressed in terms of centimetres of air under normal conditions. For example, if the emergent range of the α particle from radium C is reduced from 7 cm. to 6 cm., the absorbing sheet is said to have a stopping power for α particles of 1 cm. It was at first supposed that this stopping power was independent of the velocity of the α rays, but later researches brought out clearly that the relative stopping power depended on the velocity of the α particle and the atomic weight of the absorbing material. Taking air as a standard, the effect of velocity on relative stopping power is small for a substance like aluminium or mica, for which the average atomic weight is comparable with that of air, but becomes marked for an element of high atomic weight like gold.

The relative atomic stopping power s of an element of density ρ, thickness t and atomic weight A is given by $s = \rho_0 t_0 A / \rho t A_0$, where ρ_0, t_0, A_0 refer to the standard substance for which the stopping power is taken as unity. This is easily seen to be the case when we remember that the number of atoms in a cylinder of unit cross-section and length t contains $\rho t / m$ atoms, m being the mass of the atom, and that

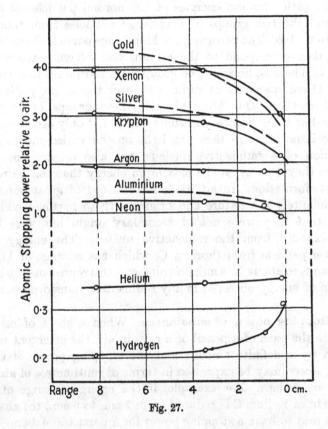

Fig. 27.

$m/m_0 = A/A_0$. In a similar way the relative molecular stopping power s of a complex molecule is given by $s = \rho_0 t_0 M / \rho t A_0$, where M is the molecular weight.

From what we have said it is clear that s is not a constant but is a complicated function of the velocity of the α particles and the atomic weight of the absorber and standard. In order therefore that s should have a definite meaning, the initial and final ranges of the α particles between which the absorption is measured should be specified.

Many observations of the stopping power of gases and solids were early made by Bragg and Kleeman, and these led to the formulation of the simple empirical rule that the stopping power of an atom is proportional to the square root of its atomic weight. Considering the complexity of the phenomena involved in the absorption of an α particle by matter, it is a matter of interest and also of practical convenience that such a simple rule should hold roughly over the whole range of the elements. A discussion of this relation will be given later (§ 26).

Observations on the change of stopping power with velocity were initially made by Bragg, Kleeman and Masek using the electric method, and by Marsden and Richardson* using the scintillation method. Recently Gurney† examined the stopping powers of the rare gases and of hydrogen and oxygen for different ranges of the α particles. His results are shown in Fig. 27 where the atomic stopping power of air is taken as unity. The corresponding values for aluminium, silver and gold are taken from the measurements of Marsden and Richardson. It is seen that for the elements of atomic weight greater than air, the relative stopping power falls off with the velocity of the α particle and the difference becomes very marked near the end of the range of the α particles. The results for hydrogen and helium are of especial interest in view of a comparison of the experimental results with theory (§ 26). It is seen that the stopping power, as first observed by T. S. Taylor‡, increases markedly near the end of the range, while the stopping power of helium shows a much smaller variation. The results are given in the following table.

Molecular stopping powers of gases relative to air for selected portions of the range.

I	II 8·6–7·6 cm.	III 3·8–3·5 cm.	IV 1·4–0 cm.	V 0·35–0 cm.	VI	
Xenon	1·98	1·95	—	1·51	1·804	(B)
Krypton	1·52	1·43	—	1·07	1·330	(B)
Argon	0·98	0·93	0·92	0·914	0·930	(B)
Neon	0·623	0·597	0·555	0·437	0·586	(B)
Helium	0·173	0·175	—	0·179	0·1757	(B)
Hydrogen	0·206	0·214	0·247	0·309	0·224	(T)
Oxygen	1·07	1·05	—	0·981	1·043	(B)

* Marsden and Richardson, *Phil. Mag.* **25**, 184, 1913.
† Gurney, *Proc. Roy. Soc.* A, **107**, 340, 1925.
‡ Taylor, *Phil. Mag.* **26**, 402, 1913.

Column VI gives the results of a careful determination by Bates* of the stopping power of the rare gases over the last 4 cm. of the range, using the scintillation method. We should expect the values to lie between those given in columns III and V, and this is seen to be the case. Gibson and Eyring† have compared the relative ionisation and stopping powers in a number of complex gases.

If we compare Fig. 27 with Fig. 19 showing corresponding results for variation of ionisation with range found by Gurney, it is seen that an increase of stopping power s in hydrogen relative to air corresponds to an increase of ionisation, while a decrease of s in neon corresponds to an increase in the relative ionisation. It has been shown (§ 18) that the ionisation observed in H_2 relative to that found in He and Ne is much less than is to be expected from our knowledge of their lowest ionisation potentials. This difference in ionisation between the monatomic and diatomic gases seems not to be connected with their stopping power, which is dependent on the result of the primary collisions of the α particles with the atoms in their path, but on the secondary ionisation due to the swift electrons liberated from the atom. It may be, as Gurney has suggested, that this secondary ionisation may be very different in monatomic and diatomic molecules, depending upon the relative amounts of energy of the α rays absorbed in ionisation, excitation, and dissociation of the molecules or a combination of these processes. It may be that much of the energy of the α rays may be expended in diatomic gases in dissociation accompanied by excitation but not by ionisation.

It is of interest to record the ranges of the α particles from radium C in H_2 and He. Taylor found a range 30·9 cm. in H_2 and 32·5 cm. in He at 15° C. and 760 mm., the range in air being 6·93 cm. By carefully purifying the helium, Bates in 1924 found a higher value for He, viz. 40 cm. This makes the range in H_2 and He 4·5 and 5·8 times respectively the range in air.

Van der Merwe‡ measured the relative lengths of the longest tracks in an expansion chamber filled with different gases. The results were in good accord with older observations.

---

* Bates, *Proc. Roy. Soc.* A, **106**, 622, 1924.
† Gibson and Eyring, *Phys. Rev.* **30**, 553, 1927.
‡ Van der Merwe, *Phil. Mag.* **45**, 379, 1923.

We have seen that Bragg and Kleeman early made a number of measurements of the stopping powers of solids and gases which led them to the conclusion that the stopping power of an element was approximately proportional to the square root of its atomic weight. The measurements of the atomic stopping power s were made on different portions of the range of the α particle, and are thus not strictly comparable. They found that $s/\sqrt{A}$, where A is the atomic weight of the absorber, was approximately a constant over the whole range of the elements. They found this rule held also for complex molecules where the stopping power is taken as the sum of the stopping powers of its constituent atoms. For example, if the atom is composed of N_1 atoms of atomic weight A_1 and N_2 of atomic weight A_2, the molecular stopping power is proportional to

$$N_1 \sqrt{A_1} + N_2 \sqrt{A_2}.$$

An excellent agreement between calculation and experiment was found in most cases. This additive rule can hold only if the α particle expends most of its energy in acting on the individual atoms and not in dissociating or exciting the molecule. It seems probable that the additive rule though very convenient for rough calculation holds only to a first approximation.

In subsequent work by Philipp* on the stopping power of some liquids and their vapours, the additive rule was found to hold for all vapours and for normal liquids like benzol and pyridin, but gave too small a value for liquids like water and alcohol where association of molecules occurs. The effect, however, is small and difficult to account for.

Von Traubenberg† made a systematic examination of the stopping power of a large number of elements by the scintillation method. The source of rays was a wire, about 1 cm. long, coated with radium C. The rays passed through an accurate wedge of the material and fell on the zinc sulphide screen placed on it. The thickness of the wedge was measured for the point where the scintillations stopped. A correction was applied for the distance traversed in air, about 5·5 mm., before entering the metal. The relative stopping power of the elements was thus compared for α particles of range between 6·4 and 0 cm. In the following table the extrapolated thickness of metal penetrated by the α rays from radium C of full range is given.

* Philipp, *Zeit. f. Phys.* 17, 23, 1923.
† von Traubenberg, *Zeit. f. Phys.* 2, 268, 1920; see also von Traubenberg und Philipp, *Zeit. f. Phys.* 5, 404, 1921.

Ranges in solid elements of the α particles of radium C.

Element	cm. $\times 10^{-3}$	Element	cm. $\times 10^{-3}$
Li	12·91	Ag	1·92
Mg	5·78	Cd	2·42
Al	4·06	Sn	2·94
Ca	7·88	Pt	1·28
Fe	1·87	Au	1·40
Ni	1·84	Tl	2·33
Cu	1·83	Pb	2·41
Zn	2·28		

The atomic stopping power s of the elements found by him is given in the following table. The results for H and He are corrected to fit the recent determination by Bates (*loc. cit.*) of the relative ranges of the α particles in those gases, compared with air. The atomic stopping power of oxygen is taken as unity.

Element	Atomic stopping power	$sA^{-\frac{1}{2}}$	Element	Atomic stopping power	$sA^{-\frac{1}{2}}$
H	0·200	0·20	Fe	1·96	0·26
He	0·308	0·154	Ni	1·89	0·25
Li	0·519	0·20	Cu	2·00	0·25
Be	0·750	0·25	Zn	2·05	0·25
C	0·814	0·25	Br	2·51	0·28
N	0·939	0·25	Ag	2·74	0·26
O	1·00	0·25	Cd	2·75	0·26
Mg	1·23	0·25	Sn	2·86	0·26
Al	1·27	0·24	I	3·55	0·31
Si	1·23	0·23	Pt	3·64	0·26
Cl	1·76	0·29	Au	3·73	0·27
A	1·80	0·29	Tl	3·76	0·27
Ca	1·69	0·27	Pb	3·86	0·27

In the third column is given the ratio $sA^{-\frac{1}{2}}$, where A is the atomic weight of the element. According to Bragg's rule, this should be constant, and this is seen to hold approximately. It will be observed that the deviations are greater for the lighter elements. Since the stopping power of an element is a function of the number of electrons in the atom and their arrangements rather than of the atomic weights, Bragg's rule is without special physical significance, but yet on the whole it fits the results of observations fairly well. Glasson[*] has shown that the rule $s \propto Z^{2/3}$, where Z is the atomic number of the element, fits the observations about as well as Bragg's rule. From the theory of absorption of α particles discussed in the next chapter, it is surprising that the variation of s with Z or A can be expressed even roughly by a simple power law over such a wide range of Z or A.

[*] Glasson, *Phil. Mag.* **43**, 477, 1922.

CHAPTER IV

SOME PROPERTIES OF THE α PARTICLE

§ 22. Retardation of the α particle. The great majority of α particles in passing through matter travel in nearly straight lines and lose energy in ionising the matter in their path. Occasionally an α particle suffers a nuclear collision with an atom and is deflected through a large angle. These occurrences, though of great interest, are so rare that they do not seriously influence the average loss of energy when a large number of α particles are under examination. The laws of retardation of the α particle are best studied by making use of the homogeneous α radiation emitted by the very thin deposits of radium C, thorium C, and polonium. It is found experimentally that the reduction of velocity in traversing normally a uniform screen is nearly the same for all the α particles, so that a homogeneous pencil of rays remains nearly homogeneous on emerging from the screen. This effect is most clearly shown with the swifter α particles, e.g. those that have a range in air between 8·6 and 3 cm. With reduction of the velocity, the "straggling" of the α particles, i.e. inequalities in the velocity and range of the emergent α particles, becomes more and more prominent and the issuing pencil of α particles becomes very heterogeneous.

The reduction of velocity is best studied by an arrangement similar to that shown in Fig. 7, where the absorbing sheet of matter is placed over the source and the deflection in a uniform magnetic field of the issuing pencil of α rays, in an exhausted chamber, is observed either by the photographic or the scintillation method. In most cases a thin wire coated with radium C has been used as a source of rays, but thin films of thorium C and polonium have in some cases been employed.

The decrease of velocity of α particles in passing through aluminium was first studied by Rutherford* using the photographic method, and in more detail by Geiger† for mica by the scintillation method. Geiger found that the reduction of the velocity of the α particles could be expressed by a simple law, viz. $V^3 = aR$, where

---

* Rutherford, *Phil. Mag.* **12**, 134, 1906.

† Geiger, *Proc. Roy. Soc.* A, **83**, 505, 1910.

V is the velocity of the issuing particles after traversing the absorbing screen, R the *emergent* range of the α particles *measured in air* and a is a constant. It will be seen that this relation holds fairly closely for ranges of α particles between the maximum and 3 cm. of air. Later, a more detailed examination by the scintillation method was made by Marsden and Taylor* with absorbing screens of mica, aluminium, copper and gold.

It was observed by Rutherford and subsequent investigators that it was unexpectedly difficult to follow the reduction of velocity below about $0 \cdot 40 \, V_0$, where V_0 is the maximum velocity of the α particles from radium C, corresponding to a range of α particles of 4 or 5 mm. in air. This difficulty was in part due to the marked heterogeneity of the issuing beam, which caused a broad diffuse band of α particles in a magnetic field, but mainly to the influence of another effect which has only recently been recognised, namely the rapid capture and loss of electrons by low velocity α particles. This effect only manifests itself clearly in a high vacuum, when the issuing beam splits up into three distinct groups of α particles corresponding to neutral, singly charged, and doubly charged particles. With the addition of a small quantity of gas in the path of the rays, these three bands disappear and give place to one main band which is less deflected than the group of doubly charged particles (see § 24).

While the early investigators all used a sufficiently good vacuum to avoid appreciable absorption of the α rays, it is probable, however, that in many cases the residual pressure of the gas was high enough to cause a marked capture and loss of electrons by the particles. Such an effect greatly complicated the interpretation of results and was undoubtedly the main factor in the failure to follow the changes of velocity for low ranges of the α particle.

To avoid this difficulty it is essential to examine the deflection of the particles in a high vacuum, where, in the case of low velocities, observations may be made of the average deflections of the singly or doubly charged group of α particles. A careful investigation fulfilling these conditions has recently been made by Briggs† for mica, using the photographic method. In order to give narrow bands on the photographic plate, fine wires coated with radium (B + C) were used as sources and the rays passed through a narrow slit. The length of path of the α rays in the vacuum and the intensity of the

* Marsden and Taylor, *Proc. Roy. Soc.* A, **88**, 443, 1913.

† G. H. Briggs, *Proc. Roy. Soc.* A, **114**, 313, 341, 1927.

magnetic field, which was kept very constant, were so adjusted that
the deflection due to the magnetic field corresponded to 3 or 4 cm.
on the photographic plate. At the same time a systematic examination was made of the broadening of the band so as to obtain an
accurate measure of the straggling of the α particles. This will be
considered later. Plate VI gives some reproductions of the photographs obtained by Briggs in these experiments. In the photographs
1 to 5, four slits were employed giving rise to four undeflected and
four deflected lines. From inspection of the photographs it will be
seen that the "midway" band due to singly charged α particles is
well marked for low velocities. The marked heterogeneity of the
issuing beam for short ranges is clearly shown. The mean reduction
of the velocity was measured after the α rays had passed nearly
normally through uniform mica sheets of known weight per sq. cm.
The stopping power of the mica sheets was measured by the scintillation and electric methods, and expressed in terms of air at
760 mm. and 15° C.

It has been shown in § 16 that the extrapolated range of the
α particles from radium C at 15° C. and 760 mm. is 6·96 cm. in air.
As this extrapolated range is difficult of interpretation, it is desirable
to express the velocity in terms of the *mean* range of the α particles
in air. This was found to be 6·92 cm. at 15° C. and 760 mm. by I. Curie
from measurements of the average length of the α ray tracks in an
expansion chamber. From consideration of the ionisation curve
and straggling of the α particles, Briggs deduced the mean range
to be 6·90 cm., and in the following table the velocity is expressed
in terms of this mean range. The ratio of the mean velocity of the
emergent particles to the initial velocity V_0 (1·922 × 10^9 cm./sec.) is
given in column 3 in the table below. The same ratio calculated from
the Geiger relation $V^3 = a(R - x)$ is given in column 4. It is seen that
this law holds fairly closely for velocities between V_0 and 0·75 V_0,
i.e. for emergent ranges greater than 3 cm. For the lowest velocity
measured, viz. 0·22 V_0, the error is more than 50 per cent. The
departure is still more marked if the extrapolated range 6·96 cm. is
taken instead of the mean range 6·90 cm.

Since the initial ranges of the α particles from the different radioactive products are generally more than 3 cm., Geiger's rule can be
applied with fair accuracy to determine the initial velocities of
expulsion from the observed ranges.

The velocity curve in air for the α particles from radium C taking the range in air as 6·96 cm. at 15° C. and 760 mm. is shown in Fig. 28. The dotted curve 3 shows the relation $V^3 = a\,(R - x)$ and curve 2 the values obtained by Marsden and Taylor. Curve 1 gives the results of Briggs.

Distance traversed in air at 15° C. and 760 mm.	Distance from end of mean range in cm. of air	V/V_0	V/V_0 calculated Geiger's rule
0	6·9	1·000	1·000
0·5	6·4	0·977	0·975
1·0	5·9	0·951	0·949
1·5	5·4	0·923	0·921
2·0	4·9	0·894	0·892
2·5	4·4	0·863	0·861
3·0	3·9	0·828	0·827
3·5	3·4	0·790	0·790
4·0	2·9	0·746	0·749
4·5	2·4	0·696	0·703
5·0	1·9	0·638	0·651
5·5	1·4	0·563	0·588
5·8	1·1	0·504	0·542
6·0	0·9	0·455	0·507
6·1	0·8	0·427	0·488
6·2	0·7	0·396	0·466
6·3	0·6	0·361	0·443
6·4	0·5	0·322	0·417
6·5	0·4	0·278	0·387
6·6	0·3	0·222	0·352

A more stringent test of the accuracy of Geiger's law can be applied by a direct comparison of the velocities of the particles whose ranges have been measured with the greatest possible precision. This can be most simply accomplished by obtaining a thin film of the active deposit of thorium and radium together on a thin wire, and comparing the deflection of the three groups of α rays, two of which arise from thorium C, in a constant magnetic field. Experiments of this kind have been made by A. Wood[*] and recently by G. H. Briggs[†] and G. C. Laurence[‡], who obtained large deflections of the α particles on a photographic plate and measured the relative deflections by means of a microphotometer. Taking the velocity of the α particles from radium C as $1\cdot923 \times 10^9$ cm./sec., Briggs found the velocities of the two groups of α particles from ThC and ThC' were $1\cdot705 \times 10^9$ and $2\cdot053 \times 10^9$ cm./sec., with an accuracy estimated

    * A. Wood, *Phil. Mag.* **30**, 702, 1915.
    † G. H. Briggs, *Proc. Roy. Soc.* A, **118**, 549, 1928.
    ‡ G. C. Laurence, *Proc. Roy. Soc.* A, **122**, 543, 1929.

to be 1 in 1000. The values found by Wood were $1 \cdot 714 \times 10^9$ and $2 \cdot 060 \times 10^9$. In a similar way, Laurence determined the velocity

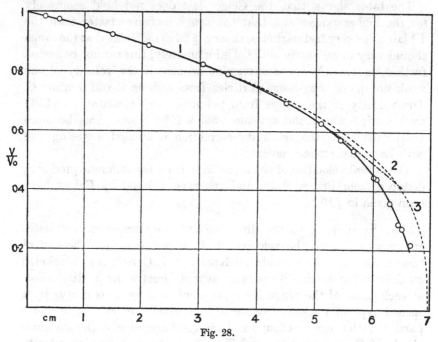

Fig. 28.

ratios for a source of radium C and polonium (radium F). The velocity ratios for the three groups obtained by Briggs and Laurence are shown in the table and compared with calculations on Geiger's rule. Laurence estimates that the relative velocities found by him are correct to 1 in 2000.

Element	Velocity ratio		
	Observed		Calculated
	Briggs	Laurence	
Th C′/Th C	1·204	1·2017	1·217
Th C′/Ra C	1·068	1·0679	1·074
Th C/Ra C	0·886	0·8885	0·882
Ra F/Ra C	—	0·8277	0·824

A slight correction for relativity has been made. Rosenblum* found the ratio of the velocities of the α particles from ThC and ThC′ to be 1·209, a somewhat higher value than that found by Briggs and

* Rosenblum, *C.R.* **180**, 1333, 1925.

Laurence. Laurence finds the ratio ThC/RaC somewhat greater than Briggs but otherwise there is good agreement.

The table shows that the Geiger law does not hold accurately for the higher ranges and that the range increases faster than the V^3 law. For very fast particles, theory (§ 26 a) indicates that the range should vary more nearly as V^4. Unfortunately, this cannot be tested further, as it is difficult to determine accurately the velocity of the weak groups of long-range particles from radium C and thorium C. The velocity of the α rays from polonium was measured carefully by I. Curie* who found a value $1 \cdot 593 \times 10^9$ cm./sec. This becomes $1 \cdot 591 \times 10^9$ when the relativity correction is applied, agreeing well with Laurence's measurements.

The actual velocities of the α particles from the different products, deduced from Briggs' data, and the ranges found by Geiger have been given in § 19.

§ 22 a. In addition to the direct velocity measurements, estimates of the variation of the velocity of the α particles near the end of their range have been made by Kapitza†, I. Curie‡, and Blackett§ by different methods. Kapitza measured directly the heating effect at each point of the range for a pencil of homogeneous α rays by a special type of radiomicrometer. The variation of energy of the α particle with range was found to be in good accord with the measurements of Geiger, Marsden and Taylor on the variation of velocity of the α particle until about 1 cm. from the end of the range, when the velocity was found to fall off more rapidly than on Geiger's rule. The results are in good accord with the direct measurements of Briggs.

In a subsequent investigation, Kapitza photographed the tracks of α particles exposed to a strong magnetic field in a small expansion chamber. Illustrations of the tracks are shown in Plate II, fig. 4. By measuring the curvature of the tracks near the end, the value of $\bar{e}/V$, where $\bar{e}$ is the average charge on the particle and V its velocity, was determined. From the data on capture and loss of α particles for different velocities (§ 24) $\bar{e}$ could be estimated and V thus determined. Assuming $\lambda_1/\lambda_2 \propto V^{4 \cdot 6}$ as found by Rutherford, where λ_1 and λ_2 are the mean free paths of the α particle for capture and loss respectively,

* I. Curie, *C.R.* **180**, 1332, 1925.
† Kapitza, *Proc. Roy. Soc.* A, **102**, 48, 1922.
‡ I. Curie, *Annales de physique*, **3**, 299, 1925.
§ Blackett, *Proc. Roy. Soc.* A, **102**, 294, 1922.

curve 3, Fig. 29 was found. Taking the index 4·3 found by Briggs (§ 24 *a*), curve 4 is obtained which agrees well with the direct determination of Briggs (curve 1).

I. Curie deduced the variation of velocity of the α particle near the end of the range from the ionisation curve obtained for an average α particle (*vide* § 16) on the assumption that the ionisation $I dx$ in a length of the trajectory dx is a measure of the loss of energy of the

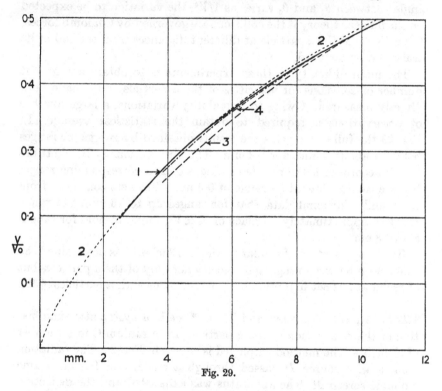

Fig. 29.

α particle, i.e. $- d(V^2) = KI dx$. Knowing the variation of I with range, the velocity V was deduced by graphical integration. It is shown in § 16 that the ionisation curve for the average single α particle is very similar in shape to the Bragg ionisation curve for a pencil of α rays. The velocity curve obtained in this way is shown in Fig. 29, curve 2.

The velocity curve has been obtained by Blackett by another distinct method. Photographs were obtained by the Wilson method of a number of collisions of the α particle in passing through air giving rise to branched forks, one due to the deflected α particle and the other to the recoiling nucleus (*vide* §§ 57, 58). These large angled deflections are comparatively numerous near the end of the range of the α particle when its velocity is much reduced. Since the velocity of the particle at the moment of the collision is unknown, it has to be estimated by indirect methods. Assuming that the statistical number of collisions for which the α particle is deflected through angles between θ_1 and θ_2 varies as $1/V^4$, the variation to be expected on the nuclear theory if the collisions are governed by Coulomb forces, the velocity of the α particle at different distances from the end of its track can be deduced.

The main difficulty in these experiments is to obtain a sufficient number of collisions for velocities of the α particle which have been directly measured. Owing to probability variations, a large number of observations is required to obtain the statistical average. In Fig. 29 the full curve gives the results obtained by Briggs for ranges between about 3 mm. and 12 mm. The observations of Blackett are in fair accord with the results of Briggs over a corresponding range, but are carried down to a range of 0·5 mm. Blackett concludes from these and additional data that for ranges up to 16 mm. the range $R \propto V^{1\cdot5}$ approximately, in place of $R \propto V^3$, found to hold for ranges above 3 cm.

By the application of similar methods, Blackett has determined the connection between range and velocity not only of the α particles but of recoil atoms set in motion by close collisions with α particles (§ 58).

§ 22 b. In 1913 Marsden and Taylor[*] made a systematic examination of the retardation of the α particles from radium C in a number of elements. The method employed is shown in Fig. 30. The radiation from a wire source D passed through a slit L and fell on a zinc sulphide screen M. The apparatus was exhausted and the deflection of the band of scintillations due to a strong magnetic field perpendicular to the plane of the paper was measured. By means of a movable ladder T, screens of known thickness of metal were interposed in front of the source. For reasons already given they were unable to follow the deflection of the pencil of α rays below a velocity $0\cdot415\ V_0$, where V_0 is the velocity of α particles from radium C. The results obtained are shown in the table below, where the weights per unit area in milligrams required to reduce the velocity

---

[*] Marsden and Taylor, *Proc. Roy. Soc.* A, **88**, 443, 1913.

of the α particles are given. The total weight required to absorb the α particles completely was deduced by the scintillation method.

The results obtained for air have already been shown in Fig. 28 for comparison with the values obtained by Briggs. The general

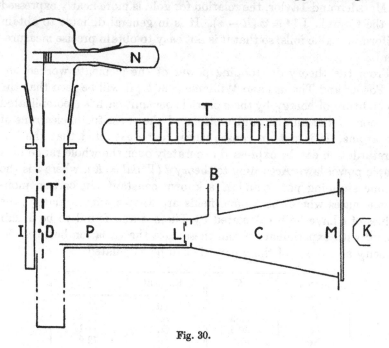

Fig. 30.

Relative velocity V/V_0	Mass per unit area				
	Gold (mg.)	Copper (mg.)	Aluminium (mg.)	Mica (mg.)	Air (mg.)
1·0	—	—	—	—	—
0·95	4·00	2·08	1·48	1·43	1·24
0·90	7·05	3·90	2·79	2·75	2·32
0·85	9·79	5·35	3·94	3·83	3·26
0·80	12·27	6·69	5·01	4·86	4·08
0·75	14·80	8·00	6·05	5·72	4·84
0·70	17·04	9·20	7·03	6·40	5·46
0·65	18·99	10·30	7·85	7·00	6·02
0·60	20·71	11·40	8·50	7·50	6·48
0·55	22·29	12·35	9·10	7·98	6·90
0·50	23·89	13·13	9·64	8·47	7·29
0·45	25·40	14·00	10·15	8·96	7·67
0·415	26·65	14·60	10·46	9·35	7·96
End	29·50	16·00	11·40	10·15	8·50

evidence indicates that the law of retardation of the α particle for velocities between V_0 and $0\cdot7\,V_0$ is very nearly the same for air as for mica, viz. $V^3 = a\,(R - x)$, where V is the velocity after traversing a distance x and R the maximum range. From the data obtained by Marsden and Taylor, the relation for gold is more nearly expressed by the formula $V^{2\cdot4} = a\,(R - x)$. It is in general difficult to obtain uniform metallic foils, so that it is not easy to obtain precise measurements.

From the theory of stopping power of the elements worked out by Fowler and Thomas and Williams (§ 26 b), it will be seen that the expenditure of energy by the α particle per unit path is a complicated function of the number and motions of the constituent electrons of the atoms. It is not, in consequence, to be expected that the law of retardation can be expressed accurately over the whole range by a simple power law. According to theory $d\,(V^4)/dl = Ks$, where s is the atomic stopping power and K a known constant. In certain monatomic gases where the atomic fields are known approximately, the values of s have been calculated by Thomas and found to be in fair accord with experiment. From these data the variation between the velocity and range of the α particle can be calculated.

Element	K	Element	K
Li	27·7	Pd	71·1
Al	39·3	Ag	70·3
Fe	50·5	Cd	71·1
Ne	52·5	Sn	72·9
Cu	54·5	Pt	99·0
Zn	55·2	Au	98·0
Mo	67·5	Pb	98·5

In some recent experiments, Rosenblum[*] has made a careful determination of the diminution of velocity of the α rays with thickness of matter traversed for a number of elements. The method was similar in principle to that used by Marsden and Taylor (loc. cit.), but the deflection of the pencil of rays in a magnetic field was measured by the photographic method. The α rays from thorium C′, initial range 8·6 cm. and velocity $2\cdot052 \times 10^9$ cm./sec., were used. He found that the relation between velocity and thickness of metal foil traversed could be expressed for all the elements by the expression

$$X = K\,(u - u^2 + 0\cdot35u^3),$$

* S. Rosenblum, C.R. 183, 851, 1927; Phys. Zeit. 29, 737, 1928.

where X is the weight in milligrams per sq. cm. of the uniform foil and $u = (V_0 - V)/V_0$, where V_0 is the initial velocity of the α particles from thorium C' and V the velocity after traversing the foil. K is a constant which varies for different elements in the way shown in the above table.

The results are compared with Bohr's theory (§ 26 a), but while there is a fair agreement for the lighter elements, the calculated values of K are about 2·5 times the observed values for heavy elements like lead (*vide* § 26 b).

§ 23. **Theory of straggling of α particles.** When an α particle from radium C is absorbed in air, it produces about 220,000 pairs of ions. Supposing that about two-thirds of these are due to secondary electrons, the number of primary ionising collisions is about 70,000. On Bohr's theory (§ 26 a), the actual number of collisions leading to a transfer of energy from the α particle is much greater than this. For illustration, consider the passage of an α particle from radium C through helium. Taking the radius of the helium atom as 1×10^{-8} cm., the α particle in its path of 38 cm. through helium makes 3×10^5 collisions. If we suppose that energy is transferred to the atom for a distance at least four times the atomic radius, the number of collisions involving appreciable transfers of energy is 5×10^6. The number in air will be smaller but of the same order of magnitude. On Bohr's theory, all transfers of energy are possible from zero to the maximum in which the escaping electron acquires twice the velocity of the α particle. On the theory of probability, it is clear that some α particles will suffer more loss of energy than others and in consequence there will be differences of range of the α particles in the matter, although the α particles are all initially liberated with identical speeds.

The ranges of the α particles will be grouped round the average range of the particles according to the well-known Gaussian law of errors. The amount of this probability variation or straggling, as named by Darwin*, can be calculated on any of the theories of the loss of energy of the α particle which are discussed in chapter v. It is clear that the number of collisions of different types involved is so numerous that the theory of probabilities for large numbers can be safely employed.

The first rough estimate of the probability variations was made by Herzfeld by assuming that the number of encounters was equal

* Darwin, *Phil. Mag.* **23**, 901, 1912.

to the number of ions produced. Elaborate calculations have been given independently by Flamm* and by Bohr† in which all encounters involving transfers of energy possible on the theory are taken into account. If R be the mean range of the α particles, the probability $W(s)\,ds$ that the α particle has a range between $R\,(1+s)$ and $R\,(1+s+ds)$ is given by

$$W(s)\,ds = \frac{1}{\rho\sqrt{\pi}}\,e^{-\frac{s^2}{\rho^2}}\,ds,$$

where ρ is a constant dependent on the particular way in which it is supposed that transfers of energy may occur. Bohr has calculated the value of ρ on his theory of the absorption of the α particle based on classical dynamics. For the α rays from radium C of range 7 cm. in air, he finds $\rho = 0\cdot0086$ and $0\cdot0116$ for hydrogen and air respectively. For the α rays from polonium of range $3\cdot9$ cm., $\rho = 0\cdot0091$ and $0\cdot0120$ for hydrogen and air respectively.

Both Flamm and Bohr showed that the straggling was only slightly affected by the occasional nuclear collisions in which the α particle is deflected through a considerable angle. The correction for this effect is small and may be neglected without serious error.

§ 23 a. It will be seen that three different methods have been employed to measure the straggling of α particles, viz. (1) the scintillation method, (2) the expansion method, and (3) the distribution in velocity of the α particles after passage through an absorbing screen. It is important to notice that different quantities are measured in each case. In the scintillation method, the actual number of α particles which are able to traverse absorbing screens of different thicknesses are counted. In the expansion method, the differences in the total lengths of the tracks of the α particles in traversing the gas of the expansion chamber are observed. In method (3), by means of a magnetic field, the distribution in velocity of the α particles in an originally homogeneous beam is measured after the rays have traversed a known thickness of matter. Methods (1) and (2) give the straggling over the whole path of the α particle, while method (3) gives a measure of the straggling at different points of the path of the average α particle.

The first experiments to estimate the straggling of the α particles were made by the scintillation method. A narrow pencil of homo-

* Flamm, *Wien Ber.* **124**, 597, 1915.
† Bohr, *Phil. Mag.* **30**, 531, 1915; **25**, 10, 1913.

geneous α rays from radium C or polonium passed through air, and the numbers of α particles at different distances from the source were counted. Experiments of this kind have been made by Geiger, Taylor, Friedmann, and Rothensteiner, with somewhat conflicting results. The variation of number of scintillations near the end of the range of the α particles from polonium is shown in Fig. 31. Curve I gives the distribution of the length of the tracks in an expansion chamber found by I. Curie (*vide infra*); Curves III and IV, obtained by Rothensteiner and Taylor respectively, the distribution found by the scintillation method. Curve II on the left shows the end of the ionisation curve for comparison with the straggling curves. There is no doubt

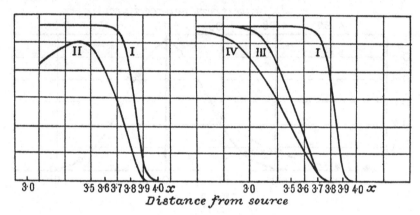

3·0   3·5 3·6 3·7 3·8 3·9 4·0 *x*   3·0   3·5 3·6 3·7 3·8 3·9 4·0 *x*
Distance from source

Fig. 31.

that the earlier observations by the scintillation method exaggerate the true straggling. This is very largely due to the fact that it is difficult to count feeble scintillations in the presence of bright ones. Unless the optical system is very good, there will be a tendency to overlook the feeble scintillations and count only the brighter ones.

A careful determination of the straggling of the α rays was made by the expansion method by I. Curie*, and by Meitner and Freitag†. By means of slits, a narrow horizontal beam of homogeneous α rays from a thin layer of active matter of small area passed into a Wilson expansion chamber and was photographed. An example of a straggling photograph obtained in this way by Feather and Nimmo, using the α rays from thorium C, is shown in Plate V, fig. 1.

* I. Curie, *Journ. de phys.* **3**, 299, 1925.
† Meitner and Freitag, *Zeit. f. Phys.* **37**, 481, 1926.

The lengths of all the tracks were measured and the average range of the α particles in the chamber was thus determined. The results obtained by I. Curie are shown in Fig. 32, where the full curve represents the probability equation $y = \dfrac{1}{\sqrt{\pi}}, e^{-x^2}$, and the circles experimental points. It is seen that the distribution of particles of lengths greater than the mean falls closely on a probability curve, but that there appears to be an excess of particles of short range. I. Curie has shown that this excess is not due to any want of homogeneity in the initial velocity of the α particles expelled from polonium, but may possibly be due to the great difficulty of obtaining a very thin and pure source of

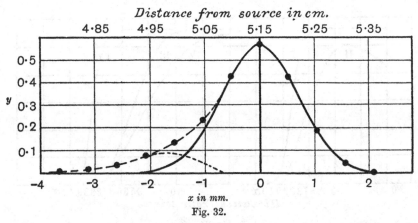

Fig. 32.

active matter which does not penetrate to some extent into the metal on which it is deposited. This is supported by the observation that the amount of the probability excess varied in different experiments. Assuming that this excess distribution is spurious, and reducing the ranges in terms of air at 760 mm. and 15° C., the straggling coefficient ρ for air was determined for polonium rays of range 3·96 cm. and found to be 0·0167. The value calculated by Bohr is somewhat lower, viz. 0·0116.

In some later experiments, I. Curie and Mercier*, using as sources radium A and radium C, obtained results in closer accord with Bohr's theory and concluded that the experiments with polonium were in error. The number of tracks examined, however, was too small to give a correct probability distribution.

Similar experiments were made independently by L. Meitner and

* I. Curie and Mercier, *Journ. de phys.* **7**, 289, 1926.

Freitag (*loc. cit.*), using thorium C and C' as sources of homogeneous α rays. In their investigations the α particles were allowed to enter the expansion chamber before the expansion was completed, so that some particles, owing to the alterations of pressure, travelled farther than others. The correction for this was large and difficult to estimate with accuracy. After correction, Meitner and Freitag found $\rho = 1\cdot55 - 1\cdot60 \times 10^{-2}$ for oxygen and $\rho = 1\cdot30 \times 10^{-2}$ for argon. The value for oxygen is considerably greater than that calculated on Bohr's theory.

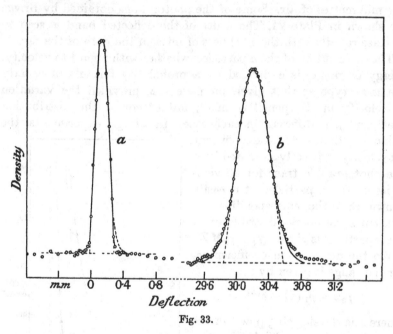

Fig. 33.

§ 23 b. In the above experiments the straggling was measured for the α particle over its whole range. It is important, however, to determine the relation between the straggling and velocity of the α particle at any point of the path of a pencil of rays. Careful experiments to test this point have been made by G. H. Briggs*, using the same general method employed to measure the retardation of the α particles (§ 22). The width of the deflected band of α rays on the photographic plate after traversing a known thickness of mica was compared with the width of the undeflected band by means of a

* G. H. Briggs, *Proc. Roy. Soc.* A, **114**, 313, 1927.

microphotometer. Narrow pencils of α rays were used and the apparatus exhausted to avoid complications arising from capture and loss of electrons by the α particles in their passage through the residual gas. As an example, the distribution in density of (a) the undeflected band and (b) the deflected band corresponding to the passage of the α particles through 2·5 cm. of mica is shown in Fig. 33. By the use of a very narrow pencil of rays, the straggling of the α particles as regards velocity could be easily detected after the particles had passed through a sheet of mica of stopping power equal to only a few millimetres of air. Some of the photographs obtained by Briggs are shown in Plate VI. The width of the deflected band is seen to increase rapidly with the thickness of mica in the path of the rays.

The distribution of the α particles, whether with regard to velocity, energy or range, is expressed by a probability formula of exactly the same type as that given on page 112, provided the variation of velocity in the pencil is small, but of course the distribution coefficient ρ is different in each case. In Briggs' experiments, the value of ρ_2, the distribution coefficient for velocity, is directly measured from the photographic trace for all velocities of the α particle. It is easily shown that the corresponding coefficient ρ_1 for energy distribution of the α particles is given by $\rho_1 = \rho_2 M V^2$, while the distribution coefficient ρ_3 for the range is given by

$$\rho_3 x = \rho_1 (dT/dx)^{-1},$$

where x is the stopping power of the absorbing screen and dT/dx the variation of energy with x, which can be deduced from the velocity curve.

The values of ρ_2, ρ_1 deduced from the broadening of bands after passing through a thickness of mica of known air equivalent are given in the table below. The variation of ρ_2 with range

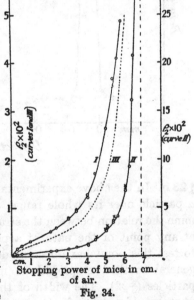

Fig. 34.

is shown in Fig. 34. Curves I and II plotted on different scales are experimental, Curve III calculated from Bohr's theory. The accuracy of measurement of ρ_2 is greatest for stopping powers between 3 and 5 cm.

PLATE VI

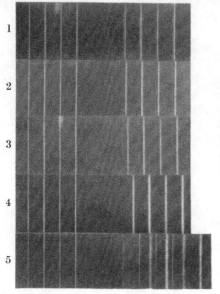

Nos. 1 to 5.

Four undeflected lines to the left and the four corresponding He_{++} lines to the right. The no-mica lines are visible to the left of the He_{++} lines. The thicknesses of mica were equivalent to 0·32, 0·55, 1·03, 3·05, and 3·49 cm. of air.

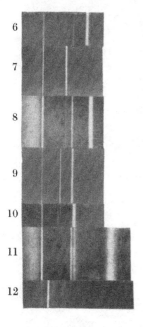

The sequence of the lines from left to right is given below and the thickness of mica (air equivalent) for each plate.

No. 6. Undeflected He_+, no-mica, He_{++}, 4·5 cm.
No. 7. Undeflected, He_+, He_{++}, 4·5 cm.
No. 8. Broad band due to β rays deflected to the left in the residual field during exposure for undeflected line (also evident in 7 and 11).

	Undeflected, He_+, no-mica, He_{++}, 4·8 cm.			
No. 9.	,,	,,	,,	,, 5·0 cm.
No. 10.	,,	,,	,,	,, 5·4 cm.
No. 11.	,,	no-mica, He_+,	He_{++},	6·0 cm.
No. 12.	,,	,,	,,	,, 6·4 cm.

The probable error increases towards the end of the range, where it is difficult to obtain very accurate results.

Air equivalent of mica (cm.)	ρ_2	ρ_1 (ergs)	Air equivalent of mica (cm.)	ρ_2	ρ_1 (ergs)
0·315	$1·53 \times 10^{-3}$	$3·63 \times 10^{-8}$	5·056	$26·3 \times 10^{-3}$	$25·6 \times 10^{-8}$
0·551	$2·12 \times 10^{-3}$	$4·94 \times 10^{-8}$	5·432	$35·0 \times 10^{-3}$	$28·3 \times 10^{-8}$
1·026	$3·06 \times 10^{-3}$	$6·80 \times 10^{-8}$	5·664	$40·9 \times 10^{-3}$	$29·0 \times 10^{-8}$
2·113	$5·08 \times 10^{-3}$	$9·83 \times 10^{-8}$	5·866	$48·4 \times 10^{-3}$	$28·1 \times 10^{-8}$
2·409	$5·86 \times 10^{-3}$	$10·9 \times 10^{-8}$	6·067	$67·3 \times 10^{-3}$	$31·8 \times 10^{-8}$
3·066	$7·64 \times 10^{-3}$	$12·7 \times 10^{-8}$	6·296	$92·4 \times 10^{-3}$	$29·5 \times 10^{-8}$
3·490	$9·56 \times 10^{-3}$	$14·6 \times 10^{-8}$	6·445	$161·0 \times 10^{-3}$	$32·9 \times 10^{-8}$
4·098	$12·0 \times 10^{-3}$	$16·4 \times 10^{-8}$	6·501	$207·0 \times 10^{-3}$	$32·6 \times 10^{-8}$
4·492	$17·0 \times 10^{-3}$	$20·2 \times 10^{-8}$	6·659	$276·0 \times 10^{-3}$	$32·5 \times 10^{-8}$

The straggling coefficient ρ_3 for range and $\rho_3 x$, where x is the air equivalent of the mica traversed, are given in the table below. The value of $\rho_3 x$ calculated on Bohr's theory of straggling and the ratio of the observed to the calculated straggling are given in the last two columns.

Stopping power of mica (cm.)	ρ_3 observed	$\rho_3 x$ observed (cm.)	$\rho_3 x$ calculated on Bohr's theory (cm.)	$\dfrac{\rho_3 x \text{ observed}}{\rho_3 x \text{ calculated}}$
0·5	$7·78 \times 10^{-2}$	$3·89 \times 10^{-2}$	$2·83 \times 10^{-2}$	1·37
1·0	$5·44 \times 10^{-2}$	$5·44 \times 10^{-2}$	$3·96 \times 10^{-2}$	1·37
1·5	$4·31 \times 10^{-2}$	$6·46 \times 10^{-2}$	$4·78 \times 10^{-2}$	1·35
2·0	$3·63 \times 10^{-2}$	$7·26 \times 10^{-2}$	$5·45 \times 10^{-2}$	1·33
2·5	$3·25 \times 10^{-2}$	$8·12 \times 10^{-2}$	$6·00 \times 10^{-2}$	1·35
3·0	$2·96 \times 10^{-2}$	$8·87 \times 10^{-2}$	$6·48 \times 10^{-2}$	1·37
3·5	$2·77 \times 10^{-2}$	$9·70 \times 10^{-2}$	$6·88 \times 10^{-2}$	1·41
4·0	$2·53 \times 10^{-2}$	$10·12 \times 10^{-2}$	$7·22 \times 10^{-2}$	1·40
4·5	$2·43 \times 10^{-2}$	$10·87 \times 10^{-2}$	$7·48 \times 10^{-2}$	1·45
5·0	$2·42 \times 10^{-2}$	$12·13 \times 10^{-2}$	$7·69 \times 10^{-2}$	1·58
5·5	$2·17 \times 10^{-2}$	$11·92 \times 10^{-2}$	$7·84 \times 10^{-2}$	1·52
6·0	$1·86 \times 10^{-2}$	$11·17 \times 10^{-2}$	$7·94 \times 10^{-2}$	1·41
6·5	$1·77 \times 10^{-2}$	$11·48 \times 10^{-2}$	$8·01 \times 10^{-2}$	1·43

It will be seen that, although there is not a quantitative agreement with theory, the general variation with thickness of mica traversed is qualitatively in good accord with the theory. On the average, the observed value of $\rho_3 x$ is about 1·4 times the calculated value. The result can be brought into accord with the theory if it be supposed that the number of the types of collision, which are mainly responsible for the straggling, is twice as numerous as those calculated by classical mechanics on Bohr's theory. This question will be discussed in more detail in § 26 b.

Bohr has shown that the quantity $(\rho_3 x)^2 = \int_0^x P \cdot (dT/dx)^{-2} dx$ where the constant $P = 16\pi e^4 N n$, where e is the charge on the electron, N the number of atoms per c.c. of the gas, and n the average number of electrons per atom. The calculated value $P = 9\cdot64 \times 10^{-16}$ for air at 15° C. and 760 mm. and about $11\cdot4 \times 10^{-16}$ for mica. The

value dT/dx can be deduced from the velocity curve, so that the integral can be evaluated graphically. The theoretical variation of the rate of increase of the straggling with thickness of air is shown in Fig. 35. It is seen that the curve passes through a minimum at about 6·5 cm. and then rises rapidly to the end of the mean range 6·90 cm. The value of $(\rho_3 x)^2$ can thus be obtained by integrating the curve corresponding to the range traversed. The total value $(\rho_3 R)^2$ of the

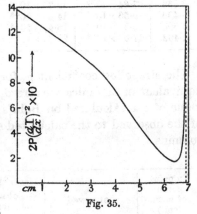

Fig. 35.

straggling due to α particles of radium C is proportional to the total area enclosed by the curve. The quantity $(\rho_3 R_1)^2$, for example, due to the α rays from polonium range R_1, is proportional to the area of the curve to the right of the abscissa $(6\cdot97 - 3\cdot92)$ cm. In a similar way the total straggling of the α particles from any product can be deduced. It will be observed that the contribution to the straggling per unit path is greatest near the beginning of the range of the particles. For example, half the total straggling for the α particles from radium C occurs in the first 2·4 cm. of the range. The rapid variation very near the end of the range produces only a small effect on the total straggling for swift α particles. The value of ρ corresponding to the full range of the α particles from radium C is for mica given by $\rho = 1\cdot73 \times 10^{-2}$. On the same data, the value of ρ for the rays from polonium is $1\cdot63 \times 10^{-2}$, while, as we have seen, I. Curie found by the expansion method a value $1\cdot67 \times 10^{-2}$ for air.

§ 24. Capture and loss of electrons by α particles. The earlier observations of the path of an α particle in a magnetic or electric field had indicated that the α particle was expelled with a double charge from a nucleus, and retained this charge for the greater part of its path through matter. It was supposed that, near the end of its path when its velocity was much diminished, the α particle ultimately captured two electrons and became a neutral atom of helium. G. H. Henderson* first observed the surprising fact that even the swiftest α particle could occasionally capture an electron

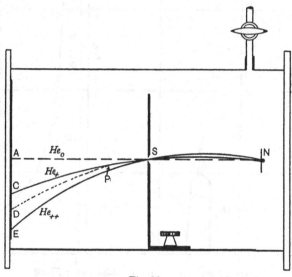

Fig. 36.

and that a pencil of α particles always consisted of a mixture of singly as well as doubly charged particles and at low speeds of neutral particles as well. This effect is most clearly shown by the arrangement shown in Fig. 36. The rays from a line source of α rays after passing through an absorbing sheet of matter N pass through a slit S and fall on a photographic plate or zinc sulphide screen placed at the end of the chamber. A uniform magnetic field is applied perpendicular to the plane of the paper. When the box is exhausted to the stage of a cathode-ray vacuum, there appears in addition to the main band E due to doubly charged particles (He_{++}) a "midway" band C which is deflected to exactly half the distance of the

* Henderson, *Proc. Roy. Soc.* A, **102**, 496, 1922.

main band. When low velocity α particles are present, a neutral undeflected band appears at A. This midway band is due to singly charged particles (He_+) which in their passage through matter have captured an electron, while the undeflected band is due to α particles (He_0) which have captured two electrons. The relative intensity of the midway to the main band varies very rapidly with the velocity of the α rays. It is small, about 1 in 200, for the swiftest rays but increases rapidly when the velocity of the α particles is reduced by

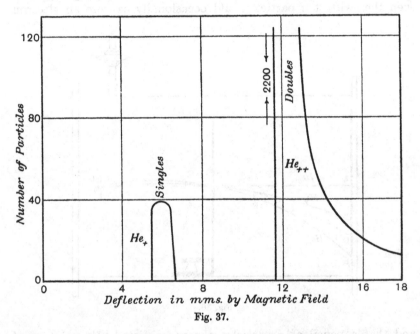

Fig. 37.

their passage through absorbing screens. Ultimately for low velocities the midway band becomes predominant while the undeflected band grows in intensity.

In the initial experiments, Henderson employed the photographic method, using Schumann plates. A detailed quantitative examination of the subject was made by Rutherford*, using the scintillation method. By deflecting the particles by an electric as well as by a magnetic field, it was shown conclusively that the midway band consisted of singly charged helium atoms having the same velocity as the main group of doubly charged α particles. The results obtained

* Rutherford, *Phil. Mag.* **47**, 277, 1924.

for different velocities using mica as an absorbing screen are shown in Figs. 37, 38, 39. The abscissae represent the actual deflection of the particles in a magnetic field which varied in the different experiments. Fig. 37 gives the result when the emergent range of the α particles was 3·6 cm., Fig. 38 1·2 cm., Fig. 39 about 0·46 cm.

In Fig. 37 the height of the He$_+$ band was 1/67 of the main band. With lowering of the velocity, the distribution varied rapidly and the width of the two bands increased markedly owing to the straggling, i.e. to inequalities in the velocities of the emergent α particles.

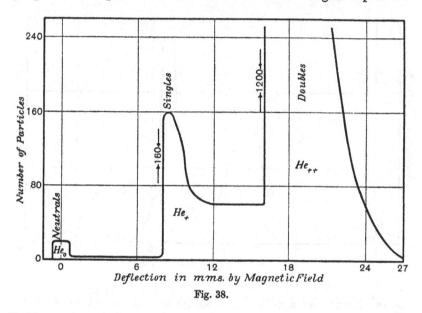

Fig. 38.

In illustration of this, the width of the band He$_0$ in Fig. 38 shows the natural width of the bands in a magnetic field if the rays were completely homogeneous. In Fig. 39 it is seen that the bands for He$_+$ and He$_{++}$ are very broad and diffuse, and that the singly charged particles are comparable in number with the doubly charged. The explanation of Curve D, Fig. 39 will be given later.

These effects are only shown in a high vacuum. If for example in the experiments illustrated in Fig. 39 a small pressure of air or other gas is introduced into the apparatus, the midway band practically disappears. This is due to the fact that the He$_+$ particles are ionised, i.e. lose an electron in their passage through the gas. For example, suppose this event happens to a He$_+$ particle at the point

P, Fig. 36, in its trajectory in a magnetic field. The particle, now He$_{++}$, more deflected by the magnetic field, reaches the screen at a point *D*, say between the two bands, outside the narrow field of view of the counting microscope. The variation in the number of He$_+$ particles as the pressure of the gas is gradually raised gives a simple method of estimating the "mean free path" of the He$_+$ particle in the gas.

If we consider a number of He$_+$ particles of the same velocity traversing a gas, some will travel farther than others before they

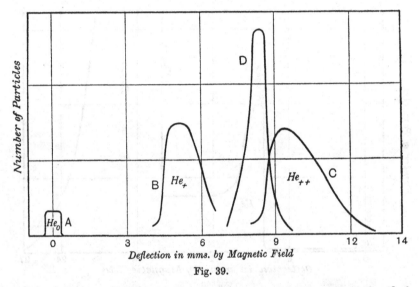

Deflection in mms. by Magnetic Field

Fig. 39.

lose an electron as a result of a collision with the atoms in their path. In considering, however, a large number of particles, the average distance of travel will be a definite quantity, and from analogy with the mean free path of a molecule in a gas will be termed "the mean free path for loss." In a similar way, a He$_{++}$ particle occasionally captures an electron as a result of collisions with the atoms in its path and becomes a He$_+$ or singly charged particle. The average distance traversed before capture will be termed "the mean free path for capture." The lengths of the mean free paths both for capture and loss will obviously depend on the nature, pressure and temperature of the gas, but are most conveniently expressed in terms of the equivalent path in the gas at normal pressure and temperature.

The number of particles in the midway band falls off nearly exponentially with increasing pressure of the gas, provided the mean free path of the He_+ particle for capture of an electron is long compared with the mean free path for loss. This is the case for high velocity α particles. From such observations the mean free path for loss of an electron can be determined.

It will be seen that a rapid interchange of charges $He_{++} \rightleftarrows He_+$ takes place along the track of the α particle, amounting to more than a thousand before the particle is brought to rest. A temporary equilibrium must consequently be set up between the number N_1 of He_{++} particles and the number N_2 of He_+ particles, such that in a given distance the number of captures is nearly equal to the number of losses in traversing a thickness dx of material. If λ_1 is the mean free path for capture $He_{++} \longrightarrow He_+$ and λ_2 is the mean free path for loss $He_+ \longrightarrow He_{++}$, then

$$N_1 dx/\lambda_1 = N_2 dx/\lambda_2$$

or
$$N_2/N_1 = \lambda_2/\lambda_1.$$

The ratio λ_2/λ_1 is thus known for a definite velocity of the α particle by measuring the ratio N_2/N_1 for that velocity. We have seen that λ_2 can be directly measured for any gas but cannot be determined directly for a mica or other solid screen. This apparent difficulty is obviated by noting that the observed ratio N_2/N_1 appears to be independent of the nature of the absorbing screen over a wide range of atomic weights. This can be directly tested by adding to the main absorbing sheet of mica a thin sheet of matter which is sufficiently thick to set up a new equilibrium for interchanges of charge, but not thick enough to alter materially the velocity of the emergent rays. This question was examined with great care by Henderson*, using an ionisation method to measure the number of α particles. No detectable difference in the ratio N_2/N_1 was noted for such different substances as aluminium, mica, copper, silver and gold. Since the average atomic weight of the constituents of mica is not much greater than that of air, we may assume without much error that N_2/N_1 for air is the same as for mica. Since N_2/N_1 and λ_2 can be directly measured, the value of λ_1 can be deduced.

The following values of λ_1 and λ_2 were obtained by Rutherford for rays of different velocities. The velocity of the particles is expressed

---

* Henderson, *Proc. Roy. Soc.* A, **109**, 157, 1925.

in terms of V_0, the maximum velocity of the α particles from radium C, viz. $1 \cdot 922 \times 10^9$ cm. per c.c.

Velocity in terms of V_0	λ_2/λ_1 for mica	Mean free path λ_2 for loss in air at N.P.T. (mm.)	Mean free path λ_1 for capture in air at N.P.T. (mm.)
0·94	1/200	0·011	2·2
0·76	1/67	0·0078	0·52
0·47	1/7·5	0·0050	0·37
0·29	1/1	0·003	0·003

It is known (§ 16) that the production of ions per unit path of a gas by an α particle varies nearly inversely as the velocity. The ions consist of the primary ions produced by the actual liberation of an electron from the atom or molecule by the α particle, and secondary ions due to swift electrons (δ rays) set in motion by the α particle. The loss of an electron from a swift He_+ particle in passing through matter may be regarded as a process of ionisation and should thus follow similar laws. This has been found to be the case, for the mean free path for loss in air λ_2 varies approximately as the velocity over the range examined. It should be pointed out that the values for λ_1, λ_2 given in the table above for a velocity $0 \cdot 29\, V_0$ have not been directly measured but deduced from this velocity relation.

The analogy between loss of an electron from a He_+ particle and ionisation of a gas by an α particle is borne out by comparing the value of λ_2 in hydrogen and helium with that observed in air. It is known that the ionisation by an α particle per unit path in hydrogen and helium is about one-fifth of its value in air for the same velocity. The observed value of λ_2 in hydrogen was 4 to 5 times greater than in air and for helium 5 to 6 times. In addition the actual magnitude of the mean free path λ_2 in air and other gases is of the order to be expected on the ionisation theory. Consider for simplicity the passage of α particles from radium C of maximum velocity V_0 through hydrogen. The α particle produces 528 ions per mm. of path and of these probably only about one-third or 170 are primary. Now the value of λ_2 for loss in hydrogen for this velocity is about 0·06 mm. In this distance 1 He_+ particle on the average is ionised, while an α particle produces about 10 primary ions. Taking into account that the ionisation potential of hydrogen, about 11·5 volts, is much smaller than that for a He_+ particle, which is supposed to be about

54 volts, the mean free path observed in hydrogen is of about the magnitude to be expected. In the present state of our knowledge of ionisation, it is difficult to give a more precise estimate.

It will be seen that for the velocities examined the ratio λ_2/λ_1 varies very rapidly with the speed of the α particle. If we suppose the variation with velocity V can be expressed by a power law V^n, the average value of n found by Rutherford was 4·6, i.e. the ratio λ_2/λ_1 varies roughly as the inverse fifth power of the velocity. Since λ_2 is roughly proportional to V, it is seen that the mean free path for capture λ_1 varies as $V^{5·6}$ or roughly as the sixth power of the velocity. Henderson*, using the electric method, found the ratio λ_2/λ_1 could not be expressed by a simple power law and concluded that the value of n decreased with increasing velocity. For velocities in the neighbourhood of $0·51\,V_0$, n was about 4·3; for a velocity about $0·70\,V_0$, n was 3·4. G. H. Briggs†, by using the photographic method and determining the relative intensity of the deflected bands by a photometer, found the value of n to be $4·3 \pm 0·1$ over a range of velocities from $0·825\,V_0$ to $0·272\,V_0$. Some of the photographs obtained are given in Plate VI. On account of the smallness of the ratio λ_2/λ_1 for high velocities, it is difficult to fix the law of variation with velocity with precision.

We have previously drawn attention to the remarkable fact that the value of λ_2/λ_1, for a given velocity of the α particle, is independent of the atomic weight of the absorber, at any rate for atomic weights between aluminium and gold, indicating a close connection or parallelism over this range between the processes of capture and loss. As the result of a mathematical analysis, Thomas‡ concluded that this relation must break down for high velocities in the case of very light elements, where the electrons are loosely bound. He found that for high speed α particles, the value of λ_1 should vary much more rapidly with velocity for hydrogen than for the heavy elements, and estimated that $\lambda_1 \propto V^{11}$. This point was examined experimentally by Jacobsen§ by observing the distribution of scintillations between the singly and doubly charged bands of α particles in a magnetic field when a small pressure of hydrogen was introduced. This was compared with the effect observed in air at a

* Henderson, *Proc. Roy. Soc.* A, **109**, 157, 1925.
† Briggs, *Proc. Roy. Soc.* A, **114**, 341, 1927.
‡ Thomas, *Proc. Camb. Phil. Soc.* **114**, 561 1927.
§ Jacobsen, *Nature*, **117**, 341, 1927.

pressure adjusted to give the same chance of loss of an electron as in hydrogen. In both cases the distribution is dependent on the ratio between the mean free path for capture and loss of electrons by the α particles in their passage through the gas. It was observed with hydrogen that the relative number of the singly charged α particles was not more than half that with air under corresponding conditions. From these results he concluded that swift α particles of velocity about $0.94 \, V_0$ captured few if any electrons in hydrogen and estimated that the mean free path for capture in hydrogen must be greater than 200 mm.

When the speed of the α particle becomes comparable with that of the electron round the hydrogen nucleus, capture in hydrogen proceeds much more freely. Rutherford (loc. cit.) found by the method described later that this was the case for velocities of the α particle about $0.3 \, V_0$.

§ 24 a. **Neutral α particles.** We have seen that neutral α particles, which are undeflected by a magnetic field, appear in numbers when the α particle is much reduced in velocity. These neutral particles which arise from the capture of an electron by a He$_+$ particle give feeble scintillations. It is difficult to make measurements on the mean free path for capture and loss in this case since, on account of the straggling of the α particles, the neutral particles vary widely in velocity. The neutral band, like the singly charged band, disappears when gas is introduced. Rough observations indicated that the mean free path in air for loss of an electron from a neutral particle was about 1/1000 mm. or about 1/5 of the mean free path for conversion of He$_+$ into He$_{++}$ for a velocity $0.47 \, V_0$.

From the data obtained for capture and loss, it can be estimated that the change of charges occurs about 600 times in air, while the velocity changes from V_0 to $0.29 \, V_0$ when the mean free path for capture and loss are equal. There seems to be little doubt that an even greater number of interchanges He$_{++} \rightleftarrows$ He$_+$ and He$_+ \rightleftarrows$ He$_0$ occurs while the velocity falls from 0.29 to zero.

It can be estimated that an α particle of radium C over a range 6.9 cm. is doubly charged for a distance about 6.4 cm. and singly charged for a distance 0.5 cm. The curvature in a magnetic field of the path of an α particle in air would consequently be less than if the α particle were doubly charged along its whole path.

§ 24 b. **Effect of gas in the path of slow α particles.** An interesting effect is observed when a small pressure of gas is introduced along the path of α particles of such low velocity that the mean free paths for capture and loss are comparable in magnitude. Consider, for example, an experiment similar to that illustrated in Fig. 39, where A, B, C represent the distribution in a vacuum of He_0, He_+, He_{++}, particles. On adding a pressure of air of $1\cdot1$ mm. in the chamber, corresponding to a stopping power of only $0\cdot23$ mm. of air, the three bands A, B, C disappear and are replaced by a single comparatively narrow band shown in D. Since the mean free path for capture for the highest velocity α particles present, viz. $0\cdot37\ V_0$, is about $0\cdot008$ mm., it is seen that the average number of interchanges of charges in passing through the air is about 36 and still more for the slower particles. When the interchanges are as numerous as this, each α particle of a given velocity is deflected by a uniform magnetic field as if it carried a charge between e and $2e$, the actual magnitude depending on the ratio of the mean free path for capture and loss.

The average charge can be calculated in the following way. Suppose that the mean free paths in the gas of α particles of velocity V for capture and loss are λ_1 and λ_2 respectively. The magnetic deflection will correspond to particles of velocity V for which the average charge is $\dfrac{\lambda_1 . 2e + \lambda_2 . e}{\lambda_1 + \lambda_2}$ or $e\left(1 + \dfrac{\lambda_1}{\lambda_1 + \lambda_2}\right)$. The total deflection of such a pencil of rays in a magnetic field will be proportional to $e/V . \left(1 + \dfrac{\lambda_1}{\lambda_1 + \lambda_2}\right)$. In this calculation, we have omitted to take into account the interchanges $He_+ \rightleftarrows He_0$ which also occur to some extent. The effect of such changes would be to diminish the value of the average charge given above.

In this particular experiment, the velocities of the α rays lay mainly between $0\cdot37$ and $0\cdot25\ V_0$. From the data obtained for capture and loss over this range of velocities, it can be calculated that the magnetic deflection will not be very different for particles of different speeds as the average charge is approximately proportional to the speed of the particles over this range. There is in consequence a marked concentration of the particles in a comparatively narrow band for which the average deflection is intermediate between the average for the broader and more diffuse bands A and B.

This concentration effect, which is so intimately connected with the frequency of capture and loss at low velocities, gives a very simple indirect method of testing whether α particles of low velocity capture electrons in their passage through hydrogen. In order to make the average mean free path in hydrogen about the same as in air, hydrogen was introduced into the chamber at a pressure of 5 mm., about equivalent to the pressure of 1 mm. of air. Under such conditions, a similar concentration of the particles in a narrow band D was observed. This affords a definite proof that the α particles capture electrons in hydrogen and that for such low velocities the ratio λ_1/λ_2 is of the same order as that observed in air. This is in contrast to the results already referred to (p. 125), where it is found that the mean free path for capture in hydrogen for very swift particles is very long compared with that for air.

The effect of capture and loss on the curvature of the track of an α particle in a uniform magnetic field has been measured by Kapitza* by a direct method. The track of an α particle was photographed in air and hydrogen in a very small expansion chamber of the Wilson type, and the curvature of the track was measured in a very strong magnetic field, produced by the passage through a coil of a very intense momentary current due to the discharge of an accumulator battery of special type. By a system of timing, it was arranged that the α particle entered into the expansion chamber at the moment when the current through the coil measured by an oscillograph was a maximum. An example of the curved tracks obtained is shown in Plate II, fig. 4, which represents the tracks in air in a field of 43,000 gauss. By careful measurement of the curvature at each point of the tracks in hydrogen at reduced pressure, it was possible to deduce directly the value of e'/V, i.e. the average charge e' of the particle divided by its velocity at each point. The results obtained are shown in Fig. 40. It is seen that the value of e'/V passes through a flat maximum for α particles between 8 and 17 mm. of hydrogen. The curvature falls off rapidly near the end of the track, but precise measurement was difficult owing to the scattering of the α particles by the gas. This effect, as we should expect, is much more marked in air than in hydrogen. This variation in the curvature of the track has been shown by Kapitza to be in good accord with the data available on the interchanges of charge of the α particles over the range of velocities involved.

This experimental method has the advantage of giving the changes of curvature in a single track and not a statistical value of a number

* Kapitza, *Proc. Roy. Soc.* A, **106**, 602, 1924.

of particles in a complex beam. The investigation was undertaken with the object of testing whether the curvature of the tracks of the particles at corresponding points of the range varied for different particles and thus be connected with the well-known phenomenon of the straggling of the α particles. The disturbances due to the scattering of the α particles by the gas, however, were too large to allow of a definite decision on this point.

Kapitza has suggested that, on certain assumptions, one α particle may be expected to differ from another in the frequency of the interchanges of its charge. The experiments on the scattering of hydrogen nuclei by α particles indicate that the helium nucleus has an unsymmetrical distribution of forces around it. It may thus be

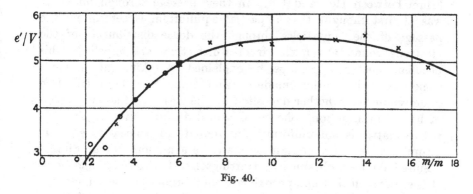

Fig. 40.

possible that an electron captured by an α particle is constrained to move in a definite orbit relative to the nuclear structure. The chance of loss of this electron, for example, may be expected to be different when the plane of motion of the electron is parallel or perpendicular to the line of flight. This question of a possible difference in the behaviour of individual α particles could be tested experimentally in this way if sufficiently strong magnetic fields were available to cause such a marked curvature of the track of an α particle that the disturbance due to scattering could be neglected.

§ 24 c. Experiments with uncovered sources. We have so far referred to observations on the distribution of He_+ and He_{++} particles after passing through mica and other absorbing screens. An unexpected effect was noted when observations were made with bare sources of platinum wire coated with the active deposit of radium by exposure to radon. Within the limit of experiment, the number of He$_+$ particles corresponded to an equilibrium distribution. This could not be due to some of the radioactive atoms being driven by recoil into the metal, for a similar effect was observed when a polished metal wire was coated with a surface deposit of radium C by electrolysis. This result is surprising when we consider that on the average the deposit is not one molecule thick and that the mean free path for capture for such velocities corresponds to 3 to 4 mm. of air. It seems probable that the distribution observed is to be ascribed to the radioactive matter itself.

It is very difficult to believe that these He$_+$ particles come directly from the nucleus and, even if this be assumed, it is a curious coincidence that the number should be about that required for equilibrium between He$_+$ and He$_{++}$ in their passage through mica. It was at first thought that a partial equilibrium is set up by the passage of the α particles through the dense distribution of the electrons round the nuclei from which they are expelled. This peculiar effect can, perhaps, be explained by taking into account some recent observations made by Mlle Chamié* on the distribution of active matter whether deposited by the action of the emanations or by chemical action. She has obtained definite evidence that the active matter is not uniformly distributed but collects together in granules which may contain as many as a million atoms. Similar results have been obtained by Harrington†. Under such conditions, it is easily seen that an approximate equilibrium between singly and doubly charged particles will be set up by the passage of the α particles from the interior of the granules to the surface. The average thickness of these granules is sufficient to lead to an approximate equilibrium between He$_+$ and He$_{++}$ in the issuing beam of particles.

§ 24 d. **Comparison with positive rays.** The phenomena of capture and loss of electrons by a flying α particle are analogous to those shown by the positive rays in a discharge tube where the interchange of charges may be very rapid. The use of α particles, however, affords us a most convenient method for studying the interchange of charges for velocities much higher than can be obtained in a discharge tube. For example, the capture and loss of electrons by the α particle have been examined for velocities between about 2×10^9 and 4×10^8 cm./sec. Detailed experiments on the mean free path of charged and

---

* Mlle Chamié, *C.R.* **184**, 1243, 1927; **185**, 770, 1277, 1927.
† E. L. Harrington, *Phil. Mag.* **6**, 685, 1928.

neutral hydrogen atoms in a discharge tube have been made by W. Wien and other investigators in his laboratory. The maximum velocity of the hydrogen particles in the experiments of Rüchardt was about 2.3×10^8 cm./sec., corresponding to an applied potential difference of 35,000 volts. In order to produce H_+ particles of velocity equal to the α particles of radium C an accelerating potential of 2·5 million volts would be required. From the results of Rüchardt on hydrogen rays, it appears that the mean free path for capture and loss in air are equal, viz. about 1/2000 mm. at N.P.T. for a velocity about 2.26×10^8 cm./sec. This corresponds to a velocity $0.12 \, V_0$ in terms of the α particle. It is to be expected that the interchange $H_+ \rightleftarrows H_0$ if governed mainly by velocity must behave very similarly to the interchange $He_+ \rightleftarrows He_0$, with the difference that the ionisation potential is higher for helium. From the data obtained from hydrogen rays, it is to be expected that the mean free paths for the interchange would be equal for a somewhat higher velocity than for hydrogen, viz. about $0.16 \, V_0$. The rough observations made with α particles are in approximate accord with this estimate, since the mean free path for the conversion of He_0 into He_+ in air was found to be about 1/1000 mm.

§ 25. **Theory of capture and loss.** We have seen that the mean free path of an He_+ or He_0 particle before losing an electron is in fair numerical agreement with the value to be expected on the ionisation theory. The problem of capture is in some respects more difficult and complicated, as it involves even in the simplest case the interaction of at least three charged bodies. From a physical point of view, it seems clear that for capture an electron must be liberated from an atom with a speed not very different from that of the α particle, and in very nearly the direction of flight of the α particle. On the ordinary theory of a collision of an α particle with a free electron, the electron projected at 60° with the direction of flight of the α particle will have the same speed as the α particle. If the electron is initially bound to an atom, the angle of escape will no doubt be dependent to some extent on the ionisation potential of the electron in its orbit. In order then that the liberated electron should have its direction altered to coincide nearly with the line of flight of the α particle, a second rectifying collision either with the nucleus or with a neighbouring nucleus must be postulated. Such a collision will have little effect on the velocity of the electron but will alter its direction. It is difficult to take into account rectifying collisions with other electrons, as the final speed depends on the closeness of the collision and the degree of binding of the electron to the nucleus, but the collisions of this type are probably unimportant in comparison with the nuclear collisions.

The problem of such a three body collision is mathematically very difficult, but approximate solutions can be obtained with certain reasonable simplifications. The question has been examined carefully by L. H. Thomas*, and he finds that the chance of capture is much influenced by the orbital velocity of the electron round the nucleus. If this orbital velocity is small compared with that of a swift α particle, as in hydrogen and helium, the chance of capture will be small but will increase rapidly with lowering of the velocity of the α particle. If V is the velocity of the α particle, and v the orbital velocity of the electron, the chance of capture varies as $(v/V)^{11}$, where v is small compared with V. We have already seen (p. 125) that experiments carried out to test this conclusion have shown that the chance of capture for swift α particles is very much less in hydrogen than in air.

In the case of the heavier elements, where v for many of the electrons may be comparable with V, the problem is very different. It involves a consideration of the variation of the electric field due to the nucleus in the outer regions of the electronic distribution. Assuming the field, as found by Hartree, to vary as r^{-3} in the region for which collisions most readily lead to capture, Thomas found that the chance of capture should vary little for the heavier elements and be proportional to $(v/V)^{5\cdot5}$, a variation with V in good accord with that observed experimentally. In addition, an actual calculation of the mean free path for capture in air for a velocity $0\cdot94\ V_0$ was of the right order of magnitude—actually about four times the value deduced from experimental observations. It is quite probable, however, that the value deduced experimentally is too great, for it involves the assumption that λ_1/λ_2 is the same for mica as for nitrogen and oxygen. On the whole it appears that the classical theory of collisions with simplifying assumptions is competent to account for the main features of the phenomena of capture and loss of electrons by α particles. A complete theory is beset with great analytical difficulties.

The consideration of capture and loss may be approached from

* Thomas, *Proc. Roy. Soc.* A, **114**, 561, 1927.

another point of view. We have previously drawn attention to the remarkable fact that the ratio λ_1/λ_2 appears to be approximately constant over the range of atomic weights from aluminium to gold. This indicates that the phenomena of capture and loss are in some way intimately connected over this range. If it be supposed that capture is the inverse process to ionisation, R. H. Fowler* has shown that from the principles of dissociative equilibrium the laws of capture may be directly deduced if the laws of ionisation are known. The qualitative and even the quantitative results deduced from this theory are in fair accord with experiment for the heavier elements, but it is doubtful whether capture and loss are true inverse processes in the sense assumed.

Oppenheimer† has investigated the problem of capture and loss of an electron by an α particle on the basis of the wave-mechanics. On certain simplifying assumptions, he finds that the mean free path for capture in air for the doubly charged α particles varies as v^{-6}, where v is the velocity of the particle. The calculated value of the mean free path for air is in fair agreement with the experimental results.

* R. H. Fowler, *Proc. Camb. Phil. Soc.* **21**, 521, 1923; **22**, 253, 1924.

† Oppenheimer, *Phys. Rev.* **31**, 349, 1928.

CHAPTER V

THEORIES OF ABSORPTION OF α RAYS

§ 26 a. Theory of absorption of α particles. We have already discussed the experimental results which have been obtained on the retardation of the α particle in passing through matter, and the variation in atomic stopping power for the different elements. It is of great interest and importance to consider in some detail how far we are able to give a theoretical explanation of the observed facts.

It is clear that the α particles have such great energy of motion that they pass freely through the electronic structure of the atoms in their path, and it is only rarely that they pass close enough to the nucleus to experience a sensible deflection. In the calculations we shall disregard the scattering of the α particles and assume that the α particle travels in nearly a straight line through the matter in its path.

In consequence of its charge, it is evident that the α particle, in passing close to an atom or in penetrating it, must disturb the motions of the electrons in the atom. The amount of energy which is communicated to the electrons by the α particle can be calculated on certain assumptions, but it is difficult to be certain of their validity. The difficulties involved may be illustrated by the distinction made to-day between an "elastic" and "non-elastic" collision of an electron with an atom. Suppose, for example, that an electron of definite energy collides with a helium atom. For an energy less than 20 volts, the electron is supposed to make an elastic collision with the atom, i.e. it is deflected with a very slight reduction of its energy. If the electron has an energy slightly greater than 20 volts, it *may* suffer an "inelastic collision" in which it gives most of its energy to the atom, resulting in the excitation of the atom by the displacement of one of the electrons to an outer quantum orbit. When the electron falls back to its original position, the energy given to the atom is radiated away. When the energy of the colliding electron is greater than 24·5 volts, it may again give most of its energy to the atom, resulting in the ionisation of the atom, i.e. the escape of an electron from the atom. Now, if the electron has an energy slightly less than the excitation energy, it must violently disturb the motions

of the electrons in the helium atom and thus momentarily give
energy to the atomic system, but this energy is in some way returned
to the colliding electron, so that finally it hàs only expended the
slight amount of energy corresponding to a perfectly elastic collision.
If this view of the conditions of transfer of energy between an electron
and an atom is substantially correct, we are faced with the question
whether or not the transfer of energy between the α particle and the
atom is controlled by similar conditions. We shall have occasion to
discuss later different theories in which these factors are taken into
consideration.

For simplicity let us at first suppose that the electrons in the
atom are at rest, and calculate the kinetic energy given to the
electrons by the passage of an α particle either through or close to
the atom. Let M, V, E be the mass, velocity, and charge of the
α particle, and m and e the mass and charge of the electron. After
a collision, the velocity u of the electron deflected through an angle
θ with the direction of the incident α particle is given by

$$u = \frac{2MV}{M+m} \cos \theta = 2V \cos \theta \quad \ldots\ldots\ldots\ldots(1),$$

since m/M is small. Suppose the forces between the α particle and
the electron are electrical and vary as the inverse square. If p be
the distance of the electron from the path of the α particle before
collision, then it can be shown that $p = \mu \tan \theta$, where

$$\mu = \frac{Ee}{V^2}\left(\frac{1}{m} + \frac{1}{M}\right) = \frac{2e^2}{mV^2} \quad \ldots\ldots\ldots\ldots\ldots(2),$$

since $E = 2e$. If $q = \tfrac{1}{2}mu^2$ be the energy communicated to the
electron, then from (1) and (2) it follows that

$$p^2 = \frac{8e^4}{mV^2}\left(\frac{1}{q} - \frac{1}{2mV^2}\right) \quad \ldots\ldots\ldots\ldots(3).$$

In passing through a thickness dx of matter containing N atoms
per unit volume, each with n electrons, the number of encounters of
the α particle with electrons for which p lies between p and $p + dp$ is

$$2\pi Nn \, p \, dp \, d\dot{x} = \pi Nn \, dx \, \frac{8e^4}{mV^2} \cdot \frac{dq}{q^2} \quad \ldots\ldots\ldots\ldots(4).$$

The loss of energy by the α particle due to encounters with the
electrons in a distance dx, in which the energy transfer lies between
q and $q + dq$, is given by

$$-MV \frac{dV}{dx} = \frac{8\pi e^4 Nn}{mV^2} \cdot \frac{dq}{q} \quad \ldots\ldots\ldots\ldots(5).$$

Integrating between the limits q_1 and q_2, the integral becomes $\log_e q_2/q_1$. From (1) the maximum energy q_2 communicated to an electron is $2mV^2$. If $q_1 = 0$ the integral becomes infinite, indicating an infinite absorption. To overcome this difficulty it is necessary to disregard distant encounters between the α particle and the electron and to give a finite value to q_1. This has been done by J. J. Thomson* in his theory of the absorption of β particles by matter. Darwin†, in his theory of the absorption of α particles, supposed that energy was taken from the α particle only if it entered the atom, and found on this assumption an approximate agreement between theory and experiment.

§ 26 b. In order to give a more physical meaning to this limitation, Bohr‡ introduced the new conception, that the electrons initially at rest are able when disturbed to vibrate with definite natural frequencies. If this time of vibration is long compared with the time taken by an α particle to traverse the atom, the energy transferred will be the same as if the electron were free. If, however, the time of vibration is short compared with the time of transit of the α particle, the energy transferred is less and becomes smaller the higher the frequency. By supposing that each electron has a natural frequency and integrating over all the electrons, the loss of energy from the α particle can be calculated. The transfer of energy to the vibrating electrons is integrated for values between $p = \infty$ and $p = a$, while the loss of energy by the closer collisions is calculated as in formula (5) for values of p between a and 0. On certain conditions which hold for light atoms, the actual calculated transfer of energy does not depend on the assumption of any particular value of a. In this way Bohr found

$$- MV \frac{dV}{dx} = \frac{4\pi e^2 E^2 N}{mV^2} \sum_1^n \log \frac{\gamma V^3 Mm}{2\pi\nu Ee\,(M+m)} \quad\ldots\ldots\ldots(6),$$

where ν is the frequency of vibration of the electron considered and n the number of electrons per atom. $\gamma = 1\cdot123$ is a numerical constant.

In order to compare theory with experiment, we shall confine ourselves to a consideration of the passage of α particles through the two simplest gases, hydrogen and helium, for which the assumptions

* J. J. Thomson, *Phil. Mag.* **23**, 449, 1912.
† Darwin, *Phil. Mag.* **23**, 901, 1912.
‡ Bohr, *Phil. Mag.* **24**, 10, 1913; **30**, 581, 1915.

made in the calculation hold closely. The hydrogen molecule contains two electrons, each of which may be supposed to have the same natural frequency ν. A similar assumption is made for the helium atom. In these cases, Bohr's formula reduces to the form convenient for calculation

$$\frac{V_0^4 - V_1^4}{X} = \frac{128\pi e^4 N}{mM} \log \frac{\gamma m \overline{V}^3}{4\pi e^2 \nu} \quad \ldots\ldots\ldots\ldots(7),$$

where V_0 and V_1 are the velocities at the ends of a short interval of length X, and $\overline{V}$ is the mean value of V for X. If V is very great, we can neglect the variation in the logarithmic term, and

$$V_0^4 - V_1^4 = aX,$$

where a is a constant which can be calculated. This is of the same form as J. J. Thomson's original theory for cathode rays and has been shown to hold experimentally by Whiddington and others. For still swifter β rays, the relation between V and X is more complicated (§ 99).

At the time Bohr advanced his theory, the quantum theory of spectra had not been developed. The natural frequency to be ascribed to the electrons was deduced from the dispersion formula of Lorentz and the experimental observations of C. and M. Cuthbertson. The frequency of the two electrons was taken as $3\cdot52 \times 10^{15}$ for hydrogen and $6\cdot0 \times 10^{15}$ for helium, corresponding to quantum energies of $14\cdot3$ and $24\cdot5$ volts respectively, values not very different from the main excitation potentials subsequently found by experiment. On the whole, there was a fair accord between calculation and experiment. For example, it was calculated from (6) that in hydrogen the value of $\frac{dV}{dx}$ is $-4\cdot9 \times 10^7$ and $-2\cdot6 \times 10^7$ for velocities of the α particle $1\cdot35 \times 10^9$ and $1\cdot75 \times 10^9$ cm./sec. respectively, while the corresponding experimental values were $-5\cdot4 \times 10^7$ and $-2\cdot7 \times 10^7$.

In order to bring the theory more into accord with modern views of spectra, it is necessary to suppose that the electron does not execute elastic vibrations as in the original theory, but rotates in a definite quantum orbit which is perturbed by the passage of the α particle. Calculations for hydrogen and helium have been made recently along these lines by R. H. Fowler*. He finds that the absorption formula is of exactly the same form as that given by Bohr; the frequency of rotation ω taking the place of ν but the constant

* R. H. Fowler, *Proc. Camb. Phil. Soc.* **22**, 793, 1925.

γ is larger, viz. 2·42 instead of the old value 1·123. This modification, which leads to a greater stopping power, appears to be connected with the fact that in elliptical motion a change of energy involves a change of frequency while in harmonic oscillations it does not. The value of ω for the hydrogen molecule is taken as 1·12 and for helium 2·9 times its value for the normal orbit of the hydrogen atom.

An excellent test of the theory is to compare the ratios of the observed and calculated stopping powers of the α particles in hydrogen and helium for different velocities. The results are given in the following table.

cm./sec.	Ratios of stopping powers H_2/He		
	Observed	Calculated $\gamma=2.42$	Calculated $\gamma=1$
1.98×10^9	1·19	1·17	1·20
1.68×10^9	1·22	1·18	1·22
1.26×10^9	1·30	1·22	1·27
1.0×10^9	1·40	1·26	1·34

The general agreement between experiment and calculation is satisfactory, but is best if $\gamma = 1$ instead of the calculated value 2·42. Fowler gives reasons why the value of γ ought to be taken smaller than that calculated. For very slow collisions, it may be that there is no energy transfer. To give a value $\gamma = 1$, the transfer of energy should cease when $2\pi p\omega/V = 0.6$, i.e. when the time taken for the α particle to describe a circle of radius p reaches about 3/5 of the time of revolution of the electron in its orbit.

Undoubtedly the classical theory of Bohr, as modified by Fowler, has proved competent to account fairly closely for the absorption of the α particle in the elements hydrogen and helium. There still, however, remains the uncertainty whether the α particle can lose energy to the atom when the collision is too slow to give sufficient energy to the electron to cause an excitation or ionisation of the atom or molecule.

This question has been attacked by Gaunt* on the basis of the new mechanics, and a partial solution obtained. The new theory deals only with statistical averages and gives a mean energy transfer in a collision without any assumption of particular energy switches. The transfer of energies by close collisions leading to ionisation, which

* Gaunt, Proc. Camb. Phil. Soc. 23, 732, 1927.

has not been computed on the new theory, is assumed to be the same as on other theories. The theory shows that occasionally energy is given to an electron by a distant collision and in this respect is analogous to Bohr's original theory. The surprising result emerges that the absorption of the α particle in hydrogen has exactly the same form as in the Bohr theory (equation 7), but the value of γ, which is difficult to calculate with certainty, may be slightly different.

The calculations required are very intricate and have so far not been applied to other atoms, but it appears that while the underlying ideas of Bohr's theory are somewhat modified, the calculated values of the absorption are essentially unchanged.

§ 26 c. We shall now consider some attempts made by Henderson* and Fowler† to calculate the stopping power on the assumption that the energy can only be lost by the α particle either in exciting or ionising the atom. The formula for this purpose has already been given in (5), where q is the energy communicated to the electron. Since dq/q is a pure number, the unit of energy employed is immaterial and it is convenient to express the energies in terms of equivalent voltages, i.e. the potential drop in volts to generate the same velocity in an electron. If W is the energy in volts of an electron having the same velocity V as the α particle and λ the ionisation potential, the limits of the integral in (5) are $4W$, which is the maximum energy transferred, and λ, then

$$- MV \frac{dV}{dx} = \frac{8\pi e^4 Nn}{mV^2} \log \frac{4W}{\lambda} \quad \ldots\ldots\ldots\ldots(8).$$

This gives the loss of energy due to ionisation of the atom.‡

It is necessary also to take into account the transference potential τ due to excitation of the atom. In general, for most atoms and molecules, the transference potentials are probably distributed at fairly close intervals after the first one. The rate of expenditure of energy due to transference is given by (8), if τ replaces λ. For convenience of calculation, the general equation can be put into the form

$$\frac{V_0^4 - V^4}{X} = \frac{32\pi e^4 N}{mM} \Sigma_s\, n_s \log \frac{4\overline{W}}{\tau_s} = \frac{32\pi e^4 N}{mM}. S,$$

* Henderson, *Phil. Mag.* **44**, 680, 1922.

† Fowler, *Proc. Camb. Phil. Soc.* **21**, 521, 1923.

‡ This calculation is based on the assumption that the electrons in an atom have a random distribution. Loeb and Condon (*Journ. Franklin Inst.* p. 95, 1923) have pointed out the difficulties in this assumption.

where X is the distance in cm. and $\overline{W}$ a suitable mean value for the velocity drop V_0 to V. The summation is extended over all electrons each with the appropriate values of τ which are obtained from optical and X ray data. All values of τ greater than $4\overline{W}$ are excluded. The quantity under the summation sign, which is the only variable from atom to atom, may be called S, the relative atomic stopping power. The values of S calculated by Fowler for velocities of the α particle between 1·0 and 0·9 of the velocity of the α particle from radium C (1.922×10^9 cm./sec.), i.e. for ranges between 7·0 and 5·2 cm. at 15° C., are given in the table.

Atomic no.	Atom	S observed	S calculated	Ratio $\dfrac{\text{calculated}}{\text{observed}}$
1	$\frac{1}{2}$ H$_2$	11·5	6·05	0·53
2	He	18·6	10·5	0·56
(7·2)	$\frac{1}{2}$ air	55·0	30·8	0·56
10	Ne	67	42·8	0·64
13	Al	81	50·2	0·62
18	A	107	63·4	0·59
29	Cu	146	94·2	0·63
36	Kr	165	102·7	0·62
47	Ag	177	124·4	0·70
54	Xe	217	131·2	0·60
79	Au	233	167·1	0·72

The experimental data for hydrogen and the monatomic gases are taken from Gurney*; the values for Al, Cu, Ag and Au from Marsden and Richardson† and from Marsden and Taylor‡. On account of the imperfection of metal foils, the results for the metals are probably not so accurate as for the gases.

The fifth column gives the ratio between the calculated and observed stopping powers. It is seen that on the average the calculated stopping power is about 60 per cent. of the observed but, considering the wide range of atomic numbers, the ratio shows little variation. It seems clear that calculations of this kind can be relied on to give the *relative* stopping power for the heavier atoms with fair accuracy.

In the original theory of Henderson and Fowler, the electrons in the atom were assumed to be at rest and to that extent the theory

* Gurney, *Proc. Roy. Soc.* A, **106**, 340, 1925.
† Marsden and Richardson, *Phil. Mag.* **25**, 184, 1913.
‡ Marsden and Taylor, *Proc. Roy. Soc.* A, **88**, 443, 1913.

is artificial. Thomas* and E. J. Williams† have carried the analysis further by taking into account the actual motions of the electrons in the atom, and have found that for the heavier atoms the discrepancies between observation and calculation shown in the above table are largely removed. The effect of the motion of the electrons is to increase the stopping power as calculated on the simple collision theory. The complete calculations are intricate and depend on a summation of the effects of all the electrons moving in a field of force which varies from point to point in the atom. In a number of cases, the results of Hartree's calculations on the field of force round certain atoms have been utilised. Thomas has compared the results of his calculations with the observations of the observed stopping power of the monatomic gases for different velocities obtained by Gurney. Considering the difficulty of the calculations, the agreement is surprisingly good over the whole range of velocities for argon and to a less extent for krypton and xenon. In the case of neon, there are considerable discrepancies especially for low velocities. This is no doubt due to the uncertainty of our knowledge of the motions of the electrons in the neon atom.

It is of interest to consider how far the different theories of stopping power of the α particle are in accord with the experimental results for the lighter gases. This is shown in the following table compiled by E. J. Williams for α rays of velocity $1 \cdot 88 \times 10^9$ cm./sec. corresponding to 1000 electron-volts.

Gas	Observed stopping power per electron	Calculated stopping power per electron		
		Bohr theory	Henderson theory	Revised Henderson theory
H_2	12·0	12·8	6·1	8·2
He	10·0	11·9	5·3	6·9
N_2	8·1	10·4	4·8	7·1
O_2	7·7	10·5	4·8	7·1
Ne	7·1	10·2	4·7	7·4

The calculations can be made with most certainty for the light elements and it is of interest to note that Bohr's theory agrees best for H_2 and He. On the other hand, the revised Henderson theory is in better accord with the experimental results for N_2, O_2 and Ne.

---

* Thomas, *Proc. Camb. Phil. Soc.* **23**, 713, 1927.
† E. J. Williams, *Proc. Manch. Lit. and Phil. Soc.* **71**, No. 4, 23, 1927.

In the case of the heavier elements, where the theory of Bohr is difficult to apply, the revised Henderson theory is in fair accord with the observations.

We have seen (§ 23 b) that while the variation of straggling of the α particles with thickness of matter traversed is in good agreement with calculation, the actual magnitude of the straggling is greater than that calculated on the classical theory of Bohr. Theory and experiment could be brought into accord if the number of close collisions, involving considerable losses of energy, were twice as numerous as those calculated. This discrepancy is considerably reduced by taking into account the velocities of the electrons in their orbits, as in the revised Henderson theory, but is not completely accounted for. We have seen, however, that in the wave mechanics there is a certain probability that even distant collisions may transfer considerable amounts of energy to the atom, and the difference may be made up in this way. It will be seen also (§ 28) that the ionisation of the α particle per unit path in gases indicates that the number of ionising collisions is about twice as great as that calculated on the simple theory.

Considering the intricacy of the problem of the passage of the α particle through matter and the uncertainty of some of the atomic data which have been used in calculation, there is a very fair agreement on the whole between theory and observation. No doubt the present theories are to some extent provisional and will have to be replaced ultimately by more elaborate calculations based on the wave mechanics.

It is clear, however, that in general a reasonable if not complete explanation can be given of the main facts of observation, viz. the laws of retardation of the α particle, the straggling of the α particles, and the variation in stopping power of different atoms and its dependence on the velocity of the α particle. It is seen that the Bragg rule for stopping power is purely empirical, although of a very convenient form for rough calculation of stopping powers. It is of interest to note that the theory of Thomas shows clearly the factors on which the variation of stopping power with velocity depends. The theory, in its present form, does not allow us to follow the variation of the velocity of the α particle near the end of its range. This no doubt is much complicated by the changes of charge of the α particle which become so prominent near the end of the range.

§ 27. Absorption of β rays. The comparison of theories of absorption of α rays with experiment can be made over a very limited range of velocity, viz. between about 0.8×10^9 and 2.0×10^9 cm./sec., corresponding in speed to electrons of energy between about 200 and 1100 volts. No α particles of higher energy are available in sufficient quantity for experiment. However, the theories of absorption of α particles and electrons of moderate speed are practically identical in form, and the comparison of theory with experiment can be carried to much higher velocities by studying the range and stopping power for β particles of speed greater than α particles but not sufficiently high to require sensible relativity corrections for mass and energy. From the requirements of the theory of Bohr, it is to be anticipated that the theory would hold more accurately the higher the speed of the particle.

When a pencil of homogeneous X-rays is passed through a gas in a Wilson expansion chamber, a number of tracks of β rays are observed of which the initial energy of projection can be calculated from X-ray data. A measurement of the length of such tracks in various gases has been made by Wilson, Nuttall and Williams, and others. E. J. Williams has made a comparison between the observed ranges and those calculated on the theory of Bohr.

Gas	Initial energy of electron (volts)	Observed range (cm.)	Calculated range on Bohr's theory (cm.)
H_2	4,800	0·36	0·22
H_2	7,560	0·73	0·50
O_2	3,790	0·049	0·022
O_2	7,560	0·119	0·072
A	20,400	0·68	0·41
Air	58,000	5·7	2·9

It is seen that the calculated ranges are only 60 to 65 per cent. of the observed values, and the discrepancy is in the same direction for all the gases including hydrogen. As in the case of the α particle, the revised Henderson theory gives a range somewhat greater than the observed values.

If these results are correct, it seems clear that the theories are still sensibly in error even for high speed particles where it is to be anticipated that the calculations would be most reliable.

§ 28. Theory of ionisation by the α particle. The ionisation produced in different gases has been discussed in §§ 16–18 and it is there shown that the amount of ionisation per unit path of a pencil of α particles is approximately proportional to the expenditure of energy by the α particle. Small differences are observed for different gases, particularly for slow velocities of the α particle. The average energy required to produce a pair of ions varies from gas to gas, but particularly in the case of diatomic gases it is very much greater than that corresponding to the ionisation potential of the gas, as ordinarily measured. The results have been given in the table on page 81.

It is known that the primary ionisation due to an α particle results from the liberation of an electron from the atom or molecule. Depending on the closeness of the collision, these electrons or δ rays as they have been termed (§ 29) may have velocities between 0 and $2V$, where V is the velocity of the α particle. If the speed of the ejected electron is sufficiently high, it may in turn produce a number of ions in the gas before it comes to rest. This is termed the secondary ionisation, and it is important in calculation to distinguish between these two modes of production of ions. If ρ is the ionisation potential of the gas and z the energy of the liberated δ particle, it is clear that the number of ions produced by the δ particle must be less than z/ρ. As a result of a detailed calculation, R. H. Fowler* finds the average number of ions to be $\dfrac{3}{4} \cdot \dfrac{z+\rho}{\rho}$, so that for a swift δ particle about three-quarters of the energy can be utilised in producing ions.

From the collision theory, it is possible to calculate at once the primary ionisation produced by the α particle when the ionisation potential is known. This has been done for the β rays by J. J. Thomson and for the α rays by Bohr†. The problem has been discussed in more detail by R. H. Fowler. We have shown in equation (4) that the number of encounters in a length of track dx which result in a total transfer of energy between q and $q + dq$ is

$$\pi N n\, dx\, \frac{8e^4}{mV^2} \cdot \frac{dq}{q^2}.$$

If ρ is the energy required to remove the electron from the atom, the number of encounters which give rise to primary electrons of energies after escape between q and $q + dq$ is

$$\pi N n\, dx\, \frac{8e^4}{mV^2} \cdot \frac{dq}{(q+\rho)^2}.$$

* R. H. Fowler, *Proc. Camb. Phil. Soc.* **21**, 531, 1923.
† Bohr, *Phil. Mag.* **24**, 10, 1913; **30**, 581, 1915.

Taking into account that there are n electrons in the atom each with a distinct value of ρ, the total amount of primary ionisation in a length dx of track is therefore

$$\pi N\,dx\,\frac{8e^4}{mV^2}\Sigma_s\,n_s\int_0^{2mV^2-\rho_s}\frac{dq}{(q+\rho_s)^2}.$$

Assuming that the average energy required to produce an ion is the same for an α particle as for an electron and is equal to the ordinary ionisation potential, it can be shown that the total ionisation is given by

$$\pi N\,dx\,\frac{8e^4}{mV^2}\Sigma_s\,n_s\int_0^{2mV^2-\rho_s}\frac{g\,(q)\,dq}{(q+\rho_s)^2},$$

where

$$g\,(q)=\frac{3}{4}\cdot\frac{q+\rho_s}{\rho_s}.$$

On these assumptions, Fowler (*loc. cit.*) has calculated the primary ionisation P and the total ionisation T per cm. of track of an α particle for different velocities. A comparison of theory with experiment is included in the following table. The values of the number of ions are taken from the data of Geiger and Taylor.

Gas	W in volts		1000	800	600	400
H$_2$	$T\times10^{-3}$	Calculated	5·28	6·35	8·04	11·13
		Observed	4·6	4·8	5·6	8·3
	T/P	Calculated	4·4	4·2	4·0	3·7
He	$T\times10^{-3}$	Calculated	2·15	2·56	3·22	4·40
		Observed	4·6	4·8	5·6	8·3
	T/P	Calculated	3·8	3·6	3·4	3·1
Air	$T\times10^{-4}$	Calculated	1·6	1·9	2·3	3·0
		Observed	2·2	2·3	2·7	3·95
	T/P	Calculated	5·8	5·4	4·9	4·4

While the calculated ionisation is in all cases of the right order of magnitude, it is noteworthy that the results for hydrogen and helium are very different. In hydrogen, the calculated ionisation is somewhat greater than that observed for all velocities; while for helium the values are about half of the observed.

In order, however, to get a clear idea of the agreement between calculation and experiment for the different gases, it is necessary to consider carefully the experimental data to be utilised in the calculations of the ionisation. In the table, the energy required to produce an ion is taken as the ordinary ionisation potential of the gases, viz. 11·5, 24·5 and about 16·5 volts for hydrogen, helium, and air respectively. We know, however, that the average energy expended by the α particle per pair of ions is 33, 28 and 35 volts for H_2, He, and air respectively (§ 18). In addition Lehmann*, Lehmann and Osgood† have directly measured the average energy required to produce an ion for electrons of energies between 400 and 1000 volts, and find values of 37·5, 33 and 45 volts for H_2, He and air respectively. Since much of the secondary ionisation due to the α particle results from electrons of this order of energy, it is evident that we cannot safely employ in calculation the actual minimum ionisation potentials. If we take Lehmann's values for the average energy in the production of ions by secondary electrons and the minimum ionisation potential for the primary ionisation, it is clear that the value of T and the ratio of T/P are much smaller than those given in the table.

Taking into account the reduction factor employed in calculating the secondary ionisation, and using the data of Lehmann, it can be estimated that the total ionisation T per cm. for an α particle of speed corresponding to 1000 electron volts in hydrogen, helium, and air are in thousands of ions 2·9, 2·15, 9·1 respectively, while the observed values are 4·6, 4·6 and 22 respectively. On the average, the ionisation calculated in this way is about one-half the observed for all three gases. This indicates that the number of collisions which lead to ionisation are about twice as numerous as those calculated on the Henderson theory, and about 1·5 times the number on the revised theory. A similar result has emerged from a comparison of theory with observation for absorption and straggling of the α rays. The analyses of these different data are thus consistent among themselves and lead to the general conclusion that for the lighter gases the number of effective collisions giving rise to the liberation of δ particles of different speed is about twice as great as that calculated from simple classical considerations, when no account is taken of the orbital velocity of electrons.

If the classical considerations of the transference of energy between the α particles and electrons are substantially correct for close encounters, the difference must be due to the probability of transfer of considerable quantities of energy in more distant collisions, which on the classical theory give only a small energy transfer. Such interchanges are possible on the new wave-mechanics and a more complete solution will no doubt ultimately be given along these lines.

* Lehmann, Prcc. Roy. Soc. A, 115, 624, 1927.
† Lehmann and Osgood, Proc. Roy. Soc. A, 115, 609, 1927.

CHAPTER VI

SECONDARY EFFECTS PRODUCED BY α RAYS

In this chapter we shall discuss some of the secondary effects accompanying the emission of α particles, including the emission of delta rays, the recoil accompanying the ejection of an α particle, the heating effect produced by the absorption of α particles, and the production of helium due to the accumulated α particles.

§ 29. Emission of delta rays. When a stream of α particles passes through a thin sheet of matter in a vacuum, a number of electrons are observed to be emitted from both sides of the plate. The energy of the great majority of these electrons is only a few volts, but the total number from each surface of the plate is of the order of 10 times the number of incident α particles. J. J. Thomson*, who first studied this emission of electrons from a polonium plate, gave them the name of δ (delta) rays. A large amount of work has been done to determine the factors involved in the liberation of these electrons, including the effect of velocity of the α particle, the nature of the bombarded material, the state of its surface, and the distribution with velocity of the escaping electrons. Before discussing these data, it is desirable to consider the origin of these electrons. It is clear that the escape of electrons, whether from a radioactive surface or a bombarded plate, is in a sense a secondary effect connected with the absorption of energy of the α rays in their passage through matter.

In fact, on modern views, the emission of δ rays is a necessary and inevitable consequence of the passage of α rays through matter. Suppose an α particle in traversing matter passes close enough to an electron to give it sufficient energy to escape from the parent atom. This is the ordinary process of ionisation by the α particle. The escaping electron may, however, have a wide range of velocity depending on the closeness of the collision. Let E, M, V be the charge, mass and velocity of the α particle; e, m, u the corresponding quantities for the electron after the collision. Suppose for simplicity of calculation that the electron is initially free and at rest. Let θ be the angle between the direction of emission of the electron relative

---

* J. J. Thomson, *Proc. Camb. Phil. Soc.* **13**, 49, 1904.

to the direction of flight of the α particle. Then by the principles of mechanics $u = V \cdot \dfrac{2M}{M + m} \cos \theta = 2V \cos \theta$, since m/M is small. Consequently the maximum velocity acquired by an electron, i.e. when $\theta = 0$, is $2V$, or the maximum velocity cannot be greater than twice the velocity of the colliding α particle. This deduction will be seen to be in good accord with observation. It can be shown (§ 26) that the number n of electrons projected within the angle θ by a single α particle in passing through a small thickness t of matter is given by $n = \pi N t \mu^2 \tan^2 \theta$, where N is the number of electrons per c.c. of matter and $\mu = Ee/mV^2$. It is clear from this calculation that the number of electrons projected nearly at right angles to the direction of the α particle, i.e. with slow speeds, will greatly preponderate. In the actual material the electron is not free but is bound to an atom, and a definite amount of energy measured by its ionisation potential is required to release it. Since the lowest ionisation potential of an atom, usually less than 20 volts, is small compared with the energy with which many of the electrons are liberated by an α particle, the above equation will not be seriously in error if we confine our attention to the swifter electrons. The number of electrons liberated increases rapidly as the velocity of the α particle is reduced but the maximum energy of emission diminishes.

Since the swifter electrons may travel several millimetres in hydrogen producing a number of ions in their track, the production of these δ rays may be studied by photographing the tracks of α particles in hydrogen or other gas at low pressure. A number of electron trails are seen to radiate from the α ray track. The slower electrons have so small a range that they appear as knobs or projections along the track, giving it a ragged edge.

Examples of such δ ray photographs by Chadwick and Emeléus are shown in Plate III, figs. 3 and 4, which illustrate the points mentioned. Photographs of this kind were first obtained by Bumstead[*] and later by C. T. R. Wilson[†], Chadwick and Emeléus, and Auger.

Using the α particles of radium C, of velocity V, the maximum velocity of the swiftest δ particle should be $2V$, corresponding in energy to about 4000 volts. Chadwick and Emeléus[‡] found that

[*] Bumstead, *Phys. Rev.* 8, 75, 1916.
[†] C. T. R. Wilson, *Proc. Camb. Phil. Soc.* 21, 405, 1922.
[‡] Chadwick and Emeléus, *Phil. Mag.* 1, 1, 1926.

the longest δ ray tracks were 2 mm. in H_2, 2·6 mm. in He, and about 0·45 mm. in air at N.T.P. The lengths of these tracks are in good accord with that to be expected for 4000 volt electrons, assuming that the length of the trail of an electron is proportional to the fourth power of the velocity, as observed by C. T. R. Wilson. Wilson noted that no visible δ ray tracks appeared for the last few centimetres of the range of the α particle in air. This is not due to a cessation of emission of δ rays but to the fact that the length of the trails falls off rapidly with the decrease of velocity of the α particle, and they are consequently difficult to observe. The swiftest δ ray emitted by an α particle of range 2·4 cm. would have a range in air of only 0·11 mm., and in consequence would only be detected under good photographic conditions.

If α particles from thorium C' of velocity $2·09 \times 10^9$ cm./sec. pass through hydrogen, the calculated number n of δ particles per cm. is given by $n = 2·4 \tan^2 \theta$, vide p. 148. Chadwick and Eméleus concluded that only those tracks were recognisable which had a minimum range of about 0·15 to 0·2 mm. Taking this factor into account, the number of δ rays to be seen per cm. of track is given by $n = 9$ for H_2, 11·5 for air, and 11 for helium. For an average velocity of the α particle 2×10^9 cm./sec. in hydrogen 157 tracks had a range greater than 0·62 mm., 51 tracks a range greater than 1 mm. and 10 greater than 1·37 mm. These numbers are in the ratio 3·1 : 1 : 0·2, while the calculated numbers are in the ratio 2·1 : 1 : 0·47, a very fair agreement with the simple theory, considering the difficulties inherent in the deductions from experiments.

Auger[*] independently made a similar comparison using the α rays from polonium of range 3·9 cm., and found an even closer agreement with theory. Over a total length of 60 cm. of a number of α ray tracks in hydrogen, he observed 246 δ tracks of range greater than 0·25 mm. or 4·1 per cm., while the calculated number was 4·9. In all, 36 tracks were observed with ranges greater than 0·67 mm.; the observed number per cm. is 0·6 and the calculated 0·64.

While there is good numerical agreement between the observed and calculated numbers of δ particles of different velocity, the direction of emission does not always agree with the theory. For δ particles in air, C. T. R. Wilson noted that the particles appear to be emitted normally to the line of motion of the α particle. Chadwick has observed the same effect both in air and in argon, but

[*] Auger, *Journ. de phys.* **7**, 65, 1926.

in hydrogen and helium, on the other hand, he observed that the
δ particles of long range have undoubtedly a velocity component
in the direction of the α particle. Chadwick concludes that these
anomalies in the direction of emission are due to the scattering of
the δ particles which is particularly marked in the heavier gases.
Since the liberation of an electron in swift motion round a nucleus
is a complicated process, it may be, however, that the simple theory
gives nearly correct results with regard to the number and energy
of the δ particles, but may be insufficient to give the final direction
of emission. Chadwick and Eméléus applied a magnetic field along
the axis of the expansion chamber to test whether any swift δ
particles were emitted in the direction of flight of the α particles.
No certain difference was noted, but the curvature given to the
δ ray track is clearly observable (Plate III, fig. 4).

It is to be expected that an α particle would occasionally remove
one of the K electrons from nitrogen or oxygen. In such a case, the
electron would escape with a slow speed and X-radiation would be
emitted when the atom returned to its normal state. This radiation
will be absorbed in the neighbouring gas and give rise to a β particle
of short range. Wilson has observed such short tracks of β rays
close to the α ray track in air and Chadwick in argon.

From these experiments on the emission of δ particles we are able
to understand in a general way the emission of electrons from a
plate bombarded by α rays. The electrons are emitted from the atoms
at all velocities between the maximum $2V$ and 0, where V is
the velocity of the α particle. These electrons are scattered and
absorbed in the matter, and the number and distribution with
velocity of those which escape from the surface will depend on a
number of complicated factors. Chadwick has deduced that the
number n of escaping electrons of energy greater than T should
be given approximately by the formula $n = a/T - b$, where a and b
can be roughly calculated.

This formula is in fair accord with the experiments of Bumstead
who determined the distribution with velocity of the particles
emitted by a surface bombarded by the α rays from polonium. An
accurate determination of this distribution is beset with many
experimental difficulties, since the swifter δ particles are able to set
up a secondary emission of electrons when they fall on a metal
surface. In general, as we should expect, the number of swift particles
is small compared with the number of slow particles. For example,

Bumstead* found that the relative number of electrons of energies greater than 20, 200, 400, 1000 volts are in the ratio 100, 14, 9, 2·4 respectively. The great majority of the electrons have a velocity between zero and a few volts. Some δ particles were observed of energy greater than 2000 volts while the theoretical maximum is about 2800 volts. Bumstead found that the graph connecting the number of electrons emitted from a metal with the range of the bombarding electrons was very similar in shape to the Bragg ionisation curve, showing a well-defined maximum near the end of the range. For the metals examined, viz. Al, Cu, Pt, Au, Pb, no certain difference was noted in the shape of the curve or in the number of escaping electrons. Bianu†, using the α rays of polonium, found the swifter δ particles were able to ionise a gas and estimated their energy to be about 2400 volts. Gurney in his experiments on the β ray spectra of radium B and radium C observed that a large number of slow electrons was emitted from the source with velocities of the order to be expected if they were δ particles arising from α particle collisions. The method used by Gurney would be very suitable for an accurate determination of the relation between the number and velocity of the escaping electrons.

On general views, we should expect a much greater emission of electrons from a surface on which the α rays fall nearly tangentially than when they fall normally. It is evident that the electrons liberated by the α particle in its track have a better chance of escaping from the surface in the former case. This no doubt is the explanation of the fact that more electrons per α particle are emitted from a surface coated with polonium than from a metal on which the α rays fall more normally.

Kapitza‡ has made calculations to estimate the number of particles emitted based on a novel point of view. As we have seen, the α particle in traversing matter gives energy to the electrons and this energy will be dissipated in the form of heat within a small distance of the track of the α particle. There will in consequence be a strong local heating and, in a sense, the local temperature will be very great. Consequently the escape of electrons from the surface of a plate may be regarded as a type of thermionic emission to which the equations of Richardson apply. In this way, Kapitza has shown that the

* Bumstead, *Phil. Mag.* **22**, 907, 1911; **26**, 233, 1913; Bumstead and McGougan, *Phil. Mag.* **24**, 462, 1912.

† Bianu, *Le Radium*, **11**, 230, 1919.   ‡ Kapitza, *Phil. Mag.* **45**, 989, 1923.

calculated electronic emission for slow velocities is of the right order of magnitude. As is well known, the amount of thermionic emission largely depends on the cleanness of the surface of a metal and the state of the vacuum. The observations of many investigators that the amount of δ ray emission is dependent to a large extent on the cleanness of the surface may receive a general explanation along these lines.

On the other hand, on this theory we should expect that the amount of electronic emission would vary greatly with the nature of the metal. No marked effect of this kind has been observed, but few experiments with cleaned metallic surfaces have so far been made. The theory applies mainly to the emission of slow velocity electrons from a bombarded surface and does not of itself set a definite limit to the energy of the escaping electrons, which appears to be characteristic of the phenomenon.

The problems connected with the emission of electrons by α particles have been the subject of numerous investigations, especially in the years 1911–1914. A review of the earlier work has been given by Hauser*.

§ 30. Secondary electrons from β rays. A swift electron (β particle) in passing through matter occasionally makes a close collision with an electron in an atom and liberates it from the atom with a velocity depending on the closeness of the collision. We should in consequence expect a β particle, like an α particle, to produce a number of secondary electrons (delta particles) in its passage through matter. On account, however, of the equality of mass of the electrons involved in the collision, the velocity u of the electron projected at an angle θ with the original direction of the β particle of velocity V is given by $u = V \cos \theta$, and the angle between the directions of the two electrons after the collision is always 90°. If the β particle has a speed comparable with that of light, a correction is required to take account of the relativity increase of mass. Assuming that the electron behaves as a point charge, the frequency of the collisions in which electrons are expelled over a definite energy range can readily be calculated by the collision theory. In general, however, on account of the high speed of the swift β particle in comparison with the α particle, the number of secondary electrons should ordinarily be much smaller for a β than for an α particle.

* Hauser, *Jahrb. Rad. u. Elect.* **10**, 145, 1913.

The presence of such secondary electrons was first clearly shown by Moseley* in his experiments on the charge carried by the β rays from radium B and C. He found that the majority of the secondary electrons had energies of only a few volts, while the number was about 1/10 of the primary β rays. Danysz and Duane† made similar observations but concluded that the secondary electrons all had energies less than 40 volts. It is clear, however, from the simple theory, that the electrons, supposed at rest, with which the β particle collides may be given any energy up to the total energy of the β particle, but of course the probability of emission falls off rapidly with increase of energy of the expelled electron.

Collisions of this type giving rise to two electron tracks perpendicular to each other have been photographed in an expansion vessel by C. T. R. Wilson‡ and Bothe‖ and an illustration is given in Plate IV, fig. 3. Bothe compared with theory the frequency of these forked tracks in which, say, more than 30,000 volts energy was given to the electrons expelled from the atom, and found the number to be of the right order of magnitude (see also p. 450).

It is well known that electrons of energy of the order of 100 volts falling on matter give rise to a copious supply of secondary electrons of slower speed. The relative numbers of such electrons fall off beyond a certain speed and, as we have seen, become comparatively small for high speed β particles.

§ 31. **Recoil atoms.** When an α particle of mass M and velocity V is ejected from an atom of mass m, from the principles of conservation of momentum it is to be expected that the recoil atom of mass $m - M$ would move with a velocity u given by $u = MV/(m - M)$.

If, for example, an α particle of mass 4 and velocity 1.69×10^9 cm./sec. is expelled from radium A of mass 218, the velocity u of recoil of the atom of radium B of mass 214 is 3.15×10^7 cm./sec., corresponding in energy to an 110,000 volt electron, and about 2 per cent. of the energy of the α particle. The recoil atom should thus behave like a swift charged atom in a discharge tube. Like the α particle, it is to be expected that it should ionise the gas and have a definite range of travel before it is brought to rest. Even if the recoil atom were expelled initially without charge, it is to be ex-

* Moseley, *Proc. Roy. Soc.* A, **87**, 230, 1912.
† Danysz and Duane, *C.R.* **155**, 500, 1912; *Sill. Journ.* **35**, 295, 1913.
‡ C. T. R. Wilson, *Proc. Roy. Soc.* A, **104**, 1, 192, 1923.
‖ Bothe, *Zeit. f. Phys.* **12**, 117, 1922.

pected that, due to collisions with the molecules in its path, it
would gain and lose a charge many times before it was stopped.

The importance of this recoil as a method of radioactive analysis
was first brought out in 1909 by Hahn and Meitner* and by Russ and
Makower†. Hahn found that a number of radioactive substances
could be separated in a pure state by application of this method.
It was found that a negatively charged plate placed near a plate
coated with the active matter collected the recoil atoms, showing
that under these conditions they carried a positive charge. By this
method, a β ray product actinium C″ was obtained in a pure state
by recoil from actinium C. Similarly thorium C″ was obtained from
thorium C, and radium B from radium A. By the application of this
new method, Hahn and Fajans found that radium C was transformed in two distinct ways.

The range of travel and the ionisation due to the recoil atoms
were carefully examined by Wertenstein‡, using the recoil of radium
D from radium C and radium G from radium F. By using a pencil
of recoil atoms, he found the number of ions remained constant
for about half the range, and then fell off, at first slowly and then
more rapidly. The ionisation of the recoil atoms fell off as their
velocity diminished, but at the beginning of the range was about
five times greater than the ionisation produced by an α particle
under the same conditions. The range of the recoil atom from
polonium was only about 1/350 of the range of the α particle although
its energy was 1/50 of that of the α particle. It is thus clear that energy
is lost by recoil atoms in traversing a gas at a much greater rate than
for an α particle. This greater expenditure of energy is accompanied
by an increase of ionisation, but on the whole the total ionisation
produced by a recoil atom is relatively less than for an α particle
for an equal expenditure of energy. This indicates that a part of
the energy of the recoil atom is lost in molecular collisions which
do not give rise to ionisation.

Similar experiments have been made by A. B. Wood§. Measurements of the range of the recoil atoms have been made by a number
of observers with concordant results. The range has been found, to
vary approximately as the range of the α particle which liberates

* Hahn and Meitner, *Verh. d. D. Phys. Ges.* **11**, 55, 1909; *Phys. Zeit.* **10**, 697, 1909.
† Russ and Makower, *Prcc. Roy. Soc.* A, **82**, 205, 1909.
‡ Wertenstein, *C.R.* **150**, 819, 1910; **151**, 469, 1910; **152**, 1657, 1911; *Theses Paris*, 1913.
§ A. B. Wood, *Phil. Mag.* **26**, 586, 1913.

the recoil atom. Wertenstein found ranges 0·7 mm. in H_2 and 0·12 mm. in air at N.T.P. for the recoil atom from radium A. A. B. Wood found the ranges in H_2 0·52 mm., 0·55 mm., 0·74 mm. for recoil atoms from actinium C, thorium C and thorium C′ for which the ranges of the α particles are 5·5 cm., 4·8 cm. and 8·6 cm. respectively. Kolhörster* used the electrical counting method of Geiger to measure the ranges of recoil atoms. For recoil from thorium C and C′, the ranges in H_2 were 0·50 mm. and 0·875 mm. and in air 0·167 mm. and 0·203 mm. respectively.

The variation of the range with nature of the gas and the velocity of recoil are in the direction to be anticipated.

The recoil of the atom due to the expulsion of an α particle is clearly observable in a Wilson photograph when the expansion chamber contains a radioactive gas. Photographs of this kind have been obtained by C. T. R. Wilson, Bose and Ghosh, Kinoshita, Akeuti and Akiyama, and Dee. At ordinary pressure, the recoil track is shown by a knob at the end of the track. As the pressure is reduced, the recoil track becomes longer and often shows evidence of a marked scattering. In some cases, the usual direction of the recoil track does not coincide with the direction of flight of the α particle, but this is no doubt due to a large deflection of the recoil atom by a collision with an atom near its point of origin. The marked ionisation of the recoil atom per unit path is clearly shown by the density of the initial part of its track. Dee† has examined the recoil tracks from actinon and actinium A in an electric field in a Wilson expansion chamber and determined the charge and mobility of the recoiling atoms; a magnified photograph obtained by Dee of a recoil track arising from the transformation of an atom of actinon is shown in Plate IV, fig. 1.

The passage of a heavy atom with its attendant electrons through another gas is obviously a very complex problem for which no theoretical solution is as yet forthcoming. It would appear necessary to suppose that at high speeds the electronic structure of two atoms can interpenetrate. At lower speeds, collision without ionisation must soon bring the recoil atom to rest. No doubt the recoil atom like the α particle suffers many changes of charge in its flight. Most of the recoil atoms from radium A, radium C, and the corresponding products of thorium and actinium carry a positive charge when

---

* Kolhörster, *Zeit. f. Phys.* 2, 257, 1920.

† Dee, *Proc. Roy. Soc.* A, **116**, 664, 1927.

they are brought to rest in air, and can thus be concentrated on the negative electrode in an electric field. This is not universal for all recoil atoms, for Briggs* has shown that the radioactive gases liberated by recoil have no charge. The effect of different gases on the final charge of the recoil atom has also been investigated. The recoil atoms behave like positive ions in a gas and lose their charge by recombination with the negative ions in the gas. Notwithstanding their heavy mass, these recoil ions have about the same mobility in air as the positive ion produced in air. In the presence of an intense α radiation, a strong electric field is required to produce saturation, i.e. to collect all the recoil atoms on the negative electrode. The effects are complicated by the strong electric wind which is set up by the motion of the ions.

Wertenstein† examined the charge carried by the recoil atoms from radium C in a high vacuum. He concluded that the recoil atoms were initially uncharged but gained a positive charge in passing through a small thickness of matter. The magnitude of the charge indicated that some of the recoil atoms under these conditions might carry more than one positive charge. Wood and Makower‡ examined photographically the deflection of a pencil of recoil atoms from radium C in a magnetic field in a good vacuum and found it to be half the deflection of the α particle under the same conditions. Since the momentum of the recoil atom is equal and opposite to that of the α particle, this showed that the recoil atoms carried a single positive charge. It is difficult to reconcile this observation with the results found by Wertenstein unless it be supposed that a fraction of the recoil atoms from a source always carry a positive charge. This is not inconsistent with the view that the recoil atom is initially uncharged, for some of the radium C atoms are driven by recoil into the wire or plate and the recoil atoms traverse a small thickness of matter before they can escape§.

When a very thin and clean deposit is obtained on a polished plate, most observers agree that the efficiency of recoil in a vacuum is 100 per cent., i.e. a recoil atom can be collected for each α particle fired into the source. When the recoil atoms are collected in air or other gas by an electric field, the efficiency of collection is not so

* Briggs, *Phil. Mag.* **41**, 357, 1921; **50**, 600, 1925.
† Wertenstein, *C.R. Soc. d. Sci. d. Varsovie*, **8**, 327, 1915.
‡ Wood and Makower, *Phil. Mag.* **30**, 811, 1915.
§ Recent experiments by McGee show that the recoil atoms may be positively or negatively charged or neutral according to conditions.

high and generally lies between 60 and 90 per cent., depending on the condition of the surface and the active deposit.

§ 31 a. β ray recoil. When a β particle is expelled from a radioactive atom, we should expect to observe the recoil of the radioactive atom as in the case of the expulsion of an α particle. On account, however, of the much smaller momentum of the β particle, the energy of recoil is minute compared with that observed in an α ray recoil. Consider, for example, the β ray recoil from radium B. The average velocity of the β rays from this substance is 1.7×10^{10} cm./sec., the average velocity of recoil of the radium C atom 4.4×10^4 cm./sec., corresponding to an electron energy of only 0·4 volt.

Observations on the β ray recoil from radium B have been made by Russ and Makower, Muszkat[*] and Barton[†] with somewhat conflicting results. It seems clear, however, that the recoil atom is initially uncharged and so cannot be collected by applying an electric field in a vacuum. Considering the small energy of the recoil atom, it is obvious that the efficiency of recoil even in a good vacuum will be greatly dependent on the purity of the thin radioactive film and the smoothness and cleanness of the surface from which recoil takes place. Barton has discussed the various factors involved and concludes that it is difficult in practice to obtain an efficiency of recoil of more than 6 per cent.

Donat and Philipp[‡] have studied the β ray recoil from thorium B and found efficiencies from 2 to 6·5 per cent., and ascribe this low efficiency to the fact that at ordinary temperatures the recoil atoms have such a low speed that many of them are scattered or reflected from the receiving surface. This effect was first noticed by Jacobsen in his experiments to estimate the period of transformation of radium C' by recoil methods. From analogy with the behaviour of atomic rays, we should anticipate that the yield of recoil atoms would be increased by lowering the temperature of the receiving surface. This has been found to be the case and Donat and Philipp obtained a yield of about 30 per cent. when the collecting surface was at − 170° C. Wertenstein[§] finds that the observed efficiency of recoil at ordinary temperatures depends essentially on two factors, namely, the nature of the radiating surface and the

[*] Muszkat, *Phil. Mag.* **39**, 690, 1920; *Journ. d. Phys.* **2**, 93, 1921.

[†] Barton, *Phil. Mag.* **1**, 835, 1926.

[‡] Donat and Philipp, *Zeit. f. Phys.* **45**, 512, 1927; *Naturwiss.* June 22, 1928.

[§] Wertenstein, *C.R.* **188**, 1045, 1929.

nature of the collecting surface. Aluminium is the best metal as radiator but the worst as receiver. Efficiencies as high as 20 per cent. were noted in the β ray recoil of radium C from radium B when aluminium was the radiator and bismuth the collector. He failed to confirm the increase of efficiency noted by Donat and Philipp when the collector is cooled to low temperature.

This question of β ray recoil acquires a special importance, as it is the only direct method of estimating the very short life of the radioactive atoms radium C' and thorium C'. Some of the atoms of radium C' are able to escape from a surface of radium C by β ray recoil, and the expulsion of α particles by these flying recoil atoms can be followed by the scintillation method. Experiments of this kind have been made by Jacobsen* and Barton† and indicate that the average life of radium C' is of the order of 10^{-6} sec. In order to obtain a reasonable efficiency of recoil, great care has to be taken in preparing and retaining clean radioactive surfaces in a high vacuum.

In his later experiments Jacobsen found evidence that a γ ray transformation precedes the expulsion of an α particle. Such experiments are very important but difficult, and complicated by the reflection of the recoil atoms from the surfaces on which they fall.

§ 32. Heating effects of the radiations. In 1903 P. Curie and Laborde‡ showed for the first time that a quantity of radium maintained itself permanently at a higher temperature than the surrounding atmosphere. They measured the heating effect in a Bunsen ice calorimeter and found that 1 gram of radium in equilibrium with its products emitted heat at the rate of about 100 gram calories per hour, or 876,000 gram calories per year. This property of radium of emitting heat at a rapid uniform rate naturally excited much attention, and several hypotheses were advanced in explanation. From subsequent investigations it is clear, however, that the emission of heat from radium and other radioactive substances is in a sense a secondary effect, for it is a measure of the energy of the radiations from the active matter, which are absorbed in the active substance or the envelope containing it.

The heating effect is very easily shown if 100 milligrams of radium or the corresponding amount of radon is available. Two small Dewar

* Jacobsen, *Phil. Mag.* **47**, 23, 1924; *Nature,* **120**, 874, 1927.
† Barton, *Phil. Mag.* **2**, 1273, 1926.
‡ P. Curie and Laborde, *C.R.* **136**, 673, 1903.

flasks provided with thermometers are placed side by side. On introducing the active matter in one flask, a rise of temperature is clearly shown in a few minutes. A very striking experiment to illustrate that the emission of heat proceeds at the same rate in liquid air or hydrogen as at ordinary temperature was early made by P. Curie and Sir James Dewar* at the Royal Institution. The apparatus is seen in Fig. 41. The small Dewar flask A containing the radium salt R is filled with the liquid and surrounded with another flask containing the same boiling liquid so that no heat is communicated to A from the outside. The gas liberated in the tube A is collected over water in the usual way.

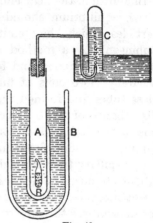

Fig. 41.

By this method the evolution of heat was found to be about the same in liquid oxygen and hydrogen as in air at ordinary temperature. The experiments in liquid hydrogen were of especial interest, for at such a low temperature ordinary chemical activity is suspended.

In later experiments P. Curie found that the heat evolution of radium depended on its time of preparation. The emission of heat, at first small, reached a steady state after about a month's interval. The experiments of Rutherford and Barnes† showed clearly that this was due to the increase of activity caused by the accumulation of radon and its subsequent products. They found that the radon a few hours after its separation from radium gave more than three-quarters of the heating effect of the original radium. It thus seemed probable that the heating effect was closely connected with the emission of α rays from the various active products. The heating effect due to absorption of the α rays can be calculated if it be supposed that the kinetic energy of the α particles is a measure of the heat evolution.

If Q be the number of α particles emitted per second by 1 gram of radium and each of its subsequent products, the kinetic energy of the expelled α particles is given by $\frac{1}{2} MQ\Sigma v^2$, where M is the mass of the α particle and v the velocity of each group of α rays present. It is also necessary to add the energy of the recoil atom which is

* Curie and Dewar, *Proc. Roy. Inst.* 1904.
† Rutherford and Barnes, *Phil. Mag.* **7**, 202, 1904.

converted into heat *in situ*. This is equal to $\dfrac{1}{2}\dfrac{M^2}{m}v^2$, where m is the mass of the recoil atom. Consequently the total energy E of the radiations is given by

$$E = \tfrac{1}{2}MQ\,\Sigma\left(1 + \frac{M}{m}\right)v^2 + E_1,$$

where E_1 is the energy of the β and γ rays absorbed under the conditions of measurement. If the heating effect is due entirely to the energy of the absorbed radiation, the value of E should agree with the observed heating effect in mechanical units.

In addition, the heat emission due to individual products in radioactive equilibrium should be proportional to the energy of the α particles which they emit. This was tested by Rutherford and Robinson* by a method which allowed rapidly changing heating effects due to radon and its products to be followed with ease and accuracy. Two coils of fine platinum wire were wound on similar glass tubes and formed the two arms of a Wheatstone bridge. The disturbance of the balance when a radon tube was introduced into one of the coils was measured with a galvanometer. The tube containing the radon fitted inside the platinum coil and was attached by a capillary tube through a stopcock to a simple pumping system. About a hundred millicuries of radon were introduced and the heating effect measured when it reached equilibrium four hours later. The emanation was then removed rapidly by means of the pump and the deflection of the galvanometer followed for an hour or more. The lag of the apparatus was sufficiently small to determine the change of heating effect due to the sudden removal of the radon, the rapid decay of the heating effect of radium A, and the curve of decay of the heating effect ultimately due to radium B and radium C. In this way it was found, taking the heating effect of radon in equilibrium as 100, that radon supplied 28·8, radium A 30·9 and radium B and radium C together 40·3. Under the experimental conditions, however, it was estimated that about 4 per cent. of the total heating effect was due to absorption of β and γ rays. Making an allowance for this, the α ray heating effect of radon, radium A and radium C were in the ratio 30·0 : 32·2 : 37·8, while the ratios for the calculated heating effect for the ranges given in § 19 are 28·0 : 31·4 : 40·6. While the heating effects are in the right order, it is seen that

* Rutherford and Robinson, *Phil. Mag.* **25**, 312, 1913; *Wien. Ber.* **121**, IIa, 1, 1912.

the extrapolated value for radium C is distinctly below the calculated value. In subsequent experiments by Watson and Henderson*, the observed and theoretical values for radium C have been found to be in good accord.

The actual value of the heating effect of radon in equilibrium was measured by replacing the source with a coil of fine manganin wire, occupying the same volume, through which a known current was sent. In this way, it was found that, subtracting the heating effect to be ascribed to the β and γ rays, one curie of radon and its products gives a heat evolution of 99·2 gram calories per hour. The heating effect of the β and γ rays under the experimental conditions was estimated at 4·3 gram calories. The value of the heating effect to be ascribed to radium itself in light of this value is 25·5 gram calories. By direct measurement of the heating effect of radium deprived of its products, Hess† found a concordant value 25·2. The heating effect to be ascribed to radium and its products and the contributions made by the β and γ rays are included in the following table. Column 2 gives the distribution calculated according to the energy of the α particles, column 3 the observed distribution. Estimates of the heating effect of the β and γ rays are included.

Heating effect in gram calories per hour corresponding to 1 gram of radium.

Product	α rays + Recoil atoms		β	γ	Total
	Calculated	Observed			
Radium	24·9	25·2	—	—	25·2
Radon	28·5	29·8	—	—	29·8
Radium A	31·2	31·9	—	—	31·9
Radium B } Radium C }	40·2	37·5	6·3	9·4	53·2
Totals	124·8	124·4	6·3	9·4	140·1

We have already seen that the value obtained by Hess for radium deprived of its products is in good accord with the calculated value deduced from the heating effect of radon. Meyer and Hess‡ made an accurate determination of the heating effect of a large quantity of pure radium salt. They found a heating effect of 132 gram calories

* Watson and Henderson, *Proc. Roy. Soc.* A, **118**, 318, 1928.
† Hess, *Wien. Ber.* **121**, 1, 1912.
‡ Meyer and Hess, *Wien. Ber.* **121**, 603, 1912.

per hour under conditions where the α and β rays were absorbed and about 15 per cent. of the γ rays. This is in good agreement with the value deduced in the table. The values of the heating effect of the β rays and the γ rays from radium B and radium C given in the table are taken from chapter XVI, where a full discussion of these measurements is given.

We should expect that the total heating effect of the radium less that due to the β and γ rays should be a measure of the energy of the α rays and the recoil atoms. Taking this as 124·4 gram calories per gram and using the values of the energy of the groups of α particles given in § 19, it can be calculated that $3·72 \times 10^{10}$ α particles are expelled per gram of radium per second by each group of α ray substances in equilibrium. It has already been pointed out (§ 13) that this value is somewhat higher than is found by several other methods of estimating this quantity. If the number of particles emitted by radium were as low as that found by Geiger, viz. $3·4 \times 10^{10}$ per gram per second, about 10 per cent. of the heat emission must be due to the liberation of energy during the disintegration in some unknown form. In order to test this important point, careful experiments were made by Watson and Henderson* on the heating effect of a number of radioactive products, viz. radon, radium (B + C), radium C, thorium (B + C), and thorium C, by a resistance thermometer method. After allowance was made for the heat emission due to β and γ rays in the experiments, it was found in all cases that the heating effect of these products was proportional to the energy of the α rays and agreed within the experimental error with a rate of emission of α particles of $3·7 \times 10^{10}$ per second per gram of radium. If an additional source of energy were present, it must be in nearly the same proportion for all three products. The experiments thus support the higher estimate of the rate of emission of α particles.

The enormous emission of energy from a radioactive substance during its disintegration is well illustrated by the case of radon and its products of rapid change. The volume of radon to be obtained from 1 gram of radium is 0·68 c.mm., and this after a few hours emits energy at the rate of 113·6 gram calories per hour. This heating effect decays with the period of radon which has an average life of 132 hours. It follows from this that the total energy liberated during the transformation of 1 c.c. of radon is $2·2 \times 10^7$ gram calories, and

* Watson and Henderson, *Proc. Roy. Soc.* A, **118**, 318, 1928.

for 1 gram of radon (atomic weight 222), $2 \cdot 2 \times 10^9$ gram calories.
Each atom of uranium in its transformation into lead emits about
$8 \cdot 1 \times 10^{-5}$ erg, corresponding in energy to a 51 million volt electron.

§ 33. **Production of helium by radioactive substances.** Following the discovery and separation of helium from the mineral cleveite
by Ramsay in 1895, it was noted that helium appeared in quantity
only in minerals containing uranium and thorium. This association
was remarkable, as there appeared to be no obvious reason why an
inert gaseous element should be present in minerals which in many
cases were impervious to the passage of water and gases. A new light
was thrown on this subject by the discovery of radioactivity. On
the transformation theory, it was to be expected that the final or
inactive products of the transformation of the radioactive elements
would be found in the radioactive minerals. Since many of the latter
are of extreme antiquity, it was reasonable to suppose that the
inactive products of the transformations would be found in some
quantity with the radioactive matter as its invariable companion.
The presence of helium in quantity in uranium and thorium minerals
was noteworthy, and it was suggested by Rutherford and Soddy*
that this helium might prove to be a disintegration product of the
radio-elements. Additional weight was lent to this suggestion by the
proof that the α particle had a ratio E/M comparable with that to
be expected for the helium atom.

This suggestion was soon verified by the experiments of Ramsay
and Soddy† in 1904, who found that helium was present in the gases
released by heating or dissolving a radium salt. This was afterwards confirmed by a number of investigators. It seemed probable
from the first that the production of helium observed was due to
the accumulated α particles expelled from the active matter, or in
other words that the α particle was a charged atom of helium. This
has been completely confirmed by subsequent experiments which
have been referred to in chapter II.

The proof of the identity of the α particle with a charged helium
atom at once gives us a method of estimating the rate of production
of helium by any active substance where the rate of expulsion of
α particles is known; for one cubic centimetre of helium at N.P.T.
corresponds to an emission of $2 \cdot 705 \times 10^{19}$ α particles. If a radium

* Rutherford and Soddy, *Phil. Mag.* **4**, 569, 1902; **5**, 441, 561, 1903.
† Ramsay and Soddy, *Proc. Roy. Soc.* A, **72**, 204, 1903; **73**, 346, 1904.

salt is sealed up in a glass tube, most of the α particles are absorbed in the material or in the walls. If some radon escapes from the salt, the α particles arising from it are embedded in the walls of the glass tube. Practically all the helium can be collected by strongly heating the whole tube.

The first definite measurement of the rate of production of helium by a pure radium salt was made by Sir James Dewar*. The radium salt, about 70 milligrams of radium chloride contained in a glass tube, was connected with a McLeod gauge and the apparatus heated at intervals to liberate the helium found in it. The helium was purified by a charcoal tube immersed in liquid hydrogen. A number of measurements were made extending over more than a year and the annual production of helium corresponding to 1 gram of radium, later corrected in terms of the International Standard, was found to be 172 c.mm. per year.

A subsequent determination was made by Boltwood and Rutherford†, using 183 milligrams of radium element in the form of chloride. In the first experiment the radium was placed in a platinum capsule contained in an exhausted Jena glass tube. After 83 days the purified radium salt was heated and the resulting gases collected and purified. In another experiment, the radium salt was set aside in a sealed vessel for 132 days. By a suitable arrangement, the radium was dissolved *in situ* and the gases pumped off. In the first experiment the amount of helium was 6·58 c.mm., in the second 10·38 c.mm. In each experiment the radium was initially completely deprived of radon. If x is the volume of helium produced per second by the radium itself in the salt, then the volume Q found over a time t is given by $Q = [t + 3(t - 1/\lambda)]x$, where λ is the radioactive constant of radon. This takes account of the growth of radon and the helium due to it, and its subsequent α ray products radium A and radium C. The two experiments were in good agreement. The annual production of helium corrected in terms of the International Standard was 163 c.mm. per gram of radium. These values are in good agreement with the calculated value based on the rate of emission of α particles from radium, and thus substantiate the view that the helium arises from the α particles.

If we reverse the process, the number Q of α particles expelled per gram of radium per second can be deduced from the experimental

* Dewar, *Proc. Roy. Soc.* A, **81**, 280, 1908; **83**, 404, 1910.
† Boltwood and Rutherford, *Wien. Ber.* **120**, 313, 1911; *Phil. Mag.* **22**, 586, 1911.

data. From Dewar's value we obtain $Q = 3.69 \times 10^{10}$; from Boltwood's value $Q = 3.50 \times 10^{10}$. These values may be compared with those obtained by other methods (§ 13).

The production of helium has been observed in all radioactive matter which emits α particles. It has been observed for ionium, radon, polonium and actinium and in amounts agreeing with the number of α particles.

The rate of production of helium by uranium and by thorium and the minerals containing them can be readily calculated from a knowledge of the number of α particles emitted. It is to be remembered that ordinary uranium (U I and U II) emits two α particles and in all one uranium atom emits eight α particles in its series of transformations.

The results are included in the following table:

Substance	Total number of α particles per gram per sec.	Calculated production of helium per gram per year (c.mm.)
Uranium	2.37×10^4	2.75×10^{-5}
Uranium in equilibrium with its products	9.7×10^4	11.0×10^{-5}
Thorium in equilibrium with its products	2.7×10^4	3.1×10^{-5}
Radium in equilibrium	14.8×10^{10}	172

The rate of production of helium in thorium and uranium minerals has been directly measured by Strutt* and found to agree approximately with the calculated values. Soddy† has measured the production of helium in purified uranium and thorium salts.

The amount of helium observed in radioactive minerals depends on a number of factors besides the age of the mineral. If the mineral is primary and compact and impervious to the passage of water, we should expect the greater part of the helium formed to be retained. In some cases the minerals are more porous and doubtless a large part of the helium has escaped. Secondary minerals like the uraninite of Joachimsthal contain relatively little helium, for they are relatively very recent in origin compared with the old primary minerals. For these reasons, it is difficult to base a reliable estimate

* Strutt, *Proc. Roy. Soc.* A, **84**, 379, 1911.
† Soddy, *Phil. Mag.* **16**, 513, 1908.

of the age of a mineral on its helium content. It is clear that the age so found is a minimum even for the most impervious primary minerals. The age of some of the old primary minerals estimated in this way cannot be less than 300 million years. For a more definite estimate of the age, however, the amount of uranium lead or thorium lead in the mineral is more reliable than its helium content.

It is now well known that helium is often found in considerable quantity, of the order of 1 per cent. by volume, in the natural gases which escape from the earth in various localities. These natural gases both in Canada and in U.S.A. have been used as sources for the separation of large quantities of helium. The general evidence indicates that the helium so obtained has its ultimate origin in the transformation of radioactive matter in the earth.

CHAPTER VII

GENERAL PROPERTIES OF THE RADIATIONS

§ 34 a. Emission of α particles and probability variations. The rate of disintegration of all radioactive substances is expressed by a simple law, namely, that the number of atoms n breaking up per second is proportional to the number N of atoms present. Consequently $n = \lambda N$, where λ is a constant characteristic for a particular radioactive substance. The rate of transformation of an element has been found to be a constant under all conditions. It is unaltered by exposing the active matter to extremes of temperature or by change of its physical or chemical state. It is independent of the age of the active matter or its concentration. It is unaffected by exposure to strong magnetic fields. Hevesy has shown that the disintegration of the primary radioactive element uranium is unaltered by exposing it to the β and γ radiation from a strong source of radium, although these rays, of great individual energy, might be expected to penetrate the atomic nucleus.

Since the expulsion of an α or β particle results from an instability of the atomic nucleus, the failure to alter the rate of transformation shows that the stability of the atomic nucleus is not influenced to an appreciable extent by the forces at our command. This is not unexpected when we consider the enormous intensity of the forces, probably both electric and magnetic, which hold the charged parts of the nucleus together in such a minute volume.

E. v. Schweidler* showed that the exponential law of decay of the radioactive bodies could be deduced without any special hypotheses of the structure of the radioactive nuclei or of the mechanism of disintegration. He assumed only that the disintegration of an atom is subject to the laws of chance, and that the probability p that an atom of a certain type shall be transformed within a given interval of time Δ is independent of the time which has elapsed since the formation of the atom and is a constant which is the same for all atoms of the same type or radioactive product.

For very small values of the time interval Δ, the chance p of transformation will be proportional to the length of the interval. There-

* Schweidler, *Congrès Internat. Radiologie*, Liège, 1905.

fore $p = \lambda\Delta$, where λ is another constant characteristic of the radio-active body.

The chance that an atom of this type shall not be transformed in a period $t = k\Delta$ will be $q = (1 - \lambda\Delta)^k$, which may be written

$$q = [(1 - \lambda\Delta)^{-1/\lambda\Delta}]^{-\lambda t}.$$

Now making Δ infinitely small while keeping the product $k\Delta$, or the time t, constant, we have

$$q = \lim_{\Delta=0} [(1 - \lambda\Delta)^{-1/\lambda\Delta}]^{-\lambda t}$$
$$= e^{-\lambda t}.$$

If at time $t = 0$ there were present a very large number N_0 of atoms of the given type, then after the time t there would remain unchanged the number N_t given by $N_t = N_0 e^{-\lambda t}$.

Comparing this with the experimental law of transformation, it is seen that the transformation constant λ is the probability that an atom shall be transformed in unit time, provided this is small compared with the half value period of the body.

On this point of view the law of decay is a statistical law, the result of a very large number of events subject to the laws of chance. If the number of atoms of the radioactive body present is N, then the number which break up in one second will, on the average, be λN, but the number which break up in any particular second will show fluctuations around this value. The magnitude of these fluctuations can be calculated from the general considerations of probability. From the theory of probability it is known that the absolute average error ϵ in a large number of observations for two events P and Q is given by $\epsilon = \pm \sqrt{Npq}$, where N is the number of observations, and p and q are the probabilities for the events P and Q respectively. Applying this formula to the radioactive changes, the number of atoms breaking up during the given time τ, small compared with the half value period of the substance, is given by $\lambda\tau N$, while the number of atoms still unchanged at the end of the interval is $(1 - \lambda\tau) N$. The probability of a single atom breaking up during the time τ is $\lambda\tau$, while the probability that the same atom will exist after that time is $1 - \lambda\tau$. Consequently the absolute average error

$$\epsilon = \pm \sqrt{N\lambda\tau (1 - \lambda\tau)}.$$

Since the square of $\lambda\tau$ is small compared with $\lambda\tau$ itself,

$$\epsilon = \pm \sqrt{N\lambda\tau} \text{ or } \pm \sqrt{Z},$$

where Z is the average number of atoms disintegrating during the time τ. The actual number of atoms breaking up in the time τ thus shows a deviation from the average value of magnitude $\pm \sqrt{\bar{Z}}$. The absolute value of the error $\sqrt{Z}$ increases with the number of atoms breaking up in the interval under consideration, but the relative error $\sqrt{\bar{Z}}/Z$ decreases.

In radioactive substances which break up with the expulsion of α particles, the number of α particles emitted is equal to the number of atoms breaking up. The variations from the average value should be detected by variations in the saturation current due to the ionisation by the α particles.

§ 34 b. **Observation of fluctuations.** The correctness of this theory was early tested by the experiments of Kohlrausch*, Meyer and Regener†, and Geiger‡, by observing the fluctuations of an electrometer when the ionisation currents due to two sources of α radiation were balanced against each other. The fluctuations were found to be of the order of magnitude to be expected from the theory. Later observations were made by Muszkat and Wertenstein§, using a string electrometer, and by E. Bormann‖, by balancing two sources of α rays under conditions where half the total radiation from each source was used. A discussion of the theory and its application to measurement of the fluctuations has been given by Campbell¶, Schrödinger**, and Kohlrausch††.

In addition to the observations of the fluctuations, a number of experiments have been made to determine by counting methods the distribution in time of the α particles falling on a small area which subtends only a small solid angle at the source. It is important to distinguish between these two types of measurement, and to recognise the difficulties of interpreting the observations in these two cases. An excellent critical discussion of this question has been given by Kohlrausch†† who has compared the experimental observations with the results obtained by 5000 drawings from an urn containing a

* Kohlrausch, *Wien. Ber.* **115**, 673, 1906.
† Meyer and Regener, *Ann. d. Phys.* **25**, 757, 1908.
‡ Geiger, *Phil. Mag.* **15**, 539, 1908.
§ Muszkat and Wertenstein, *Journ. de phys.* **6**, 2, 119, 1921.
‖ Bormann, *Wien. Ber.* **127**, 232, 1918.
¶ Campbell, *Proc. Camb. Phil. Soc.* **15**, 117, 1909.
** Schrödinger, *Wien. Ber.* **127**, 237, 1918.
†† Kohlrausch, *Naturwiss,* **5**, 192, 1926.

hundred numbered counters, replaced after each drawing. The number of the counter drawn was noted in each case and the observations compared with theoretical calculations of probability.

In order to test adequately the random emission of α particles from a radioactive substance, it is essential to use a large solid angle so as to be as independent as possible of the variations in the emission in space of the particles. The ideal method would be to count directly all the particles emitted from a particle of radioactive matter over a solid angle 4π and analyse their distribution in time.

Using the fluctuation method, the experiments of Bormann in some respects best fulfil the theoretical conditions, for half of the

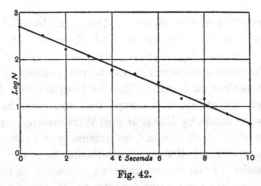

Fig. 42.

α rays from each of the balancing sources were used. However, it is not easy to interpret accurately the fluctuations observed in the measuring instrument and to compare them with the theory. A number of factors are involved between the primary act of disintegration and the reading of the deflection, viz. the emission of the α particle, the ionisation in its path, and the characteristics of the measuring instrument. Notwithstanding these difficulties, the evidence obtained in this way strongly supports the view that the disintegration of atoms follows the laws of chance as regards emission in time.

A more direct test of the probability law of disintegration of active matter can be made by observing the distribution in time of *all* the α particles emitted by a source. This has been done by Constable and Pollard in some unpublished experiments using the scintillation method. A trace of polonium was mixed with the zinc sulphide crystals of a small screen about 2 sq. mm. in area which was placed between two counting microscopes. The scintillations were counted by the two observers and recorded on a moving tape. About 1300 scintillations were counted and the time intervals examined as in the work of Marsden and Barratt (page 172). The results are shown in Fig. 42, where the ordinates represent logarithms of the number of intervals and the abscissae time intervals in seconds. It will be seen from the theory given on page 172 that for a very large number of particles the curve should be a straight line. This is seen to be the case approximately, supporting the theory that the α particles from a source are emitted at random. Considering the small number counted, the agreement with the probability law is as close as could be expected. Similar observations have been made by N. Feather* with results in close accord with theory.

§ 34 c. In the other type of experiments, the α particles emitted in a small solid angle were counted. Such experiments do not test the random distribution of the particles in time but only in space. For example, even if the α particles were emitted from the source at a uniform rate without random variations in time, we should still observe a random distribution in the counting experiments. Such experiments, however, which test the random emission in space of the α particles, have a certain interest and importance, as so much work has been done in counting α and β particles by the scintillation and electric methods. In counting scintillations from a constant source, the variations of the number in a given time are occasionally so marked that it is difficult at first sight to credit that they are mere examples of random distribution and are not necessarily to be ascribed to the inefficiency of the counter.

Experiments of this kind have been made by Rutherford and Geiger, Marsden and Barratt, using the scintillation method of counting α particles, and by Mme Curie, using the electric method.

In the experiments of Rutherford and Geiger†, about 10,000 scintillations produced by α particles emitted by a film of polonium were counted, and the time of appearance of each scintillation was recorded on a moving tape by pressing an electric key. The number appearing in equal intervals of time was then deduced. For a very large number of particles, the probability that n α particles should appear in a given interval is given by $\frac{x^n}{n!}.e^{-x}$, where x is the average number during the interval and n may have any positive integral

---

* N. Feather, *Phys. Rev.* **35**, 705, 1930.
† Rutherford, Geiger and Bateman, *Phil. Mag.* **20**, 698, 1910.

value from 0 to ∞. The agreement between theory and experiment is shown in Fig. 43, where the full line gives the theoretical curve and the circles observed numbers. The number of α particles counted was 10,097 and the average number appearing in the interval under consideration, namely 1/8 minute, was 3·87. The analysis of the data is given below:

Number of α particles observed in interval:

0	1	2	3	4	5	6	7	8	9	10	11	12	13	14

Observed number of occurrences:

57	203	383	525	532	408	273	139	45	27	10	4	0	1	1

Theoretical number of occurrences:

54	210	407	525	508	394	254	140	68	29	11	4	1	1	1

Considering the comparatively small number of α particles counted, the agreement between theory and experiment is quite as close as could be expected.

Marsden and Barratt* have used another method of comparing the distribution in time of the α particles with the theory of probability. If the *average* time between successive α particles from a constant source is $1/\mu$, the probability that no scintillation is observed on the screen in any interval t is $e^{-\mu t}$. The probability that a scintillation occurs in the interval between t and $t + dt$ is μdt. Consequently the probability of an interval greater than t and smaller than $t + dt$ is $\mu e^{-\mu t} dt$. In a large number N of intervals, the pro-

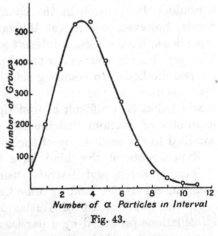

Fig. 43.

bable number of intervals larger than t and smaller than $t + dt$ is $N\mu e^{-\mu t} dt$. This deduction shows that *small* intervals are more probable than *large* ones. If a curve be plotted with the number of intervals as ordinates and the interval in seconds as abscissae, the curve is exponential (*v.* Fig. 42).

The number of intervals greater than t_1 and less than t_2 is given by

$$N \left(e^{-\mu t_1} - e^{-\mu t_2} \right).$$

* Marsden and Barratt, *Proc. Phys. Soc.* **23**, 367, 1911; **24**, 50, 1911.

Marsden and Barratt used as data the record of the scintillations on a chronographic tape obtained by Rutherford and Geiger in the experiments mentioned above. They found that the analysis of the intervals was in close agreement with the theory.

A careful analysis by the same method has been made by Mme Curie* to test the random emission of α rays from polonium. The electrical method of detection was used with a string electrometer and the time of emission of the α particles was recorded photographically on a moving film. About 10,000 particles were counted and the analysis was found to be in very close accord with the theory. The experiments were continued over several months during which the activity of the polonium had sensibly decayed. Taking this factor into account, the curve connecting the number of particles and time intervals was very nearly exponential over a wide range. The agreement between theory and experiment is thus excellent, indicating that the α particles are emitted at random in space and that the variations accord with the theory of probability.

§ 34 d. **Double scintillations.** The theory of probability, with which the experimental results have been compared, refers only to cases where each atom in breaking up emits one α particle and where the time of transformation is long compared with the duration of the experiments. In certain cases, however, it appeared at first as if a single product emitted two α particles simultaneously. This was observed in special experiments with the radioactive gases thoron and actinon. It is now known that the former gives rise to an α ray product thorium A

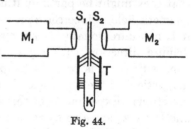

Fig. 44.

of average life 1/5 sec. and the latter to actinium A of average life 1/350 sec. The radioactive gas under consideration is allowed to diffuse in a narrow cell formed by two zinc sulphide screens S_1, S_2 (Fig. 44) which are observed through two microscopes M_1, M_2. Each observer sees the scintillations on one screen but not on the other. In this way, Geiger and Marsden† found that more than half of the scintillations from actinon appeared as pairs, sometimes one on each screen, and in other cases both on one screen. Such a result is to

---

* Mme Curie, *Journ. de phys.* **6**, 1, 12, 1920.
† Geiger and Marsden, *Phys. Zeit.* **11**, 7, 1910.

be expected since the eye is not able to distinguish the short interval between the emission of α particles from the successive products. By introducing some actinon into an expansion chamber, C. T. R. Wilson[*] and Dee[†] have obtained clear photographs of the α particles expelled with a very short time interval from the two products.

The frequency of these scintillation pairs led to experiments which disclosed the presence of a new product of rapid transformation. In a similar way, a number of pairs of scintillations have been observed with the radioactive gas thoron. In this case, however, the life of the subsequent product is longer and many of the apparent pairs are separated by an appreciable time interval.

By a similar method, it can be shown that polonium, ionium and other substances emit one α particle per disintegrating atom. It is thus clear that a study of the distribution of α particles with regard to time has been instrumental in bringing to light a number of important facts.

§ 34 e. Direction of emission of α rays. We shall now refer to certain experiments which have been made to test whether α particles are emitted equally in all directions under special conditions, viz. in a crystal or under the influence of an intense magnetic field. If the nuclei of radioactive atoms have a magnetic moment, it is possible that they might be partially if not completely orientated by a strong magnetic field. Since the α particles are all ejected at identical speeds, it is not unreasonable to suppose that the α particle is ejected in a definite direction with regard to the nuclear structure. On these suppositions, it is to be anticipated that under the influence of a magnetic field the α particles would be emitted unequally in different directions. Experiments of this kind have been made by Wertenstein and Danysz[‡] and also by Henderson[§]. The former, using an α ray tube as a source of α rays, observed a slight difference in the ionisation current due to the α rays—about 1 in a 1000—by the application to the source of a magnetic field of 35,000 gauss. This effect, however, was so small that it might well be due to experimental disturbances. As far as observation has gone, we may conclude that no definite asymmetry in the emission of α rays has been observed when the source is exposed in an intense magnetic field.

In a similar way it is possible that the nuclei of radioactive atoms

[*] C. T. R. Wilson, *Proc. Camb. Phil. Soc.* **21**, 405, 1923.

[†] Dee, *Proc. Roy. Soc.* A, **116**, 664, 1927.

[‡] Wertenstein and Danysz, *C.R. d. Soc. d. Sci. d. Varsovie*, **7**, 555, 1914.

[§] Henderson, *Proc. Camb. Phil. Soc.* **21**, 56, 1922.

composing a crystal may be orientated in space under their mutual forces. If this be the case, some asymmetry in the emission of α particles might be expected. This interesting question has been examined by several investigators, using large crystals of uranium nitrate as a source of α rays. Merton* measured the ionisation due to the α rays from the various faces but was unable to observe any certain difference. Mühlestein†, on the other hand, using a similar crystal, found positive results both by the ionisation and scintillation methods. He found a ratio of activities from three different faces of the crystal 1·00 : 1·05 : 0·85 by the electric method, and 1·00 : 1·09 : 0·68 by the scintillation method. On account of the smallness of the effects, experiments of this kind are very difficult, and great care must be taken to obtain definite results. If these results are substantiated by further observations, it will afford definite proof that the nuclei are in some way orientated in the crystal and that the α particles are ejected more freely in one direction than in another relative to the nuclear structure.

§ 35. **Relation between range and period of transformation.** It was early observed that there appeared to be a connection between the period of transformation of a product and the velocity of the α particles expelled from it. The speed of the α particle was greater the shorter the period of transformation. Geiger and Nuttall‡, who made a determination of the ranges of many of the α particles, found that such a relation existed in the majority of cases. This relation, generally known as the Geiger-Nuttall rule, is shown in its simplest form by plotting the logarithms of the range of the α particle against the logarithm of λ, the transformation constant of the element. This is shown in Fig. 45§, where the ranges refer to those obtained by Geiger (§ 19). It is seen that the products of each of three radioactive series lie nearly on straight lines which are approximately parallel to one another. This linear relation is only approximate in character and there are certain marked disagreements, notably the product actinium X which lies completely off the line for the actinium series. Various methods have been proposed for expressing the general relation. Swinne‖ showed that the relation

---

* Merton, *Phil. Mag.* **38**, 463, 1919.   † Mühlestein, *Arch. d. Sci.* **2**, 240, 1920.

‡ Geiger and Nuttall, *Phil. Mag.* **22**, 613, 1911; **23**, 439, 1912.

§ The figure is taken from Geiger (*Zeit. f. Phys.* **8**, 45, 1921) with the nomenclature adopted at that time. The horizontal line at Ur shows the uncertainty as to the ranges of uranium I and II at that period. See also Fig. 93, p. 332.

‖ Swinne, *Phys. Zeit.* **13**, 14, 1912.

$\log \lambda = a + bv^n$, where a and b are constants, v the velocity of the α particle and $n = 1$ or 2, is in fair accord with observation.

While the existence of such a relation between speed of emission of α particles and period of transformation is of great importance and interest, it is doubtful whether we can rely on any simple relation between these quantities. The extrapolated curves of Fig. 45 show that the period of transformation of radium C′ should be of the order 10^{-8} sec., while Jacobsen and Barton find experimentally

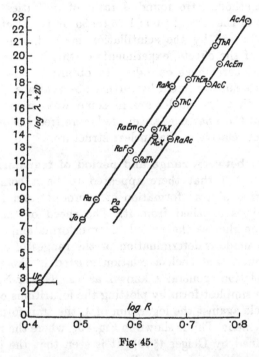

Fig. 45.

a period of about 10^{-6} sec. In fact it is unlikely that the data for any individual series can be expressed at all accurately by a simple relation. At the same time, we are justified in using the relation to form a rough idea of the period of transformation of a product which cannot be measured directly. In this way the period of transformation of the product uranium II, which emits rays of range 3.28 cm., has been estimated to be about 10^6 years.

Various attempts have been made to give a theoretical explanation of this relation, notably by Lindemann*, who on certain assumptions

* Lindemann, *Phil. Mag.* **30**, 560, 1915.

found an expression agreeing in general form with the Geiger-Nuttall rule. Other theories have been proposed by van den Broek, Brossler and others.

Recently the problem has been attacked in another way based on the wave-mechanics. On certain general assumptions of the variation of the electric field in the neighbourhood of a radioactive nucleus, Gamow* and Gurney and Condon† have been able to show not only why a relation should exist between the velocity of emission of α particles and the transformation constant, but also to account for the experimental results in an even more satisfactory way than is given by the Geiger-Nuttall relation. On account of its interest and importance, this theory will be considered in some detail in § 77b, when the problem of nuclear structure is under discussion.

§ 36. The distribution of radiation. The general results of experiment show that α and β particles are expelled equally in all directions from a thin layer of active matter. By exposing solid bodies in the presence of the emanation of radium, thorium or actinium, they become coated with an exceedingly thin film of the active deposit from which particles escape at their maximum initial velocity. For these reasons the distribution of the radiation is in these cases very different from that observed for the same body when heated to a sufficient temperature to emit ordinary light‡. In the case of light, the distribution is controlled by Lambert's Law, which supposes that the intensity of radiation in any direction is proportional to the cosine of the angle between the normal and the direction of the emitted light. In a thin film of radioactive matter, Lambert's Law does not hold. For example, if a pinhole photograph be taken of a luminous sphere, e.g. the sun, the intensity of the photographic effect is nearly uniform over the cross-section. If, however, we take a pinhole photograph of a small sphere uniformly coated with a thin film of active matter by the aid of the radiations emitted from it, the intensity of the radiation, and consequently of the photographic impression, is greatest at the extreme edge of the image and shades off rapidly towards the centre. In a similar way, the photographic impression due to a cylinder coated with a thin film of active matter, after the rays have passed through a

* Gamow, *Zeit. f. Phys.* **51**, 204, 1928.
† Gurney and Condon, *Nature*, Sept. 22, 1928; *Phys. Rev.* **33**, 127, 1929.
‡ Rutherford, *Phil. Mag.* **12**, 152, 1906.

narrow slit parallel to the cylinder, is greatest at the edges of the geometrical image, and falls off rapidly towards the centre. If the photographic effect is weak, the image of the cylinder often appears to consist only of two straight lines corresponding to the edges of the image.

The effects of distribution can be simply and strikingly illustrated in another way. Suppose, for example, that a rod of triangular or square cross-section, several centimetres long, is exposed in the presence of radon. The active deposit is then nearly uniformly distributed over the surface. After removal, the rod is placed vertically on a photographic plate. Examples of the positives of the photographs produced in this way by the α rays are shown in Plate VII, figs. 1, 2. In the case of the rod of square cross-section, the photograph is marked by four dark bands at right angles to each other, of the same width as the rod. In each of these dark bands, the photographic effect is due to the radiation from one side of the rod only. Immediately outside the dark bands, the radiations from two sides produce their effect, so that the intensity of the radiation varies very nearly in the ratio of one to two in crossing the boundary. By using very active bodies, this effect of distribution can be very clearly seen on a screen of zinc sulphide. The results obtained are to be expected theoretically if the α rays are expelled equally in all directions and escape from the metal without appreciable absorption.

In the experiments described the photographic effect is due almost entirely to the α rays; but no doubt somewhat similar effects would be observed if the thin film of active matter gave out only β rays.

Greinacher* has investigated the distribution of radiation around a prism of uranium metal, which is placed on a photographic plate. In this case the radiation is emitted from the whole volume of the material, and the photographic action is due mainly to the β rays of various velocities which emerge from the uranium. The distribution of the radiation was very different from that already shown. For thick films of active matter the photographic effect was greatest perpendicular to the surface of the triangle and small in the space formed by the continuation of the sides, where in the case of a thin film the intensity would be strongest.

In another experiment, a radium preparation was placed in a cubical lead box, which was placed on a photographic plate. The

* Greinacher, *Phys. Zeit.* **9**, 385, 1908; **10**, 145, 1909.

radiation was most intense opposite the sides, and least where the radiations from two sides of the box produced their effect.

This is no doubt due to the observed fact that the β radiation from a thick layer of active matter is greatest in the normal direction.

§ 37. **Light emission of gases exposed to α rays.** When strong sources of radium C, thorium C, or polonium deposited on a small flat plate are viewed in a dark room with a rested eye, the source is seen to be surrounded by a weakly luminous hemispherical zone extending over the range of the α particles. This luminosity excited by the α rays in gases was first studied by the late Sir William and Lady Huggins*, who examined the spectrum of the weak light emitted by radium preparations in air and found it to coincide with the band spectrum of nitrogen. This appears to be due to the α ray bombardment of nitrogen occluded in the source. These experiments have been confirmed by a number of observers including Himstedt and Meyer†, Walter‡, Pohl§, and Bosch ||. The spectrum of luminous radium bromide is in part continuous, due to the fluorescence of the crystal. In an atmosphere of helium, a few helium lines were noted but no lines were observed in hydrogen, carbon monoxide and carbon dioxide. The luminosity excited in gases can be best studied by using a source of polonium which emits only α rays, and screening the photographic plate from the direct α radiation. There seems to be no doubt that the spectrum of the weak luminosity produced by α rays is in general similar to that produced by the electron discharge under special conditions in gases at low pressure. Such an effect is to be expected, since the α rays cause the same general type of ionisation and molecular dissociation produced in a vacuum tube by cathode rays and positive rays. The luminosity produced by α rays, however, under ordinary experimental conditons is exceedingly feeble compared with that produced in a discharge tube.

§ 38. **Pleochroic haloes.** Plates of mica are rapidly coloured brown or black by α rays, the rapidity of coloration depending on the composition of the mica. Geologists long regarded with curiosity

* Sir William and Lady Huggins, *Proc. Roy. Soc.* A, **72**, 196, 409, 1903; **76**, 488, 1905; **77**, 130, 1905.
† Himstedt and Meyer, *Phys. Zeit.* **6**, 688, 1905; **7**, 672, 1906.
‡ Walter, *Ann. d. Phys.* **17**, 367, 1905; **20**, 327, 1906.
§ Pohl, *Ann. d. Phys.* **17**, 375, 1905; **18**, 406, 1905.
|| P. Bosch, *Arch. Néerl.* **8**, 163, 1925.

small coloured areas in certain kinds of mica, for example, in biotite, cordierite, and muscovite. The sections of these coloured areas are usually circular in shape, and exhibit the property of pleochroism under polarised light, and for this reason were called "pleochroic haloes." The centres of these areas usually contain a minute crystal of foreign matter. Joly* first pointed out that these haloes were of radioactive origin, and were due to the coloration of the mica by the α rays expelled from a nucleus which contains radioactive matter. This subject has been investigated in detail by Joly and his pupils, by Mügge and many others. In homogeneous material, the haloes are spherical with a radioactive inclusion in the centre and often show a well-marked structure consisting of a number of rings. The radii of these rings are closely connected with the ranges in mica of the different groups of α rays expelled from the active material. In the complete uranium halo, the outer ring marks the range of the α particle from radium C' (7 cm. in air), and in the thorium halo the range of the α particles from thorium C' (8·6 cm. in air).

The detailed structure sometimes seen in a halo is shown in Plate VII, fig. 3. The central blackened area or pupil is produced by the α particles emitted from all the products contained in the uranium mineral, and its limit is defined by the range of the α particles from radium itself. The limit of the second ring, which is much lighter in colour, marks the range of the swifter α particles from radium A, while the edges of the outer ring or corona mark the limit of action of the α rays from radium C. It was observed in all haloes that the darkening due to the α particles from radium C is most intense near the end of their range. The reason of this is clear when we take into consideration the variation of the ionisation along the path of the α particle. The ionisation due to an α particle shows a sharp maximum near the end of its range and then falls rapidly to zero. Assuming that the coloration of the mica at any point depends upon the density of ionisation, it is to be expected that the darkening of the mica would show the type of gradation illustrated in the figure.

Assuming that the uranium in the inclusion is in equilibrium with radium and its other products, the integrated ionisation to be expected in air at various distances from the centre is shown in Fig. 46. These are drawn from the typical ionisation curve found by Henderson (§ 16) for the α particles from radium C. The darkened areas at the top of the figure illustrate the positions of rings in the

* Joly, *Phil. Mag.* **13**, 381, 1907.

PLATE VII

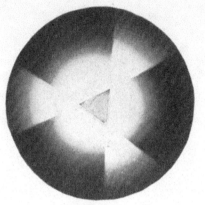

Fig. 1.

Fig. 2.

Distribution of α ray activity.

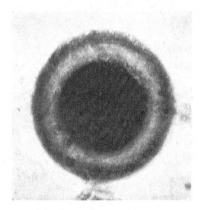

Fig. 3. Halo in mica.

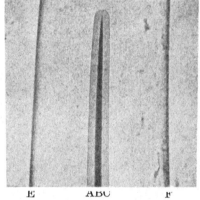

E   ABC   F

Fig. 4. Halo in glass.

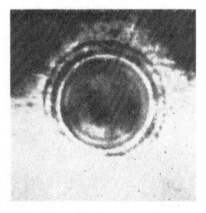

Fig. 5.

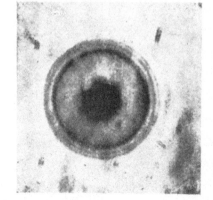

Fig. 6.

Haloes in fluorspar.

halo which correspond to some of the humps on the compound ionisation curve. Accounts of the development of haloes in mica have been given by Joly* in 1916 and 1923, including the phenomena of reversal analogous to over-exposure in the case of the photographic plate. All stages of the development of haloes are sometimes seen in a single specimen of mica. If the radioactive inclusion is very small and feebly active, only the inner ring shows coloration and the rings due to α particles from radium A and

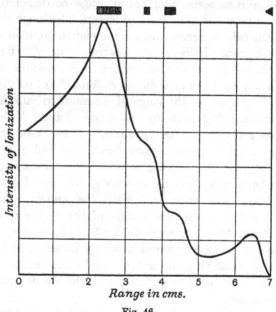

Fig. 46.

radium C do not appear. The coloration of the mica is due to the integrated effects of the α particles extending over geological ages. By measuring the diameter of the inclusion which gives a fully developed halo, Joly has estimated that not more than 5×10^{-10} gram of uranium is required. It can be calculated that such a nucleus would on the average emit only one α particle in 10 hours. This is a maximum estimate, and a much smaller rate of emission of α particles is required to form an embryonic halo. By comparing the coloration observed in haloes in mica with the coloration of the same mica due to the action of a known number of α particles, Joly† estimates

* Joly, *Phil. Trans.* A, p. 217, 1916; *Proc. Roy. Soc.* A, **102**, 682, 1923.
† Joly and Rutherford, *Phil. Mag.* **25**, 644, 1913.

from the size of the inclusion that some of the haloes would require several hundred million years to develop to the stage observed to-day.

Mügge* has made a systematic examination of a number of materials for haloes and found that the blue fluorspar from Wolsendorf shows haloes in greater frequency and variety than any other substance. The rings due to different groups of α rays show up even more clearly than in mica. There are eight groups of α particles from the uranium-radium series and seven rings corresponding to the ranges of the α particles have been observed. The inner ring is twice the width of the others, accounting for the eighth group of α particles. Examples of fluorspar haloes† are shown in Plate VII, figs. 5 and 6. In fig. 5 the rings due to the α ray bodies from uranium to radium A are shown clearly, and the ring due to radium C' is just visible in the upper half. Fig. 6 shows the rings of uranium to radium, the ring due to polonium being indicated at a few points. The haloes in some respects are different in appearance from those in mica and the rings apparently represent more nearly the end of the range of the α particles than in the case of mica. The dimensions of these haloes have been measured by Gudden†, who has found the radii of most of the rings agree well with the known ranges of the α particles in air. The observed ranges of the α rays from U I and U II were found to correspond to 2·83, 2·93 cm. respectively at 15° C., while the ranges measured in air by Laurence (§ 19) are 2·73 and 3·28 cm. Considering the agreement in the case of the ranges of the other six groups, this difference in the shorter ranges is not easy to account for.

While the main observations on the detailed structure of haloes are in good accord with the theory of their radioactive origin, certain divergences have been noted which require explanation. Joly initially pointed out that the inner ring due to the α rays of U I and U II was somewhat greater in radius than was to be expected on Geiger's original data for the ranges. This difficulty has been at any rate in part removed by the newer measurements of the ranges by Laurence which are substantially higher than the old. In addition to this, Joly (loc. cit.) noted that the radius of the inner ring was not constant but appeared to be greater the higher the geological age of the mica. Measurements are difficult in old micas but

* Mügge, Centralb. f. Min. pp. 65, 113, 142, 1909.
† Gudden, Zeit. f. Phys. 26, 110, 1924.

definite observations of this kind may help to throw light on the early history of the radioactive matter in the inclusions. Besides the possibility suggested at one time by Joly that the rate of disintegration of uranium and range of the α particles has progressively changed in geological time, there is always the question whether one or more radioactive isotopes of uranium, which have a much shorter life than the long-lived isotope, were originally present in the inclusion. If this were the case, the activity due to such an isotope might be small to-day and difficult to detect, but the radiations from it should have left a record in the darkening of the mica. In particular it has been suggested by T. R. Wilkins* and others that actinium may not be, as generally supposed, a branch product of the uranium series but may arise from an isotope of uranium which has a much shorter average life than the main isotope. In such a case, in very old haloes some evidence may be obtained to throw light on this point. We have seen (§ 8 a) that recent evidence strongly indicates that actinium does arise from an isotope of uranium, but the range of the α particle expelled from the isotope is not known. Lotze† has suggested that owing to the physical and chemical changes produced in mica by continuous bombardment, the actual measured radius of the inner ring of intense haloes may not represent the true ranges of the α particles in unaltered mica. Joly‡ does not consider such an effect important. Poole concludes that the colour changes in mica are due to chemical decomposition of water molecules by the α rays and oxidation of the iron in the mica. Joly has observed in some specimens of mica very small "X" haloes which resemble in structure the ordinary halo but correspond to ranges of α particles between 1·5 and 2·25 cm. In a similar way, Iimori and Yoshimura,§ who have made a number of observations on samples of mica from Japan, have noted the presence of certain haloes called "Z" haloes whose radii are not in agreement with known ranges of the uranium and thorium series. Their exact origin is unknown, but they correspond to ranges of α particles between 1·2 and 2·1 cm. and are different in some respects from those noted by Joly.

While it is very difficult to be certain of the exact origin of these minute haloes, a systematic examination of varied materials is

* T. R. Wilkins, *Nature*, **117**, 719, 1926.
† Lotze, *Nature*, Jan. 21, p. 90. 1928.
‡ Joly, *Nature*, Feb. 11, p. 207, 1928; *Proc. Roy. Soc.* A, **102**, 682, 1923.
§ Iimori and Yoshimura, *Instit. Phys. and Chem. Res.* Tokyo, **5**, 11, 1926.

required to see whether any definite evidence can be obtained of the existence of additional types of radioactive matter in the early history of the earth. The subject is one of great interest and importance, as the study of haloes gives us a method of detection of possible radioactive changes in matter which does not survive in detectable amounts to-day.

Types of haloes may be produced artificially by radioactive matter in the laboratory. Rutherford showed that an effect similar in character to the coloration produced in mica could be observed in glass. A large quantity of radon was enclosed in a capillary tube and allowed to decay *in situ*. A microscopic examination showed that the glass was coloured to a depth of about 0·04 mm. from the inner surface of the tube corresponding to the range in glass of the α particles from radium C. A microphotograph is shown in Plate VII, fig. 4, where EF is the glass tube, B is the base of the capillary and A,C are the boundaries of the coloration.

Artificial haloes may also be easily produced in a photographic plate by infecting the plate at a point by a trace of radioactive matter. On development, the magnified photograph is similar to that shown in Plate II, fig. 1. Mühlestein* showed that a more definite halo effect is produced by pouring a layer of mercury over the surface of the plate. In ordinary air, some of the α particles have part of their range in air and are then scattered back into the photographic plate. The addition of mercury confines the photographic effect to the range of the α particles in the gelatine. An example of a halo due to radium C photographed by Harrington is shown in Plate II, fig. 2.

§ 39. **Luminous effects.** Besides their power of acting on a photographic plate and ionising gases, the radiations from active bodies are able to produce marked luminous effects in various substances. On account of the ease of absorption of the α rays, the effects produced by them are superficial but the effects of penetrating β rays, which arise directly from the source and from the conversion of γ rays, extend to a considerable depth. As the β rays are identical with swift cathode rays in a vacuum tube, it is to be expected that they would produce similar luminous effects.

We have already referred (§ 12 a) to the scintillations produced in phosphorescent zinc sulphide and their importance in affording

* Mühlestein, *Arch. d. Phys.* Geneve, **44**, 63, 1917; **46**, 48, 1918; **2**, 423, 1920.

a simple method for counting individual α particles. Intense luminous effects are produced when a strong beam of α particles falls on a zinc sulphide screen. The amount of luminosity and its rate of disappearance after the α rays are removed depends to some extent on the impurities present in the zinc sulphide. In general the effect due to α rays is very much more intense than that due to β and γ rays. The mineral willemite (zinc silicate) gives a beautiful greenish colour under the influence of both α and β rays. It shows to a slight extent the scintillation effect of zinc sulphide, but the luminosity due to β and γ rays is much more marked than in zinc sulphide. The platino-cyanides glow brilliantly under the action of radium rays. The calcium and barium salts glow with a deep green light and the sodium compounds with a lemon yellow. They are especially sensitive to β and γ rays and in this respect are superior even to willemite. The mineral kunzite, a variety of spodumene discovered by Kunz, is insensitive to α rays but glows with a reddish colour under the influence of β and γ rays. The different actions of the rays on these substances can be illustrated very simply and beautifully by the following experiment. A tube filled with fragments of the minerals arranged in layers is immersed in liquid air and the emanation from about 20 milligrams of radium condensed in the tube. On removing the liquid air, the emanation distributes itself uniformly in the tube and the different shades of colour are clearly seen. The luminosity of kunzite is small at first but increases with the gradual growth of β and γ rays due to the successive transformations of the emanation.

Some diamonds phosphoresce strongly under radium rays and others only to a slight extent. Glew and Regener found that some diamonds give scintillations, but these are not as bright as those in zinc sulphide.

The action of continued bombardment of zinc sulphide by α particles is not only of theoretical but of practical importance. Zinc sulphide mixed with radioactive matter is used to coat the hands and figures of clocks, watches and gun-sights, to render them permanently luminous in the dark. Any radioactive substance which emits α rays and has a reasonably long life may be used for this purpose. Radiothorium, on account of its limited life, is more suitable and cheaper for this purpose than radium itself.

Marsden* examined the effect of continued bombardment of zinc

* Marsden, *Proc. Roy. Soc.* A, **83**, 548, 1910.

sulphide by α particles. It was found that the luminosity decreased with time, depending on the intensity of the bombardment. The number of scintillations observed from a given source was unchanged by bombardment of the screen but the intensity of the scintillations became weaker. After very intense bombardment the luminosity sinks to a minimum of a few per cent. and there is often a change of colour of the emitted light. Similar effects are observed in willemite but the decay of luminosity was somewhat slower. In the case of barium platino-cyanide the decay of luminosity is very rapid. In explanation of these results, Rutherford* supposed that each of these phosphorescent substances contained a number of "active centres" which were destroyed by the radiations with the emission of light. The decay of luminosity was ascribed to the gradual destruction of these active centres throughout the volume exposed to the α radiation. From a knowledge of the range of the α particles in the material, it was calculated that a single α particle on the average destroyed all active centres in its path within a cylinder of diameter $1\cdot3 \times 10^{-7}$ cm. The corresponding diameter for willemite was $2\cdot5 \times 10^{-7}$ and for barium platino-cyanide $1\cdot6 \times 10^{-5}$ cm. This suggests that the disturbance due to an α particle extends to a considerable area round its path.

The commercial use of self-luminous radioactive paint has led to very detailed investigations† of the change of intensity and tint of the luminosity with time when different quantities of radioactive material are mixed with the zinc sulphide. The change of luminosity due to continued bombardment by β and γ rays varies widely in different materials and is more complicated than in the case of α rays.

§ 40. Physical and chemical changes due to the radiations. The rays from radioactive substances produce marked coloration and physical and chemical changes in a number of substances. As we should anticipate, the effect of the α rays extends only to a short depth while the effect of the more penetrating β and γ rays extends throughout the volume of the substance. The coloration is very marked in glass, but the rapidity and tint of the coloration depends on the nature of the glass. Ordinary soda glass is coloured a deep violet and with long-continued action the glass becomes almost

---

* Rutherford, *Proc. Roy. Soc.* A, **83**, 561, 1910.

† See Patterson, Walsh and Higgins, *Proc. Phys. Soc.* (Lond.), **29**, 215, 1917.

black. Other kinds of glass are coloured brown or yellow. Giesel found that rock-salt and fluorspar are rapidly coloured and the effect extends much deeper than that due to ordinary cathode rays. Quartz shows varied effects due to the action of the rays depending on the presence of impurities. These have been discussed by Salomonsen and Dreyer, Hönigschmid and Berthelot. Rutherford observed that a fused quartz tube after long exposure to the rays from radon became brittle and cracked, and the whole tube on examination showed a multitude of approximately circular marks. Similarly Mme Curie observed that quartz tubes cracked when exposed to the action of the α rays from a strong source of polonium. For these reasons, quartz is a very unsuitable substance in which to enclose strong radioactive preparations.

Doelter* has examined the coloration produced by radium rays in a number of minerals. In many cases the coloration rapidly disappeared under the action of ultra-violet light. It is well known that the colour produced in glass tubes is to a very large extent destroyed by heating them to the temperature of luminescence.

Under intense radiation, obvious physical and chemical changes occur in many solid substances. For example, glass becomes devitrified and brittle and difficult to work; a thin film of collodion which is sometimes stretched over a radioactive source to prevent contamination soon loses its strength and becomes useless for its purpose. Glew long ago observed that mica sheets exposed to strong radiation were bent, and similar effects have been observed in other crystals.

It is well known that the α rays rapidly decompose water and produce marked chemical changes in many materials. This important subject will not be discussed here as a complete account of our knowledge of it has been given by Lind in his book *The chemical effects of α particles and electrons* (2nd ed. 1928). The amount of chemical action is in many cases closely connected with the ionisation produced by the α and β rays.

It may be of interest to record a striking observation made by G. H. Henderson†. He found that nitrogen iodide ($N_2H_3I_3$) explodes when a strong radioactive source emitting α rays is brought near it. From 10^7 to 10^8 α particles falling on the material are required to

---

* Doelter, *Centralbl. f. Min.* p. 161, 1922; *Das Radium und die Farben*, Steinkopffs, Dresden, 1910.

† Henderson, *Nature*, **109**, 749, 1922.

produce an explosion. Poole* examined other explosives, including fulminate of mercury, silver azide, dynamite, nitroglycerine and potassium picrate, but found no effect after bombardment by 10^{12} α particles. It appears probable that the explosion of nitrogen iodide is the result of an α particle collision of a rare type with one of the constituent atoms.

§ 41. **Physiological actions and precautions.** It was early observed that the rays from radium and other radioactive substances produce burns of much the same character as those caused by X rays. Some time after exposure, there is a painful irritation followed by inflammation which lasts for some weeks. After continued exposure the skin may break, giving rise to sores which are difficult to heal.

On account of the small penetrating power of the α rays, their effect is mainly confined to the skin exposed to the rays; but the effect of the more penetrating β rays extends much deeper, while the penetrating γ rays traverse the whole body. It is now well established that a controlled exposure to the β and γ rays from radium and other radioactive bodies is of much utility in the treatment of certain diseases. The effects produced are similar to those observed with X rays, but the use of radium has in some cases marked advantages. In many of the large hospitals there are radium departments for preparation and standardisation of the radioactive sources employed. In some cases, sealed radium preparations are used, in others small glass tubes containing radon. A short time after preparation, such radon tubes emit the characteristic β and γ rays emitted by radium in equilibrium with its products. The activity of these radon tubes is not permanent but decays with time, falling to half value in 3·82 days.

For the preparation of radon tubes the radium is kept in solution. The radon, mixed with hydrogen and oxygen, is pumped off and then purified. Finally it is compressed into the fine glass tubes which are sealed off. In many institutions elaborate apparatus has been installed for removing and purifying the radon, with a minimum of handling and exposure for the operator.

It would be out of place here to describe the results of numerous experiments that have been made on the therapeutic effect of the radiations and their action on different organisms. It is desirable,

---

* Poole, *Nature*, 110, 148, 1922; *Proc. Roy. Dubl. Soc.* 17, 93, 1922.

however, to draw attention to the precautions that should be taken in the preparation and use of radioactive products for direct scientific purposes.

Reference should be made to the action of the β and γ rays on the eye, first noted by Giesel. On bringing up a radium preparation to the closed eye, in a dark room, a sensation of diffuse light is observed which increases with the intensity of the radiation. This appears to be due to a fluorescence produced by the rays in the eye itself. The blind are able to perceive this luminosity if the retina is intact but not if the retina is diseased. Hardy and Anderson*, who examined this effect in some detail, conclude that it is produced both by the β and γ rays. The eyelid absorbs a large fraction of the β rays so that the luminosity observed with the closed eye is mostly due to the γ rays. The lens and retina of the eye phosphoresce under the action of the β and γ rays. The luminosity observed with the open eye in a dark room, the light from the active preparation being cut off by black paper, is mainly due to the phosphorescence of the eyeball. The γ rays for the most part produce the sensation of light when they strike the retina.

In counting scintillations where strong sources of radium C and thorium C are employed, for example in studying the artificial disintegration of the elements (chapter x), the intense β and γ rays produce a disturbing effect on the eye. Even if the β rays are largely bent away by a strong magnetic field, the γ rays are always present. To avoid this difficulty, it is desirable that the light emerging from the objective of the counting microscope should be bent through a right angle by a totally reflecting prism before entering the eyepiece. This arrangement allows suitable screens of lead or other absorbing material to be interposed between the source and the eye and head of the observer, so as to reduce greatly the intensity of the γ radiation. Experience has shown that counting under these conditions is much less tiring and more reliable than when the observer is exposed to the full intensity of the γ radiation.

For many experiments on the α, β and γ rays it is necessary to obtain strong sources of radium (A + B + C) on wires or plates and occasionally to use "α ray tubes", in which almost pure radon is compressed into such a thin-walled glass tube that the α rays emerge freely into the air. For preparation of such sources, it is desirable that the radium should be kept in solution and the radon pumped

* Hardy and Anderson, *Proc. Roy. Soc.* A, 72, 393, 1903.

off at intervals and collected in a tube over mercury and subsequently purified to the degree required.

These operations should be carried through as expeditiously as possible before the β and γ rays from the radon and its products reach an appreciable value. There is little β and γ radiation for the first 10 minutes after removal of the radon from the solution, but it then grows rapidly, reaching half its maximum value in 40 to 50 minutes. In the transfer operations over mercury, rubber gloves should be worn to protect the fingers from the small traces of the active deposit which are always present in the mercury through which the radon has bubbled. This prevents irritation of the skin of the fingers by the α rays and also keeps the hands free from radioactive contamination.

The radium in solution should be surrounded by sufficient lead to absorb the greater part of the γ rays. The radium room should be well ventilated so that any traces of radon which may escape into the air are rapidly removed. It is very undesirable to remain for long in a room in which free radon is present in appreciable quantity. During respiration, the radon enters the body and some of it is transformed *in situ*.

For these reasons, it is of great importance that all radium preparations should be kept in sealed tubes to prevent the escape of the emanation into the air, and that every precaution should be taken against the liberation of emanation in the various operations of preparing sources.

In all chemical operations involving radioactive bodies, the greatest care should be taken to keep the hands free from contamination and to prevent the dissemination of active matter. Similar precautions should be taken with radioactive sources, for it must be borne in mind that sources of the active deposit of radium after decay are always coated with the long-lived deposit of radium D and radium F (polonium). To prevent the radioactive contamination of a laboratory, no material which has been exposed in the presence of the radium emanation should be made use of in the workshop. It is surprising how easy it is to contaminate the tools in a workshop, resulting in a radioactive contamination of all apparatus under construction. This is obviously a great disadvantage in all ionisation experiments where a very low natural leak is desired.

CHAPTER VIII

THE SCATTERING OF α AND β PARTICLES

§ 42. The scattering of α particles and the nuclear theory of the atom. When α particles pass through matter, some of them are deviated from their original direction of motion and undergo the process known as scattering. The presence of this scattering was first shown by Rutherford* by a photographic method. He found that the image of a narrow slit produced by a beam of α particles had sharply defined edges when the experiment was performed in an evacuated vessel. If air was admitted into the apparatus, or if the slit was covered with a thin sheet of matter, the photographic trace of the pencil of α rays was broadened and the intensity of the photographic effect faded off slowly on either side of the centre.

A detailed examination of the amount and character of the scattering of the α particles in passing through matter was first made by Geiger†, using the scintillation method of detecting the particles. These experiments will be described later. It will be sufficient to say here that they showed that the scattering suffered by α particles in penetrating the atoms of matter is relatively very small. The average angle of scattering even by comparatively thick sheets of matter was only a few degrees. About the same time Geiger and Marsden‡ made the very striking observation that some of the α particles in a beam incident on a sheet of matter have their directions changed to such an extent that they emerge again on the side of incidence, that is, they are deflected through angles greater than 90°. The amount of this "reflection" of α particles increased at first with the thickness of the foil on which the α particles were incident, and then became constant. The fraction of α particles thus reflected increased rapidly with the atomic weight of the radiator. An estimate of the number of α particles reflected from a platinum radiator showed that about 1 particle in 8000 incident on the plate was scattered through more than a right angle.

This fraction was very much greater than that to be expected from the experiments of Geiger on the scattering at small angles.

* Rutherford, *Phil. Mag.* **11**, 166, 1906.
† Geiger, *Proc. Roy. Soc.* A, **81**, 174, 1908.
‡ Geiger and Marsden, *Proc. Roy. Soc.* A, **82**, 495, 1909.

On the views then current the electrical fields in the atom were not of sufficient intensity to deflect such an energetic projectile as an α particle through more than a very small angle in a single collision with an atom. A large deflection of an α particle was ascribed to the combined effect of a very great number of atomic encounters. Rutherford*, however, pointed out that the probability that an α particle should be scattered in this way through more than a right angle was very much smaller than that found experimentally by Geiger and Marsden. He concluded, therefore, that the deflection of the α particle through a large angle was due to a single atomic encounter, for the chance of a second encounter of a kind to produce a large deflection must usually be very small. It was then necessary to suppose that the atom must be the seat of an intense electrical field, and Rutherford suggested a simple structure of the atom which provided such a field. He supposed that the positive charge associated with an atom is concentrated into a minute centre or nucleus and that the

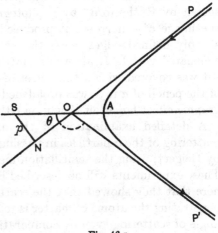

Fig. 46 a.

compensating negative charge is distributed over a sphere of radius comparable with the radius of an atom. For the purposes of calculation he assumed that the central or nuclear charge of the atom and also the charge on the α particle behaved as point charges. He was then able to show that for all deflections of the α particle greater than one degree the field due to the negative charge could be neglected and the deflection due to the field of the central charge need alone be considered. The deflections due to the nuclear field can be calculated very simply if we assume that the electrical force between the nucleus and the α particle is given by Coulomb's Law. We shall consider first the case of a heavy atom such that the nucleus may be assumed to remain at rest during the collision. The mass of the α particle will be taken as constant, since its velocity is always small in comparison to the velocity of light. The path of the

* Rutherford, *Phil. Mag.* 21, 669, 1911.

α particle under these conditions will then be a hyperbola with the nucleus S of the atom as the external focus (Fig. 46 a).

Let PO be the initial direction of motion of the α particle and OP' its final direction after deflection in the nuclear field. Let $p = SN$ be the perpendicular distance from the nucleus to the initial line of motion of the α particle.

If E, M, V be the charge, mass and initial velocity of the α particle, v its velocity at A, and Ze the charge of the nucleus, we have from the conservation of angular momentum

$$pV = SA \cdot v \qquad \ldots\ldots(1)$$

and from conservation of energy

$$\tfrac{1}{2}MV^2 = \tfrac{1}{2}Mv^2 + \frac{Ze \cdot E}{SA}$$

or

$$v^2 = V^2\left(1 - \frac{b}{SA}\right) \qquad \ldots\ldots(2),$$

where $b = \dfrac{2Ze \cdot E}{MV^2}$. b is therefore the closest possible distance of approach of an α particle of velocity V to a nucleus of charge Ze, which occurs of course when the collision is "head on" or $p = 0$, and the α particle is deflected through 180°.

Since the eccentricity is $\sec\theta$ and the focal distance SO is the eccentricity times half the major axis, i.e. $SO = OA \sec\theta$, we have from the geometry of the hyperbola

$$SA = SO + OA = p \operatorname{cosec}\theta\,(1 + \cos\theta)$$
$$= p \cot \tfrac{1}{2}\theta.$$

From (1) and (2)

$$p^2 = SA\,(SA - b) = p \cot \tfrac{1}{2}\theta\,(p \cot \tfrac{1}{2}\theta - b)$$

or

$$b = 2p \cot\theta.$$

The angle of deflection ϕ of the α particle is $\pi - 2\theta$, and hence

$$\cot \tfrac{1}{2}\phi = \frac{2p}{b} \qquad \ldots\ldots(3).$$

This gives the deflection of the particle in terms of the impact parameter p and b, the closest possible distance of approach.

It is of interest to see how the angle of deflection varies with the value of p/b. This is shown in the following table:

p/b	10	5	2	1	0·5	0·25	0·125
ϕ	5·7°	11·4°	28°	53°	90°	127°	152°

We have now to calculate the probability of deflection through an angle ϕ. Suppose that a pencil of α particles falls normally on a thin screen of matter of thickness t, containing n atoms per unit volume. We assume that the screen is so thin that the particles pass through without appreciable change of velocity and that the number of particles deflected from their path is small, so that the pencil traverses the screen normally. Then the chance that a particle shall pass within a distance p of a nucleus is

$$q = \pi p^2 nt.$$

A particle which moves so as to pass within a distance p of the nucleus is deflected through an angle greater than ϕ, where ϕ is given by equation (3).

Hence the probability of deflection through an angle greater than ϕ, or the fraction of the total number of α particles which are deflected through an angle greater than this, is

$$q = \tfrac{1}{4}\pi ntb^2 \cot^2 \tfrac{1}{2}\phi \qquad \dots\dots(4).$$

Similarly the chance of deflection through an angle between ϕ and $\phi + d\phi$ is equal to the chance of striking between the radii p and $p + dp$ and is given by

$$dq = 2\pi pnt \,.\, dp$$
$$= \tfrac{1}{4}\pi ntb^2 \cot \tfrac{1}{2}\phi \,.\, \operatorname{cosec}^2 \tfrac{1}{2}\phi \,.\, d\phi \qquad \dots\dots(5).$$

For comparison with the experiments to be described later, it will be convenient to express the equation (5) in another form. In these experiments the scattering was determined by counting the number of α particles falling normally on a constant area of a zinc sulphide screen placed at a constant distance r from the radiator. If Q be the total number of α particles incident on the radiator, then the number of particles scattered to unit area of the zinc sulphide screen placed at an angle ϕ to the original direction of the particles is given by

$$y = \frac{Qntb^2 \operatorname{cosec}^4 \tfrac{1}{2}\phi}{16r^2} \qquad \dots\dots(6).$$

Since $b = \dfrac{2Ze \,.\, E}{MV^2}$, we see that on Rutherford's theory the number of α particles falling on unit area at a distance r from the point of scattering must be proportional to

(a) $\operatorname{cosec}^4 \tfrac{1}{2}\phi$;

(b) t, the thickness of scattering material (provided that single scattering prevails);

(c) the square of the central charge Ze;

and inversely proportional to

    (d) $(MV^2)^2$, the square of the energy of the particle, or the fourth
    power of the velocity if M be constant.

§ 43. Experimental test of the nuclear theory. The first point,
the angular distribution of the scattered particles, was tested by
Geiger* in some preliminary experiments. He found that the distribution between 30° and 150° of the particles scattered by a thin
gold foil was in agreement with this theory. Later, a beautiful series
of experiments was carried out by Geiger and Marsden†, in which
the above conclusions drawn by Rutherford from his theory were

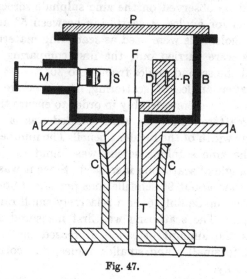

Fig. 47.

tested point by point. They first investigated the variation of
scattering of α particles over a wide range of angles, using the
apparatus shown in Fig. 47. The apparatus consisted in essentials
of a strong cylindrical metal box B which contained the source of
α particles R, the scattering foil F, and a zinc sulphide screen S
attached rigidly to a microscope M. The box was fixed to a graduated
circular platform A, which could be rotated in the airtight joint C.
The microscope and zinc sulphide screen rotated with the box,
while the scattering foil and source remained fixed. The box was

* Geiger, *Proc. Manch. Lit. Phil. Soc.* **55**, 20, 1911.
† Geiger and Marsden, *Phil. Mag.* **25**, 604, 1913.

closed by a ground glass plate P and could be exhausted through the tube T.

The source of α particles was an α ray tube filled with radon. The beam of α particles was therefore not homogeneous, for it contained particles from the products radium A and radium C in addition to those from the radon. This fact did not interfere with the investigation of the law of scattering with angle, since each group of α particles is scattered according to the same law. A narrow pencil of α particles from the source R was directed through the diaphragm D to fall normally on the scattering foil F.

By rotating the platform A the α particles scattered in different directions could be observed on the zinc sulphide screen. Observations were taken for angles of scattering between 5° and 150° and both silver and gold foils were used as scattering material. Two sets of experiments were carried out, the first comparing angles from 15° to 150° and the second angles from 5° to 30°.

For the smaller angles of scattering, the aperture of the diaphragm D was reduced considerably in order to ensure that the angle at which the scattered particles were counted was large compared with the angular width of the incident pencil. The number of particles scattered to the zinc sulphide screen was found to decrease very rapidly as the angle of scattering increased. Since it was not feasible to count more than about 90 scintillations per minute or less than 5, observations were made only over a relatively small range of angles at the same time. The scattering was first measured at the larger angles, and as the amount of radon decreased the measurements were extended to smaller and smaller angles, due correction being made for the decay.

Even when no scattering foil was in position at F a few scintillations were observed. These were due to particles scattered from the edge of the diaphragm limiting the incident pencil and from the walls of the vessel. The number of these extraneous particles was reduced by lining the vessel with paper and by using aluminium for the material of the diaphragm, for the scattering, as will be seen, is least for substances of low atomic weight. This extraneous effect was determined at different angles and allowed for in the subsequent measurements of the true scattered particles.

The collected results of these experiments are given in the following table. The first column gives the values of the angles ϕ between the direction of the incident pencil of α particles and the direction in which the scattered particles were counted, and the second column gives the corresponding values of $\cosec^4 \frac{1}{2}\phi$. Columns III and V give the observed numbers N of scintillations for silver and gold respectively. Columns IV and VI show the ratio of N to $\cosec^4 \frac{1}{2}\phi$. This ratio is seen to be approximately constant for both sets of experiments.

Variation of scattering with angle.

I	II	III	IV	V	VI
		SILVER		GOLD	
Angle of deflection, ϕ	$\cosec^4 \frac{1}{2}\phi$	Number of scintillations, N	$\dfrac{N}{\cosec^4 \frac{1}{2}\phi}$	Number of scintillations, N	$\dfrac{N}{\cosec^4 \frac{1}{2}\phi}$
150°	1·15	22·2	19·3	33·1	28·8
135	1·38	27·4	19·8	43·0	31·2
120	1·79	33·0	18·4	51·9	29·0
105	2·53	47·3	18·7	69·5	27·5
75	7·25	136	18·8	211	29·1
60	16·0	320	20·0	477	29·8
45	46·6	989	21·2	1435	30·8
37·5	93·7	1760	18·8	3300	35·3
30	223	5260	23·6	7800	35·0
22·5	690	20300	29·4	27300	39·6
15	3445	105400	30·6	132000	38·4
30	223	5·3	0·024	3·1	0·014
22·5	690	16·6	0·024	8·4	0·012
15	3445	93·0	0·027	48·2	0·014
10	17330	508	0·029	200	0·0115
7·5	54650	1710	0·031	607	0·011
5	276300	—	—	3320	0·012

Fitting the experiments at the smaller angles to those at larger angles, the numbers of scattered particles are proportional to $\cosec^4 \frac{1}{2}\phi$ over the whole range investigated, where $\cosec^4 \frac{1}{2}\phi$ varies from 1 to 250,000. These experiments thus afford abundant proof of the law of scattering with angle deduced by Rutherford from the nuclear theory of the atom.

Geiger and Marsden next examined the variation of scattering with the thickness of the scattering material. In these and most subsequent experiments it was necessary to use a source of homogeneous α particles, for it was found—and it was predicted by the theory—that the scattering increased very rapidly as the velocity of the α particle was reduced. In this series of experiments the angle of scattering was kept constant, while the thickness of the scattering foil was varied. The α particles from a source R of radium (B + C) passed through the diaphragm D and impinged on the scattering foil F, Fig. 48. The scattered particles were observed on a zinc sulphide screen fixed at Z. The angle of scattering in most experiments was about 25°. The main part of the apparatus was enclosed in a brass ring A closed by the glass plates B, C. The opening O in the plate B, through which the α particles entered the chamber, was closed by a thin sheet of mica. The chamber was evacuated during the experiments. The scattering foils were carried on a disc S which could be rotated by means of a ground joint so that different foils could be brought to F to intercept the incident pencil of α particles. As in the previous experiments, some difficulty was experienced with particles scattered from the edges of the diaphragm D. The number of these was very large even when the diaphragm was made of paper. They were finally eliminated

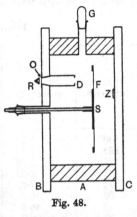

Fig. 48.

by introducing a subsidiary screen, the position of which could be adjusted by means of the ground-joint G.

Observations were made with foils of different materials and the results showed that for each material the amount of scattering was proportional to the thickness of the scattering foil, provided this was small. When the thickness of the foil was so great as to cause an appreciable decrease in velocity of the α particle, the amount of scattering increased rather more rapidly than the thickness. This increase was due to the decrease in velocity suffered by the particles in their passage through the foil. A correction for this effect was made from the results of measurements of the variation of scattering with the velocity of the α particles. The final results, after this correction, are shown in Fig. 49.

For the four metals examined, the number of scattered particles was proportional to the thickness of the scattering foil. This may be taken to show that the scattering was single, i.e. that the particles observed under the conditions of these experiments had undergone only one deflection in passing through the scattering foil.

With the same experimental arrangement as shown in Fig. 48 the amount of scattering by different atoms was compared. Thin foils of the different materials were carried on the disc S. The thicknesses of the foils were chosen to give about the same numbers of scattered

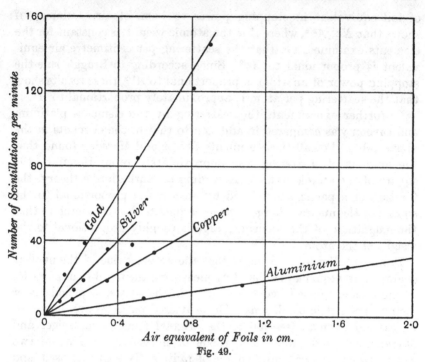

Fig. 49.

particles. The results of one experiment are given in the following table. The source of α particles was radium (B + C).

Variation of scattering with atomic weight.

I	II	III	IV	V	VI	VII
Substance	Atomic weight, A	Air equivalent in cm.	Number of scintillations per minute corrected for decay	Number N of scintillations per cm. air equivalent	$A^{3/2}$	$N \times A^{2/3}$
Gold	197	0·229	133	581	2770	0·21
Tin	119	0·441	119	270	1300	0·21
Silver	107·9	0·262	51·7	198	1120	0·18
Copper	63·6	0·616	71	115	507	0·23
Aluminium	27·1	2·05	71	34·6	141	0·24

The numbers given in column IV are the numbers of scintillations observed corrected for the variation in activity of the source, and also for the change in velocity of the α particles in the scattering foil. In column V are the numbers N of scintillations for a thickness of foil equivalent in stopping power to 1 cm. of air. Column VII shows that $N \times A^{2/3}$, where A is the atomic weight, is constant for the elements examined, i.e. that the scattering per centimetre air equivalent is proportional to $A^{3/2}$. Since according to Bragg's rule the stopping power of an atom is proportional to $A^{1/2}$, these results show that the scattering per atom is approximately proportional to A^2.

In further experiments the scattering by the elements platinum and carbon was compared in addition to that of the elements in the above table. For all these elements Geiger and Marsden found that the amount of scattering was approximately proportional to the square of the atomic weight. According to Rutherford's theory, the fraction of α particles scattered by an atom is proportional to the square of the nuclear charge. The experiments showed therefore that the magnitude of the nuclear charge is roughly proportional to the weight of the atom.

From equation (6) it is clear that the actual value of the nuclear charge of an atom can be found by measuring the number of particles in the incident pencil and the number in the scattered pencil under definite geometric conditions. The scattered particles are, however, an extremely small fraction of the original pencil, and Geiger and Marsden adopted different methods of measurement for the two cases. In one experiment a source of radium (B + C) was used, and the number of particles in the incident pencil was calculated from its γ ray activity and the geometric conditions. The number of particles scattered by a thin gold foil was observed at an angle of 45°, and a correction was applied for the fact that only 85 per cent. of the particles falling on the zinc sulphide screen produced scintillations. In another experiment the source of α particles was a radon tube and the number of particles in the main beam was determined directly by the scintillation method after an interval of some weeks, in which time the radon had decayed to a suitably small amount. These experiments, combined with the observed variation of scattering with atomic weight, showed that for elements of weight greater than aluminium, the positive charge on the nucleus of the atom is roughly about $\frac{1}{2}A \cdot e$, where A is the atomic weight and e the electronic charge.

Another series of observations was carried out to investigate the way in which the scattering of the α particles depended on their velocity. A source of radium (B + C) was used, and beams of different velocities were obtained by placing absorbing screens of mica between the source and the window O (Fig. 48). The particles were scattered by a thin foil of gold or silver placed at F. The results are shown in the following table.

Variation of scattering with velocity.

I Range of particles	II Relative values of $1/V^4$	III Number N of scintillations	IV NV^4
5·5	1·0	24·7	25
4·76	1·21	29·0	24
4·05	1·50	33·4	22
3·32	1·91	44	23
2·51	2·84	81	28
1·84	4·32	101	23
1·04	9·22	255	28

Column III gives the number N of scintillations per minute under fixed conditions when α particles of the ranges given in column I were used. The relative values of $1/V^4$ are entered in column II, the velocity V being calculated from the range by Geiger's rule $V^3 = aR$. Over the whole range examined the product NV^4 remains sensibly constant, showing that the scattering varies inversely as the fourth power of the velocity of the α particle. This accords with the predictions of the theory.

§ 44. **The magnitude of the charge on the nucleus.** These experiments of Geiger and Marsden thus verified the conclusions developed by Rutherford in a most remarkable way, and left no doubt as to the essential truth of his conception of the atom as containing a small highly charged nucleus surrounded by an inverse square field of force. As is well known, this nuclear atom formed the basis of all subsequent speculations on the intimate structure of the atom, and in the hands of Bohr led to the remarkable development of the quantum theory of spectra.

On this conception of atomic structure the fundamental constant of an atom is not the atomic weight but the magnitude of the nuclear charge. The charge on the nucleus fixes the number of the external electrons in the atomic system and controls their arrangement. The mass of the nucleus influences the arrangement of the electrons only to a very small degree. The physical and chemical properties of an element are determined by the charge on the nucleus, and the question of its actual magnitude thus becomes of great importance.

The experiments of Geiger and Marsden, as mentioned already, gave a rough estimate of the value of the nuclear charge. To within about 20 per cent. the charge on the nucleus was equal to $\frac{1}{2}A \cdot e$, where A is the atomic weight, e the value of the electronic charge. Later, van den Broek* made the interesting suggestion that the nuclear charge might be equal to the atomic number of the element, and he showed that his suggestion was in good agreement with the scattering experiments. The importance and bearing of this suggestion were quickly recognised, and it was adopted by Bohr in his papers on the structure of atoms and the origin of spectra.

The first clear experimental evidence in its favour was given by Moseley's investigations of the X ray spectra of the elements. Moseley found that the X ray spectra depended on the square of a number which increased by unity in passing from one element to the next of higher atomic weight. This number was not exactly equal to the atomic number but could be written as $N - a$, where N is the atomic number and a a constant for the spectral series. Moseley supposed that this constant a represented a disturbing effect due to the electrons in the atom, and concluded that the atomic number gave in electronic units the actual value of the charge on the nucleus. This generalisation has been of fundamental importance in modern atomic theory, and gave a starting point for the development of the external structure of the atom. It has been subjected to direct test by Chadwick†, who has measured the nuclear charge of the three elements copper, silver and platinum.

§ 44 a. Measurement of nuclear charge. The only direct method we have of measuring the nuclear charge is from the single scattering of a particles. The method consists in the measurement of the number of a particles in the original pencil of particles and the number in the scattered pencil under well-defined geometric conditions.

The arrangement adopted in these experiments has been much used in subsequent observations of scattering and will therefore be described in some detail.

In order to increase the fraction of scattered particles relative to the incident beam, a scattering foil in the form of an annular ring subtending a fairly wide cone was used. In Fig. 50 let R be the source of particles, AA' the scattering foil, and let a zinc sulphide

* van den Broek, *Phys. Zeit.* 14, 32, 1913.
† Chadwick, *Phil. Mag.* 40, 734, 1920.

screen be placed at a point S on the axis of the cone RAA' such that $RA = AS$.

The solid angle subtended at R by an elementary annular ring at P is

$$2\pi \sin \tfrac{1}{2}\phi . d\tfrac{1}{2}\phi.$$

If Q is the number of α particles emitted per second by the source, the number falling per second on this elementary ring is

$$\tfrac{1}{2}Q . \sin \tfrac{1}{2}\phi . d\tfrac{1}{2}\phi.$$

The number scattered to unit area at S placed at right angles to RS is then

$$\tfrac{1}{2}Q . \sin \tfrac{1}{2}\phi . d\tfrac{1}{2}\phi . \cos \tfrac{1}{2}\phi . nt \sec \tfrac{1}{2}\phi . \frac{b^2 \operatorname{cosec}^4 \tfrac{1}{2}\phi}{16r^2}.$$

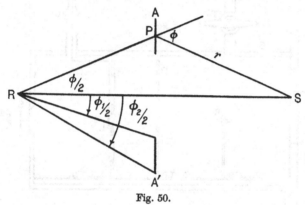

Fig. 50.

For the whole scattering foil of angular limits $\tfrac{1}{2}\phi_1$, $\tfrac{1}{2}\phi_2$, the number of particles falling per second on unit area at S is

$$\frac{Qntb^2}{32\bar{r}^2} \int_{\tfrac{1}{2}\phi_1}^{\tfrac{1}{2}\phi_1} \operatorname{cosec}^3 \tfrac{1}{2}\phi . d\tfrac{1}{2}\phi,$$

or

$$\frac{Qntb^2}{64\bar{r}^2} [\log \tan \tfrac{1}{4}\phi_2 - \log \tan \tfrac{1}{4}\phi_1 + \cot \tfrac{1}{2}\phi_1 \operatorname{cosec} \tfrac{1}{2}\phi_1 - \cot \tfrac{1}{2}\phi_2 \operatorname{cosec} \tfrac{1}{2}\phi_2],$$

where $\bar{r}^2$ is a mean value.

The number of α particles falling per second directly on unit area at S is $Q/4\pi l^2$, where $l = RS$.

With this arrangement the number of particles scattered by a suitable foil, even of a heavy atom like platinum, is still about 1/500 to 1/1000 of the number in the direct pencil. Thus if the scattered particles were falling on the screen at a convenient rate, say about 30 per minute, the direct number would be about 20,000 per minute.

The counting of so large a number was accomplished by rotating in the path of the direct pencil a wheel containing a slit. By adjusting the ratio of the width of the slit to the circumference of the wheel, the number of particles falling on the screen could be reduced to the desired extent. Thus the direct and scattered particles were counted on the same zinc-sulphide screen under similar conditions, and the accuracy of the measurement of the nuclear charge depended chiefly on the total number of α particles counted in the experiments.

The apparatus used in these experiments is shown in Fig. 51. The diaphragm D served to define the angular limits of the beam of

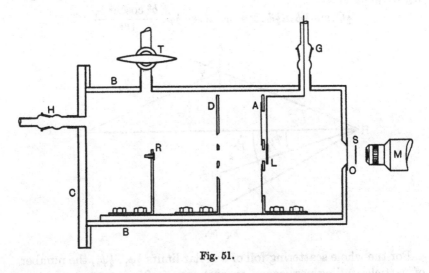

Fig. 51.

α particles striking the scattering foil held on the support A. It was arranged by suitable disposition of D and A that no particles could be scattered directly to the screen S from the edges of the diaphragm. The direct pencil of α particles passed through holes in the central discs of D and A, and could be cut off at will by the lead screen L carried by a glass joint G. The wheel containing the slit was rotated between the end of the box and the screen S. The source of α particles, R, was a disc of 2 to 3 mm. diameter coated on its face with radium (B + C).

Three series of experiments were carried out in which foils of copper, silver and platinum were used as scattering materials. The results of the measurements gave for the values of the nuclear charge 29·3e, 46·3e and 77·4e, for copper, silver and platinum respectively, with an estimated accuracy of between 1 and 2 per cent. The atomic numbers of these elements are 29, 47 and 78, so that these experiments gave a direct proof of Moseley's generalisation that the charge on the nucleus is given by the atomic number.

§ 45. The size of the nucleus and the law of force. In his theory of the scattering of α particles, Rutherford assumed that the force between the α particle and the nucleus of an atom was given by Coulomb's Law. It was pointed out by Darwin* that a direct test of the law of force is given by the dependence of scattering on the velocity of the α particle. If the force around the nucleus vary with the distance as $1/r^p$, then the number of scattered particles should be proportional to $\left(\dfrac{1}{V^2}\right)^{\frac{2}{p-1}}$, other factors being kept constant. We have seen that the experiments of Geiger and Marsden showed that the number of scattered particles varied as the fourth power of the velocity of the α particle, to the nearest integral power (§ 43). Combined with the observed law of angular distribution of the scattered particles, this leaves no doubt as to the general validity of the Coulomb law of force in these collisions.

This test of the law of force was repeated more strictly, though over a narrower range of velocity, by Chadwick†, using the apparatus shown in Fig. 51. By means of a ground joint H, mica sheets of known stopping power were brought in front of the source to cut down the velocity of the α particles incident on the scattering foil. The number of scattered particles was counted for three velocities, with a scattering foil of platinum. The velocity was calculated from the emergent range by Geiger's rule. The results are given below.

Mica sheet	Relative V^4	NV^4
0	1	100
1	0·549	101
2	0·232	103

The values of NV^4 are constant within the experimental error of about 4 per cent. It may be noted that if, instead of using Geiger's rule to calculate the velocity of the particles, we use Brigg's later measurements (§ 22), the values of NV^4 in the table will show still better agreement.

---

* Darwin, *Phil. Mag.* 27, 499, 1914.
† Chadwick, *loc. cit.*

The distance of approach of the fastest α particles used in these experiments was about 7×10^{-12} cm., and of the slowest about 14×10^{-12} cm. Over this range of distance the force between the platinum nucleus and the α particle obeys Coulomb's Law very closely.

These results were extended by Rutherford and Chadwick* in one direction, by Rose† in the other. Rose observed the single scattering of α particles by gold at angles between 1° and 8°, with the object

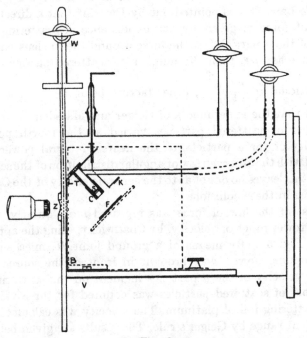

Fig. 52.

of investigating the effect of the K shell of electrons on the field around the nucleus. No such effect was found, and the results showed that the inverse square law of force held up to 1.7×10^{-10} cm. from the nucleus, within the accuracy of experiment. Rutherford and Chadwick examined the scattering of α particles by several elements. The most accurate experiments were performed with gold as scattering element. Using the annular ring method previously described, they measured the scattering by a thin gold foil between 40° and 70° and

* Rutherford and Chadwick, *Phil. Mag.* **50**, 889, 1925.
† Rose, *Proc. Roy. Soc.* A, **111**, 677, 1926.

between 65° and 100°. In each set of experiments the scattering was observed for α particles of two ranges, 6·7 cm. and 4·4 cm. in air, so that a direct check of the law of force was obtained. The results gave a nuclear charge for gold less than 2 per cent. different from the atomic number, and the comparison of the scattering of particles of different ranges showed that the inverse square law held very closely.

In another series of experiments the scattering angle was kept constant at an average value of 135° and the amount of scattering observed for α particles of different ranges. The apparatus used in these experiments is shown in Fig. 52. The source S was carried in a narrow tube T, which was closed by a thin sheet of collodion. The diaphragm C, made of graphite to reduce extraneous scattering, served to limit the beam of α particles. The particles fell normally

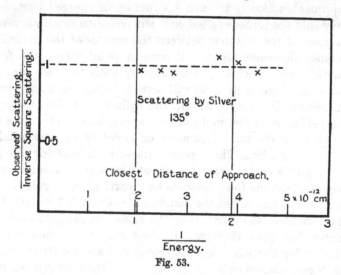

Fig. 53.

on F, a thin foil of the element under examination. The scattered α particles were observed on the zinc sulphide screen Z.

The source of α particles was radium (B + C). By placing absorbing screens of mica over the source, α particles of different ranges were obtained, varying from 2·4 cm. to 6·7 cm. in air, and their scattering observed. Foils of gold, platinum, silver, and copper were used as scattering materials. No divergence from the inverse square law of force was found with these elements. The experimental results obtained with silver are illustrated by Fig. 53, where the ratio of the observed to the theoretical scattering is plotted against the reciprocal of the energy of the α particle. The closest distance of approach of an α particle to the nucleus is, for a given angle of scattering, inversely proportional to the energy of the particle. The abscissae therefore represent also the region around the nucleus which was investigated, and a scale of distances is given in the figure. The dotted line gives the amount of scattering calculated from theory. It is seen that there is no evidence of any departure from theory, either in the variation of scattering with velocity of the α particle or in the actual fraction of particles scattered, within the limits of error of the experiments of about 10 per cent.

On the generally accepted views of atomic structure, all nuclei consist ultimately of protons and electrons and must occupy a finite volume. When the α particle, itself a helium nucleus and a complex body, approaches closely to such a structure of charged particles it is unlikely that the forces exerted on it will continue to vary inversely as the square of the distance between the centres of the nuclei. As this distance decreases there will come a point where the forces between the two nuclei vary very rapidly with the distance, and the scattering of the α particle will depart widely from the predictions of theory. We may compare the collisions of the two nuclei with the collisions of the molecules of a gas, and in the same way that we speak of the size of an atom or molecule we may speak of the "size" of a nucleus. The experiments on the scattering of α particles then give some information about the size of atomic nuclei. We have seen that the force between an α particle and a gold nucleus obeys Coulomb's Law when the distance of collision is 3.2×10^{-12} cm. We may conclude, therefore, that the radius of the gold nucleus is not greater than 3.2×10^{-12} cm. In the same way we find that the upper limits for the silver and copper nuclei are 2×10^{-12} cm. and 1.2×10^{-12} cm. respectively. The nucleus is thus minute in comparison with the atom, which has a radius of roughly 10^{-8} cm.

It will be seen later that experiments on the collisions of α particles with light elements have yielded more definite information about the size of nuclei, for in certain cases marked divergences from the scattering laws have been found for very close distances of approach between the colliding nuclei. No departure from inverse square forces has been observed in the collisions of α particles with the atoms of copper or of any element of higher atomic number. In all cases investigated, from copper to uranium, the scattering of α particles has been found to obey the simple rules formulated by Rutherford.

§ 46. Scattering at small angles and multiple scattering. The experiments so far described have dealt with the scattering of a particles through comparatively large angles by thin sheets of matter. Under these conditions the scattering is single, the deflection of the particle is the result of a single atomic encounter. We have seen that the phenomena of single scattering are essentially simple and that it is easy to interpret the results of experiment in terms of the structure of the atom.

When the scattering of a particles is examined at very small angles the conditions for single scattering will only be attained if very thin foils of matter are used as scattering agent. In general, the particles which are observed may have been scattered more than once. The relations which hold for such conditions will usually be extremely complicated. When, however, the final deflection of the particle is the resultant of a very large number of small deflections of the same

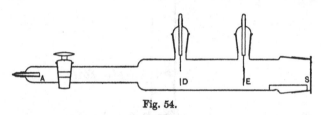

Fig. 54.

order of magnitude, statistical methods can be applied to the calculation of the distribution. This scattering process is known as multiple scattering. The relations which describe multiple scattering are not so simple as those for single scattering, nor are they based upon so definite a theoretical foundation. A thorough discussion of the application of statistical methods to the problem of multiple scattering has been given by Bothe*. It is clear that since the particles do not penetrate so deeply into the structure of the atoms through which they pass we cannot expect the phenomena of multiple scattering to provide much information about the structure of the atoms. The experiments on multiple scattering have indeed led to no definite conclusions about atomic structure and, speaking generally, the best that can be done is to show that the results are in accord with the nuclear theory.

The conditions for the multiple scattering of a particles are most nearly realised when the scattering is observed at small angles from

* Bothe, *Zeit. f. Phys.* 4, 300, 1921; 5, 63, 1921.

relatively thick sheets of matter. The first experiments of this kind were those of Geiger*, to which we have already referred in § 42. His experimental arrangement is shown in Fig. 54.

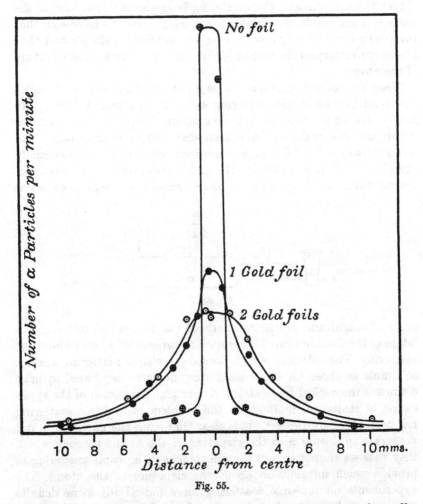

Fig. 55.

The source of α particles was radium (B + C) deposited on the walls of a narrow conical tube A, which was closed by a thin sheet of mica. A narrow pencil of particles passed through a circular hole D and was received on the zinc sulphide screen S. Scattering foils could be brought into the path of the rays at E or at D.

* Geiger, *Proc. Roy. Soc.* A, **81**, 174, 1908; **83**, 492, 1910; **86**, 235, 1912.

When the apparatus was exhausted and no foil was placed in the path of the particles, a small and very bright spot of scintillations was observed on the screen. When a thin foil of a few mm. air equivalent was introduced at D or E the bright spot disappeared, for, owing to scattering, the scintillations were distributed over a much greater area. The scintillations were counted at different distances from the centre with the result shown in Fig. 55. From these measurements the most probable angle of scattering for the given foil can be deduced. Let n be the number of scintillations observed

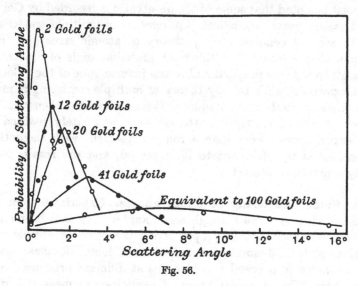

Fig. 56.

per unit area at a distance r from the centre of the spot and s be the distance from the scattering foil to the zinc sulphide screen. Then $2\pi r n \, dr$ represents the total number of α particles scattered through an angle θ given by $\tan \theta = r/s$ and the most probable angle of scattering is given when $2\pi r n \, dr$ is a maximum. Fig. 56 shows Geiger's results for scattering by different thicknesses of gold foil, plotted in this way. Geiger also investigated the dependence of the most probable angle of scattering on the thickness of the scattering foil, on the atomic weight of the material, and on the velocity of the α particles.

Experiments on the multiple scattering of α particles have also been carried out by Mayer* with, in principle, the same method, and with much the same results.

* Mayer, *Ann. d. Phys.* **41**, 931, 1913.

These experiments of Geiger were at the time taken to show that the scattering of an α particle in any encounter with a single atom was small and that a large deflection could only result from a very large number of encounters. Rutherford, however, showed that the results could be accounted for equally well on the theory of the nuclear atom. A closer examination of these experiments has been made recently by Bothe*, who developed statistical methods to deal with them. Bothe concluded that the experiments were in good agreement with his calculations.

It must be noted that some of the observations recorded by Geiger do not satisfy purely statistical requirements of multiple scattering, such as are independent of any theory of atomic structure. For example, Geiger found that the most probable angle of scattering with a given foil was proportional to the inverse cube of the velocity of the α particle, while on any theory of multiple scattering it must be proportional to the inverse square. It seems doubtful then whether the conditions for multiple scattering were adequately realised in these experiments. More than a rough agreement with theoretical calculations is therefore not to be expected, and the results have mainly a practical interest.

§ 46 a. **Single scattering at small angles.** Experiments on the scattering of α particles through small angles have also been made by Rutherford and Nuttall†. A pencil of α particles was sent between two glass plates 0·3 mm. apart and 14·5 cm. long. The glass plates were contained in a vessel to which gas at different pressures could be admitted. The emergent beam of particles was measured in an ionisation chamber of sufficient depth to absorb the particles completely. The conditions of scattering are obviously not very definite, but it was possible by this method to compare the scattering produced by various gases—hydrogen, helium, oxygen, etc. The results certainly seemed to agree with the law of single scattering and led to fairly correct relative values of the nuclear charges of the atoms used as scatterers. It is, however, somewhat doubtful if much importance can be attached to this agreement. In view of the small angle of scattering and the magnitude of the gas pressures which were used (about 7 cm. on the average), it does not seem likely that the conditions were suitable for the examination of single scattering.

* Bothe, *loc. cit.*
† Rutherford and Nuttall, *Phil. Mag.* 26, 702, 1913.

More definite are the experiments of Rose*, to which some reference has already been made. In these experiments the arrangement shown diagrammatically in Fig. 57 was used. A very thin foil of gold was placed at F in the path of a narrow pencil of α particles. The particles scattered through the annular opening AA' in the diaphragm P were observed. The number of particles in the direct pencil could be measured by allowing the particles to pass through the hole O. In most experiments the particles were counted by means of a Geiger electrical counter, in others the scintillation method was used. By changing the distance between the foil and the diaphragm, the scattering was examined over a range of angles from 1·2° to over 8°.

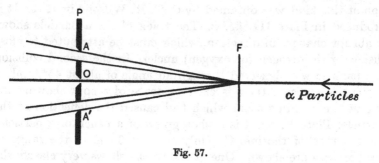

Fig. 57.

The results agreed very well with the angular distribution given by the theory of single scattering, except at the smaller angles. For example, with a foil of gold consisting of two layers of the thinnest leaf and for which the value of nt was $1·46 \times 10^{18}$, the observed scattering was in good accord with theory when the angle of scattering was greater than about 2·5°. We may conclude that under these experimental conditions the scattering was at least mainly single, but that at smaller angles than 2·5° the scattered particles consisted in an appreciable part of particles which had suffered more than one atomic encounter. Scattering of this latter type, in which the final deflection of the particle is the resultant of a small number of atomic encounters, has been called "plural scattering." It is the transition stage between single scattering and multiple scattering. The importance of plural scattering will be discussed later (§ 50) in connection with the scattering of β rays. The experiments of Rose are of value in specifying the conditions under which the effects of plural scattering begin

* Rose, loc. cit.

to appear and to affect the comparison of experiment with the theory of single scattering. In addition, these experiments showed that the field of force around the nucleus of the atom of gold obeys the inverse square law fairly closely at distances between $0\cdot4 \times 10^{-10}$ cm. to $1\cdot7 \times 10^{-10}$ cm., from inside the K shell of electrons to well outside it. The measurements were not sufficiently accurate to reveal any effect due to the electrons of the K shell.

§ 47. Scattering from Wilson photographs. The fact that an α particle can suffer a large deflection from its path in a single encounter with an atom is most vividly illustrated by photographs of the tracks of α particles in an expansion chamber. The first photograph of this kind was obtained by C. T. R. Wilson* in 1912. It is reproduced in Plate III, fig. 2. The track of the α particle shows two abrupt changes in direction, which must be attributed to close collisions with nitrogen (or oxygen) nuclei. In the second collision the α particle was deflected through an angle of about 45°, and at the point of collision there is clear evidence of a spur showing the track of the recoiling atom which had caused the deflection of the α particle. Plate V, fig. 1 is a photograph of a pencil of α particles from a source of thorium C. Only the last 3 cm. of the range of the α particles are shown. One of the tracks shows very clearly the result of a close collision with a nitrogen atom. The α particle is deflected through about 70° and the nitrogen atom has received sufficient momentum to travel a distance of about $0\cdot6$ mm. in the gas. Close inspection of the photograph will show that most of the tracks suffer changes of direction, which are small until near the end of the range. Since the chance of deflection through a given angle is inversely proportional to the fourth power of the velocity, the deflections should occur more frequently near the end rather than the beginning of the range.

An investigation of the laws of scattering by measurements of the tracks is, of course, a difficult and a tedious task, which has been attempted in only a few cases. Auger and Perrin† measured 52 tracks in argon and used their results to deduce a value for the atomic number of argon, which they found to be 19, as close to the actual value (18) as can be expected from the small number of observations. Blackett‡

* C. T. R. Wilson, *Proc. Roy. Soc.* A, **87**, 277, 1912.
† Auger and F. Perrin, *C.R.* **175**, 340, 1922.
‡ Blackett, *Proc. Roy. Soc.* A, **102**, 294, 1922.

made a statistical analysis of a large number of tracks in argon and in air. In these experiments the track was photographed from two directions at right angles. From the two images thus obtained any deflection suffered by the particle could be measured with reasonable accuracy. Blackett examined chiefly the ends of the tracks when the velocity of the particle was small and where deflections were comparatively frequent. The velocity of the particles was not known, for Geiger's rule loses its validity when the velocity is small and no direct measurements of range and velocity have been made in the region concerned in these experiments. The test of the nuclear theory was therefore restricted to observation of the angular distribution of the deflections. The angular distribution is quite sensitive to modifications in the law of force*. For example, if the law of force is the inverse square, the ratio of the number of particles scattered at 10° to that at 90° is 4330, while if the force varies as the inverse cube the ratio is 525. The number of deflections examined was 281 in argon, and 1524 in air. In both cases the distribution of the deflections was in agreement with the inverse square law of force. The distances of approach in the collisions studied varied from $7\cdot4 \times 10^{-12}$ cm. to $5\cdot4 \times 10^{-10}$ cm. in the case of argon, and from $3\cdot2 \times 10^{-12}$ cm. to $3\cdot7 \times 10^{-10}$ cm. in the case of air. It will be noted that the upper limit with argon is well outside the K shell of electrons. The experimental data are not sufficiently precise to show any effect of an apparent change in the nuclear charge due to shielding by the K electrons.

§ 48. **The scattering of β rays.** In comparison with the agreement between the predictions of the nuclear theory of the atom and the results of experiments on the scattering of α particles, the experiments on the scattering of β rays leave much to be desired. The conclusions derived from the measurements which have so far been made are not only less precise and definite than in the corresponding experiments with α particles but indeed some observers have found it difficult to reconcile their results with the accepted views of atomic structure.

The reasons for this unsatisfactory state of our knowledge are mainly connected with the greater technical difficulties of experiments with β particles. In contrast to the α radiation, the β radiation emitted by a radioactive source is not homogeneous but consists

---

* Cf. Darwin, *Phil. Mag.* **27**, 499, 1914.

of particles of widely different velocities. For a detailed study of scattering it is important to use a primary beam of particles of approximately the same velocity. To obtain such a beam with the available sources of rays it is then necessary to adopt some kind of resolving device, such as a magnetic field, to sort out particles of a suitably narrow range of velocity, with the result that the primary beam is of low intensity. A further complication is introduced by the fact that there is generally a strong γ radiation present, emitted by the source of β rays. The measurement of the scattered beam of β rays, which must usually be of very small intensity, in the presence of a strong γ radiation is a matter of some difficulty.

The definiteness of the experiments on the scattering of α particles is partly due to the simplicity and certainty of the scintillation method of counting the particles, even in the presence of strong β and γ radiations. For the β particles no such method is available; while it is possible to record β particles by the electrical counter the method is subject to various limitations, particularly in the presence of an appreciable γ radiation.

The chief defect of the earlier experiments, and the one most responsible for the unsatisfactory agreement with theory, was the failure to fulfil the experimental conditions under which true single scattering can be obtained with β particles. The difficulty of attaining such conditions is a necessary consequence of the small energy of the average β particle. According to the theory of single scattering the deflection experienced by a particle in a given encounter is inversely proportional to the square of the kinetic energy of the particle. For a given distance of approach between the nucleus and the particle the β particle will therefore be deflected through a much greater angle than the α particle, or to produce the same deflection in both cases a much greater distance of collision will suffice for the β particle. Thus the effective radius of the scattering atom is very much greater and there is a correspondingly greater likelihood that the final deflection of the β particle under the usual conditions of experiment may be due to two or more encounters, instead of a single one. We shall return to this question in § 50.

In addition, the theory of scattering of β rays is complicated by various factors, such as the change in mass of the β particle with its velocity, and the possible loss of energy by radiation as the β particle accelerates around the nucleus.

§ 49. Early experiments. It was early shown by many workers*
that the β rays were much more easily scattered by matter than
the α rays. For example, Crowther showed, in 1908, that the β particles were strongly scattered in passing through sheets of matter
so thin that the absorption was inappreciable. The early experiments
dealt mainly with the scattering of a complex beam of β radiation by
comparatively thick sheets of material, and it was not to be expected
that the results would show any simple relation either with the
average velocity of the β particles or with the nature of the scattering
material. The first experiments in which the scattering of β particles
was studied in detail were carried out by Crowther† in 1910. His
experimental arrangement is shown in Fig. 58.

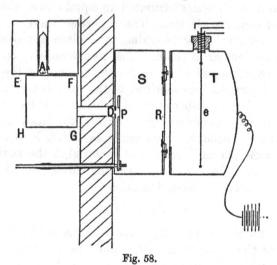

Fig. 58.

A nearly homogeneous pencil of β rays from a radium preparation A was obtained by passing the rays between the poles of an
electromagnet $EFGH$. This homogeneous pencil passed through the
opening D into the scattering chamber S. The rays after passing
through an opening R covered with thin aluminium foil entered the
testing vessel T where the ionisation current due to them was
measured. In order to avoid the scattering of the β rays by the

---

* Cf. Becquerel, *C.R.* **130**, 979 and 1154, 1900; Eve, *Phil. Mag.* **8**, 669, 1904;
McClelland and Hackett, *Trans. Roy. Soc. Dubl.* (2) **8**, 169, 1905; Crowther, *Proc.
Roy. Soc.* A, **80**, 186, 1908; Madsen, *Phil. Mag.* **18**, 909, 1909.
    † Crowther, *Proc. Roy. Soc.* A, **84**, 226, 1910.

air their path to R was through an exhausted space. On introducing a thin foil P the rays are scattered, and the effect measured in T is then due to those rays which have been scattered through an angle less than ϕ, where ϕ is half the angle subtended by the opening R. In order to obtain as homogeneous a pencil of β rays as possible the opening D was small, with the result that the effect of the γ rays from the radium was comparable with that due to the β rays. A balance method was therefore employed to obtain accurate measurements of the ionisation currents.

The results obtained were compared with the theory of scattering advanced by Sir J. J. Thomson*. On this theory the atom of matter was supposed to consist of a sphere of uniform positive electrification in which was distributed an equal and opposite charge in the form of negative electrons. The β particle traverses the atom and suffers irregular deflections due to a close approach to the constituent electrons or to the field due to the positive sphere. The deflections were supposed to be in general small, and the observed deflection of β particles was ascribed to the probability distribution resulting from a large number of small deflections. If θ is the average deflection due to an encounter with an atom, the mean deflection of a particle after N encounters was taken as $\sqrt{N}.\theta$. Now if n be the number of atoms per unit volume of material, b the radius of the atom, the number of encounters in passing through a thickness t is $n\pi b^2 t$. Consequently the mean deflection

$$\phi_m = \sqrt{n\pi b^2 t}.\theta,$$

or $\phi_m/\sqrt{t}$ should be constant for a given material. The theory also indicated that the value of $mv^2/\sqrt{t_m}$ should be constant for a given material for a given value of ϕ, where t_m is the thickness of foil required to scatter half the particles through the angle ϕ, and m, v are the initial mass and velocity of the β particle.

Both these relations were confirmed by Crowther, and his experiments seemed therefore to support the Thomson theory of the atom. Later, Rutherford was able to show that the same relations, differing only by a numerical constant, should also hold on the theory of single scattering by the nuclear atom. These tests were therefore not sufficient to decide between the two theories of atomic structure. As we have seen, the decisive evidence was provided by the experiments on the scattering of α particles in favour of the nuclear theory.

* J. J. Thomson, Proc. Camb. Phil. Soc. 15, 465, 1910.

In view of the small angles of scattering and comparatively large thicknesses of the scattering foils, there is no doubt that the experiments of Crowther are to be interpreted on the laws of multiple scattering. If the scattering is multiple, the probability of a deflection greater than ϕ will be $e^{-\phi^2/n\theta^2}$ or $e^{-\phi^2/c t\theta^2}$, where c is a constant. For a given angle of scattering the relation between the intensity of the current in the ionisation chamber and the thickness of the scattering foil will therefore be given by $I/I_0 = 1 - e^{-k/t}$, where k is a constant for that angle. Crowther's measurements were in accord with this relation. On single scattering the relation would be, for small values of t only, $I/I_0 = 1 - at$, where a is a constant for a given angle of scattering.

In a discussion of this and later experiments on the scattering of β particles Bothe* calculated the relations which should hold between the "angle of half scattering" and the thickness of scattering foil in the cases of single and of multiple scattering, on the basis of the nuclear theory of the atom. For purposes of calculation he considered the atom as a nucleus with a continuous distribution of electrons in the outer shell of the atom. He supposed the conditions to be such that the angle of scattering ϕ was so small that $\tan \frac{1}{2}\phi$ could be replaced by $\frac{1}{2}\phi$. He then showed that if the scattering is single a foil of thickness t containing n atoms per c.c. of atomic number Z will scatter half the incident particles through more than Φ, where Φ is given by

$$\Phi = \sqrt{2} \cdot \frac{2e^2}{mv^2} \cdot Z \cdot \sqrt{\pi nt}$$

and where m, v, e are the mass, velocity, and charge of the β particles. If the scattering is multiple, the corresponding relation is with certain approximations

$$\Phi = 2 \cdot 6 \cdot \frac{2e^2}{mv^2} \cdot Z \cdot \sqrt{\pi nt}.$$

Both relations are of the same form, and differ only by a numerical constant of $1 \cdot 8$. Bothe showed that Crowther's results were in very fair accord with theory; in particular the value of $\Phi/Z\sqrt{t}$ calculated from the measurements was constant within the error of experiment.

Substantially the same experimental method was used later by Crowther and Schonland†. The earlier results were confirmed but they found the linear relation between I/I_0 and t which is indicated

* Bothe, *Zeit. f. Phys.* 13, 368, 1923.
† Crowther and Schonland, *Proc. Roy. Soc.* A, 100, 526, 1922.

by the theory of single scattering. The actual amount of scattering was, however, found to be much too large, and to account for this they suggested that either the law of force close to the nucleus was not that of the inverse square or the electron might have a magnetic moment.

Certain difficulties arose with either of these explanations, and although Schonland* in a later paper dealt with the sources of error in the method and obtained results in fair accord with theory, it seems that the true explanation of the discrepancy between the results of Crowther and Schonland and the predictions of the nuclear theory is that their experiments did not satisfy the conditions for single scattering. The linear relation found by them to hold between I/I_0 and t was not a true one, but depended on the fact that the theoretical curve connecting the amount of scattering with thickness of foil is not far from linear about the middle of the range in which they were working, and their experimental error was sufficient to allow a linear relation to be suggested.

Meanwhile the multiple scattering of β rays had been investigated by Geiger and Bothe†, using a photographic method for measuring the intensity of the β radiation. On the whole their results were in good accord with the corresponding experiments on α rays, and with Crowther's early experiments. For small thicknesses of scattering foil, they found that the scattering was proportional to the thickness of the foil rather than to the square root. This linear relation, though taken to indicate that single scattering conditions were beginning to be realised, was however only an approximation to linearity similar to that found by Crowther and Schonland.

§ 50. "Plural" scattering. The discrepancies in these experimental results led to several attempts‡ to provide a more complete theory of scattering, and in particular to the question of the scattering to be expected under conditions between the two extremes of multiple and true single scattering. In discussing Crowther and Schonland's experiments, Bothe§ pointed out that any method which measures the "half scattering angle" must of necessity deal to a notable extent with particles which have been scattered more than once.

* Schonland, *Proc. Roy. Soc.* A, **101**, 299, 1922.

† Geiger and Bothe, *Zeit. f. Phys.* **6**, 204, 1921.

‡ Cf. Jeans, *Proc. Roy. Soc.* A, **102**, 437, 1923; H. A. Wilson, *Proc. Roy. Soc.* A, **102**, 9, 1923; and the papers of Bothe and Wentzel cited below.

§ Bothe, *Zeit. f. Phys.* **13**, 368, 1923.

This is easily seen to be the case, for if the probability that a particle be scattered through the angle ϕ is $\frac{1}{2}$ the chance that it will be scattered twice through that angle is $\frac{1}{4}$. When the variation in scattering with angle is taken into account, it is clear that the particles scattered through more than ϕ must consist to a large extent of particles which have suffered more than one deflection. The relations which should hold approximately between the "angle of half scattering" Φ and the thickness t of foil have been given above for the cases of single and multiple scattering. Bothe showed that most of Crowther and Schonland's results were in a region intermediate between these relations, but that those obtained under conditions of multiple scattering were in agreement with the results of Geiger and Bothe.

The problem of dealing with a type of scattering between the two extremes of single and multiple scattering, that is, with scattering in which the final deflection of the particle is the result of a few fairly large deflections, has been considered in detail by Wentzel*. Scattering of this type has been termed, for want of a better word, "plural" scattering. Wentzel showed that under certain conditions "plural" scattering may be more important than single scattering, or, in other words, that the deflection of a particle through a certain angle is more likely to occur in a small number of medium steps than in a single large step. The scattering observed under such conditions will therefore be greater than that predicted by the theory of single scattering.

Wentzel showed that Crowther and Schonland's experiments should not be interpreted on the basis of single scattering, but that plural scattering and in some cases multiple scattering was effective. He developed the theory of plural scattering in great detail and applied his calculations to some of the results of Crowther and Schonland. He was able to show that in these cases the theory provided an adequate explanation of the experiments, on the basis of the nuclear atom and the Coulomb law of force. In many of the observations of Crowther and Schonland the scattered particles had suffered so many deflections that the calculations became too cumbrous. One would expect that in these cases the relations of multiple scattering should begin to hold, and Bothe did in fact find, as already noted, that the results could be interpreted in this way.

* Wentzel, *Ann. d. Phys.* 69, 335, 1922.

An example of the application of Wentzel's calculations to a particular case studied by Crowther and Schonland is shown in Fig. 59. The angle of scattering was 6·3°, the scattering element was gold, and the velocity of the incident β particles was 0·85 c, corresponding to an energy of 4·6 × 10^5 electron volts. The gold foils used as scattering material were each 8·5 × 10^{-6} cm. thick. The relation calculated by Wentzel between the amount of scattering and the number of scattering foils is given by the curve, while the circles represent the experimental observations. The variation in amount of scattering with number of foils given by the theory of single scattering is

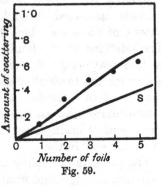

Fig. 59.

shown by the straight line S. As the thickness of foil is increased, the curve for plural scattering rises more rapidly than the single scattering line, with which it is identical for sufficiently thin foils. It is of interest to note that an appreciable portion of the plural scattering curve is approximately linear; a linear relation between amount of scattering and thickness of foil, if determined by experiments over a short range of thickness or with considerable error, does not therefore necessarily imply the presence of purely single scattering.

§ 51. Criterion for single scattering. In the course of this work Wentzel (*loc. cit.*) developed an adequate criterion for the presence of true single scattering. The argument by which he arrives at this is briefly as follows.

If the scattering foil is of thickness t and contains n atoms per c.c., then Wentzel shows that the most probable number of atomic encounters made by the particle in passing through the foil, when the number is small, is the whole number between $\pi p^2 nt$ and $\pi p^2 nt - 1$, where p is the effective radius of the atom. That is to say, the scattering will be mainly single so long as $\pi p^2 nt < 2$. The greatest atomic radius which still fulfils this condition is $p_{max} = \sqrt{2/\pi nt}$. Fixing the effective atomic radius in this way means that we regard all particles which do not penetrate within this distance of the centre of the atom as undeflected. A particle which just penetrates into this region will undergo a deflection ω given, on the nuclear theory, by cot $\frac{1}{2}\omega = mv^2 p_{max}/Ze^2$, where m, v are the mass and velocity of the particle. For single scattering to hold, deflections less than ω must not contribute appreciably to the measured scattering. The conditions of experiment must therefore be so chosen that the angle of scattering ϕ at which the particles are observed is several times greater than ω. Otherwise the possibility arises that an appreciable number of the particles attain their final deflection as a result of several deflections each less than ω.

The relation between ϕ and ω is obtained from the following considerations. The greatest of the neglected deflections are those which correspond to impact parameters between p and $p + \Delta p$, and the number of such deflections suffered by a particle in passing through the foil of thickness t is

$$[\pi (p_{max} + \Delta p)^2 - \pi p^2] nt \cong 2\pi p_{max}\Delta p nt = \frac{4\Delta p}{p_{max}}.$$

Each of the deflections will be less than ω and therefore the resultant deflection of the particle will be less than $4\omega . \dfrac{\Delta p}{p_{max}}$. So long as this is small compared with the observed angle of scattering ϕ, the deflections less than ω will not make an appreciable contribution to the scattering. This condition will be fulfilled even when Δp is of the same order as p_{max}, provided that ϕ is at least a small multiple of 4ω. This is then a sufficient criterion for the presence of single scattering under the given experimental conditions—viz. a thickness t of scattering foil containing n atoms per c.c., and an angle of scattering ϕ.

The criterion may be made more precise by comparison with experimental data. Taking, for example, the α ray experiments of Chadwick (§ 44), where the results show that the conditions for single scattering were adequately realised, we find that $\phi/4\omega$ is at least 7. In the β ray experiments of Crowther and Schonland the highest value of $\phi/4\omega$ was about 3. In this case Wentzel's calculations showed that the scattering was mainly single but that for smaller values of $\phi/4\omega$ plural scattering began to be important. This estimate of the limiting value of $\phi/4\omega$ was confirmed in the experiments of Rose on the scattering of α particles through small angles, from which it appeared that the scattering was chiefly single so long as ϕ was not less than 10ω, and in the experiments of Schonland described below (§ 53).

§ 52. Calculation of the single scattering of β particles*. Before describing experiments in which the single scattering of β rays has been studied, we must proceed to consider what modifications to the expressions developed in § 42 for the scattering of α particles are necessary when they are applied to β particle scattering.

The relations of § 42 were obtained on the assumption that the mass of the particle remained constant and that its kinetic energy was given by $\tfrac{1}{2}MV^2$. These assumptions are sufficiently correct in the case of the α particles, but they cannot be retained in considering the scattering of the β particles. Since the mass of the β particle increases with its velocity, some of the energy acquired by the particle in moving under the attracting force of the nucleus will be used in increasing the mass, and the velocity will not increase to the extent it would if the mass remained constant. Consequently the particle spends a longer time in the nuclear field and its deflec-

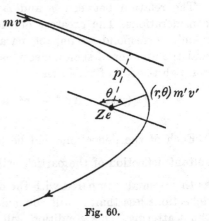

Fig. 60.

tion is greater than that given by the simpler theory. The variation of mass with velocity changes the character of the orbit of the particle, and in certain cases the orbit becomes a spiral going right into the deflecting centre, representing a capture of the β particle by the nucleus.

The modifications in the orbits caused by the change of mass have been considered by Darwin†, whose calculation has been given in a convenient form by Schonland‡.

Since the mass of the electron is very small except when its velocity is close to that of light, the motion of the nucleus during the encounter may be neglected.

Let m, v be the initial mass and initial velocity of the β particle, m', v' the mass and velocity at the point (r, θ) (Fig. 60). Then

$$m = m_0 \left(1 - v^2/c^2\right)^{-\tfrac{1}{2}}, \quad m' = m_0 \left(1 - v'^2/c^2\right)^{-\tfrac{1}{2}}.$$

---

* Mott, *Proc. Roy. Soc.* A, **124**, 425, 1929, has calculated the scattering of β particles by an atomic nucleus on the wave mechanics. Some points may be noted: the spiral orbits of the classical theory do not occur, and the beam of scattered particles is polarised. The scattering calculated in this way is rather less than that observed experimentally, but the difference may perhaps be accounted for by the effects of radiation.

† Darwin, *Phil. Mag.* **25**, 201, 1913.   ‡ Schonland, *Proc. Roy. Soc.* A, **101**, 299, 1922.

From the conservation of angular momentum we have

$$m'r^2\dot\theta = mvp,$$

and from the conservation of energy

$$m_0c^2\left\{\frac{1}{\sqrt{(1-v'^2/c^2)}}-1\right\} = m_0c^2\left\{\frac{1}{\sqrt{(1-v^2/c^2)}}-1\right\}+\frac{Ze^2}{r}.$$

From these two equations we can deduce that the equation of the orbit is

$$p\gamma/r = \sin\gamma\theta + (1-\cos\gamma\theta)\,p_0/(p\gamma\beta) \qquad \ldots\ldots(1),$$

where $\gamma = \sqrt{1-p_0{}^2/p^2}, \quad \beta = v/c,$

and $p_0 = Ze^2\sqrt{1-\beta^2}/m_0c^2\beta.$

If we write $p = p_0 \operatorname{cosec}\psi$, this becomes

$$p\gamma/r = \sin(\theta\cos\psi) + [1-\cos(\theta\cos\psi)]\tan\psi\,.\,1/\beta \quad\ldots(2).$$

If the deflection of the particle is ϕ, then $\pi + \phi$ is the angle between the asymptotes and $\frac{1}{2}(\pi+\phi)$ is the vectorial angle of minimum r. Here $\theta = \frac{1}{2}(\pi+\phi)$, and therefore

$$\beta\cot\psi = \tan[\pi - \tfrac{1}{2}(\pi+\phi)\cos\psi] \qquad \ldots\ldots(3),$$

and from this ψ can be found in terms of ϕ and β.

If the scattering foil contains n atoms per c.c. and has a thickness t cm., the chance that the particle will be scattered through an angle greater than ϕ will be

$$q = \pi p^2 n t = \pi n t\,(p_0 \operatorname{cosec}\psi)^2$$

$$= \pi n t\left[Ze^2\frac{\sqrt{1-\beta^2}}{m_0c^2\beta}\,.\operatorname{cosec}\psi\right]^2 \qquad \ldots\ldots(4),$$

where ψ is given by the above relation.

This expression has often been written in the following form:

$$q = \pi n t\,.\left(\frac{Ze^2}{2T}\right)^2.\cot^2\tfrac{1}{2}\phi\,.f^2\,(\phi,\beta),$$

where T is the kinetic energy of the β particle, and where

$$f(\phi,\beta) = \frac{2\,(1-\sqrt{1-\beta^2})\operatorname{cosec}\psi}{\beta\cot\tfrac{1}{2}\phi}.$$

The term $f^2\,(\phi,\beta)$ has been there regarded as the correction to be applied to the simpler theory in which the variation of mass of the scattered particle is neglected. This is hardly justified, for it is equivalent to calculating the initial velocity of the particle from its energy by the relation $\tfrac{1}{2}mv^2 = $ Energy.

It is perhaps preferable to use the form

$$q = \pi n t . \left(\frac{Ze^2}{mv^2}\right)^2 \cot^2 \tfrac{1}{2}\phi . \left(\frac{\beta \cosec \psi}{\cot \tfrac{1}{2}\phi}\right)^2 \qquad \ldots\ldots(5).$$

The correcting term, $\left(\dfrac{\beta \cosec \psi}{\cot \tfrac{1}{2}\phi}\right)^2$, is then nearly unity for moderate values of ϕ and β, and in the comparison of theory with experimental results it can often be neglected.

The orbits represented by equation (1) may be divided into two kinds, according as p is greater or less than p_0.

When p is greater than p_0 the orbit is a curve modified from an hyperbola having the nucleus in the focus, the modification consisting in increasing every element of vectorial angle in the ratio of 1 to $\sqrt{1 - p_0{}^2/p^2}$. For large values of p or high velocity of the particle, the orbit is nearly a true hyperbola. For smaller values of p or smaller velocity, the path of the particle tends to wind round the nucleus and when p/p_0 approaches unity the particle may describe several turns before it can escape.

When p is less than p_0 the orbits have the form of a spiral and the particle winds round and round the nucleus and is finally captured. It was pointed out by Darwin that these spiral orbits cannot in fact result in a capture of the particle, for otherwise a transmutation of the elements under β particle or cathode ray bombardment would be easily observed. It seems probable that the particles which should describe these spiral orbits escape from the atom without serious loss of energy. It is of interest to note in this connection that if the electron possesses a magnetic moment, however small, no capture of the electron can occur*.

In these calculations it has been assumed that the particles suffer no loss of energy by radiation processes. We know from experiment that some radiation does take place in collisions between β particles and matter, and the theoretical work of Kramers† on the quantum theory of "white" X radiation suggests that radiation occurs from orbits of the kind considered above. While a direct calculation of the result of a radiation of energy in modifying the orbits seems at present impossible, it is easy to see that the general effect must be to distort the orbits still farther from the hyperbola and to increase the amount of scattering. At the same time the critical impact

---

* Bothe, *Zeit. f. Phys.* **44**, 543, 1927.

† Kramers, *Phil. Mag.* **46**, 845, 1923. According to Kramers, the effects of radiation will be inversely proportional to the atomic number of the scattering nucleus. They are thus most serious for the light elements.

parameter p_0 would increase and with it the number of particles which should describe spiral orbits. We should expect, on general grounds, that when p is large compared with p_0 the radiation loss would be small and that the equations developed above would be valid.

Another factor which has been neglected in the above calculation is the influence of a possible polarisation of the nucleus by the approaching β particle (cf. § 62). The polarisation will be less than in the case of the α particle, on account of the smaller charge carried by the β particle. The polarisation forces will in general be small,

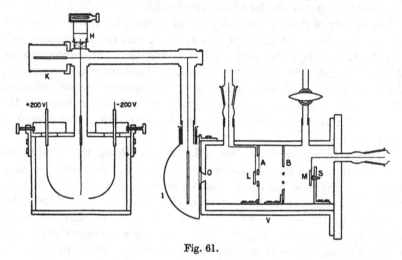

Fig. 61.

except in the case of scattering of fast β particles through large angles. Their effect will be to increase the critical impact parameter p_0, and the increase in p_0 will be the more pronounced the greater the velocity of the particle. The effect of polarisation is thus in the same direction as that of radiation processes, to distort the orbits from the hyperbola and to increase the number of spiral orbits. An exact calculation has not yet been given.

§ 53. **Experiments on single scattering.** Experiments in which the conditions for single scattering have been realised have been carried out by Chadwick and Mercier* and by Schonland.

In the former series of experiments the method developed earlier by Chadwick for measuring the scattering of α rays was used. The apparatus is shown in Fig. 61 and its general arrangement will be

* Chadwick and Mercier, *Phil. Mag.* **50**, 208, 1925.

clear from the previous description. The annular limits of the scattering foil, held on the support A, were 12° and 20°, so that the scattering was measured between angles of 24° and 40°. The intensities of both the scattered particles and the direct beam from the source S were measured in the hemispherical ionisation chamber I. Thus the scattered and the direct particles had the same length of path in the chamber, and the ionisation currents gave to a first approximation the ratio of the numbers of particles in the scattered and direct beams. The currents were measured by means of the electrometer shown on the left of the figure.

The source of β rays was a deposit of radium E, which emits very little γ radiation. No attempt was made to sort out a pencil of rays of definite velocity. The primary beam was therefore heterogeneous. While this lack of homogeneity is of little moment when comparing the scattering by different elements, it becomes very important when the actual fraction of scattered particles is to be estimated, for the chance of scattering varies inversely as the square of the energy of the particles.

The scattering produced by foils of four elements—aluminium, copper, silver and gold—was examined. For the thinnest foils used Wentzel's criterion for single scattering was fulfilled, and the scattering was observed to be proportional to the thickness of foil. This is shown by the results obtained with foils of aluminium, which are given in the curve of Fig. 62. The ordinates represent the ratio of the ionisation currents due to the scattered particles to those for the direct particles and the abscissae the value of nt for the foil used. If the conditions for single scattering are fulfilled the curve should be a straight line passing through the origin. It is seen that this is true only for small values of nt. The effect of plural scattering begins to be appreciable for an aluminium foil of nt about $1\cdot2 \times 10^{19}$, corresponding to a weight of $0\cdot54$ mg./sq. cm. or a thickness of $2\,\mu$.

The amounts of scattering shown by thin foils of the four elements examined are compared in the following table.

Element	Atomic number Z	$\dfrac{\text{Scattered}}{\text{Direct}}$	Ratio	Corrected	Z^2/Z^2_{Al}
Aluminium	13	0·00234	1	1	1
Copper	29	0·0113	4·83	5·03	4·98
Silver	47	0·0302	12·9	13·5	13·1
Gold	79	0·075	32·1	34·1	37·0

The ratio of the scattered to the direct particles given in the third column is calculated for a foil of $nt = 10^{18}$ of each element from measurements similar to those shown in Fig. 62 in the case of aluminium. In the fourth column the amounts are compared taking the amount for aluminium as unity. Strictly, in order to compare the nuclear scattering for the different elements an allowance should be made for the scattering effect of the electrons in the atom. In this way the corrected ratios of the next column are obtained. It will be seen that the correction is very small. Comparison with

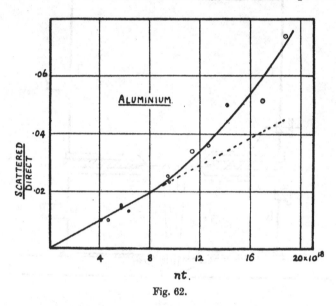

Fig. 62.

column 5 shows that the scattering effect of an atomic nucleus is proportional to the square of its charge, as required by the theory, except in the case of gold. The discrepancy in this case was explained by a calculation of the orbit of the β particle round the gold nucleus. Under the experimental conditions the path of the particle lies outside the K shell of the atom, and thus the effective nuclear charge is less than the actual charge by about two units.

The actual amount of scattering observed was compared with that to be expected on the theory. This comparison could not be made with much accuracy owing to the heterogeneity of the primary pencil of β rays. The β rays of radium E differ widely in velocity; the distribution in the magnetic spectrum extends up to 5000 $H\rho$ but has a pronounced. maximum at 2200 $H\rho$. Since the chance of scattering under given conditions is inversely proportional to the fourth power of the velocity of the particle, the calculation of the scattering for such a distribution of particles cannot be satisfactory. Such comparison as was possible showed that there was no marked discrepancy between theory and experiment. The observed scattering was about 25 per cent. greater than the calculated scattering, a divergence within the errors of calculation. It should be noted that under the

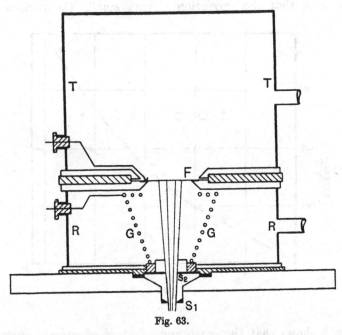

Fig. 63.

conditions of these experiments the impact parameter p of the orbits had a comparatively large value. The spiral orbits do not arise, and the effects of loss of energy as the particles describe their orbits in the nuclear field are not likely to be serious.

A more extensive series of measurements was made by Schonland[*]. The essential part of his apparatus is shown in Fig. 63.

A homogeneous beam of fast cathode rays entering through the slits S_1, S_2 fell on the scattering foil F situated at the junction of two evacuated brass cylinders R and T. The top cylinder T received the transmitted beam of particles and the bottom one R received the

* Schonland, *Proc. Roy. Soc.* A, **113**, 87, 1926.

particles scattered by the foil through more than 90°. The currents passing to the cylinders were measured by a balance method. The negatively charged grid G of fine wires prevented the disturbing effect of slow secondary rays emitted from the foil. Some of the scattered particles strike the ebonite base of this grid so that the scattering was actually measured between the angles 90° and 161°. The energy of the primary particles was varied from 30,000 to 80,000 volts.

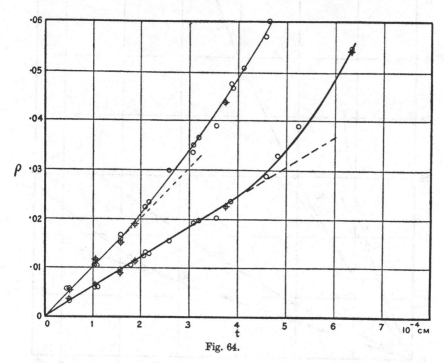

Fig. 64.

The results obtained showed that the conditions for single scattering were fulfilled for the thinnest foils used. Fig. 64 shows the results obtained with various thicknesses of aluminium foil, the ordinates giving the fraction of scattered particles. The upper curve was obtained with particles of energy 59,900 volts, or $\beta = 0.447$, and the lower curve with particles of energy 77,300 volts, $\beta = 0.497$. It will be seen that the linear relation between the amount of scattering and the thickness of material holds only over a certain range. This range, in which the conditions for single scattering are fulfilled, is shorter in the upper curve than in the lower, as we should expect.

The upward bend of the curves indicates the appearance of plural scattering. It is of interest to note the value of Wentzel's criterion for single scattering at the points where the experiments show that single scattering is giving place to plural scattering. From the upper curve

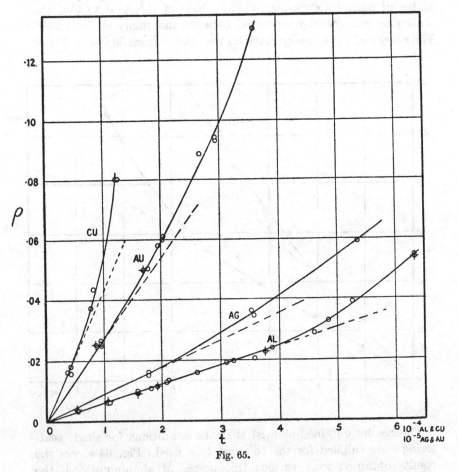

Fig. 65.

we find $\phi/4\omega = 2.7$ and from the lower curve we find $\phi/4\omega = 3.1$. The average value $\phi/4\omega = 2.9$ is in fair agreement with the previous estimate of Rose (§ 51).

The scattering powers of four elements, viz. aluminium, copper, silver and gold, were compared, with the results shown in Fig. 65. (Two scales of thickness are employed, one for aluminium and copper, and one for silver and gold, as shown in the figure.) It will be noted that the curves for copper, silver and gold are linear only over the initial portion of the range of thickness examined, indicating the early incidence of plural scattering. The dotted lines were taken to represent the effects of pure single scattering.

The comparison of these measurements with the theory of single scattering is shown in the following table.

Element	ρ	n	Z	ρ/nZ^2
Aluminium	0·000612	6·04 × 10²²	13	6·0 × 10⁻²⁹
Copper	0·00043	8·51	29	6·0
Silver	0·00087	5·90	47	6·7
Gold	0·00272	5·93	79	7·3

The second column gives the fraction ρ of particles scattered by a foil of thickness 1×10^{-5} cm. of the material in the first column. The third column gives the value of n, the number of atoms per c.c. in the foil, the fourth Z the atomic number of the element. The figures in column 5 show that the quantity ρ/nZ^2 is constant within the experimental error, as is required by the theory, except in the case of gold. The discrepancy in the case of gold was attributed to the presence of secondary rays emitted by the gold foil under the bombardment of the primary rays.

From these results Schonland calculated the values of the nuclear charges of the scattering elements and obtained a satisfactory agreement with the atomic numbers.

He also investigated the variation of scattering with the energy of the particles. He used an aluminium foil of 5×10^{-5} cm. thickness as scattering agent, and varied the energy of the incident particles from 39 kilovolts to 77·3 kilovolts. The conditions prevailing in these experiments were, however, unsuitable for an adequate test of the theory in respect of the absolute amount of scattering and its variation with the energy of the particles, for the impact parameter p had too small a value. For example, the measurements with aluminium foils were made under conditions such that all the scattered particles approached a nucleus in an orbit with a parameter p less than $2·9p_0$. An appreciable fraction of the scattered particles should therefore have described looped orbits and the forbidden spiral orbits. The probability of radiation processes was also thereby increased.

In a later paper Schonland* modified his conclusions to allow for these conditions. On the reasonable assumption that the particles

* Schonland, *Proc. Roy. Soc.* A, **119**, 673, 1928.

which should theoretically describe spiral orbits are in effect scattered equally in all directions, and, neglecting any radiation losses, he showed that the variation of scattering with the energy of the particles was as given by the theory of the previous section, and that the fraction of scattering was about 20 per cent. greater than the calculated amount. This difference may be due to the effect of radiation processes.

It is clear from what has been said that an adequate test of the theory of β ray scattering has not yet been carried out. We can, however, say that there is no discrepancy between the experimental results and the predictions of the theory which cannot be reasonably ascribed to the conditions of experiment or to factors the influence of which cannot be exactly calculated. The scattering varies with the thickness of the scattering material and its atomic number in the way demanded by the theory, but the amount of scattering is probably a little too great. This difference has been attributed to the effect of radiation of energy by the particles in their orbits. The scattering varies with the energy of the incident particles in approximately the expected way. No examination of the angular distribution of the scattered particles has been made.

It may be pointed out here that for a strict test of the theory the following conditions must be satisfied—the scattering must be single, particles homogeneous in velocity must be used, and the impact parameter p must be large compared with the critical value p_0. This means in practice the use of the thinnest foils of scattering material, moderate angles of scattering, and particles of an energy of at least 200,000 volts.

§ 54. **Scattering of β particles by electrons***. In the previous calculation only the scattering effect of the nucleus of the atom has been considered, not that of the electrons. The electronic scattering may, however, be appreciable in the case of scattering by light elements. If we consider the electron in the atom to be at rest before the encounter and take its mass equal to that of the incident β particle, then the chance that the β particle will be deflected through an angle greater than ϕ is, neglecting the relativity correction,

$$q_e = \left(\frac{2e^2}{mv^2}\right)^2 . \pi n t \cot^2 \phi .$$

* See, however, Mott, *Proc. Roy. Soc.* A, **126**, 259, 1930, where the collision of two electrons is considered from the standpoint of quantum mechanics.

If the atomic number of the scattering atom is Z, then the total electronic effect of an atom is given by*

$$q_{\bullet} = Z . \left(\frac{2e^2}{mv^2}\right)^2 . \pi n t \cot^2 \phi.$$

The correction for the variation in mass will not generally be of importance and will diminish the scattering, for in these collisions the β particle is retarded in its motion, not accelerated as in a collision with the positive nucleus.

Comparing this expression with that for the nuclear scattering, we see that the effect of the electrons in the atom is the more marked the smaller the atomic number of the atom. In fact, in the case of hydrogen the electron is almost as efficient in scattering as the nucleus when the angle is small, while for gold the amount of the electronic scattering is only about 1 per cent. of that of the nucleus.

An investigation of the electronic scattering of β rays has recently been made by M. C. Henderson†, by comparing under the same experimental conditions the scattering of hydrogen and helium with that of heavier atoms such as oxygen, argon, bromine and iodine. Two experimental arrangements were used, one similar to that of Crowther, the other the annular ring method adapted for scattering in a gas. The results showed that between angles of 10° and 30° the electronic scattering was about three times that given by theory. This excess scattering may arise from the presence of magnetic forces around the electron, in addition to the ordinary electrostatic force. The suggestion that the electron possesses a magnetic moment has been made from time to time by various workers, but it has received its strongest support from considerations developed in quite another field of work.

Uhlenbeck and Goudsmit‡, in 1925, suggested that if the electron could be considered to be rotating on its axis, another degree of freedom would be provided that could be quantised in the usual way to yield a fourth quantum number, needed to interpret the anomalous Zeeman effect and multiplet structure. Their hypothesis has had a considerable degree of success in explaining various anomalies in optical spectra. On this hypothesis the rate of spin of the electron

---

* At a given angle ϕ there will be present also electrons projected by the collision of the β particles which are scattered through the angle $\frac{1}{2}\pi - \phi$. These will have the same velocity as the β particles scattered at angle ϕ.

† M. C. Henderson, *Phil. Mag.* **8**, 847, 1929. Cf. also Williams and Terroux, *Proc. Roy. Soc.* A, **126**, 289, 1930.

‡ Uhlenbeck and Goudsmit, *Naturwiss.* **13**, 953, 1925; *Nature*, **117**, 264, 1926.

is such as to give the electron a magnetic moment of one Bohr magneton ($= eh/4\pi m_0 c$). The magnetic force should therefore be much greater than the electrostatic force for a comparatively large distance around the electron, and we should expect the collisions between electrons to be profoundly modified in consequence. The exact calculation of the effect of magnetic forces is very complicated and has not so far been accomplished. Henderson's experiments certainly point to a departure from the normal electrostatic force, but until the variation of scattering with angle has been more closely investigated it is not justifiable to draw any more specific conclusion than this. Perhaps, as Dirac* has suggested, it may be possible to account for all the observed facts without the assumption of a magnetic electron, by taking account in the equations of the wave-mechanics of the change of mass of the electron with its velocity. He has already shown that such a method can lead to a solution of the problem for which the spinning electron was postulated.

A further point which must be mentioned in discussing the question of electronic scattering is the possibility that the interaction of a β particle with an atomic electron may not be adequately described by the laws of classical mechanics. It is well known that the collisions of slow electrons with the outer electrons of an atom follow the laws of quantum mechanics. In analogy with these, we should expect that the classical laws would suffice for the collisions of fast β particles with weakly bound electrons and that the failure of ordinary mechanics would be the more marked the closer the energy of the β particle approaches to the ionisation or resonance energy of the atomic electron. This supposition is supported by some experiments of Webster† on the excitation of characteristic X rays by cathode particles, in which he found that the probability of a quantum absorption of the cathode particle jumped to a maximum when its energy just exceeded the excitation energy and then decreased continuously as the energy of the particle was further increased. In the case of β rays the energy of the particles is in general much greater than the energies of the electron levels in the scattering atom, except in the case of heavy elements. Inelastic collisions should therefore be comparatively rare, and should occur to an appreciable extent only with the K electrons of heavy atoms, such as gold. It is difficult to predict the effect of an inelastic collision on the scattering of the

---

* Dirac, *Proc. Roy. Soc.* A, **117**, 610, 1928; **118**, 351, 1928. Cf. also Iwanenko and Landau, *Zeit. f. Phys.* **48**, 340, 1928.

† D. L. Webster, *Phys. Rev.* **7**, 599, 1916.

particle, but it may perhaps appear as an excess scattering super-imposed on the ordinary scattering.

Some experimental evidence of the existence of these inelastic collisions has been obtained by Bothe*, in observations of the scattering of the β rays of radium E by a thin gold foil. He found that the scattering was greater than was to be expected from calculation, and that the excess particles consisted at least in the greater part of particles of very small velocity. These were not present in the particles scattered from aluminium under the same conditions. Bothe, therefore, considered them to be particles which had suffered a quantum absorption in the K level of gold, in which the binding energy is about 80,000 volts. Inelastic collisions of a similar kind may perhaps account for the excess scattering observed by Schonland in the case of gold, although the energy of the incident particles used by him was 77,300 volts, rather less than the binding energy of the K level and much greater than that of the L level, which is about 14,000 volts.

§ 55. **Evidence of scattering in β ray tracks.** It is of interest to note that photographs of the tracks of β particles in an expansion chamber show very clearly the types of scattering discussed in the previous sections. Both Bothe† and C. T. R. Wilson‡ have published such photographs and have pointed out that the tracks show the abrupt changes in direction, sometimes through more than a right angle, due to single scattering, and the slower curvature due to the cumulative effect of a large number of smaller deflections. Bothe observed, in all, 17 cases in which a β particle was deflected in a single encounter through more than 45°, 6 of these being through more than 90°. He calculated that under the conditions of experiment the numbers to be expected on an average were 14·4 and 2·5 respectively. Wilson found that out of 503 β ray tracks 44 showed deflections greater than 90° in the portion between 0·5 cm. and 1·5 cm. from the end of their path. The calculated number was about 54. We may conclude that the frequency of the large single deflections is roughly in accord with theory.

It has occasionally been pointed out that some β ray tracks show a gradual curvature, which often appears to persist in one direction only. To explain this, somewhat strained hypotheses have from time to time been made. Bothe, after a careful study of such tracks, came

* Bothe, *Zeit. f. Phys.* **13**, 368, 1923.
† Bothe, *Zeit. f. Phys.* **12**, 117, 1922.
‡ C. T. R. Wilson, *Proc. Roy. Soc.* A, **104**, 192, 1923.

to the conclusion that the mechanism of multiple scattering was able to account for all the observations. The persistent curvature in one direction was more apparent than real and due to a tendency of an observer to see a succession of random points as a smooth curve.

These photographs also give evidence of the scattering of β rays by electrons, shown by the occurrence of forked tracks. These are due to a close collision of the β particle with an atomic electron, which acquires sufficient energy to escape from its orbit and to produce a track comparable with that of the original β particle. If the masses of the atomic electron and the β particle were equal, and if energy and momentum are conserved in the collision, the angle of the fork should be exactly 90° and the three arms should be coplanar. The relativity change of mass of the β particle will make the angle of the fork rather less than 90°; for example, if the initial velocity of the β particle is 0·9c (700,000 volts), the angle should be about 76° in the special case when the two branches make equal angles with the stem of the fork, that is, when the β particle transfers half its energy to the struck electron.

These forks are of comparatively rare occurrence. Bothe* found on the average one fork for every metre of β ray track. He obtained in his photographs only eight clear examples of β ray forks, a number sufficient to give only a very rough estimate of the frequency of electron scattering, and to show that there was no serious departure from theoretical expectations. It is impossible to distinguish after the collision between the recoil electron track and that of the incident β particle. The chances are that the least deflected prong of the fork is due to the β particle, for a small deflection is more likely than a large one.

Examples of β ray tracks obtained in a Wilson chamber are shown in Plates IV and VIII. Plate VIII, fig. 2, shows a straight track due to a fast β particle, and fig. 1 of the same plate a curved track due to a slow particle which has suffered many small deviations in its passage through the air. Plate VIII, fig. 3, shows a track in which the β particle has suffered a large abrupt deflection due to collision with a single nucleus. In Plate IV, fig. 3, a stereoscopic pair of photographs is reproduced; the β particle has here collided with an atomic electron and it will be seen that the angle between the directions of motion of the β particle and of the recoiling electron immediately after the collision is very close to a right angle.

* Bothe, *Zeit. f. Phys.* 12, 117, 1922.

PLATE VIII

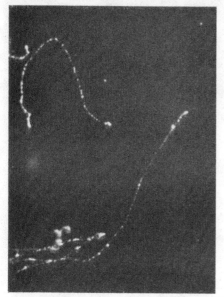

Fig. 1. Curved β tracks.

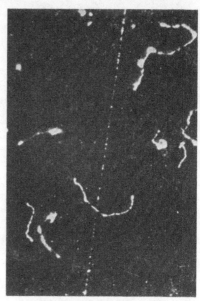

Fig. 2. Straight β track.

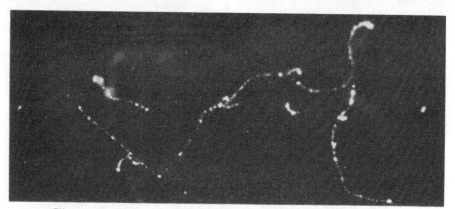

Fig. 3. β track showing nuclear deflection and electronic collision.

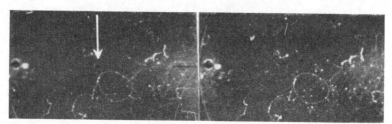

Fig. 4. Stereoscopic pair showing loss of energy of β after collision
with a nucleus.

In some experiments of Skobelzyn* a strong magnetic field was applied to bend the β rays while in the expansion chamber and so to reveal their velocities. Skobelzyn was thus able to measure the velocity of the particle both before and after a collision with a nucleus as shown by a large single deflection. He found that in a few cases the velocity of the particle was definitely less after the collision than before. In one of these cases the particle was deflected through an angle of 30° in a collision with, presumably, a nitrogen nucleus. The curvature of the track before the collision gave a value for $H\rho$ of 3330, corresponding to an energy of 610 kilovolts; after the collision an $H\rho$ of 2700, corresponding to 450 kilovolts. In another case, a picture of which is given in Plate VIII, fig. 4, the particle had initially an energy of about 900 kilovolts, corresponding to a range in air of some metres. After the collision with a nucleus resulting in a deflection of about 90° the energy of the particle was much less. The whole of the subsequent path of the particle was visible in the photograph and had a length of about 13 cm., which corresponds to an energy of less than 100 kilovolts. The β particle thus suffered a considerable loss of energy in passing through the nuclear field. This energy was presumably emitted in the form of radiation, by the processes considered by Kramers† and Wentzel‡. These observations of Skobelzyn bring out very clearly a point which has been mentioned previously, that radiation processes must be taken into account in an adequate theory of the scattering of β particles.

In the experiments of Wilson (*loc. cit.*) it was noted that in a fair proportion of the forked tracks a branch of considerable range occurred when the primary β ray suffered only a very small deviation, and that momentum was not conserved in the motion of the particles. These observations have as yet received no satisfactory explanation. Williams and Terroux§, working with β rays of much greater initial energy, found no evidence of such anomalous collisions.

* Skobelzyn, *Zeit. f. Phys.* **43**, 354, 1927.
† Kramers, *Phil. Mag.* **46**, 845, 1923.
‡ Wentzel, *Zeit. f. Phys.* **27**, 257, 1924.
§ Williams and Terroux, *Proc. Roy. Soc.* A, **126**, 289, 1930.

CHAPTER IX

THE COLLISIONS OF α PARTICLES WITH LIGHT ATOMS

§ 56. In the previous chapter those experiments on the scattering of α particles in passing through matter were described which confirm the essential assumption of the present theory of atomic structure, that the atoms of matter contain a positively charged nucleus having practically the whole mass of the atom associated with it. The experiments showed that the magnitude of the positive charge of the nucleus was fixed by the atomic number of the atom, or its position in the ascending series of the chemical elements. In the collisions examined in these experiments, in which no elements of smaller atomic number than copper were investigated, the atomic nucleus and the α particle behaved as point charges repelling each other with a force varying inversely as the square of the distance between them.

It is to be anticipated that divergences from the Coulomb law of force would appear when the collisions between the atomic nucleus and the α particle are sufficiently close, for the nuclei are generally supposed to have a complex structure and to be built up in some way from two common units, the electron and the hydrogen nucleus or proton. Since the closest distance of approach to a nucleus of an α particle of given energy is proportional to the charge of the nucleus, it is in the collisions of fast α particles with the nuclei of the lighter elements that deviations from the normal law of force might be expected to be most readily discovered. In this chapter we shall give an account of experiments in which such collisions have been studied and which provide definite evidence of a departure from the Coulomb law of force at very small distances between the colliding particles. Before proceeding to these experiments it will be convenient to discuss the question of the collision of an α particle with a nucleus in more detail than in the preceding chapter.

The general scattering relation. In considering the collisions of an α particle with an atomic nucleus in § 42, it was assumed that the mass of the nucleus was infinitely large compared with that of the α particle. The nucleus remained at rest during the collision, and the α particle preserved its original kinetic energy. While this assumption was justifiable in describing the scattering of α particles by heavy atoms, it is obviously untrue for the collisions with light atoms.

We shall now deal with the general case and include the fact that the nucleus is set in motion in collision. The calculations for the general case have been given by Darwin*. Let the charge and mass be E, M for the α particle, Ze, m for the nucleus. Let V be the initial velocity of the α particle, v its velocity after impact. The nucleus is at rest before the collision. There are four unknown quantities to be determined by the collision, the velocity v and deflection ϕ of the α particle, and the velocity u and direction of motion θ of the struck nucleus (Fig. 66). If momentum and energy are conserved in the

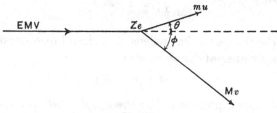

Fig. 66.

translational motion—that is, there is no energy lost by radiation or transferred to rotational motion—the direction of motion of the nucleus will be in the plane fixed by the initial and final directions of the α particle, and we shall have the following relations:

$$\left.\begin{array}{l} MV = Mv \cos \phi + mu \cos \theta \\ 0 = Mv \sin \phi - mu \sin \theta \\ MV^2 = Mv^2 + mu^2 \end{array}\right\} \quad \ldots\ldots(1).$$

From these we derive

$$\left.\begin{array}{l} u = 2V \dfrac{M}{M+m} \cos \theta \\[2mm] \tan \phi = \dfrac{m \sin 2\theta}{M - m \cos 2\theta} \end{array}\right\} \quad \ldots\ldots(2),$$

$$\left.\begin{array}{l} v = \dfrac{V}{M+m}[M \cos \phi \pm \sqrt{m^2 - M^2 \sin^2 \phi}] \\[2mm] \tan \theta = \dfrac{m \cot \phi \pm \sqrt{m^2 \operatorname{cosec}^2 \phi - M^2}}{M + m} \end{array}\right\} \quad \ldots\ldots(3).$$

* Darwin, *Phil. Mag.* 27, 499, 1914.

In the latter equations the positive sign is to be taken when $m \geq M$. The negative sign therefore applies only in the case of collisions of α particles with hydrogen nuclei, or with electrons.

Certain aspects of the collisions of an α particle with an electron have been considered in chapter v. Since in such a collision $m^2 - M^2 \sin^2 \phi$ becomes negative when the value of ϕ is greater than about 30 seconds, when we substitute for m the mass of the electron, we see that an α particle cannot be deflected in a collision with an electron through an angle greater than this.

In the calculation of the orbit of an α particle deflected by an infinitely heavy nucleus surrounded by a Coulomb field of force the following relation was deduced between the impact parameter p and the deflection ϕ':

$$\tan \tfrac{1}{2}\phi' = \frac{1}{p} \cdot \frac{ZeE}{MV^2},$$

where Ze is the charge on the nucleus. The closest distance of approach in this orbit, or apsidal distance, was

$$p \cot \tfrac{1}{4}(\pi - \phi').$$

By a well-known process in the theory of orbits the more general case in which the nucleus is allowed to move during the collision can be obtained from this particular case by substituting $1/M + 1/m$ for $1/M$ in the relations. The results then apply to the relative motion of the α particle and a nucleus of mass m.

Thus if we write
$$\tan \tfrac{1}{2}\phi' = \frac{1}{p} \cdot \frac{ZeE}{V^2} \cdot \left(\frac{1}{M} + \frac{1}{m}\right),$$

the deflection ϕ' of the α particle is relative to the moving nucleus. Now

$$\tan \phi' = \frac{v \sin \phi + u \sin \theta}{v \cos \phi - u \cos \theta},$$

and using the relations (1) and (2) this can be reduced to

$$\tan \phi' = -\tan 2\theta.$$

Hence
$$\theta = \tfrac{1}{2}(\pi - \phi').$$

From this relation and (2) we obtain, by eliminating θ,

$$\cot \phi = \cot \phi' + \frac{M}{m} \operatorname{cosec} \phi',$$

from which the actual deflection ϕ is easily obtained from the relative deflection ϕ'.

Substituting the value of ϕ' in terms of θ we get the relation

$$p = \frac{ZeE}{V^2}\left(\frac{1}{M} + \frac{1}{m}\right)\tan\theta \qquad \ldots\ldots(4).$$

The apsidal distance is given by

$$D = \frac{ZeE}{V^2}\left(\frac{1}{M} + \frac{1}{m}\right)(1 + \sec\theta) \qquad \ldots\ldots(5).$$

The fraction of α particles scattered through an angle greater than ϕ by a foil of very small thickness t (so that the scattering is single) and containing n atoms per c.c. will be

$$q = \pi p^2 nt = \pi nt\,\mu^2\tan^2\theta \qquad \ldots\ldots(6),$$

writing $$\mu = \frac{ZeE}{V^2}\left(\frac{1}{M} + \frac{1}{m}\right),$$

and where $\tan\theta$ is given in terms of ϕ by (3) above.

If Q α particles per second fall on the scattering foil, then the number scattered between ϕ and $\phi + d\phi$ will be $Qnt\,.\,2\pi p\,.\,dp$ and the number of these observed within a solid angle ω will be

$$q = Qnt\omega\,\frac{p\,dp}{\sin\phi\,d\phi}.$$

Substituting the value of p in terms of ϕ deduced from (4) and (3) we find

$$q = Qnt\omega\,\frac{Z^2 e^2 E^2}{M^2 V^4}\,\mathrm{cosec}^3\,\phi\,\frac{[\cot\phi \pm \sqrt{\mathrm{cosec}^2\,\phi - (M/m)^2}]^2}{\sqrt{\mathrm{cosec}^2\,\phi - (M/m)^2}}$$
$$\ldots\ldots(7).$$

This is the general scattering relation when the motion of the nucleus is taken into account. For all collisions except those in helium or hydrogén $m > M$ and the positive sign is to be taken. The above expression can then be expanded in powers of M/m and, writing $b = \dfrac{2ZeE}{MV^2}$ as before,

$$q = \tfrac{1}{16}Qntb^2\omega\,\{\mathrm{cosec}^4\,\tfrac{1}{2}\phi - 2\,(M/m)^2 + (1 - \tfrac{3}{2}\sin^2\phi)\,(M/m)^4\}$$
$$\ldots\ldots(8).$$

The first term of this expansion is that given by equation (6) of § 42, where the motion of the nucleus was neglected. Since $\mathrm{cosec}^4\,\tfrac{1}{2}\phi$ is always greater than 1, the effect of the correcting terms due to motion of the nucleus is very small except for the light atoms. Even in the case of scattering by aluminium the correction is never greater than 4 per cent.

In the following table are given the closest possible distances of approach, i.e. the apsidal distances in the case of a head-on collision, of the α particle of radium C, of velocity $1\cdot922 \times 10^9$ cm./sec., to the nuclei of certain atoms. These distances of approach are calculated on the assumption that the α particle and the nucleus behave as point charges repelling each other with a force varying inversely as the square of the distance between them. It will be seen later that this assumption is not true in the case of close collisions with the nuclei of the lighter elements. The values given in the table for the first four elements, therefore, do not represent the distances of approach realised experimentally.

Element	Atomic number	Atomic weight	Closest distance of approach on Coulomb's Law
Hydrogen	1	1·008	$1\cdot9 \times 10^{-13}$ cm.
Helium	2	4·0	1·5
Carbon	6	12·0	3·0
Aluminium	13	27·1	5·5
Copper	29	63·57	11·5
Silver	47	107·9	18·2
Gold	79	197·2	30·1
Uranium	92	238·2	38·0

In this table the atomic weight has been taken to give the mass of the nucleus. This is of course only approximately true in the case of elements which consist of isotopes. Since the distance of approach depends on the mass of a nucleus as well as its charge, an α particle of given velocity will approach more closely to a heavier isotope of an element than to a lighter one. The difference is, however, very small except in the case of very light elements; in the case of lithium, for example, the closest distance of approach of the α particle of radium C is $1\cdot88 \times 10^{-13}$ cm. for the isotope of mass 6 and $1\cdot77 \times 10^{-13}$ cm. for the isotope of mass 7.

§ 57. **Deductions from energy and momentum relations.** We shall now consider certain consequences which do not depend on any hypothesis of a law of force but that follow from the assumption that the collisions of α particles and atomic nuclei are perfectly elastic.

The relation (3) of the previous section gives the velocity of the α particle after a collision resulting in a deflection ϕ. If we write $m/M = k$, we get

$$v/V = \frac{\cos\phi}{k+1} + \frac{\sqrt{k^2 - \sin^2\phi}}{k+1}.$$

Assuming Geiger's rule that $V^3 = aR$, we can calculate the range of the α particle after the collision. The following table gives the range in cm. of air of the α particle of radium C, of initial range 7 cm., after scattering through various angles by some atoms.

	30°	60°	90°	120°	150°	180°
He	4·5	0·9	0	—	—	—
C	6·1	4·2	2·5	1·4	1·0	0·9
Al	6·6	5·6	4·5	3·6	3·0	2·9
Cu	6·8	6·4	5·8	5·3	5·0	4·8
Ag	6·9	6·6	6·3	5·9	5·7	5·6
Au	7·0	7·0	6·8	6·4	6·2	6·2

It is seen that an α particle when scattered by a light atom through a large angle suffers a reduction in range so marked as to be readily detected by experiment. So far as measurements have been made they are in good agreement with the above calculations*.

Considering now the motion of the struck nucleus, we see from relations (2) above that the light atomic nuclei may attain considerable velocities if the collision is close. In the case of a head-on collision, a hydrogen nucleus will receive a velocity of $1\cdot6V$, a helium nucleus V, carbon $\frac{1}{2}V$, and oxygen $0\cdot4V$, for example. It is to be expected that the shock of the collision will free the hydrogen and the helium nuclei of their electrons, but it is not possible to say *a priori* what will be the charge carried by the recoiling carbon or oxygen atom. We can therefore make a reliable estimate of the ranges of the projected atoms only for hydrogen and for helium. To find the range of the projected hydrogen nucleus, Darwin† applied Bohr's theory of the motion of electrified particles and he deduced that the range should be about the same as that of an α particle of the same velocity. This result is confirmed by experiment (cf. § 59).

It has already been mentioned that in the expansion photographs of α ray tracks a large deflection of the α particle is accompanied by a "spur" due to the recoiling atom. By photographing a large number of α ray tracks, Blackett has obtained many cases of "spurs" or "forks." In his earlier experiments two photographs of the track, viewed from mutually perpendicular directions, were taken, by means of an arrangement of mirrors, through the same lens on the same plate. In later experiments two independent cameras were

* Cf. Geiger and Marsden, *Phil. Mag.* **25**, 604, 1913; Rutherford and Chadwick, *Phil. Mag.* **50**, 889, 1925.
† Darwin, *loc. cit.*

used, their axes being at right angles. From the two projections of the track thus obtained it is possible to deduce the true angles between the forks and the lengths of the forks. In this way Blackett* has made a study of the forked tracks obtained in hydrogen, helium, air and argon. For the present, we are concerned only with their bearing on the assumption of the conservation of energy and momentum in the collisions. It follows from equations (2) and (3) of § 56 that

$$m/M = \sin \phi / \sin (2\theta + \phi),$$

a relation which can be tested directly by measurement of the angles between the arms of the fork and the stem.

Of the collisions examined in this way none, except those resulting in a disintegration of the struck nucleus (see § 71), showed a departure from this relation greater than the experimental error. For illustration, we may take the collisions shown in Plate IX, figs. 1, 2, 3. Fig. 1 shows a collision of an α particle with a hydrogen atom, the least deflected arm of the fork is the path of the α particle. Fig. 2 is a collision with a helium atom. In this case it is impossible to say which of the arms is due to the α particle, though the probability is that it is the least deflected arm. In Fig. 3 the collision is with an oxygen atom, which has produced the shorter of the two branches. The measurements of these forks gave the following results:

Atom	ϕ	θ	m/M observed	m/M calculated
Hydrogen	8° 27′	68° 0′	0·253	0·252
Helium	38° 34′	50° 53′	0·981	1·000
Oxygen	76° 6′	45° 12′	4·18	4·00

In some later and more accurate experiments Blackett and Hudson† obtained two very good photographs of collisions with hydrogen atoms. The measurements of these showed that any kinetic energy lost in the collision could not be greater than 85,000 electron-volts in the one case, or 51,000 electron-volts in the other, while the energy of the incident α particle was about 6,000,000 electron-volts. These experiments thus confirm very strongly the assumption that within the experimental error both momentum and energy are conserved in the majority of the collisions of α particles with atomic nuclei.

* Blackett, Proc. Roy. Soc. A, 103, 62, 1923.
† Blackett and Hudson, Proc. Roy. Soc. A, 117, 124, 1927.

PLATE IX

Fig. 1. Collision of α particle with hydrogen atom.

Fig. 2. Collision of α particle with helium atom.

Fig. 3. Collision of α particle with oxygen atom.

On the other hand, some evidence has been brought forward which indicates that in some collisions the momentum may be conserved and not the energy, in others neither momentum nor energy. For example, Auger and Perrin* state that although in most of the cases of collision observed the energy and momentum relations were satisfied, a few were found in which the energy relation failed. Momentum was still conserved, as was shown by the coplanarity of the three branches of the track. Akiyama† has obtained stereoscopic photographs of seven forked α ray tracks in air. Of these only two appeared to satisfy both momentum and energy relations, while in three cases the momentum relation was not fulfilled.

It may perhaps be possible to explain these anomalous cases as due to the difficulties of measurement inherent in the method. Not only may the tracks be distorted by instrumental errors of many kinds, but the errors involved in the calculation of the angles may take large values under certain circumstances. For these reasons only a very small proportion of the forked tracks observed may be suitable for a test of the energy and momentum relations. It is indeed highly improbable that the momentum criterion, the coplanarity of the fork, should fail to hold within the accuracy of experiment. If there is a loss of momentum in the collision, it must appear in the motion of a third body or as radiation. In the former case the track of the third body should be observed, while in the latter, owing to the small momentum associated with a given energy of radiation, it is easy to show that the energy lost by radiation would be a large fraction of the total energy of the α particle even when the departure from coplanarity was small.

It is, however, not unlikely that in very close collisions the atomic nucleus may suffer a deformation of such magnitude that an appreciable amount of kinetic energy disappears. Such inelastic collisions have been discussed by Smekal‡ and are also suggested by considerations developed to explain the anomalous scattering of α particles (see § 62). It is generally assumed that the nucleus of an atom is a system of positively and negatively charged particles. When the α particle approaches very close to such a system, there may be a considerable displacement of the particles relative to each other, produced by the electric field of the α particle. The energy

* Auger and Perrin, *C.R.* **175**, 340, 1922.
† Akiyama, *Jap. Journ. Phys.* **2**, 279 and 287, 1923.
‡ Smekal, *Phys. Zeit.* **27**, 383, 1926.

transferred to the nucleus to cause this distortion may not all be returned to the α particle as it recedes from the nucleus and may remain as vibration energy of the displaced particles, to be dissipated later in some form. It is possible to make a very rough estimate of the amount of energy which may be lost in this way, by assuming that the polarisation of the nucleus is due to the displacement of an elastically bound particle with a frequency of free vibration ν. Kuhn* has shown that if the particle is a proton or an α particle, ν may be, for some nuclei such as those of magnesium or aluminium, of the order of the γ ray frequencies, about 10^{20}. Now the time of collision is of the order of 10^{-21} sec., so that the time of collision is short compared with a period of free vibration of the particle. In such a case the displacement of the particle will lag behind the electric field due to the α particle and the energy of distortion will not all be returned to the α particle as it moves away; the energy transferred to the nucleus may be a few per cent. of the energy of the incident α particle. If ν is much greater, as it may be in very stable nuclei, e.g. the helium nucleus, or if the displaced particles are electrons, the amount of energy remaining with the nucleus will be very small. In this case, when the time of collision is long compared with a period of free vibration of the particle, the displacement of the particle will correspond exactly to the external field, and no energy will remain with the nucleus when the α particle has moved away. The amount of energy lost in the collision will be the work done in displacing the particle and will be only a very small fraction of the energy of the α particle. Since the α particle must be regarded as a very stable nucleus, the distortion suffered by it during the collision is not likely to absorb any appreciable amount of energy.

It is evident that very accurate measurement will be required to give definite information on this point. The only cases in which inelastic collisions have been established with certainty are those in which a nitrogen nucleus is disintegrated and the α particle appears to have been captured by the nucleus. In this connection, however, the experiments of Slater† are of interest. Slater found that when heavy atoms, such as gold, are bombarded by α particles a small amount of γ radiation is emitted which from its penetrating power he judged to have its origin in the nucleus. This observation would indicate that some of the collisions of α particles with gold nuclei are inelastic, but the evidence cannot be regarded as definite.

* Kuhn, *Zeit. f. Phys.* **44**, 32, 1927.   † Slater, *Phil. Mag.* **42**, 904, 1921.

We may conclude therefore that at least in the great majority of the collisions of an α particle with atomic nuclei momentum and energy are conserved in the motion of the particles.

§ 58. **Range and velocity of recoil atoms.** The investigation by Blackett* of branched α ray tracks has already been mentioned on account of its bearing on the conservation of energy and momentum in the collisions of α particles with atoms. These experiments also provided information about the relation between the range and velocity of the recoiling atom, set in motion by the α particle. The range of the recoil atom was given immediately by the length of its track and the velocity was obtained from measurement of the angles of the fork.

If in Fig. 66 the ranges of the α particle after the collision and of the recoil atom are R_M and R_m respectively, then we may write $v = f_1 (R_M)$ and $u = f_2 (R_m)$. This assumes that both the α particle and the recoil atom travel a given distance for a given initial velocity; in other words, that no straggling occurs. If the straggling should be considerable, these functions may be regarded as giving an average value for the velocity at a given distance from the end of the tracks.

The equations (1) and (2) of § 56 give

$$\frac{u}{v} = \frac{M \sin \phi}{m \sin \theta}$$

and

$$\frac{m}{M} = \frac{\sin \phi}{\sin (2\theta + \phi)}.$$

Since R_M, the range of the α particle after collision, can be measured from the fork, the velocity v can be found at once if the relation $v = f_1 (R_M)$ is known. This relation was determined by Blackett† from a statistical analysis of the deflections shown by α particle tracks (cf. § 22 a). He concluded that for small values of the range, between 0·5 mm. and 16 mm. in air, the relation was approximately $v = R_M{}^{\frac{2}{3}}$. When the range R_M of the deflected α particle is greater than a few millimetres, the velocity of the particle may be estimated from its range by using the results of Briggs (§ 22). In some cases another estimate of v may be obtained by measuring the distance of the point of collision from the origin of the α particle and a knowledge of the initial velocity of the particle. When v is thus determined and if θ and ϕ are measured, the velocity u of the recoil atom

* Blackett, *Proc. Roy. Soc.* A, **103**, 62, 1923.
† Blackett, *Proc. Roy. Soc.* A, **102**, 294, 1922.

can be calculated from the above equations. The range of the recoil atom may be so short that θ cannot be measured with the desired accuracy, but if the nature of the recoil atom and therefore the value of m is known it is not necessary to know θ. In this way the values of u and R_m can be determined for the recoil atom in each branched track.

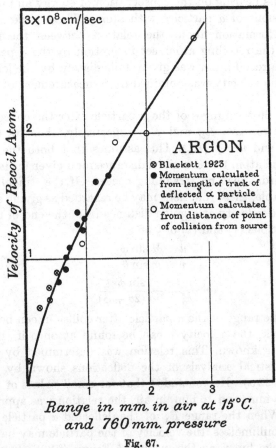

Range in mm. in air at 15°C.
and 760 mm. pressure

Fig. 67.

This procedure was carried out with a number of forked tracks obtained in air, hydrogen, helium and argon. The results found for recoiling atoms of argon are shown in Fig. 67 together with some later results obtained in the same way by Lees*. The range of the recoil atom is given in terms of air at 15° C. and 760 mm. pressure.

* Unpublished.

Both R_m and R_M were of course in this case measured in argon, and the reduction to air was made by measuring the range R' of the α particles in the argon used in the expansion chamber as well as the range R in air. The measured values of R_m and R_M were then multiplied by R/R'. It will be noted that there is no very marked variation in range for particles of similar velocities. It seems certain that the particle must change its charge during its motion through

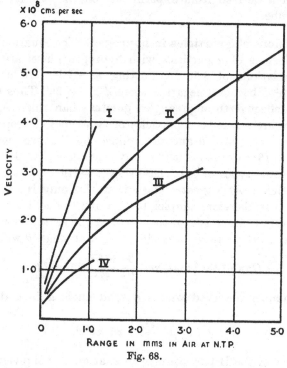

Fig. 68.

the gas, and therefore one must conclude that the change of charge is so frequent that the average state of each particle is much the same.

Blackett's results for the four kinds of recoiling atoms which he investigated are shown in smoothed curves in Fig. 68. From these curves it appears that the relative ranges of recoil atoms of hydrogen, helium, nitrogen and argon of the same initial velocity are as 1·0 : 1·9 : 3·2 : 5·1 respectively. Since the energies of atoms of the same velocity are proportional to their masses, the average loss of energy per unit length of track varies as 1·0 : 2·1 : 4·5 : 7·9 for these four particles. Now the rate of loss of energy of a charged particle in its passage through matter is to a first approximation proportional to the square of its charge. Hence the relative effective charges of the hydrogen, helium, nitrogen and argon recoil atoms are as 1·0 : 1·5 : 2·1 : 2·8. The average charge of the recoil atom of hydrogen must be very nearly $+ 1e$, and therefore the above numbers represent the average charge carried by the four kinds of recoil atoms. These results seem reasonable and are in general agreement with other evidence derived from experiments on the positive rays in a discharge tube.

§ 59. **Collisions of α particles in hydrogen.** The equations describing the collisions of α particles with hydrogen nuclei are obtained from the relations of § 56 by putting $m = \frac{1}{4}M$. We notice that $m^2 - M^2 \sin^2 \phi$ becomes negative when $\phi > 14° 29'$. Thus no α particles are deflected through angles greater than this. At smaller angles there are scattered α particles of two types corresponding to two different values of θ and determined by the two signs of the ambiguity in (3). Consequently at any angle less than $14° 29'$ there will occur three types of particles—slow α particles, fast α particles, and H particles. At a greater angle than this only H particles will be found. It is therefore simpler to describe the collisions in terms of the projected H particles rather than of the scattered α particles. The number of H particles projected within an angle θ will be

$$Q\pi n t \mu^2 \tan^2 \theta = Q\pi n t . \frac{1}{V^4} . \frac{e^2 E^2}{M^2} . 25 \tan^2 \theta,$$

and the number observed within a solid angle ω in a direction θ will be

$$Q n t \omega . \frac{1}{V^4} . \frac{e^2 E^2}{M^2} . 25 \sec^3 \theta.$$

The velocity of the H particle projected at angle θ is given by

$$u = \tfrac{8}{5} V \cos \theta.$$

In all but the most distant collisions the velocity imparted to the hydrogen atom is so great that the moving particle must be the free hydrogen nucleus, or proton. The range of the projected protons was calculated by Darwin, on the basis of Bohr's theory of the motion of electrified particles through matter. On this theory the range of a moving proton should be slightly less than that of an α particle of the same velocity. Hence, assuming that Geiger's rule connecting range and velocity still applies, the range of the proton struck straight on by an α particle of radium C′ should be nearly 30 cm. in air or about 120 cm. in hydrogen.

These projected H nuclei were first observed by Marsden*, who found that the passage of α particles through hydrogen gave rise to particles which produced scintillations on a zinc sulphide screen placed far beyond the range of the α particles. The range of the particles was more than 100 cm. in hydrogen, in agreement with Darwin's calculation. The first detailed investigation of the collisions was made by Rutherford†. His experimental arrangement is shown in Fig. 69. The source of α rays was radium (B + C) deposited on

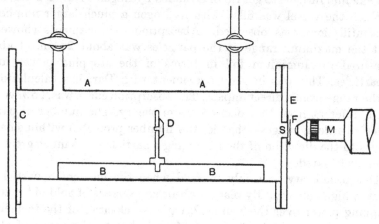

Fig. 69.

the face of a disc D. This was mounted on the slide BB, and enclosed in a rectangular box AA which could be exhausted or filled with a suitable gas. In the end E of the box was cut an opening which was covered by a thin silver foil of a stopping power of 6 cm. of air. The zinc sulphide screen was placed at F about 2 mm. from the window, so that further absorbing screens could be inserted between the window and screen. The apparatus was placed in a strong magnetic field in order to bend away the β rays, which otherwise caused a strong luminosity of the zinc sulphide screen.

The observations consisted in counting the number of scintillations produced on the screen as successive thicknesses of absorbing sheet were introduced into the path of the particles. In this way the

* Marsden, *Phil. Mag.* 27, 824, 1914.
† Rutherford, *Phil. Mag.* 37, 537, 1919.

distribution of the H particles with range was found. If R_0 is the range given to an H particle by a direct impact, the range R of a particle projected at an angle is given by $R = R_0 \cos^3 \theta$. The results could therefore be readily compared with the calculations given above.

When the vessel was exhausted of air, a few scintillations were always observed on the screen, when the absorptions in the path corresponded to less than about 28 cm. of air. These were shown to be due to H particles arising from hydrogen occluded partly in the source and partly in the absorbing screens exposed to the α particles. All the foils used in the path of the α particles were therefore heated in a vacuum furnace to get rid of occluded hydrogen as far as possible.

When the vessel was filled with hydrogen a much larger number of scintillations was observed. Absorption measurements showed that the maximum range of the particles was about 28 cm. of air, measured in aluminium foil in terms of the stopping power for α particles. This was in good agreement with Darwin's calculation of the range from a direct impact. The absorption curve was, however, quite different from that deduced from theory. The number of particles with long ranges—that is, the number projected within small angles to the direction of the impinging α particle—was much greater than anticipated.

This peculiarity was only marked when the incident α particles were of high velocity. By placing absorbing screens of gold of known stopping power over the source D, various velocities of the incident α particles were obtained. It was found that as the range of the α particle was reduced, the absorption curve of the H particles became more like the theoretical curve and for an α particle of 3·9 cm. range was in fairly good agreement. On the other hand, the number of H particles observed was always greater than the number to be expected on the theory, and the variation in number with the velocity of the α particle was in the opposite direction.

In interpreting these results, Rutherford showed that although the observations suggested that the forces between the α particle and H nucleus varied more rapidly than on Coulomb's Law, it was unlikely that any theory of purely central forces would suffice. He attributed the divergences to the effects of the complex structure of the α particle or helium nucleus. If we suppose that the α particle is built up in some way of four H nuclei and two electrons, we should expect a complicated field of force round the nucleus and rapid variation in magnitude, possibly also in direction, of the forces at

distances of the order of the diameter of the electron. This agrees
with the observations, for in the anomalous collisions in hydrogen
the distance between the α particle and the hydrogen nucleus was
about 3×10^{-13} cm., while the accepted value of the diameter of the
electron is $3 \cdot 6 \times 10^{-13}$ cm. Assuming for simplicity that the hydrogen
nucleus acts as a point charge, the helium nucleus must have dimensions of about 3 to 4×10^{-13} cm. Rutherford suggested that a general
agreement with the experiments could be obtained if the α particle
behaved in these collisions as a plate of radius $3 \cdot 5 \times 10^{-13}$ cm. moving
in the direction of its axis.

In the above account it has been assumed that the scintillations
observed when α particles passed through hydrogen were in fact due
to projected H nuclei. This assumption was subjected to direct test
by Rutherford, who measured the magnetic and electrostatic deflections of the particles. Owing to the small number of H particles
obtained under the experimental conditions and to the fact that
particles of various velocities were necessarily present, the measurements were very difficult and great accuracy could not be
attained.

The magnetic deflection of a pencil of H particles was compared
with that of the α particles of radium C under similar conditions of
experiment. The mean value of mu_0/e obtained from these determinations was $3 \cdot 15 \times 10^5$ E.M. units, u_0 being the maximum velocity
of the H particles arising from the impact of the α particles of
radium C. On the collision theory $u_0 = 1 \cdot 6V$ and the value of mu_0/e
if the particles were H nuclei should be $3 \cdot 2 \times 10^5$ E.M. units. The
agreement was therefore very satisfactory.

The electrostatic deflection could not be measured directly, but
the deflection in a magnetic field was compared with that due to a
combined magnetic and electric field. In this way the electric field
equivalent in deflecting the particles to a magnetic field was found,
and this then gave the velocity of the H particles directly. The value
found was $u_0 = 3 \cdot 12 \times 10^9$ cm./sec., compared with the calculated
value $3 \cdot 07 \times 10^9$ cm./sec. Combining this result with the value found
for mu_0/e above, we have $e/m = 10^4$ E.M. units. The agreement with
the value of e/m for the hydrogen ion in electrolysis, $e/m = 9570$, is
sufficiently close to show that the scintillations observed when α rays
pass through hydrogen are due to hydrogen nuclei, and that, within
the margin of experimental error, energy and momentum are conserved in these collisions.

§ 59 a. By a more direct method Chadwick and Bieler* made a very close study of the collisions of α particles with hydrogen nuclei, and determined the angular distribution of the projected H particles for α rays of various velocities. The arrangement was similar to that used by Chadwick for the scattering of α particles, the foil in Fig. 50 being replaced by a thin sheet of paraffin wax to provide the source of hydrogen particles. Let the number of H particles projected within an angle θ by a single α particle in passing through 1 cm. of hydrogen gas at N.T.P. be $q = F(\theta)$. Then it is easy to show that the number of H particles falling per second on unit area of a zinc sulphide screen at S will be

$$\frac{Q.\bar{t}}{16\pi\bar{r}^2}.[F(\theta_2)-F(\theta_1)],$$

where Q is the number of α particles emitted per second by the source, $\bar{r}^2$ is the mean value of r^2 and $\bar{t}$ is the mean value of t sec $\frac{1}{2}\theta$, t being the thickness of the paraffin wax sheet expressed as the thickness of hydrogen gas containing the same number of atoms of hydrogen.

If the paraffin wax is in the form of a circular sheet, the number of H particles observed on the screen is directly proportional to $F(\theta_2)$. Thus the simplest method of experiment is to use circular sheets of different angular limits $\frac{1}{2}\theta_2$, $\frac{1}{2}\theta_2'$, etc. The observations then give points on the curve $q = F(\theta)$ corresponding to the different angles θ_2, θ_2', etc. This method was used for smaller angles only. For angles greater than 20°, annular rings of wax of suitable angular limits were used. Scattered α particles were prevented from reaching the screen by means of suitable absorbing screens.

The results obtained in these experiments are shown in Fig. 70, in which q is plotted against θ for various velocities of the α particle, expressed in the diagram as ranges in air.

The observed numbers of H particles at small angles were greatly in excess of those given by the theory of point charges and Coulomb's Law. On this theory the value of q for $\theta = 30°$ and an α particle of range 8·2 cm. is $4·4 \times 10^{-7}$, and for an α particle of 2·9 cm. range it is $15·8 \times 10^{-7}$. The corresponding observed values were $4·3 \times 10^{-5}$ and $0·6 \times 10^{-5}$; in the first case 100 times as great as the inverse square number, in the second not quite 4 times as great.

There is thus a rapid approach to inverse square conditions as the

* Chadwick and Bieler, *Phil. Mag.* 42, 923, 1921.

range of the α particle is reduced, that is, for large distances of collision. Further experiments, the results of which are not shown in Fig. 70, showed that with α particles of 2 cm. range the number and distribution of the projected H particles were in fair agreement with the simple theory.

A comparison of the experimental curve B with the curve B' calculated on the inverse square law shows the approach to normal

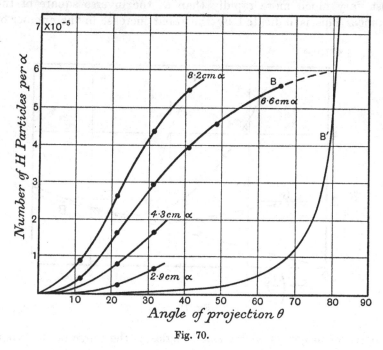

Fig. 70.

collisions as the value of $θ$, and with it the distance of collision, increases. It would appear that inverse square conditions would be reached with α particles of 6·6 cm. range at a distance of collision corresponding to a value of $θ$ of about 80°, i.e. at about 7×10^{-13} cm.

These results confirmed the general conclusions reached by Rutherford, that within a certain distance of approach of the particles the force between them varied much more rapidly than on Coulomb's law and that no system of central forces between two point charges would explain the observations. A closer analysis led to an estimate of the dimensions of the region over which the forces between the α particle and the H nucleus became abnormal. This region is not spherical, as the experiments show that for central collisions, i.e. small values of θ, the law of force is approximately the inverse square until the distance of approach between the centres of the two nuclei is less than about 4×10^{-13} cm., while for oblique collisions of fast α particles the distance of approach must not be less than about 8×10^{-13} cm.

At closer distances than these the force between the particles must vary much more rapidly than as the inverse square of the distance. This is indicated by the rapid increase in the number of

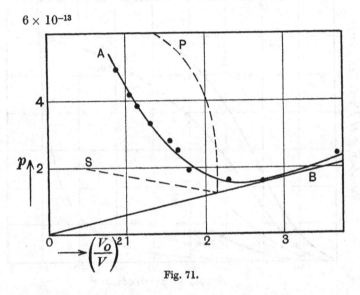

Fig. 71.

H particles projected within small angles as the range of the α particle increases.

The experimental results were compared with the collision relations calculated by Darwin* in a consideration of certain models of α particles. The models which apply here are the elastic sphere and elastic plate. The elastic sphere may be considered to afford the simplest representation of a rapidly varying field of force. An H particle projected towards the sphere moves under Coulomb forces until it strikes the sphere, when it rebounds elastically. Similarly, the elastic plate repels the H particle with a force varying as the inverse square of the distance from the centre. Fig. 71, curve A,

* Darwin, *Phil. Mag.* 41, 486, 1921.

COLLISIONS OF α PARTICLES IN HYDROGEN 259

shows the variation observed experimentally in the number of H particles projected between 0° and 31·3° with the velocity of the α particle. In this figure the abscissae are values of $(V_0/V)^2$, where V_0 is the velocity of the α particle of radium C, and the ordinates are values of p, which is defined in the following way. If P is the probability of a collision of a single α particle with a single H nucleus resulting in the projection of the H nucleus within an angle θ, the quantity p is given by $P = \pi p^2$; hence $p = \sqrt{q/\pi N}$ (see p. 256), where $N = 5\cdot41 \times 10^{19}$, the number of H atoms per c.c. of hydrogen gas at N.T.P. Curve S is calculated for an elastic sphere of radius 4×10^{-13} cm. and curve P for an elastic plate of radius 8×10^{-13} cm. At the point of intersection they turn into the inverse square law line B. The dimensions of the sphere and plate are so chosen as to give a deviation from the inverse square law in the region given by the experiments. The experimental curve lies midway between the curves S and P. Chadwick and Bieler concluded that the α particle behaves in these collisions like an elastic oblate spheroid of semi-axes about 8×10^{-13} cm. and 4×10^{-13} cm. respectively, moving in the direction of its minor axis, and interpreted the experiments in the following way. An H particle projected towards an α particle moves under the normal forces given by the inverse square law so long as it remains outside the spheroidal region of the above dimensions; if the particle penetrates the surface it encounters an extremely powerful force varying rapidly with the distance, and the collisions become abnormal. Assuming that the H particle can be regarded as a point charge, then the deviations of experiment from the simple theory may be ascribed to the complex structure of the α particle. The spheroidal region of abnormal forces gives some idea of the "dimensions" of the helium nucleus. It may be supposed that the nucleus is normally of this flattened or plate-like shape or that this shape is due to deformation by the intense forces acting on the nucleus in the collisions.

§ 59 b. An interesting application of the mass spectroscope to the measurement of the mass of the projected H particles has been made by Stetter*. It is well known that in Aston's mass spectroscope the particles are first deflected in a strong electric field and then pass into a magnetic field where they are deflected in the opposite direction. By suitable adjustment of the strengths of the electric and

* Stetter, Zeit. f. Phys. 34, 158, 1925.

magnetic fields particles which have different velocities but the same value of e/m are brought to a focus in a plane. The mass spectroscope thus splits up a beam of particles according to the value of e/m. The application of this principle to the examination of the projected H particles is beset with two difficulties as compared with its application to the positive rays of a discharge tube. The velocity of the H particles is about ten times greater than that of the positive rays, and, since the deflection of a charged particle in an electric field varies inversely as the square of the velocity, very intense electric fields are required; and the number of H particles available is of the order of 10^{-8} of the number of positive rays, necessitating the use of a zinc sulphide

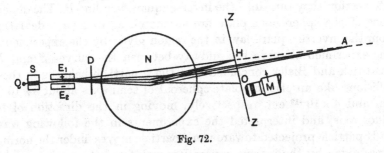

Fig. 72.

screen to locate the particles instead of a photographic plate. It is clear that the measurements cannot be carried out with the same accuracy as in Aston's experiments.

The arrangement as adopted by Stetter is shown in diagram in Fig. 72. The particles from the source Q passed through a slit system S, consisting of six parallel slits each 0·14 mm. wide, and thence into the electric field between the plates E_1 and E_2. These were 2 mm. apart and a difference of potential of 15,000 volts could be applied between them, from a transformer with a rectifying system. The deflected beam then entered through a diaphragm D, which cut out the particles of extreme velocities, into the magnetic field N, in which they were deflected in the opposite direction. The plane in which the particles of different values of e/m were brought to a focus is shown by the dotted line in the figure; H particles were focussed at H and α particles at A. The zinc sulphide screen could not be placed in the position of this focal plane, for, owing to the small angle of incidence and the width of the original pencil of particles, the band on the screen would have been wider than the field of view of the microscope. The screen was therefore placed in the position ZZ, at right angles to the direction of the H particles. The microscope M could be moved along the screen by an arrangement with vernier reading and the position of a band was found directly or by counting the number of scintillations observed at different positions.

The apparatus was tested by using as source a preparation of polonium. When no fields were applied, the position of the undeflected band was at O. The deflections of the α particles under electric field alone, and with combined electric and magnetic fields, were measured and found to agree within 1 per cent. with the calculated deflections. For the measurement of the H particles the source was a thin-walled glass tube filled with radon, the inner wall exposed to the α particles being coated with a thin layer of paraffin wax. The spectrum of the particles was a single band with a well-defined maximum. The measurements gave for the mass of the particle a value $1 \cdot 017 \pm 0 \cdot 03$, which is to be compared with the mass of the H nucleus $1 \cdot 007$.

§ 60. Collisions of α particles in helium. Further information on this question of the field of force in the immediate neighbourhood of the He nucleus has been given by a study of the collisions of α particles in helium made by Rutherford and Chadwick*. In these collisions the two particles are the same, and putting $m = M$ we have from the energy and momentum relations

$$u = V \cos \theta, \quad v = V \cos \phi, \quad \phi = \tfrac{1}{2}\pi - \theta.$$

If the forces between the particles are given by Coulomb's law, the impact parameter p is given by

$$p = \frac{1}{V^2} \cdot \frac{2E^2}{M} . \tan \theta,$$

and the closest distance of approach during the collision is

$$D = \frac{1}{V^2} \cdot \frac{2E^2}{M} (1 + \sec \theta).$$

It may be noted that this distance is less than the corresponding distance for collisions in hydrogen.

If, as before, ω is the solid angle subtended by a zinc sulphide

* Rutherford and Chadwick, *Phil. Mag.* 4, 605, 1927.

screen placed at an angle ϕ, then the number of α particles scattered to the screen is

$$Qnt\omega \cdot \frac{1}{V^4} \cdot \frac{E^4}{M^2} \cdot 4 \cot \phi \, \mathrm{cosec}^3 \phi,$$

where n is the number of He atoms per c.c., t is the length in cm. of the scattering path, and Q is the number of particles passing through the defining aperture.

The number of helium atoms set in motion in the direction ϕ under the same conditions is

$$Qnt\omega \cdot \frac{1}{V^4} \cdot \frac{E^4}{M^2} \cdot 4 \sec^3 \phi.$$

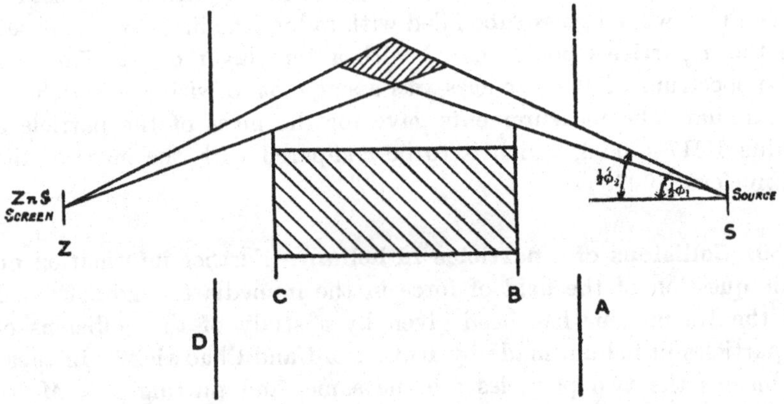

Fig. 73.

The projected helium nuclei should be quite indistinguishable from α particles, for, neglecting the energy required to remove the electrons, the velocity of the nucleus recoiling in the direction ϕ is $V \cos \phi$, the same as the velocity of the α particle scattered in that direction. The scintillations observed on the zinc sulphide screen will be due therefore to both sets of particles and there is no means of distinguishing between a scattered α particle and a projected He nucleus. The two types correspond, in general, to collisions of different kinds, for the He nuclei projected at an angle ϕ are produced in the collisions in which the α particles are scattered through an angle $\frac{1}{2}\pi - \phi$. When $\phi = 45°$ both sets arise from the same type of collision.

The experimental arrangement used by Rutherford and Chadwick was the annular ring method, adapted for scattering in a gas, as shown in the diagram of Fig. 73.

The diaphragms A and D were graphite sheets containing circular holes, diaphragms B and C were graphite discs. A and B served to define the incident beam of α particles, an annular ring of limits $\frac{1}{2}\phi_1$ and $\frac{1}{2}\phi_2$ from the source S, and C and D defined the scattered beam received by the zinc sulphide screen Z. The volume of gas which is effective in scattering is shown shaded in the figure. If the gas used is helium, the number of α particles scattered per second to unit area of Z is given by

$$\frac{Qnt}{4r^2} \cdot \frac{E^4}{M^2V^4} (\operatorname{cosec}^2 \phi_1 - \operatorname{cosec}^2 \phi_2),$$

where r is the mean distance from the source, and Q the number of particles emitted per second by the source.

The number of He nuclei projected per second to unit area of Z is

$$\frac{Qnt}{4r^2} \cdot \frac{E^4}{M^2V^4} (\sec^2 \phi_2 - \sec^2 \phi_1).$$

These particles correspond to the α particles scattered between the angles $\frac{1}{2}\pi - \phi_1$ and $\frac{1}{2}\pi - \phi_2$.

On account of the impossibility of distinguishing between the scattered α particles and the projected He nuclei, the results are subject to a certain ambiguity. We shall consider first the observations made when ϕ_1 and ϕ_2 were 40° and 50° respectively. In this case the scattered α particles and the projected He nuclei are equal in number, and arise from the same collisions. The curve showing the variation in number of the particles observed for different velocities of the incident α particles is shown in Fig. 74. The ordinates represent the ratio of the observed number to the number calculated on inverse square forces, and the abscissae are proportional to $1/V^2$. The dotted line for ordinate unity thus represents the numbers calculated on the inverse square law. It will be noted that the ratio is large for the swift α particles, falls very rapidly to a value just below unity as the velocity of the α particle is decreased, and then rises again to the inverse square value.

The results are very similar to those found for the collisions in hydrogen, not only in form but also in magnitude.

In the collisions between angles of 40° and 50° the forces are given by Coulomb's law when the incident α particle has a range less than 3 cm., that is, until the distance of approach between the centres of the particles is less than about $3\cdot5 \times 10^{-13}$ cm. A comparison of the curve of Fig. 74 with the corresponding curve for an elastic sphere shows that the region of abnormal forces is not spherical but

has a plate-like shape. This indication is confirmed by a consideration of experiments performed at small angles of scattering. These
again can be interpreted without ambiguity, for at small angles the
number of projected particles is, if Coulomb's law holds, very small
compared with the number of scattered particles. It can therefore

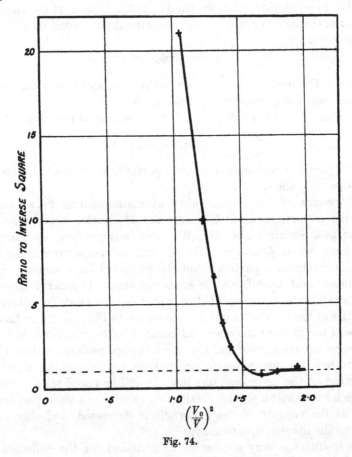

Fig. 74.

be assumed without serious error that, except in the collisions of
very fast α particles, the particles observed are scattered α particles
only.

The experiments at small angles showed that, as the velocity of
the α particle is reduced, the ratio of the number of scattered particles
observed to the number calculated on inverse square forces falls,
sometimes to as low as one-third, and then rises to roughly unity.

This deficiency in scattering at the small angles is the necessary consequence of the excess scattering shown at other angles by α particles of the given velocity. From the results of these experiments the distance of approach between the centres of the colliding particles at which inverse square numbers are finally reached is about 12 to 15×10^{-13} cm.

Thus while for central collisions between an α particle and a helium nucleus the law of force remains normal until the distance of approach is less than about $3 \cdot 5 \times 10^{-13}$ cm., for glancing collisions the critical distance is about 14×10^{-13} cm.

§ 61. Scattering of α particles by aluminium and magnesium. The natural development of the experiments just described to a study of the collisions of α particles with the elements next in atomic number—lithium, beryllium, etc.—might be expected to provide further evidence of a departure from Coulomb forces and so to help in the interpretation of the origin of the forces which come into action in close collisions. A general investigation is, unfortunately, attended with many difficulties arising from the chemical properties of the light elements, and up to the present time the only elements for which the scattering of α particles has been examined in any detail are magnesium and aluminium, which can be readily obtained in the form of thin foil. It will appear from the account which follows that the experimental data which have been obtained are neither very accurate nor very comprehensive. This defect is to be ascribed mainly to the comparatively large atomic number of aluminium (or magnesium). We have seen in the case of collisions in hydrogen and helium that the inverse square law of force fails to hold when the α particle approaches the nucleus within, roughly, 5×10^{-13} cm. The closest possible distance of approach of the α particle of radium C to an aluminium nucleus is, on inverse square forces, $5 \cdot 5 \times 10^{-13}$ cm. While we may with reason expect to find in collisions with the aluminium nucleus a failure of Coulomb's law at greater distances than with the helium nucleus, yet the margin is not large, and the distance of approach of the particles, even in close collisions, is still so great that the departure from inverse square scattering is small.

A further difficulty must be mentioned. Some of the collisions with aluminium and magnesium nuclei result in the ejection of a proton from the nucleus. The fate of the α particle is not known, it may possibly be captured by the nucleus, but in any case such collisions are anomalous, for kinetic energy is not conserved. The type of collision which causes a disintegration is probably a very close collision of a fast α particle, but what fraction of these collisions is of this nature it is not possible to say definitely from our present knowledge. It is, however, likely to be small. The scattering of fast α particles through large angles by aluminium or magnesium must therefore be interpreted with this reservation.

The first experiments which are of interest in showing a departure from theory in the scattering of α particles by aluminium and magnesium are those of Bieler*.

In these experiments the annular ring arrangement of scattering foil was again adopted. By the use of a series of diaphragms made so that the maximum scattering angle, ϕ_2, of one arrangement was the same as the minimum scattering angle, ϕ_1, of the next, Bieler investigated the scattering by thin aluminium and magnesium foils over the whole range between the angles 18° and 100° in a series of four steps.

At the large angles the ratio of the number of particles in the direct to that in the scattered beam was so great that it was not practicable to count the direct beam by the wheel and slit method. The number of scattered particles to be expected on the inverse square law was therefore deduced, in all the experiments, from the scattering produced by a thin gold foil under the same conditions. It was assumed that, over the range of angles used, 20° to 100°, the scattering of gold was normal. Later experiments by Rutherford and Chadwick, described in § 45, have shown that this assumption was justified.

Two sets of experiments were carried out. In the first the source of α particles was radium (B + C), and in the second polonium, so that the scattering was observed for two velocities of the α particle. The results of the experiments on the scattering by aluminium are shown in Fig. 75. The curves show that the scattering at small angles is very close to that which would be expected on an inverse square law of force. As the angle of scattering increases, the ratio of the observed scattering to that given by inverse square forces diminishes rapidly for the fast α particles, more gradually for the slower α particles. The experiments indicate therefore a departure from the inverse square law of force, which increases rapidly as the distance between the nuclei diminishes.

* Bieler, *Proc. Roy. Soc.* A, **105**, 434, 1924.

These results of Bieler were confirmed and extended by Rutherford and Chadwick* in an examination of the scattering of α particles by several elements, some of the results of which have already been mentioned. In these experiments the scattering was examined at large angles only and the distance of collision was varied by using α particles of different velocities. Apart from the fact that it is at large angles of scattering that any divergence from inverse square forces should be expected, this method has the further advantage that a change in the law of force should show itself more clearly than when the scattering is examined at small angles. A consideration of the shape of the hyperbolic orbits shows that the deflection

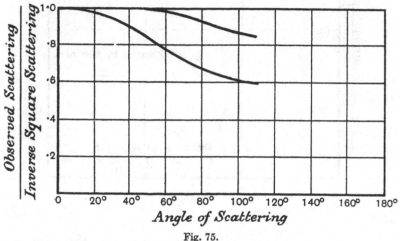

Fig. 75.

of the α particle occurs mainly in the region near the apse of the orbit, and this region is the more confined the larger the angle of scattering. Two series of experiments were carried out at average scattering angles of 135° and 90°. In both the scattering angle was kept fixed, while the velocity of the incident α particles was varied by placing sheets of mica of known stopping power in their path. We have seen that, if Coulomb's law of force is true, the number of particles scattered by a foil between fixed angles should be inversely proportional to V^4, where V is the velocity of the particles. Consequently, any change in the law of force would be immediately evident from a comparison of the numbers of scattered particles obtained with different velocities of the incident particles. In addition, the numbers

* Rutherford and Chadwick, *Phil. Mag.* **50**, 889, 1925.

to be expected on inverse square forces were deduced by measuring the scattering produced by a gold foil under the same conditions. The results of the experiments on the scattering by aluminium at 135° are shown in Fig. 76, where the ratio of the observed number of particles to the inverse square number is plotted against $1/V^2$. Since for a given angle of scattering the closest distance of approach of an α particle to the nucleus is proportional to $1/V^2$, the abscissae also represent the distances of collision, calculated on the assumption of inverse square forces. The dotted line gives the number of particles

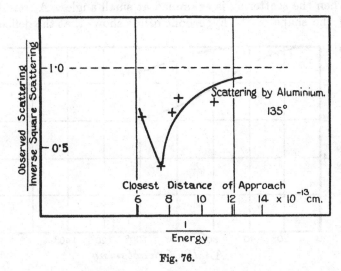

Fig. 76.

calculated from the simple theory. It is seen that for low energies of the incident α particles the scattering is approximately normal. As the energy of the particles is increased, the scattering diminishes to a minimum corresponding to α particles of 5 cm. range, and then rises sharply.

We may compare these results with those obtained in the collisions with helium which have been given in Fig. 74. The aluminium curve corresponds to the small portion of the helium curve which lies below the dotted line representing the inverse square numbers. The comparison suggests that if α particles of greater energy were available the scattering by aluminium at 135° would increase perhaps to several times the normal amount.

The experiments at an average angle of scattering of 90° showed only a diminution in scattering as the energy of the particles was increased and not a subsequent rise. (Cf. Fig. 78, where the observations are represented by the points.) The absence of the latter is to be attributed to the greater distance of collision due to the change in scattering angle, and on the whole the results confirm those obtained at 135°.

In both Bieler's experiments and those of Rutherford and Chadwick the departure from the calculated scattering is in one sense only, the scattering is always less than normal. If we consider the scattering of α particles of a given velocity, it is clear that if the scattering is less than normal over one range of angles it must be greater than normal over another range. Changes in the law of force between the particles will affect the angular distribution of the scattered particles, but the chance that a particle shall be scattered between 0° and 180° must always be unity, whatever the law of force, provided only that the particle cannot be captured by the nucleus*. Over the whole of the range examined in the above experiments the observed scattering is less than the inverse square scattering. This must be compensated by an excess of scattering in another region. The curve of Fig. 75 indicates that this region may be at angles larger than 135°. Some as yet unpublished experiments support this view, for they show that the scattering at 155° is several times the inverse square value. It is, however, difficult to say whether the excess in this region is sufficient to compensate the deficiency at smaller angles. If not, there must be either an excess of scattering at angles less than 20° or a sufficient number of α particles must be captured by the nucleus. Now it is known that protons are ejected from aluminium (and magnesium) by bombardment of α particles and it is probable that the collision which leads to the ejection of a proton also involves the capture of the α particle. The evidence available at present is, however, not sufficient to give even a rough estimate of the responsibility of such captures for the observed deficiency in scattering.

The data given by the above experiments are not sufficient to allow a definite estimate to be made of the size and shape of the region in which the forces between the aluminium (or magnesium) nucleus and the α particle depart from Coulomb's law. It is, however, of interest to observe that the dimensions of this region are not very different from those which were found from the collisions in hydrogen and in helium. The experiments show that the inverse square law is approximately true when the α particle does not

* Cf. Rutherford, *Proc. Phys. Soc. Lond.* **39**, 359, 1927.

approach closer to the aluminium nucleus than about 13×10^{-13} cm. and that it certainly no longer holds at a distance of roughly 6 or 8×10^{-13} cm.

§ 62. Discussion of the anomalous scattering. We have now to consider the suggestions which have been made to explain the conclusion derived from the experiments described in the last three sections, that the force between an α particle and light atomic nuclei is not given by Coulomb's law when the distance between them is very small, but varies much more rapidly with the distance. Either, or both, of two courses lies open; the first, to suppose that the force between point charges is not given by Coulomb's law at very small distances; the second, to ascribe the divergence to the complex structure of the particles. Both courses have been followed.

In seeking an explanation of his results, Bieler* examined the effect of an attractive force, varying as a higher power of the distance than the second, and published curves showing the distribution of scattered particles when the additional attractive force varied inversely as the cube and as the fourth power of the distance†. Comparing these curves with the observed distribution of α particles scattered by aluminium, he concluded that the assumption of an additional attractive force varying as the inverse fourth power of the distance would give a satisfactory explanation of his results. The magnitude of this additional force was such as to balance the Coulomb force of repulsion at a distance of about $3\cdot44 \times 10^{-13}$ cm. from the centre of the aluminium nucleus; within this distance, which on this view may be considered as the effective radius of the nucleus, the attracting force would be greater than the electrostatic repulsion. Such an additional attractive force would diminish very rapidly with the distance from the nucleus and would have, for example, no noticeable effect upon the motions of the electrons of the atom.

This assumption of an attractive force varying inversely as the fourth power of the distance, in addition to the Coulomb force, seems sufficient to account also for the later results of Rutherford and Chadwick. No detailed comparison of calculations on this basis

---

* Bieler, *Proc. Camb. Phil. Soc.* 21, 686, 1923.

† It should be noted that the theoretical curves published by Bieler are not accurate; they show an amount of scattering which is at all angles less than that given by Coulomb forces. Actually the defect in the scattering at large angles is compensated by an excess scattering at small angles. Since the scattering at small angles is relatively great, a very small percentage excess is sufficient to compensate a marked defect in the scattering at the large angles.

with their experimental results has yet been made. The calculations must however be closely analogous to those made by Debye and Hardmeier, which deal in effect with an additional attractive force varying as the inverse fifth power of the distance. As will be seen below, these give a fairly good agreement with experiment; it seems probable that the assumption of an inverse fourth power force of suitable magnitude will also give a fair account of the anomalous scattering.

Bieler's hypothesis of an additional inverse fourth power term in the law of force is an assumption solely made to account for the anomalous scattering. If the charged particles composing the nucleus and α particle describe orbits, it is natural to suppose that the nucleus and α particle may have magnetic moments. Two magnetic doublets, if in the proper relative orientation, will attract each other with a force varying as the inverse fourth power of the distance. The possibility of explaining the assumed inverse fourth power term in this way was examined by Bieler, and rejected on account of the artificial mechanisms which had to be postulated to obtain the proper orientation of the particles and also because large magnetic moments were necessary. Detailed calculations of the effects of magnetic forces in the scattering of α particles were made by Wessel*, and lead to the same conclusion.

A more general explanation† has been given by Debye and Hardmeier‡ on the following lines. According to our present views the aluminium nucleus must be a complex structure of positive and negative charges held in equilibrium under their mutual forces. When the α particle approaches close to the nucleus, the powerful external force due to it distorts the arrangement of the charges and the nucleus becomes polarised. The case is analogous to the polarisation of an atom or molecule by the approach of a charged particle. As in that case, it is assumed that the electric moment of the nucleus due to the displacement of the charges is proportional to the external force.

If P be the electric moment of the nucleus, we have

$$P = \alpha \cdot \frac{E}{r^2}$$

* Wessel, *Ann. d. Phys.* **78**, 757, 1925.
† The first suggestion of this explanation was given by Pettersson, *Wien. Ber.* **133**, 510, 1924.
‡ Debye and Hardmeier, *Phys. Zeit.* **27**, 196, 1926.

where r is the distance between nucleus and α particle, and α is the moment acquired by the nucleus in a field of unit strength. Since the force due to a dipole of moment P at a point on its axis is $2P/r^3$, the total force on the α particle will be

$$F = \frac{ZeE}{r^2} - 2\alpha\,\frac{E^2}{r^5}.$$

Thus the polarisation of the nucleus gives rise to an attractive force on the α particle, which varies inversely as the fifth power of the distance. By giving a suitable value to α, Debye and Hardmeier, and later Hardmeier* in greater detail, were able to show that this

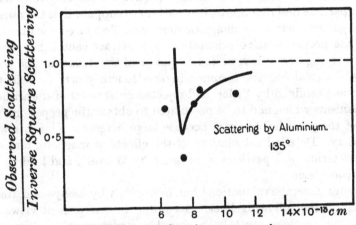

Distance of approach on inverse square force

Fig. 77.

hypothesis would account very fairly for the experimental results. The value of α was roughly 0.4×10^{-36} cm.³ for both aluminium and magnesium, that is, of the order of magnitude of the volume of the nucleus. According to Debye this may be taken to mean that the charged particles of which the nucleus is composed are held together by Coulomb forces.

The curves of Figs. 77 and 78 show the calculated scattering at angles of 135° and 90° respectively on this hypothesis, while the circles represent the observations of Rutherford and Chadwick. The agreement is good for the collisions of slow α particles, but there is a discrepancy in the case of the fast α particles which is particularly marked in the scattering at 90°. A similar result holds for the com-

* Hardmeier, *Phys. Zeit.* 27, 574, 1926; 28, 181, 1927.

parison between Bieler's results and the scattering calculated on this hypothesis; the theoretical curve agrees very well with the observations made with slow α particles, but for the faster particles gives values which are higher than are found. On the whole, the agreement is satisfactory and it seems possible to ascribe the anomalous scattering to the polarisation of the nucleus in the electric field of the α particle.

The divergences between the calculations on this hypothesis and the experimental results may be partly due to the difficulties of the experiments, for these cannot be carried out under sufficiently defined conditions, and partly to the assumptions which have been

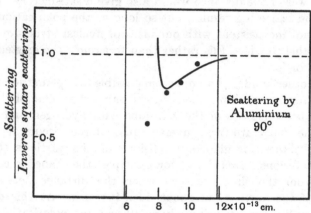

Distance of approach on inverse square force

Fig. 78.

made implicitly in the calculations, that the forces are central and that the α particle behaves as a point charge. The distances of approach in these collisions are so small that the structure of the nucleus, which may not be symmetrical, may be of importance, and the forces due to polarisation may not vary as some simple power of the distance. The α particle must also be considered to be a complex structure and must therefore acquire an electric moment, though perhaps a very small one, in the strong field of the aluminium nucleus.

Further, it has been assumed that the polarisation of the nucleus at any instant is proportional to the field due to the α particle at that instant. Owing to the inertia of the charged particles in the nucleus the polarisation will lag behind the field, and in consequence the additional term to the Coulomb force which arises from the polarisation will be, according to Kuhn*, proportional to r^{-3} over a portion of the orbit of the α particle.

Such considerations as these are however to be regarded as approximations which will not change the general agreement between the experiments and the deductions from the hypothesis of the polarisation of the nucleus.

It appears then that the hypothesis of the polarisation of the nucleus is able to account for the divergences from Coulomb's law which have so far been observed in the collisions of α particles with aluminium and magnesium nuclei. There can be little doubt that Bieler's assumption of an attractive force varying inversely as the fourth power of the distance would also give a satisfactory explanation of the scattering results, but so long as the polarisation hypothesis is not inconsistent with our ideas of nuclear structure it may be preferable to adopt this rather than to assume the existence of a new type of force.

On the other hand, it does not seem possible to explain the collisions with hydrogen or helium nuclei in the same way. One of the most striking characteristics of the collisions with hydrogen and helium is that the departure from inverse square forces occurs at a much greater distance for glancing collisions of the particles than for central collisions. In helium, for example, the glancing collisions depart from the simple theory when the distance between the centres of the particles is about 14×10^{-13} cm. At this distance the Coulomb forces are relatively small and any polarisation of the nuclei must correspondingly have a negligible effect. It is of no avail to assume a particular structure of the α particle, such as will give an asymmetrical field, if the inverse square law of force is retained, for this cannot provide forces of the necessary magnitude. A similar conclusion is arrived at by Hardmeier†, who has made recently a calculation of the effects of polarisation in the case of collisions in helium resulting in a deflection of 45°. He found that it was possible in this way to explain only the small initial fall below inverse square scattering of the experimental curve of Fig. 74 and that other causes must be sought to account for the very large and rapid rise at shorter distances of collision. Assuming that the polarisation forces are indeed responsible for the departure at large distances, he deduced that the polarisability α of the helium nucleus is about

* Kuhn, *Zeit. f. Phys.* **44**, 32, 1927.
† Hardmeier, *Helvetica Phys. Acta*, **1**, 193, 1928.

$1 \cdot 6 \times 10^{-39}$ cm.^{3}. This is extremely small compared with that found above for the aluminium nucleus, viz. 400×10^{-39} cm.^{3}, and it points to the conclusion that the forces which hold the helium nucleus together are much stronger than the inverse square forces.

Whatever may be the origin of these forces, it seems necessary to assume some plate-like "shape" for the region around the α particle in which they act. The experiments on the collisions of α particles in hydrogen suggested, regarding the H nucleus as a point charge, that the abnormal forces come into action over a spheroidal surface of semi-axes 8×10^{-13} cm. and 4×10^{-13} cm., the short axis being in the line of motion of the α particle. We should expect to find from the collisions in helium that the forces become abnormal over a region of about twice these dimensions. We have seen, however, that while for glancing collisions the critical distance is about 14×10^{-13} cm., for central collisions it is about $3 \cdot 5 \times 10^{-13}$ cm., actually rather less than the corresponding distance for collisions in hydrogen. If we suppose that the effects of distortion are small in these collisions, we may bring the hydrogen and helium results into agreement by assuming that the forces are also abnormal around the H nucleus or proton. In previous sections we have used a rough mechanical model, analogous to the model of a molecule used in the early kinetic theory of gases, in which the unknown forces around the α particle were likened to elastic forces. Interpreting the results in this way, we may say that the proton behaves like an elastic sphere of radius 2×10^{-13} cm., and the α particle like an oblate spheroid of semi-axes 7×10^{-13} cm. and 2×10^{-13} cm., the short axis being in the line of motion of the α particle.

In the interpretation of these collisions it has been assumed that the α particle has a definite orientation. No adequate proof of this assumption has been obtained, but it is strongly suggested by the experiments both in hydrogen and in helium. From the experiments in helium, indeed, it seems almost necessary to assume further that the helium nucleus does not present all possible orientations in the collisions, and to suppose that the abnormal forces also provide a turning couple controlling the relative orientations of the colliding nuclei.

We have seen that it seems impossible to account for the presence of strong forces over such an extensive region around the α particle solely on the basis of Coulomb forces, even when modified by the effects of polarisation.

It is natural, before postulating the existence of a force of a new type, to examine the possibility that the abnormal forces may be

276   ANOMALOUS SCATTERING IN HYDROGEN

of magnetic origin. The hypothesis of magnetic forces might help to explain in a general way many of the phenomena observed in the collisions of an α particle with hydrogen and helium nuclei, such as the rapid variation in scattering with the velocity of the α particle and the apparent orientations of the particles during the collision. It has been assumed in recent years that the electron has a magnetic moment of one Bohr magneton, and it was suggested by Frenkel* that the proton also may have a magnetic moment, of magnitude 1/1800 of a Bohr magneton. Experimental evidence in favour of this suggestion that the proton possesses a spin is afforded by the work of Dennison† on the specific heat of the hydrogen molecule, based on Hund's investigations of the band spectra of hydrogen. On the other hand, there is very strong evidence that the α particle has not merely a small magnetic moment but has none at all. This evidence is based on the fact that in the band spectrum of helium every alternate line is missing; this is interpreted to mean that no vector quantity, however small in magnitude, can be associated with the helium nucleus‡. The helium nucleus can therefore possess no spin and no magnetic moment. It is perhaps still possible to suppose that the slight distortion of the α particle produced in the collision causes it to acquire an appreciable magnetic moment; the alternative is to assume the existence of a hitherto unknown, though not altogether unsuspected, force in the immediate neighbourhood of the α particle §.

In the above examination and interpretation of the results of experiments on the collisions of α particles with atomic nuclei we have assumed that the laws of classical mechanics are valid. It is well known that the failure of classical mechanics in molecular events occurs when the wave-length associated with the particles is not small compared with molecular dimensions. Now the wave-length associated with the α particle is about 10^{-13} cm. and the distances of collision are only a few times greater. One may with reason inquire whether the classical mechanics is adequate to deal with these very close collisions and whether we should not rather adopt the methods of the new wave mechanics. This question has been examined by Wentzel‖, Oppenheimer¶, and most closely by Mott**, who has shown

---

* Frenkel, *Zeit. f. Phys.* 37, 243, 1926.   † Dennison, *Proc. Roy. Soc.* A, 114, 483, 1927.
‡ Cf. Solvay Report, "Electrons et Photons," p. 271, 1928.
§ See Mott, *Proc. Roy. Soc.* A, 126, 259, 1930; Chadwick, *Proc. Roy. Soc.* A, 128, 114, 1930. It is now clear that the α particle has a perfectly spherical field of force and that the "shape" deduced from the collision experiments is due to the distortion of the particle in the collision.   ‖ Wentzel, *Zeit. f. Phys.* 40, 590, 1926.
¶ Oppenheimer, *Zeit. f. Phys.* 43, 413, 1927.   ** Mott, *Proc. Roy. Soc.* A, 118, 542, 1928.

that the scattering of particles by a Coulomb centre of force follows the same laws on the new mechanics as on the classical theory, whatever the velocity of the incident particles. It might still arise that the anomalous scattering observed in the experiments is a consequence of a central field of force, not that of the inverse square, combined with the wave mechanics. Certain criteria have been given by Schrödinger and de Broglie for the conditions under which marked diffraction patterns may be expected to appear in the collisions of particles. These criteria have been interpreted by Blackett* in terms of collision relations founded on classical mechanics; they are, however, not precisely defined, and until some scattering phenomena have been completely explained in this way, so as to furnish comparative data, it is difficult to suggest where diffraction patterns may be expected to appear in collisions of a particles with atomic nuclei †.

It may, however, be mentioned here that one very important application of the new mechanics in the interaction of a nucleus and an a particle has been made by Gamow and by Gurney and Condon to explain the emission of a particles from radioactive nuclei and later, by Gamow, to estimate the probability of disintegration of nuclei. This application will be discussed in §§ 72 and 77 a.

§ **63. The potential of an a particle in the nuclear field.** If the abnormal forces around the helium nucleus extend over such distances as the experimental evidence suggests, it is possible that the marked divergences from Coulomb forces observed in the collisions of a particles with the aluminium nucleus may be due partly to a real departure from the Coulomb law of force as well as to the effects of polarisation. For the purpose of obtaining a general picture of the forces acting between an a particle and the nucleus of a light atom it is sufficient to suppose only that in addition to the Coulomb force of repulsion there is also a force of attraction which varies inversely as a high power of the distance, without making any additional hypothesis about the nature or origin of this force.

The potential energy of an a particle at a distance r from the centre of the nucleus will then be

$$V(r) = \frac{2Ze^2}{r} - \frac{a}{r^n},$$

---

* Blackett, *Proc. Camb. Phil. Soc.* **23**, 698, 1927.

† Attempts to apply quantum mechanics to the scattering of a particles by aluminium and magnesium have been made by Banerji, *Phil. Mag.* **9**, 273, 1930; Sexl, *Naturwiss.* **18**, 247, 1930; and Massey, *Proc. Roy. Soc.* A, **127**, 671, 1930.

where a is some constant which fixes the magnitude of the attractive force, and where $n = 3$ or $n = 4$ will agree with experiment. The potential energy of the α particle will be a maximum at a distance r_0 given by

$$r_0 = \sqrt[n-1]{\frac{na}{2Ze^2}},$$

and the value of the maximum potential will be

$$V(r_0) = \frac{2Ze^2}{r_0}\left(1 - \frac{1}{n}\right).$$

An estimate of the values of r_0 and $V(r_0)$ is provided by the scattering experiments. Thus Bieler, assuming $n = 3$, obtained a fair

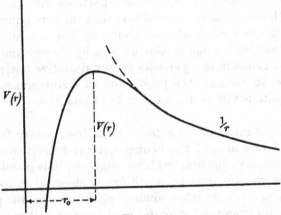

Fig. 79.

agreement between his observations with aluminium and the calculated scattering when r_0 was put equal to $3 \cdot 44 \times 10^{-13}$ cm. If $n = 4$, as is the case when the attractive field is due to polarisation, then Hardmeier's calculations lead to a value of r_0 of about 4×10^{-13} cm. Both assumptions lead to nearly the same value for the maximum potential, about $7 \cdot 4 \times 10^6$ electron-volts.

Although the maximum potential energy of an α particle in the field around the aluminium nucleus is on this view less than the kinetic energy of the α particle of radium C, yet on account of the motion of the nucleus during the collision the closest possible distance of approach of this α particle is about $4 \cdot 6 \times 10^{-13}$ cm.

The paths of α particles of velocity rather higher than those of thorium C' in such a potential field may be represented by the curves

of Fig. 80, given by Hardmeier* for the case $n = 4$. The curves show
the orbits corresponding to different impact parameters. If the
impact parameter is large, the orbit of the particle will be practically
the same as in a Coulomb field of force. As the impact parameter
decreases the angle of scattering will increase, but not so rapidly as
in a Coulomb field, that is, to attain the same angle of scattering the
particle must have a smaller p; there will therefore be more particles
scattered through the smaller angles and fewer through the greater
angles. As p decreases, the attractive force tends more and more to
compensate the repulsive force and the angle of scattering may
decrease instead of increasing; in this region the amount of scattering
may be greater than on the inverse square force. There will thus be
a maximum angle of scattering under these conditions. For a narrow

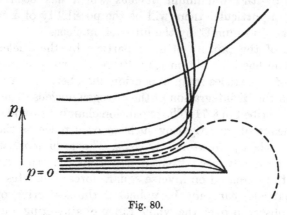

Fig. 80.

range of values of p the particle may describe orbits which loop
round the nucleus and may be in effect scattered at random. Finally,
for values of p smaller than a certain limiting value for which the
orbit is shown by the dotted curve, the particle will penetrate into
the region within the critical distance r_0 in which the resultant force
is one of attraction. It is to be supposed that such particles are
either re-ejected in all directions or are captured by the nucleus, and
such cases of capture are perhaps responsible for the disintegration
of certain nuclei under the bombardment of α particles.

The distribution of the scattered particles with angle compared
with that on a pure Coulomb field will be as follows: the scattering
will rise slightly above the normal amount at small angles and then
fall below it as the angle increases, afterwards rising rapidly in the

* Hardmeier, *Phys. Zeit.* **28**, 181, 1927.

neighbourhood of the maximum angle of deflection; at angles greater than this there will be present only those particles which are just not captured and describe looped orbits round the nucleus before escaping in random directions.

This view of the potential field between an aluminium nucleus and an α particle may be extended to other light elements. For a rough approximation we may suppose that the volume of a nucleus is proportional to its weight A and we may write for the value of r_0, the distance at which the potential energy of the α particle is a maximum,
$$r_0 = \text{ca. } 1 \cdot 2 \times 10^{-13} . A^{\frac{1}{3}}.$$

On this view the scattering of the α particles of radium C' by the nitrogen nucleus will be very similar to the scattering of the faster α particles by the aluminium nucleus which has been described above, and in particular there will be the possibility of a capture of an α particle of radium C' by the nitrogen nucleus.

Evidence of the capture of an α particle by the nucleus of the nitrogen atom has been obtained by Blackett[*], when photographing the tracks of α particles in an expansion chamber. The result of the capture was the disintegration of the nitrogen nucleus. These experiments are described in § 71. The expansion chamber provides perhaps the only means of investigating such an occurrence as the capture of an α particle by an atomic nucleus. The argument, which has sometimes been advanced, that a defect in the scattering below the amount anticipated on inverse square forces indicates a capture of the α particle, can only be upheld if the scattering of a given particle is observed over the whole range of scattering angles, for a defect over a certain range of angles may be compensated by an excess over another; if the excess occurs at small angles, where the scattering is relatively very large, a very small increase above the normal scattering, of such an amount as to be difficult to detect experimentally, will be sufficient to compensate a large defect at the greater angles.

While this picture of the potential field between an α particle and a nucleus is able to reproduce the salient points of the scattering experiments, it must be remembered that the evidence for it is by no means complete, and that the assumptions which have been made may not hold even approximately for very close distances of approach.

* Blackett, *Proc. Roy. Soc.* A, **107**, 349, 1925.

CHAPTER X

THE ARTIFICIAL DISINTEGRATION OF THE LIGHT ELEMENTS

§ 64. The idea of the artificial disintegration or transmutation of an element is one which has persisted since the Middle Ages. In the times of the alchemists the search for the "philosopher's stone," by the help of which one form of matter could be converted into another, was pursued with confidence and hope under the direct patronage of rulers and princes, who expected in this way to restore their finances and to repay the debts of the state. In spite of this encouragement the successful transmutation of some common matter into gold was but seldom reported. Even in these cases the transmutation could not be repeated; either the alchemist had vanished or his supply of the "philosopher's stone" had been exhausted. The failures were many and the natural disappointment of the patron usually vented itself on the person of the alchemist; the search sometimes ended on a gibbet gilt with tinsel. But when the confidence of the patrons departed the hope of the alchemists still remained, for the idea of transmutation not only accorded with the desires of the man but was founded on the conceptions of the philosopher. According to Aristotle all bodies are formed from a fundamental substance, "primordial matter," and the four elements—water, earth, air, and fire—differed from each other only by possession of different combinations of the properties of cold, wet, warm, and dry. By changing the properties one element should be changed into another. On this view, it was almost self-evident that bodies so closely allied as the metals could interchange their qualities. The experimental evidence was never more than meagre, and consisted mainly of such facts as that a trace of arsenic mixed with copper changed its red colour to grey-white. This was a first step towards the conversion of the copper into silver. Perhaps the most convincing evidence was that provided by an experiment in which a vessel of iron was changed into one of copper. This experiment consisted in immersing an iron vessel for several hours in certain natural springs. When removed the shape of the vessel was the same but the iron had changed into copper. The water of the spring contained traces of a copper salt, in an amount which could not be detected by the methods of those times, and when the vessel was immersed in this solution metallic copper was deposited on the surface of the iron, the less noble metal. This experiment is often quoted in alchemistical treatises.

With the development of experimental science and the growth of modern chemistry the authority of Aristotle was broken and the methods of the alchemist fell into disrepute. The idea of a primordial matter and the hope of a transmutation of the elements was, however, never completely abandoned. While Mendelejev himself was violently convinced of the independent nature and stability of the chemical elements, yet to some the discovery of the periodic system and the harmonious order of the elements suggested a common bond and relationship and revived the belief in transmutation. Faraday said: "To decompose the metals, to re-form them, and to realize the once absurd notion of transmutation—these are the problems now given to the chemist for solution." The weapons at the disposal of the chemist were however not sufficiently powerful to overcome the forces holding the atoms together, and it seemed for some time that the transmutation of elements was a theoretical possibility rather than a practical one.

The evidence provided by the radioactive elements of the spontaneous disintegration of atoms led naturally to renewed speculation and experiment concerning the disintegration of the ordinary inactive elements. Although the physical and chemical forces available in the laboratory were unable to influence the rate of transformation of the radioactive atoms, these themselves provided, in their radiations, new and energetic weapons of attack. It was thought that the radiations must occasionally penetrate very deeply into the structure of the atoms in their path, and it seemed possible that in some cases a disintegration of the atom might occur. At the same time it was generally realised that such cases of disintegration must be comparatively rare. Of the three types of radiation from the radioactive bodies the α particle was likely to be the most effective in promoting disintegration, owing to its enormous energy and momentum, corresponding to that of helium atoms at a temperature of 6×10^{10} degrees.

This question of the artificial disintegration of the ordinary chemical elements was attacked by numerous investigators. Perhaps the best-known of the early experiments are those of Cameron and Ramsay in 1907. Their method was to add radon to the substance to be examined, thus exposing it to the action of the α particles from radon, radium A, and radium C, and to test chemically for the products of disintegration. They announced some positive results, as, for example, that water treated in this way produced small quantities of neon and argon, and that copper produced traces of lithium. These experiments, however, did not stand the test of more rigorous examination, and it is clear that their results, and those announced later by Ramsay and Usher, were due not to the disintegration of matter exposed to the bombardment of α particles, but to the liberation of chemical elements already present as such in the substances examined.

The first successful experiment of this kind was carried out by Rutherford in 1919*. The method of experiment was extremely

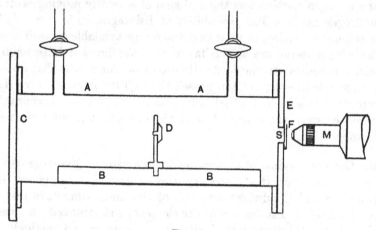

Fig. 81.

simple and at the same time very sensitive. The apparatus was the same as that used in his experiments on hydrogen collisions, and is shown again in Fig. 81. The α particles emitted by the source D of radium active deposit were absorbed in the gas used to fill the box. The opening S in the face of the box was covered with a sheet of silver and outside this opening was placed a zinc sulphide screen. When the gas through which the α particles passed was nitrogen, scintillations were observed on the screen even when the absorbing matter between the source and the screen had an air equivalent much greater than that which corresponded to the range of the α particles. When the nitrogen was replaced by oxygen or carbon dioxide these scintillations were not observed. Rutherford concluded therefore that

* Rutherford, *Phil. Mag.* **37**, 581, 1919.

the scintillations were due to particles ejected with great speed from the nitrogen nucleus by the impact of an α particle. The particles had approximately the same range as the protons set in motion when the α particles were sent through hydrogen, that is, about 28 cm. in air. He was able to show that the particles could not be due to any appreciable extent to hydrogen impurity in the nitrogen used in the experiment. Rough measurements of the magnetic deflection of the particles suggested that they were hydrogen nuclei or protons.

These experiments provided the first evidence of the artificial disintegration of elements. The instrument of disintegration was the α particle emitted by radium C, and the result of the disintegration of the nitrogen nucleus was the emission of a swiftly moving proton or hydrogen nucleus. The probability of disintegration is very small. The amount of hydrogen produced when any available quantities of radioactive material are used is far beyond the limits of detection by chemical methods. Rutherford estimated from his results that about one α particle in every million passed through the nitrogen gave rise to a proton. This indicates that only a very close or a very favourable collision between the α particle and the nitrogen nucleus results in a disintegration.

§ 65. General review of experiments on artificial disintegration. A few years after this discovery the subject was re-opened by Rutherford and Chadwick* and one of the main difficulties in the way of a general examination of the elements was removed. A closer study of the disintegration of nitrogen with improved methods of observing scintillations showed that the protons ejected from nitrogen had a much greater maximum range than the protons set in motion in hydrogen under similar conditions. For example, when the α rays of radium C of range 7 cm. in air were passed through hydrogen, no H particles could be detected when the total absorption in the path of the particles was equivalent to more than 29 cm. of air. When these α rays passed through nitrogen, however, particles were observed which had ranges greater than this, up to a maximum of 40 cm. of air. This observation showed at once that the H particles observed with nitrogen could not possibly be due to the presence of free hydrogen or hydrogen in combination as a chemical impurity, and it opened a safe and certain way to the examination of other elements for artificial disintegration. The element under test was

* Rutherford and Chadwick, *Phil. Mag.* **42**, 809, 1921; **44**, 417, 1922.

bombarded by the α particles from radium (B + C), and scintillations were looked for at absorptions greater than 32 cm. of air, that is, beyond the range of any protons due to the presence of hydrogen, whether free or in combination, in the material exposed to the α rays. In this way, unmistakable evidence was obtained of the disintegration of six elements, viz. boron, nitrogen, fluorine, sodium, aluminium, and phosphorus. Many other elements were examined, but no certain evidence of a disintegration resulting in the expulsion of protons of long range was found. It was, of course, possible that in some elements the maximum range of the liberated proton might be less than 32 cm. in air. The investigation for particles of shorter range than this led to no conclusive result. It was found that some protons were always present under such conditions, whatever the material examined. Some of these were emitted from the source of α particles (cf. § 59), and some were due to hydrogen impurity in the substances exposed to bombardment of the α particles. The examination at small absorptions was further complicated by the presence of long range α particles from the source of radium C (cf. § 20) which are emitted in such numbers as to interfere seriously with the detection of the small number of particles to be expected from disintegration of the material under examination.

These experiments had, however, shown that the protons liberated in disintegration were emitted in all directions. It was therefore legitimate to search for particles ejected from the bombarded material in a direction at right angles to the direction of the incident α particles. This method had the advantage that hydrogen impurity in the material no longer affected the results and that the number of long range α particles scattered at 90° was so small as to be negligible. The search for particles of disintegration could thus be pursued with safety down to an absorption corresponding to 7 cm. of air, the range of the α particles from the source of radium (B + C), and with special precautions to even smaller absorptions in the case of the lighter elements. This method was adopted at the same time by Kirsch and Pettersson in their experiments on artificial disintegration.

In this way Rutherford and Chadwick* found evidence of the disintegration of all elements from boron to potassium, with the two exceptions of carbon and oxygen. No certain evidence of the disintegration of lithium, beryllium, or the elements beyond potassium

* Rutherford and Chadwick, *Nature*, **113**, 457, 1924; *Proc. Phys. Soc. Lond.* **36**, 417, 1924.

was obtained. In all cases the product of the disintegration was a swiftly moving proton or H nucleus. No evidence of the liberation of particles of other types was obtained.

Parallel with the later experiments of Rutherford and Chadwick progressed the investigations of Kirsch and Pettersson and their colleagues in Vienna. These workers have carried out a large number of experiments on the artificial disintegration of elements and points connected therewith. Their results are in most cases very different from those obtained by Rutherford and Chadwick. Put shortly, the main point of difference is that the Vienna results indicate that the disintegration of an element is a much more frequent occurrence than would appear from the Cambridge results. Kirsch and Pettersson find that nearly every element can be disintegrated by bombardment with α particles and that the chance of disintegration is much greater than is suggested by Rutherford and Chadwick's experiments*.

It seems difficult to put forward any satisfactory explanation of these and other divergences between the two series of investigations, and this is not the place to attempt it. Many of the doubtful points would be most clearly decided by attacking the problem in a different way, by obtaining photographs of the disintegration in a Wilson expansion chamber. The first step in this direction has already been taken and pictures of the disintegration of nitrogen have been obtained by Blackett and by Harkins and his co-workers. These experiments will be discussed later (§ 71). It would also be of advantage to repeat the experiments using, instead of the scintillation method, an electrical method of detecting the particles of disintegration. Investigations of this kind have been begun by Bothe and Fränz†, and some results have been published.

In this account of artificial disintegration we shall, in order to avoid fruitless discussion, refer only to the experiments carried out in the Cavendish Laboratory and we shall deal as far as possible with those points on which there is little conflict of opinion. For the full statement of the Vienna observations the reader is directed to the original papers of Kirsch and Pettersson and to their book *Atomzertrümmerung* ‡.

---

* A discussion of some of the points at issue has been given by Chadwick, *Phil. Mag.* 2, 1056, 1926, and by Kirsch, Pettersson, and others, *Zeit. f. Phys.* 42, 641, 1927.

† Bothe and Fränz, *Zeit. f. Phys.* 43, 456, 1927; 49, 1, 1928.

‡ Kirsch and Pettersson, *Atomzertrümmerung*, Akademische Verlagsgesellschaft m.b.H., Leipzig, 1926.

§ 66. Disintegration of nitrogen and other elements. The experimental arrangement used by Rutherford and Chadwick is shown in Fig. 82. The face of the brass tube T was provided with a hole 5 mm. in diameter, which was closed by a silver foil of 6 cm. air equivalent. The source R of α particles was a disc coated with radium (B + C). The zinc sulphide screen S was fixed to the face of the tube so as to leave a slot in which absorbing screens could be inserted.

When a stream of dry air was passed through the apparatus scintillations were observed on the zinc sulphide screen. The number of these was counted for different thicknesses of mica inserted in the path of the particles. The result of such an experiment is shown in curve A, Fig. 83. The ordinates represent the number of scintillations observed per minute per milligram of activity of radium C, and

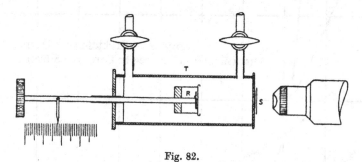

Fig. 82.

the abscissae give the air equivalent of the absorbing screens. It is seen that scintillations were observed up to an absorption corresponding to 40 cm. of air. When the air in the apparatus was displaced by a mixture of CO_2 and H_2 which had the same stopping power for α rays as air, the results shown in curve B were obtained. At small absorptions the number of particles from this mixture was very much greater than the number of particles from nitrogen, but no particles were observed at absorptions greater than 29 cm. of air. The particles observed with nitrogen at absorptions greater than this could therefore not arise from any hydrogen impurity in the gas or in the materials exposed to the impact of the α particles. They must be ascribed to the disintegration of the nitrogen nucleus.

When dry oxygen was passed through the tube only a few scintillations were observed and none were found at absorptions greater than 29 cm. of air (curve C). There was thus no evidence of a disintegration of the oxygen nucleus. The scintillations observed in this case were

ascribed to slight hydrogen impurities in the surfaces exposed to the
α rays and to the "natural H particles" from the active source.

The effects observed when a thin foil of aluminium was placed
over the source (the air in the apparatus being displaced by oxygen)
are shown in curve D. It is seen that particles are liberated from
aluminium and that some are able to penetrate relatively large thicknesses of matter, up to an absorption equivalent to about 90 cm.
of air.

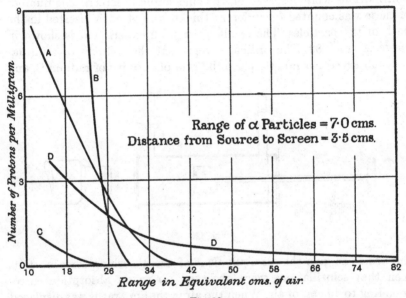

Fig. 83.

In this simple way many elements were exposed to the bombardment of α particles and tested for disintegration. The element to be
examined was placed immediately in front of the source R. When
possible thin foils of the element were used. In some cases the
material was in the form of powder, generally an oxide; a film of
such material was prepared by dusting the powder on to a thin gold
foil smeared with alcohol. A stream of dry oxygen was passed
through the apparatus and a search was made for the liberation of
particles with ranges greater than 32 cm. of air. It is clear that such
particles could not arise from hydrogen contamination but must be
ascribed to disintegration of the atoms of the element exposed to
the impact of the α particles. It was shown in these first experiments

that, in addition to nitrogen and aluminium, the elements boron, fluorine, sodium, and phosphorus were also disintegrated. The maximum ranges of the H particles liberated in the disintegration were in each case greater than 40 cm. of air.

§ 66 a. Experiments by the right angle method. It was observed in the course of the above experiments that the H particles were liberated from aluminium not only in the same direction as the motion of the

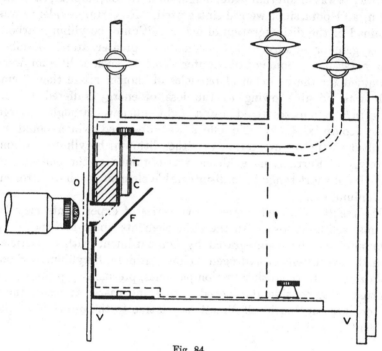

Fig. 84.

beam of α particles but also in the opposite direction, and later experiments showed that this held also for other cases of disintegration. It appeared that the H particles were emitted roughly equally in all directions relative to the incident α rays. This observation led to an extension of the experiments to much smaller absorptions than could be used in the above method.

The beam of α rays from the source C (Fig. 84) fell on the material to be examined placed at F and the liberated particles were observed on the zinc sulphide screen O, placed at an angle of 90° or more to the direction of the incident α particles. The source and the material under examination were contained in an evacuated brass box. The zinc sulphide screen was outside the box, opposite a window covered with a sheet of mica of 7 cm. air equivalent. In this arrangement any scintillations observed on the screen could be due only to particles ejected in the disintegration of the bombarded atoms. The presence of hydrogen had no effect and the number of long range α particles scattered through 90° was too small to disturb the observations. In this way it was found that neon, magnesium, silicon, sulphur, chlorine, argon, and potassium were disintegrated. No certain evidence was obtained of the disintegration of helium, lithium, beryllium, carbon, or oxygen, or of any of the elements of greater atomic number than potassium. Some of the lighter elements, such as lithium, were examined for the emission of particles of shorter range than 7 cm. This was possible owing to the loss of energy suffered by an α particle when scattered by a light nucleus through a large angle (compare § 57). No effect was observed which could be ascribed to the disintegration of either lithium, beryllium, carbon, or oxygen. Further, no evidence was obtained of the emission of particles of short range from disintegrable elements such as nitrogen, aluminium, etc.

The results obtained in the above series of experiments may be summarised as follows. All the light elements up to and including potassium can be disintegrated by bombardment with α particles with the exceptions of hydrogen, helium, lithium, beryllium, carbon, and oxygen. In the disintegration particles, presumably protons, are emitted with considerable speeds. In some cases, at least, these protons are emitted in all directions relative to the direction of the incident α particle.

§ 67. **The magnetic deflection of the particles.** The number of the particles emitted in these disintegrations is too small to permit a determination of the ratio of the charge to the mass by the usual methods. An approximate estimate of the deflection of the particles in a magnetic field has been obtained by the following method, similar in principle to that described in § 20.

The source of α rays was placed at R and was inclined at an angle of 20° to the horizontal face of the brass plate S (Fig. 85). Source and plate were contained in a box placed between the poles of an electromagnet, the field being perpendicular to the length of the plate.

The material, the particles from which were to be examined, was laid directly over the source.

The estimate of the magnetic deflection of the particles was obtained by observing the effect of a magnetic field on the number of scintillations near the line E, the shadow of the edge of the plate S which was formed on the zinc sulphide screen Z.

By allowing the α rays from the source to strike the screen, the position of the line E was clearly defined as the edge of the band of scintillations. The microscope was adjusted so that this edge was a little above a horizontal cross-wire in the eyepiece, which marked the centre of the field of view. The material to be examined, say aluminium, was now placed over the source and dry oxygen was

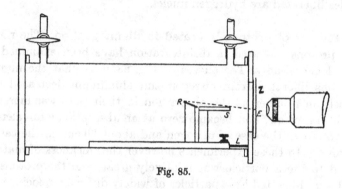

Fig. 85.

circulated through the box. The scintillations appearing on the screen were now due to the long range particles resulting from the disintegration of aluminium. Their number was far too small to give a band of scintillations with a definite edge. It is clear that if a magnetic field be applied in such a direction as to bend the particles upwards the edge of the shadow should move upwards and the scintillations should appear only in the lower half of the field of view of the microscope. If the field be applied in the opposite direction, the edge E should move downwards and the scintillations should occur over the greater part of the field of view. The ratio of the numbers of scintillations observed in this way clearly gives a measure of the amount of deflection suffered by the particles. This ratio was determined for the long range particles liberated from aluminium, phosphorus, and fluorine, and it was compared with that for projected H particles of known velocity and also for the α particles of 8·6 cm. range emitted by thorium C'. It was shown in each case that the long range particles carried a positive charge, and that they were deflected in a magnetic field to the degree to be expected if they were H nuclei moving with a velocity estimated from their range as measured in absorption experiments. There can be little doubt therefore that the particles are H nuclei, though it is difficult to obtain conclusive proof without an actual determination of the velocity and the value of e/m of the particles.

While the measurement of the magnetic deflection of the particles liberated in disintegration has been examined only for four elements, viz. nitrogen, fluorine, aluminium, and phosphorus, there is reason to suppose that the other elements would give similar results and it is very probable that in all the known cases of disintegration the particles liberated are hydrogen nuclei.

§ 68. **Ranges of protons liberated in disintegration.** The ranges of the protons liberated in disintegration have been examined for a few elements only. The curves of Fig. 83 show that the number of protons liberated from nitrogen and aluminium decreased continuously in number as the absorption in their path was increased from 30 cm. of air, and became zero at an absorption equivalent to 40 cm. of air in the case of nitrogen and about 90 cm. in the case of aluminium. In these experiments however the thickness of material exposed to the α particle was relatively great, and the protons observed were liberated by α particles of widely differing velocities and over a wide range of angle to the direction of the effective α particle.

The results were therefore of value mainly in giving the maximum range of the protons liberated under bombardment by α particles of a known initial range.

Experiments under similar conditions were made to measure the maximum range of the protons liberated in the opposite direction to that of the incident particles. This was called the maximum range in the backward direction. Since the material was placed immediately behind the source, it was possible for protons liberated at all angles from 180° to nearly 90° to be observed. It is therefore doubtful whether this "backward" range should be attributed to the protons emitted in directions making angles close to 180° with that of the α particle, and it may be that it really corresponds to a direction of emission not much more than 90°.

In the later experiments in which the protons were examined at right angles to the direction of the α particles the maximum ranges were determined very roughly, except in the two cases of nitrogen and magnesium.

The results of these measurements are given in the following table. The initial range of the bombarding α particles was 7 cm. The ranges of the protons are expressed in equivalent cm. of air at 15°.

Element	Atomic number	Atomic masses	Maximum range of proton		
			Forward	Right angles	Back-ward
Boron	5	10, 11	58	—	38
Nitrogen	7	14	40	24	18
Fluorine	9	19	65	—	48
Neon	10	20, 22	—	16	—
Sodium	11	23	58	—	36
Magnesium	12	24, 25, 26	40	24	—
Aluminium	13	27	90	—	67
Silicon	14	28, 29, 30	—	18–30*	—
Phosphorus	15	31	65	—	49
Sulphur	16	32, 33, 34	—	18–30	—
Chlorine	17	35, 37	—	18–30	—
Argon	18	36, 40	—	18–30	—
Potassium	19	39, 41	—	18–30	—

* In this and other cases where the range is given as 18–30, it is to be taken that the range has not been measured but probably lies between 18 cm. and 30 cm. of air.

It will be noted that the maximum range of the proton depends on the direction in which it is emitted relative to the direction of the incident α particles. This is not unexpected, for one would suppose on general grounds that momentum must be at least approximately conserved in these collisions. The nucleus struck by the α particle will receive a momentum in the forward direction and the proton liberated from the moving nucleus may be expected to have a greater velocity in the forward direction than in the backward.

On certain hypotheses as to the mechanism of the disintegration it is possible to calculate the relation between the range of the incident α particle and the range of the proton emitted in the forward and in the backward direction (cf. § 72), but in view of the uncertainty regarding the "backward" ranges it is scarcely justifiable to draw any definite conclusions from the comparison of these calculations with the observations.

Some striking similarities in the ranges of the protons from different elements may be pointed out. The seven elements which have been examined most closely may be grouped in three pairs, with aluminium as the exception. The maximum ranges of the protons from boron and sodium are about the same, as are those from nitrogen and magnesium, and fluorine and phosphorus.

While the data are still rather meagre we may say that, in general, the protons emitted by the elements of odd atomic number have a greater range than those emitted by the elements of even atomic number.

It is also of interest to note that the kinetic energies of some of the protons liberated from some elements may be greater than the energy of the incident α particle. Assuming as a rough approximation that the range of the proton is proportional to the cube of its velocity, and making use of the fact that a proton of velocity 1·6 times the velocity of the α particle of radium C has a range in air of about 29 cm., we find that a proton which has a kinetic energy equal to that of the α particle of radium C will have a range of about 57 cm. in air. We see therefore that in several cases of disintegration the kinetic energy of the emitted proton may be greater than that of the incident α particle; for example, the maximum range of the protons ejected from aluminium is 90 cm., corresponding to an energy 1·37 times that of the α particle of radium C responsible for the disintegration. The nucleus which was struck by the α particle and disintegrated will in general also have an appreciable kinetic energy. It follows therefore that in several cases of disintegration the sum of the kinetic energies after impact of the α particle, the released proton, and the residual nucleus may be distinctly greater than the initial kinetic energy of the α particle.

§ 68 a. Minimum range. Some experiments, full details of which were not published*, were made under more definite conditions to examine the protons liberated with short ranges. A thin foil of aluminium, of about 5 mm. air equivalent, was bombarded by α particles of 7 cm. range from radium (B + C) and the ejected protons were examined in a direction at right angles to the α ray beam. An absorption curve was taken in the usual way with the result shown in Fig. 86, where the ordinates represent the number of protons observed at absorptions in equivalent cm. of air given by the abscissae. It is seen that the gradual increase in the number of protons observed as the absorption in their path is decreased does not continue down to the smallest ranges. The particles appear to possess a minimum range of about 10 to 12 cm. of air.

* Cf. Rutherford and Chadwick, *Proc. Camb. Phil. Soc.* 25, 186, 1929.

A similar absorption curve was determined for the protons ejected from sulphur. The curve was similar in shape and gave a value for the minimum range of the proton of about 15 cm.

These experiments suggest that the protons liberated in a disintegration possess a certain minimum energy as well as a maximum energy. On the classical mechanics one would expect that no proton could escape from a nucleus with a kinetic energy less than that corresponding to a fall through the maximum potential in the field of the nucleus. The experiments on the scattering of α particles by aluminium suggest that the potential of a proton in the field around the aluminium nucleus would have a maximum value of about

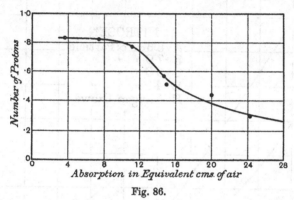

Fig. 86.

$3\cdot7 \times 10^6$ electron volts (cf. § 63). The energy of the proton of minimum range released from aluminium corresponds to a fall through about 3×10^6 volts, which is roughly of the same magnitude. On the other hand, considerations which will be discussed in § 72 indicate that the laws of classical mechanics are probably not applicable in these disintegrations, and it seems possible that protons may escape from a nucleus with smaller energy than corresponds to the maximum potential. A test of this hypothesis could probably be obtained by investigating how the minimum range of the proton depends on the range of the incident α particle; on classical mechanics the minimum range should be characteristic of the disintegrating nucleus and independent of the range of the α particle. As yet however no such measurements have been made.

§ 69. The effect of the velocity of the α particle on disintegration. In two cases only has the disintegration been studied in any detail, viz. in the cases of aluminium and nitrogen. With these elements some experiments were made to find how the number and range of the emitted protons depend on the velocity of the bombarding α particle.

The experimental arrangement was that shown in Fig. 82, and the method of experiment was as already described. Sources of thorium C were used to obtain α particles of 8·6 cm. range, and sources of α rays of less than 7 cm. range were obtained by placing gold or silver foil

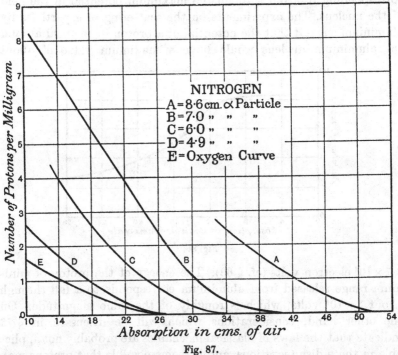

Fig. 87.

of the requisite stopping power over a source of radium (B + C). The results of the experiments with nitrogen (in the form of air) are shown in Fig. 87, where the ordinates give the observed number of protons per milligram activity of the source, and the abscissae the absorption in the path of the particles expressed in equivalent cm. of air. Only the end part of the curve A, for particles of range 8·6 cm., is given, since the initial portion is complicated by the effects of the α particles of 4·8 cm. range which are also emitted from a source of thorium C. The maximum ranges of the protons were 51, 40, 34 and 27 cm. for α particles of range 8·6, 7·0, 6·0 and 4·9 cm.

respectively. To a rough approximation, therefore, the maximum range of the proton was proportional to the range of the incident α particle.

It will be seen from the curves that the number of protons observed increased rapidly with the range of the α particle. It is perhaps not justifiable to use these results for a strict comparison of the effects of α particles of different ranges, for the experimental conditions were not suitable. The α particles passed through a length of 3·5 cm. of gas and, in the absence of information about the angular distribution of the protons, there is no means of estimating the effects of different parts of the gas. Further, owing to the presence of the long range α particles, observations were not taken below an absorption of 12 cm. of air, and it appears from the curves that the number of protons was still increasing with diminishing absorption. We may however make a rough comparison of the effects of α particles of different ranges. After allowing for the "natural" scintillations as given by curve E, we find that the number of protons observed at 12 cm. absorption was 6·5, 3·4, 0·8 per mg. activity of the source for particles of range 7·0, 6·0, and 4·9 cm. respectively. The chance of producing a disintegration therefore decreases very rapidly as the velocity of the α particle is reduced.

Similar observations were made in the case of aluminium, though the results were in some respects less definite owing to the smaller number of protons and the shape of the absorption curve.

In these experiments the effects of the long range α particles and of the "natural" scintillations from the source and materials near it prevented a thorough examination and an extension of the observations to the effect of α particles of shorter range. For example, when α rays of range 4·9 cm. were used a small but definite number of particles was observed to be liberated from nitrogen, but with aluminium it was doubtful whether any residual effect existed.

In some later observations the right angle method of experiment was used. The relative numbers of protons observed for incident α rays of different ranges are shown in the curve of Fig. 88. The element examined was nitrogen and the total absorption in the path of the protons was equivalent to 5 cm. of air. The source of α particles was radium (B + C) and the ranges shorter than 7 cm. were obtained by placing absorbing screens of mica over the source. Since a wide cone of particles must be used in order to obtain a measurable number of protons, the beam of incident particles is necessarily heterogeneous.

In the figure an average value of the range of the incident particles
has been taken. It will be seen from the figure that the chance of
disintegration increases rapidly with the velocity of the α particle.
It is very small for a velocity corresponding to a range of 3 cm. in
air and would appear to fall to zero at a range not much less than
this. The minimum range of the α particle effective in producing
disintegration could not be fixed with any accuracy for the reason
mentioned above.

Similar results were obtained in experiments in which the disintegration of aluminium was examined. Few if any particles of

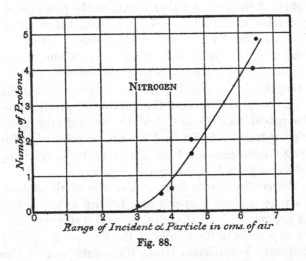

Fig. 88.

disintegration could be observed when the range of the incident
α particles was less than 3 cm., and as the range of the α particles
was increased above this value the number of protons rose rapidly.

**§ 70. The probability of disintegration. The numbers and distri-
bution of the protons.** It is difficult to make a reliable estimate
of the probability of the disintegration of a nucleus by the collision
of an α particle, for no comprehensive study has so far been made
with any element. This question is intimately connected with the
angular distribution of the protons emitted during disintegration. It
is known that the protons are emitted in all directions, for they have
been observed in the same direction, at 90°, and at 135° to the
direction of the incident α particles, but as yet there is very little
information about the variation in the number of protons with the

direction of emission. An inspection of the curves of Fig. 89, which show the numbers of protons observed from aluminium in the forward and backward directions, suggests that the number in the backward direction is decidedly less than in the forward. The observations were not carried to small absorptions on account of the effects of the long range particles and of free hydrogen, so that they provide no definite information about the total numbers of protons released under these conditions. The suggestion that there is a preponderance of protons in the forward direction is however supported by comparison of the number of protons observed in these experiments

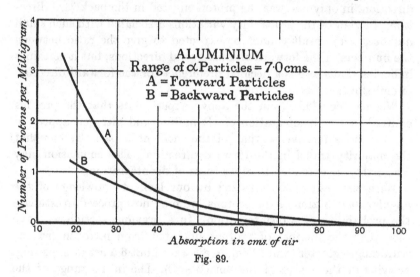

Fig. 89.

with those observed in others where the angle of observation was 90° and 135°. In some experiments, not yet published, a polonium source was enclosed between a sheet of aluminium foil and a sheet of platinum foil and placed in the centre of a Wilson expansion chamber. The tracks of the protons liberated from the aluminium were visible over a very large solid angle on each side of the source. They were photographed by two cameras at right angles as in Blackett's experiments. The effects of free hydrogen in the materials cannot be eliminated in this experiment but are not likely to be large. The tracks due to hydrogen collisions are as likely to appear in the forward direction, i.e. from hydrogen in the aluminium, as in the backward, from hydrogen in the platinum; and, since the total absorption in the path of the particles was 8 cm. and the range of the α particle from polonium is 3·9 cm., they should be few in number. The analysis of a number of photographs, in which the tracks of nearly 200 protons were obtained, showed that about two to three times as many protons appeared in the forward direction as in the backward.

Further evidence on this point is provided by the experiments of Blackett, which are described in the next section. In an examination of the tracks of α particles in nitrogen Blackett observed eight cases in which the collision resulted in the disintegration of the nitrogen nucleus. In seven of these the proton was emitted in the forward direction, in only one was the proton emitted in the backward direction. In view of the probability variations associated with such small numbers, this result cannot be expected to give the ratio between the numbers in the forward and backward directions, but it certainly indicates that there is a decided preference for the forward direction of emission.

We may conclude from the above experiments that the protons emitted in the disintegration of aluminium and nitrogen appear in all directions relative to that of the incident α particle, but that the majority travel in the forward direction. This conclusion may reasonably be extended to other cases of disintegration.

With the reservation imposed by our lack of knowledge of the angular distribution of the protons, we may now proceed to estimate the probability of a disintegration. In a previous section we have seen that the number of protons liberated from nitrogen (as air) with ranges greater than 12 cm. in air was about 6·5 per min. per mg. activity of the source of radium (B + C). The initial range of the incident α particles was 7 cm. and the protons were observed on a zinc sulphide screen of 8·3 sq. mm. area placed at a distance of 3·5 cm. from the source. The efficiency of the screen was about 0·75. Assuming for the purpose of calculation that the protons are emitted equally in all directions, we find that about 18 protons are liberated for every million α particles of 7 cm. range passing through the nitrogen. To take into account the preponderance of protons in the forward direction, this result should be reduced. On the other hand, the observations probably did not record the total number of protons, for those with ranges less than 12 cm. did not reach the screen, and the conditions of experiment were such that the whole path of the α particle was not effective. As a very rough estimate of the probability of disintegration we may say that when one million α particles of 7 cm. range are totally absorbed in nitrogen about 20 nitrogen nuclei will be disintegrated.

Similar calculations may be made for the case of aluminium. In this case about 8 nuclei are disintegrated when one million α particles of 7 cm. range are totally absorbed.

Only the roughest estimates of the probability of disintegration can be given for the remaining elements. In general it may be said that the probability is greatest for the lighter elements and least for the heavier, and greater for the elements of odd atomic number than for those of even atomic number. For example, the chance of disintegrating a potassium nucleus by bombardment with an α particle is less than that of disintegrating a sodium nucleus, and the chance for a silicon nucleus is less than for an aluminium or a phosphorus nucleus. The probability of disintegration for the heavier elements, such as chlorine, argon, potassium, is very small, about one nucleus being disintegrated for every million α particles of 7 cm. range. It is not justifiable to make more definite comparisons of the results obtained for one element with those for another owing to the fact that the experiments were not carried out under the same conditions.

It appears therefore that the chance of disintegration may depend not only on the charge of the nucleus but on the internal structure. It also depends greatly on the velocity of the incident α particle. We have seen in § 69 that the chance of disintegration is very small for small velocities of the α particle and increases rapidly as the velocity of the α particle is increased.

§ 71. **Experiments in the expansion chamber.** The experiments so far described have consisted in the detection by the scintillation method of the proton emitted in the disintegration of an atom. This method can give no direct information about the motion of the atom or of the α particle after the disintegrating collision. The Wilson expansion chamber provides an obvious and perhaps the only certain means of observing the motion of these particles. Since the probability of a collision resulting in disintegration is small, for example about 20 protons are liberated when one million α particles of 7 cm. range are passed through nitrogen, it is necessary to photograph a very large number of tracks. The expansion chamber must therefore be worked automatically in order to provide the required information in a reasonable time (see § 12 d). Experiments of this kind have been carried out by Blackett* and by Harkins†. In both cases the tracks were photographed in two directions at right angles, both pictures being received on the same photographic film.

In Blackett's experiments the apparatus made one expansion and took one photograph every 10 or 15 seconds. The source of α rays was thorium (B + C), giving particles of 8·6 cm. range and 4·8 cm. range. The expansion chamber was filled with nitrogen containing about 10 per cent. of oxygen, which was added to improve the sharpness of the tracks. The average number of tracks on each photograph was 18 and about 23,000 photographs were taken. Thus the tracks of about 270,000 α particles of 8·6 cm. range and 145,000 of 4·8 cm. range were obtained.

Among these tracks a large number of forks was found corresponding to elastic collisions between the α particles and nitrogen atoms. In addition to these normal collisions, however, eight were obtained of a very different type. Two of these are reproduced on Plate X, figs. 1 and 2. These eight forks represent the ejection of a proton from a nitrogen nucleus. In each, however, there are only two branches. The path of the ejected proton, as an inspection of the photographs will show, is obvious. It consists of a fine straight track along which the ionisation is clearly less than along the α tracks near it. The second of the two branches of the fork is a short track similar in appearance to the track of the nitrogen nucleus in a normal forked track. There is no sign of a third branch to correspond to the track of the α particle itself after the collision. It is difficult to see how an α particle which has penetrated into the structure of the nitrogen nucleus can escape without acquiring sufficient energy to produce a visible track. Since no track appears on the photographs, we must conclude that the α particle does not escape but that in ejecting the proton from the nitrogen nucleus it is itself absorbed into the nuclear structure. The resulting nucleus must have a mass of 17 and, if no electrons are gained or lost in the process, an atomic number of 8 (cf. p. 313).

The conclusions based on the appearance of the anomalous tracks were confirmed by measurement. Blackett showed that the angles between the components of each of these anomalous forks were not consistent with an elastic collision. Kinetic energy was not conserved in these collisions. The anomalous forks were however coplanar, indicating that momentum was conserved. If ψ and ω are the angles

---

* Blackett, *Proc. Roy. Soc.* A, **107**, 349, 1925.

† Harkins, *Zeit. f. Phys.* **50**, 97, 1928.

PLATE X

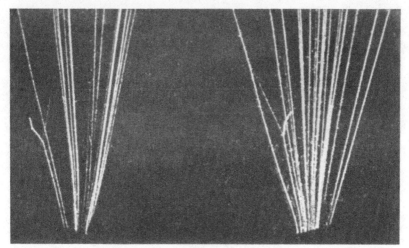

Fig. 1.

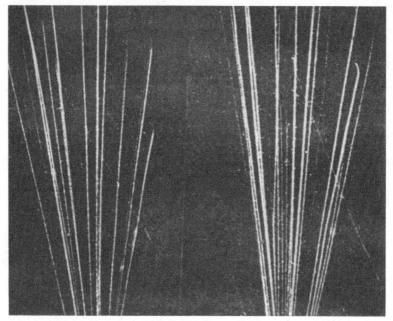

Fig. 2.

The ejection of protons from nitrogen nuclei by α particles. In Fig. 1 the proton is emitted in the forward direction, in Fig. 2 in the backward direction.

between the initial track of the α particle and the track of the proton and the new nucleus respectively, we have, if momentum is conserved,

$$m_p v_p \sin \psi - m_n v_n \sin \omega = 0,$$

$$m_p v_p \cos \psi + m_n v_n \cos \omega = MV,$$

where m_p, m_n are the masses and v_p, v_n the velocities of the proton and final nucleus, and M, V the mass and velocity of the α particle at impact.

From these we find that the momenta of the proton and nucleus are given by

$$m_p v_p = MV \sin \omega / \sin (\psi + \omega),$$

$$m_n v_n = MV \sin \psi / \sin (\psi + \omega).$$

The velocity V of the α particle was calculated from the distance of the fork from the source, ψ and ω were measured from the photographs. Thus the momenta of the particles making the branches of the fork could be found. Assuming that the fine track was due to a proton, the velocity of the proton was deduced. The results obtained were in fair agreement with the measurements of Rutherford and Chadwick on the ranges of the protons ejected from nitrogen nuclei.

The momentum of the residual nucleus was also found from the angles of the fork, and its range was measured and compared with the ranges observed for recoiling nitrogen atoms (§ 58). Blackett concluded that the results were consistent with the assumption that the residual nucleus had an atomic number 8 and a mass 17.

If we assume that the mass of the residual nucleus is 17, then the kinetic energy of the particles after collision can be calculated and compared with that of the incident α particle. From the measurements of six disintegration forks, Blackett found that the ratio of the sum of the kinetic energies of the nucleus and proton to that of the α particle at the point of impact was 0·83, 1·02, 0·72, 0·68, 0·88, 0·93. On the average therefore there was a loss of kinetic energy of 19 per cent. of the energy of the incident α particle.

It is preferable to express the energy change in the disintegration in terms of some definite unit, say as a fraction of the energy of an α particle of 7 cm. range. Blackett's observations then give for the energy change in the six disintegrations obtained $- 0·16$, $+ 0·02$, $- 0·30$, $- 0·33$, $- 0·11$, and $- 0·06$ times the energy of an α particle of 7 cm. range.

The frequency of the disintegrations was in rough agreement with the estimates of § 70, when the velocity of the α particles was taken into account, for 8 cases of disintegration were observed for 270,000 α particles of 8·6 cm. range and 145,000 of 4·8 cm. range.

Some collisions were observed in which the α particle was deflected through a large angle (110° to 120°) without the ejection of a proton. If these collisions were in fact with nitrogen and not with oxygen and if the α particle concerned in the collision was one of the group of 8·6 cm. initial range, this observation would show that the disintegration of a nitrogen nucleus does not necessarily occur when the α particle penetrates within a certain distance of the centre of the nucleus, but that some other factor determines whether an elastic collision is to be made or whether the α particle is to be captured and a proton ejected; for the distance of approach of the α particle was on this assumption about $3·5 \times 10^{-13}$ cm. on the Coulomb law of force, and the experiments described in § 69 show that disintegration of the nitrogen nucleus may occur when the distance of approach is less than about 6×10^{-13} cm. If, however, the α particle concerned in the collision belonged to the shorter group of particles, the distance of collision may have been a little greater than the critical distance. Curtiss* also has reported a collision of an α particle with a nitrogen nucleus in which the distance of collision was estimated to be about 5×10^{-13} cm. but in which there was no evidence of disintegration.

§ 72. **Discussion of the results†.** We have seen in the foregoing sections that all the elements from boron to potassium can be disintegrated by the bombardment of α particles, with the two exceptions of carbon and oxygen. The result of the disintegration is in every case the emission of a hydrogen nucleus or proton. No evidence has been obtained of the ejection of particles of any other type. So far as has been ascertained, only one proton is emitted from the disintegrating nucleus; this appears most clearly from the experiments in the expansion chamber.

While in some cases, e.g. aluminium, the kinetic energy of the particles after disintegration may be greater than the kinetic energy of the incident α particle, in other cases, e.g. nitrogen, it is less. Thus some disintegrations take place with liberation of energy, others with an absorption of kinetic energy. The probability of producing a disintegration with an α particle of given energy varies from one element to another. While the information so far obtained on both these points is meagre and indefinite, yet there emerges from the

---

* Curtiss, *Phys. Rev.* **31**, 1128, 1928.   † See also Appendix, p. 572.

results a distinct difference in behaviour between the elements of odd atomic number and those of even atomic number. In general the protons emitted by the odd-numbered elements are faster and more numerous than those from the even-numbered elements. It will be recalled that the early experiments were successful in showing the disintegration of the odd elements as far as phosphorus, while the disintegration of the even elements was only observed later.

This difference between elements of odd and even number manifests itself in other ways. Harkins* has shown that elements of even number are on the average much more abundant in the earth's crust than those of odd number. Aston† has found that odd elements may have at most two isotopes differing in mass by two units, while even elements may have a large number with a considerable range in mass. The conclusion seems to be that the even elements are more firmly built than the odd elements. In the case of the lighter elements the difference between odd and even has been shown by Aston‡ to extend to the masses of the nuclei. Aston's measurements of the masses of the lighter elements are sufficiently accurate to give some indication of the relative stabilities of nuclei. It is generally assumed that the nuclei of all elements are built up from two units, the proton and the electron. The mass of a nucleus is not given by the sum of the masses of the protons and electrons contained in it, as measured in free space, for inside the nucleus the protons and electrons are packed so closely together that their electromagnetic fields interfere and a fraction of their combined mass is destroyed. The mass destroyed appears as a release of energy in the formation of the nucleus, the greater the loss of mass the more firmly are the charged particles bound together and the more stable is the nucleus so formed. The atomic number and the mass number of the element give the numbers of protons and electrons in the nucleus, and an accurate measurement of the mass of the nucleus will give the loss of mass or release of energy in its formation. Aston expressed his measurements of the masses of the elements in terms of the "packing fraction," the divergence of the mass of the atom from the whole number rule divided by its mass number. The packing fraction is therefore "the mean gain or loss of mass per proton when the nuclear packing is changed from that of oxygen to that of the atom in question." A high packing fraction indicates looseness of packing of the proton and

---

* Harkins, *Phil. Mag.* **42**, 305, 1921.   † Aston, *Isotopes*. Arnold & Co., London.
‡ Aston, *Proc. Roy. Soc.* A, **115**, 487, 1927.

electron, and, therefore, low stability. We are here concerned not so much with the actual value of the packing fraction as in the change of the packing fraction from one element to the next. If we assume that the disintegration of an element consists solely in the ejection of a proton from the nuclear structure, the residual nucleus will be the next lower in the scale. If, as seems very probable from Blackett's experiments, some disintegrations consist in the binding of the α particle and the emission of a proton, the newly formed nucleus will be the next higher in the scale of elements. In either case an element of odd number will change into one of even number. Now Aston's results show that the light elements of odd number have in general a much higher "packing fraction" than those of even number. Thus when an element of odd number changes into one of even number there will be in general a disappearance of mass. This mass may appear as kinetic energy associated with the emitted proton. Thus Aston's measurements afford a general explanation of the high speeds associated with the protons emitted in the disintegration of the odd-numbered elements.

In the above argument we have neglected to take into account the gain in mass of the proton on its release from the nuclear binding and the kinetic energy of the incident α particle. These energies nearly balance, for in the nucleus the mass of the proton is very little greater than 1 and outside it is 1·00724, while the kinetic energy of the α particle of radium C corresponds to a mass of 0·0082.

The information at present available is not sufficient to test accurately the balance between the energy and mass changes in the disintegrations. In many of the lighter elements the atomic masses are not known with the necessary accuracy, nor have the disintegration experiments provided any information about the motions of the α particle and residual nucleus after disintegration except in the one case of nitrogen. As an example, we may consider the disintegration of fluorine. If we assume that the disintegration of fluorine consists in the absorption of the α particle into the fluorine nucleus and the ejection of one proton, the residual nucleus will have a mass number 22 and an atomic number 10, that is, it is an isotope of neon. The mass of the fluorine atom is given by Aston as 19·000, the mass of the α particle is 4·0011 (subtracting the mass of the electrons from the mass of the helium atom) and the energy of an α particle of 7 cm. range is, in mass, 0·0082. The mass (plus kinetic energy) of the nucleus formed by absorption of the α particle will therefore be 23·0093 minus the mass of 9 electrons. The mass of the proton is 1·00724. The maximum range of the proton liberated from fluorine by an α particle of 7 cm. range is 65 cm., corresponding in energy to a mass of 0·0089. The kinetic energy of the residual nucleus is not known, but neglecting this we can conclude that the mass of the residual atom, adding the mass of 10 electrons to the mass of the nucleus, is not greater than $23·0104 - (1·00724 + 0·0089) + 0·00054 = 21·9937$. The mass of the neon isotope of mass number 22 is given by Aston as a provisional estimate as 22·0048. The agreement is not good, and the value deduced appears to lie outside the range of error of Aston's measurements.

Some information* about the energy relations in the disintegration processes may be derived from a consideration of the experiments in which the maximum ranges of the protons emitted when nitrogen and aluminium were bombarded by α particles of different ranges were measured. If it is assumed that the α particle is captured by the nucleus which is disintegrated, then on the conservation of momentum we have

$$MV = m_p v_p \cos \psi + m_n v_n \cos \omega,$$
$$0 = m_p v_p \sin \psi - m_n v_n \cos \omega,$$

where M, V; m_p, v_p; and m_n, v_n are the masses and velocities of the α particle, proton, and residual nucleus respectively, and ψ, ω are the angles between the initial track of the α particle and the directions of motion of the proton and nucleus respectively.

If each disintegration of a nucleus of a given type takes place with the same change of energy Q, we shall have

$$\tfrac{1}{2} m_p v_p^2 + \tfrac{1}{2} m_n v_n^2 = \tfrac{1}{2} MV^2 + Q.$$

For the comparison with experiment only the maximum value of v_p is required and under the above energy condition this will be given when $\psi = 0$. The equations then lead to a relation between the maximum value of v_p and V in terms of the masses of the particles and the energy change Q. If we write as a first approximation $(v_p/V)^3 = r/R$, where r is the range of a proton of velocity v_p and R is the range of an α particle of velocity V, and express Q as a fraction ρ of the energy of the α particle of radium C' of 7 cm. range, this relation becomes

$$r = 64R \left[\frac{1}{m_n + 1} + \sqrt{\frac{m_n}{m_n + 1} \left\{ \frac{1}{4} - \frac{1}{m_n + 1} + \frac{\rho}{4} \left(\frac{7}{R} \right)^{\frac{2}{3}} \right\}} \right]^3.$$

* Cf. Bothe, *Zeit. f. Phys.* **51**, 613, 1928; and Rutherford and Chadwick, *Proc. Camb. Phil. Soc.* **25**, 186, 1929.

This gives the maximum range of the proton in terms of the range of the incident α particle.

If the experimental results for nitrogen shown in the curves of Fig. 87 are inserted, this relation leads to a consistent value for ρ of -0.13. Thus the observations with nitrogen are consistent with the assumption that a definite quantity of energy, equal to 0.13 of the kinetic energy of the α particle of radium C', is absorbed in the process of disintegration of the nitrogen nucleus.

This agreement must be compared with the results of Blackett, who found from the measurement of six disintegration forks in nitrogen that the change of kinetic energy varied from -0.33 to $+0.02$ of the energy of the α particle of radium C'. This result depends only on the measurement of angles and the assumption of the conservation of momentum, an assumption confirmed by the coplanarity of the forks. It does not seem possible to ascribe such a large variation in energy to errors in the measurements, and it appears more probable that the agreement shown between the maximum ranges of the protons found in the scintillation experiments and in the above calculations is fortuitous.

If the observations with aluminium are examined in the same way, the relation between the maximum range of the protons and the range of the incident α particle cannot be satisfied by assuming a definite value for Q. It is perhaps possible to regard this result as a suggestion that in the aluminium disintegration the α particle is not absorbed into the nucleus, but it seems more probable that it indicates that the energy change is not the same for each disintegrated nucleus.

This view is strengthened by an examination of the results given in Fig. 86. In this experiment a thin foil of aluminium of 5 mm. air equivalent was bombarded by the α particles of radium C', and the figure gives the absorption curve of the protons ejected in a direction at right angles to the incident particles. Thus no considerable absorption, either of the α particles or of the emitted protons, could take place in the foil, and since the protons were examined over a comparatively narrow range of angle not much variation in range can be attributed to variations in the direction of emission. Nevertheless protons were emitted which had ranges varying from about 12 cm. of air to from 70 to 80 cm., corresponding, on the assumption that the range is proportional to the cube of the velocity, to energies varying from 0.33 to about 1.1 or 1.2 of the energy of the α particle of radium C′. The possibility that the particles of very long range might not be protons was tested by examining their deflection in a magnetic field. This experiment showed definitely that the long range particles were in fact protons. If it be granted that the experimental conditions were sufficiently defined, this result indicates very clearly that the energy change in the disintegration of aluminium is not constant for each nucleus, independently of any assumptions about the mechanism of disintegration.

A necessary consequence of this conclusion is the suggestion that either the mass of the aluminium nucleus or that of the new "integrated" nucleus is not constant, or both. If this can be extended to the other light elements, a test of its truth might be obtained with the mass spectroscope, although the suggested variations in mass are small. The amount of variation may be illustrated by the case already taken of the disintegration of fluorine. If the proton emitted in this disintegration when an α particle of radium C′ is captured may have energies varying from 0·003 to 0·0089 in mass units, corresponding to minimum and maximum ranges of 12 cm. and 65 cm., then, again neglecting the kinetic energy associated with the residual nucleus, the mass of the neon isotope formed will vary from 21·994 to 22·000.

In this discussion of the energy relations in the disintegrations it has been assumed that all the energy is associated with the three particles—the α particle, the nucleus, and the proton—and that no other particle or radiation is emitted in the process. The best evidence that no positive particle is emitted is provided by Blackett's experiments, which refer only to nitrogen, but the emission of an electron or of radiation by the nucleus would most probably not have been detected in any of the observations which have yet been made. If one quantum of γ ray energy were emitted in each disintegration, the detection of the γ radiation would be a matter of some difficulty even if polonium, which does not itself emit γ radiation, were used as the source of α particles. The above discussion of the energy relations is therefore subject to certain reservations.

A very general description of the phenomena of disintegration can be given in terms of the picture of the potential field between an α particle and the nucleus of a light atom obtained from the scattering experiments (§ 63). To explain those experiments it was necessary to assume, in addition to the Coulomb force of repulsion, an attractive force varying as the inverse fourth or fifth power of the distance between the α particle and the nucleus. There is then a certain critical distance r_0 at which the potential of the α particle is a maximum, and within this distance the resultant force is attractive. If the velocity of the incident α particle is sufficiently great, it will be possible for an α particle to penetrate within this region and to be captured by the nucleus. It is probable that in order that a disintegration of the nucleus should occur such a capture of an α particle must be made, but it may be that every capture does not result in the emission of a proton and that certain nuclei cannot retain the particle. If the potential field around a nucleus were known, it would be possible to calculate the probability of capture for α particles of different velocities and to compare the calculations with the observations of the number of protons emitted under the same conditions. The scattering experiments have, however, been carried out only with two elements, magnesium and aluminium, and even in these cases the data obtained are such as to give only an approximate idea of the magnitude of the attractive force and its variation with distance. In the discussion of the field around the aluminium nucleus given in § 63 it was concluded that the critical distance r_0 is about $3 \cdot 4 \times 10^{-13}$ cm. if the attractive force is assumed to vary as the inverse fourth power of the distance, while the closest possible distance of approach of the α particle of radium C′ is about $4 \cdot 6 \times 10^{-13}$ cm., allowing for the motion of the nucleus during the collision. On this view, penetration of the α particle into the aluminium nucleus and capture would only be possible when its velocity is somewhat greater than 2×10^9 cm./sec. The disintegration experiments show, on the other hand, that the aluminium nucleus is disintegrated by α particles of much smaller velocity than this; that is, the disintegration experiments suggest that the nucleus, or value of r_0, is much larger than is deduced from the scattering experiments. A similar conflict of evidence arises in the consideration of the size of the uranium nucleus (cf. § 75), and it has been resolved by Gamow and by Gurney and Condon on the conceptions of the wave mechanics. The same explanation will apply to the present case; while on classical mechanics an α particle of energy E (or rather an effective energy E, for the motion of the nucleus in the collision must be allowed for) is unable to cross the region, shown shaded in Fig. 90, in which it would have a negative kinetic energy, on the wave mechanics there will always be a certain probability that the particle will be able to enter the nucleus. This probability will be greater the nearer the energy of the particle is to that necessary on the classical mechanics. An approximate calculation on the basis of the wave mechanical equations of the probability of capture by an aluminium nucleus has been made by Gamow* for α particles of different energies. The agreement with experiment was as good as could be expected, in view of the simplifications which were necessary to make calculation possible. He has also calculated the probability of disintegration of other light elements when bombarded by α particles of radium C′

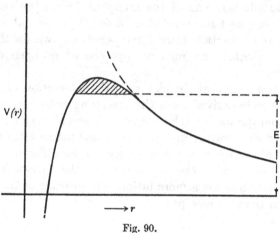

Fig. 90.

and by α particles of polonium. In this he assumed that the critical distance r_0 varied from element to element proportionally to the cube root of the atomic weight. A close comparison of these calculations with the experimental results cannot be made, since the latter are very uncertain and obtained under varied conditions, but it is possible to see a general agreement. In particular, the calculations show that the chance of penetration decreases very rapidly as the atomic number of the element increases, and is small for numbers greater than 20; for example, only about one α particle of radium C′ in 10^{11} is able to penetrate into the nucleus of iron.

There is the further possibility that the entrance of an α particle into the nucleus does not result in a disintegration. Such a case

* Gamow, *Nature*, **122**, 805, 1928; *Zeit. f. Phys.* **52**, 510, 1928.

may be supplied by carbon and oxygen, but the present experimental results do not exclude the possibility that these nuclei disintegrate with the emission of protons or of α particles of very short range. The agreement between Gamow's calculations of the chance of penetration and the experimental numbers of emitted protons is sufficient to show that for elements such as nitrogen or aluminium the probability that a capture of the α particle results in disintegration must be fairly high. In this connection it is of interest to refer again to the collision observed by Blackett in which the α particle was deflected through 110° and no disintegration occurred. If the α particle was one of the group of 8·6 cm. range, then on the assumptions we have here adopted for the potential field round the nucleus the α particle must have penetrated within the critical distance r_0, whether the nucleus was one of nitrogen or one of oxygen.

We may conclude that the picture of the interaction of a nucleus and an α particle derived from the scattering experiments is, when interpreted on the wave mechanics, able to provide a general description of the disintegration and to give a reasonable estimate of the probability of disintegration of a nucleus under different conditions. To extend this view to the explanation of the energy relations in the disintegration needs a more intimate knowledge of the structure of the nuclei than we now possess.

§ 73. The stability of the residual nucleus. It has so far been assumed without question that the new nucleus formed in a disintegration is stable. The evidence for this assumption rests partly on Blackett's experiments and partly on an unpublished investigation of Rutherford and Ahmad. Blackett's photographs show that if the nucleus formed in the disintegration of nitrogen breaks up with the emission of a positively charged particle it must have a life greater than 0·001 sec., for otherwise the track of the particle would be obtained in the photographs. Rutherford and Ahmad bombarded nitrogen contained in a closed vessel by α particles from a strong source of radon and its products. After removal of the source the conductivity of the nitrogen was examined; none was found that could not be attributed to unrecombined ions due to the initial ionisation produced by the α particles.

If we disregard the possibility of the emission of an electron, which might not have been detected in either experiment, this evidence suggests very strongly that the nucleus formed from nitrogen is stable. It must then be an isotope of oxygen with a mass 17, and should exist on the earth. If it does exist it can only be present in such small quantities as to escape detection in the mass spectroscope of Aston, or as not to influence to any appreciable degree the chemical atomic weight*.

An attempt to detect the break up of the nucleus formed in a disintegration was made by Shenstone† in 1922. In some experiments it was arranged that the substance, after several minutes bombardment by a strong source of α particles, was brought beneath a zinc sulphide screen within an interval of 1/100 sec. No residual activity of the bombarded materials was observed. In the main experiments the substance to be examined was fixed on the face of a disc. By rotating the disc at a speed of about 250 revolutions per sec. the substance could be brought from beneath the source of α particles in less than 1/10,000 sec. to a zinc sulphide screen where its activity could be observed. Aluminium, carbon, iron, and lead were examined in this way. The result of the observations was negative in all cases, and Shenstone concluded that in the case of the first three elements no α particles or protons of range greater than 2 mm. are emitted after an interval of 8×10^{-5} sec. from the bombardment, and none of range greater than 6 mm. after an interval of 3×10^{-5} sec. In the case of lead the corresponding intervals are 10^{-4} sec. and 5×10^{-5} sec. It will be noted that Shenstone examined only one element, aluminium, which is known with certainty to be disintegrated under the impact of α particles. His results suggest that the nucleus formed in the disintegration of aluminium is stable, and that the act of disintegration takes place within an interval of 10^{-5} sec. after the impact of the α particle.

§ 74. The transmutation of elements. There have been in recent years numerous claims that one element has been changed into another by processes in which forces of very ordinary quality were

---

* From an examination of the absorption bands of oxygen in the spectrum of the sun, Giauque and Johnston (*Nature*, March 2, 1929) claim to have obtained definite evidence of the presence of an isotope O 18 in the atmosphere, and in a later communication (*Nature*, June 1, 1929) evidence also of an isotope O 17. Babcock (*Nature*, May 18, 1929) estimates from the intensity of the bands that O 18 is present in about 1/1250 of the amount of O 16. The amount of O 17 must be much less than this.

† Shenstone, *Phil. Mag.* **43**, 938, 1922.

employed and in which the amount of transmuted element was so large as to be capable of detection by the usual chemical methods. The most important of these claims is perhaps that of Miethe*, who reported the transmutation of mercury into gold in a mercury vapour lamp. Miethe's claim was supported by Nagaoka†, who passed a high-tension discharge between electrodes of mercury and tungsten immersed in transformer oil, and found both silver and gold in the carbonised residues of the oil. These results were upheld on the ground that the transmutation of one element into the next of lower atomic number, for example of mercury into gold, requires merely the addition of one electron to the nucleus; and since the nucleus is positively charged, this operation may be easy to bring about by electrical means and may need very little energy. This argument is unsound. Unless some mechanism exists which prevents the absorption of casual electrons driven into the neighbourhood of the atomic nucleus, matter could not exist for any reasonable time. Moreover, this process must result in the formation of an atom one unit less in number but of the same weight. Now the measurements of Aston of the masses of atoms show that every pair of isobars known with certainty differs by two units in atomic number, and that elements of odd and of even number differ fundamentally in the grouping of their isotopes. There is therefore no evidence in nature for the type of transmutation suggested.

Miethe was able to collect from old mercury lamps so much gold that a determination of its atomic weight could be made. This was carried out by Hönigschmid and Zintl‡. The value obtained was not distinguishable from that of ordinary gold, viz. 197·2. At the same time Aston§ showed that mercury contained no appreciable quantity of an isotope of mass number less than 198. Although this evidence was not conclusive, since the gold examined by Hönigschmid and Zintl was not prepared by the actual process devised by Miethe, it was nevertheless significant. The claims of Miethe and of Nagaoka have been the subject of numerous investigations, notably by Haber, Jaenicke and Matthias‖, by Garrett¶, and by Sheldon and Estey**,

---

* Miethe, *Naturwiss.* **12**, 597, 1924; **13**, 635, 1925.

† Nagaoka, *Nature*, **116**, 95, 1925.

‡ Hönigschmid and Zintl, *Zeit. f. anorg. Chem.* **147**, 262, 1925.

§ Aston, *Nature*, **116**, 208, 1925.

‖ Haber, Jaenicke, and Matthias, *Zeit. f. anorg. Chem.* **153**, 153, 1926.

¶ Garrett, *Proc. Roy. Soc.* A, **112**, 391, 1926.

** Sheldon and Estey, *Phys. Rev.* **27**, 515, 1926.

with the result that there is now little if any support for this process of transmutation.

A claim that lead could be transmuted to mercury and thallium in an arc was made by Smits*, but after further experiment† withdrawn.

Reports of transmutations of even more remarkable type were made by Wendt and Irion‡, who stated that tungsten was decomposed and partly transmuted into helium when, in the form of a fine wire, it was electrically deflagrated; by Riding and Baly§, who claimed that helium and neon were formed from nitrogen subjected to cathode ray bombardment; and by Paneth and Peters‖, who stated that small quantities of helium could be synthesised from hydrogen by dissolving the latter in palladium. The experiments of Wendt and Irion were repeated very carefully by Allison and Harkins¶, but no evidence of the production of helium was found. The claim of Paneth and Peters was soon withdrawn and the appearance of helium in the experiments received a more prosaic explanation. It was found** that a glass surface which has been completely freed from helium by prolonged heating in hydrogen is able to absorb a detectable amount of helium, almost free from neon, from the atmosphere during only a single day's contact with the air; and also that glass (and asbestos) surfaces which do not give off detectable quantities of helium when heated in a vacuum or in oxygen release quantities of the order of 10^{-9} c.c. when heated in hydrogen. In later experiments†† precautions were taken to exclude the escape of helium absorbed in the glass surfaces, or the diffusion of helium from the air through hot glass, and the possibility was examined of the formation of helium from certain substances under the action of various agents. Under these conditions the results were entirely negative, and no quantity of helium larger than 10^{-10} c.c. was obtained.

---

* Smits, *Nature*, **117**, 13, 1926; Smits and Karssen, *Zeit. f. Elektrochem.* **32**, 577, 1926.

† Smits, *Nature*, **120**, 475, 1927; Smits and Frederikse, *Zeit. f. Elektrochem.* **34**, 350, 1928.

‡ Wendt and Irion, *Journ. Amer. Chem. Soc.* **44**, 1887, 1922.

§ Riding and Baly, *Proc. Roy. Soc.* A, **109**, 186, 1925.

‖ Paneth and Peters, *Ber. d. D. chem. Ges.* **59**, 2039, 1926.

¶ Allison and Harkins, *Journ. Amer. Chem. Soc.* **46**, 814, 1924.

** Paneth, *Nature*, **119**, 706, 1927; Paneth, Peters, and Günther, *Ber. d. D. chem. Ges.* **60**, 808, 1927.

†† Paneth and Peters, *Zeit. f. phys. Chem.* **134**, 353, 1928.

There can be little doubt that the claims which have so far been made of the transmutation of elements by the use of agents other than swift α particles are untrustworthy. In all the cases reported the presence of an element has been mistaken for its creation. On the present views of atomic structure the nucleus of an atom is extremely small and the forces binding together its component parts are of great intensity, corresponding to several million volts. It seems unlikely that the transmutation of elements on a chemical scale, as opposed to the radioactive scale, will be possible until such potentials are also available in the laboratory.

CHAPTER XI

THE RADIOACTIVE NUCLEI

§ 75. The investigation of radioactivity during the last twenty-five years has led to the accumulation of a wealth of data concerning the emission of energy in the form of α, β, and γ rays from the radioactive nuclei and in nearly all cases the rate of disintegration of the element has been determined. This information must have an intimate bearing on the structure of the radioactive nuclei and it provides a variety of quantitative tests which can be applied to any hypothesis of this structure. It is, however, only within very recent times that a picture, even of the most general type, has been given which will explain satisfactorily the spontaneous disintegration of a nucleus with the emission of an α particle, and no application of this has yet been made to the emission of the β and γ rays.

The difficulty in the way of an adequate theory of the structure of the radioactive nuclei has been twofold. In the first place, early in the study of radioactivity it became clear that the time of disintegration of an atom was independent of its previous history and depended only on chance. Since a nuclear particle, say an α particle, must be held in the nucleus by an attractive field, it seemed necessary, in order to explain its ejection, to invent some mechanism which would provide a spontaneous revulsion to a repulsive field. This mechanism must be of such a nature as to provide degrees of instability corresponding to a range of half-value periods from 10^{10} years to 10^{-6} sec., and the energy change in the disintegration must be the same for nuclei which have just been formed as for nuclei which have existed for some time, in the case of uranium, for example, for some thousand million years. In addition any theory of the radioactive disintegration must offer some explanation of the Geiger-Nuttall relation, which reveals a connection between the chance of disintegration and the energy of the α particle which is emitted.

The second point of difficulty arises from a consideration of the dimensions of the radioactive nuclei, in particular of the uranium nucleus. The experiments on the scattering of α particles have shown that deviations from the Coulomb law of force occur in the close collisions of α particles with the nuclei of light atoms, but no such deviation has been found in collisions with atoms as heavy as copper. For example, the scattering of swift α particles by the nucleus of gold has been tested very closely and the results are in complete accord with the Coulomb law of force. Experiments are described later in this chapter (§ 76 a) in which the scattering of α particles by uranium has been examined with the same result, that no indication of a failure of the inverse square law of force was found. Since we should anticipate that deviations from this law of force must occur when the α particle approaches closely to the nucleus, we conclude that the radius of the nucleus of uranium is less than the closest distance of approach in these experiments, viz. about 4×10^{-12} cm. On the other hand, the radioactive evidence suggests just as clearly that the nuclear structure of uranium extends to a distance greater than 6×10^{-12} cm. This deduction is based on the following argument. When an α particle is released from a nucleus it must gain energy in escaping through the repulsive field due to the residual nucleus. If the force is given by Coulomb's law and the particle starts at a distance r from the centre of the nucleus, the energy gained by the particle in its escape will be $2Ze^2/r$, where Ze is the charge of the residual nucleus. Now the α particle of uranium I has an energy corresponding to $4 \cdot 07 \times 10^6$ electron-volts. The charge on the nucleus after release of the α particle will be $90e$, therefore the value of r is $6 \cdot 3 \times 10^{-12}$ cm. Since the value of r will be greater if the α particle is released with some initial velocity, this is a minimum estimate for the radius of the uranium nucleus. Thus the dimension of the uranium nucleus deduced from the energy of the α particle emitted by it is nearly twice as great as that indicated by the scattering experiments in which α particles are fired at the nucleus.

This conflict of evidence about the size of the uranium nucleus is difficult to reconcile on any theory based on the classical mechanics, but it has received a simple explanation on the wave mechanical equations, an explanation which has succeeded not only in removing this difficulty but also in providing a general idea of the instability of a radioactive nucleus and an understanding of the connection between the rate of disintegration and the energy of the emitted α particles which is contained in the Geiger-Nuttall relation.

In the following sections of this chapter we shall give a brief account of experiments concerning the radioactive nuclei and of some hypotheses which have been put forward about their structure. Chiefly those experiments have been selected which have a direct bearing on the fundamental difficulty which has been pointed out above, that α particles of greater energy than those emitted by a nucleus are unable to penetrate it from the outside.

§ 76. **Experiments concerning the radioactive nuclei.** Since the early days of radioactivity attempts have been made to influence the rate of disintegration of the radioactive bodies. The effects of temperatures up to 3000° C., of pressures of 20,000 atmospheres, of powerful electric, magnetic, and gravitational fields have all been looked for, and all in the final analysis without a positive result. The stability of the radioactive nuclei against forces of these types received an immediate explanation on the nuclear theory of atomic structure, according to which the radioactive processes take place in the nucleus of the atom where the electrostatic and perhaps also the magnetic forces are of a different order of magnitude from those which have been applied in the laboratory. A more promising instrument of attack on the radioactive nuclei seemed to be provided by their radiations, and this method has been used in many experiments and by various investigators. In most cases the radioactive substance was bombarded by α particles, for these might be expected to have the greatest chance of success, but the effects of β and of γ radiations have also been investigated.

If the bombardment of the radioactive nuclei by α particles caused a very marked disintegration, the rate of decay of a product which emits α particles would depend upon its concentration. Such an effect has never been observed. Rutherford* found that the rate of decay of radon remained constant to within 0·05 per cent. while its concentration decreased more than a thousand times. This observation was confirmed by the later measurements of Mme Curie† on the rate of decay of radon. We can conclude from these observations that the effect of bombarding radioactive nuclei is not large, but an effect of the same order as that found in the case of the lighter elements could not be detected in this way.

An attempt to accelerate the rate of decay of a radioactive substance by bombardment with α particles was made by Danysz and Wertenstein‡. A thin layer of uranium oxide, U_3O_8, was exposed for six days to the α particles emitted from a thin-walled tube containing

---

* Rutherford, *Wien. Ber.* **120**, 303, 1911.

† Mme Curie, *Ann. de physique*, **2**, 405, 1924.

‡ Danysz and Wertenstein, *C.R.* **161**, 784, 1925.

18 millicuries of radon. If an α particle which penetrates into or very close to the uranium nucleus were able to cause a disintegration with the emission of an α particle, the rate of decay of the uranium would be accelerated and the amount of uranium X present in the preparation would increase above the equilibrium amount. The β ray activity of the layer of uranium oxide was therefore measured before and after the exposure to the α particles. No change in activity could be detected, although a growth in the amount of the uranium X corresponding to the amount in equilibrium with 3 mg. of U_3O_8 could have been measured. During the course of the experiment about 6×10^{14} α particles fell on the uranium oxide, while the measurements would have revealed the formation of 1.2×10^8 atoms of uranium X_1. Thus under the conditions of this experiment not more than one α particle in five million could have been effective in accelerating the disintegration of uranium.

Danysz and Wertenstein made similar experiments in which the bombarded substance was mesothorium. In this case the attempt was to influence the β ray disintegration of mesothorium 1 into mesothorium 2. The effect of the α particle bombardment was observed by measuring the γ ray activity of the preparation, which is due to the mesothorium 2. The result of this experiment also was negative. It was concluded that not more than one α particle in 26,000 was effective in promoting the disintegration of mesothorium 1 into mesothorium 2.

This question has recently been re-examined by Herszfinkiel and Wertenstein*. According to a theory suggested by Rutherford (§ 77) the α particles in a radioactive nucleus circulate as neutralised satellites round a central core. In the spontaneous disintegration of the nucleus the neutral satellite is supposed to break up, the two neutralising electrons go to the central core and the α particle escapes. If external action brings about an α transformation, the satellite would again break up but it might be that in this case the electrons would escape as well as the α particle. Thus a favourable collision of an α particle with a radioactive nucleus might on this view result in a triple transformation of one α and two β particles. Such a transformation would explain the negative result obtained by Danysz and Wertenstein when uranium was bombarded by α particles, for the disintegration would result not in the formation of an atom of uranium X_1 but of an atom of uranium II. This effect would not have

* Herszfinkiel and Wertenstein, *Nature*, 122, 504, 1928.

been detected by the method employed, and on account of the long period of uranium II would be difficult to detect by any method unless the disintegration was of very frequent occurrence.

In the case of thorium, however, the result of this triple transformation might be radiothorium. This has a half-value period of less than two years, so that a change from thorium to radiothorium can be easily detected. Herszfinkiel and Wertenstein accordingly exposed thin layers of thorium oxide, containing about 1 mg. ThO_2, to bombardment of the α particles from a tube containing 28 millicuries of radon. The α ray activity of the thorium oxide layer was measured before and after six days' exposure to the α particles. No change in activity of any of the preparations was found, although a difference corresponding to 1/20 mg. of thorium could be detected. Assuming that the result of an artificial disintegration of the thorium nucleus is to produce radiothorium, it was calculated from these data that not more than one α particle in eight million was effective.

In some experiments radioactive substances have been subjected to bombardment by cathode rays in a discharge tube. It will be sufficient to mention here the experiments of Jorissen and Vollgraf*, in which uranium oxide was bombarded for some hours without any change in activity which could be detected. Walter† exposed uranium oxide to the X rays from a tube operated at about 200 kilovolts. He observed no appreciable change in the β ray activity of the preparation, that is, in its content of uranium X.

A very careful examination was made by Hevesy‡, who exposed uranium and radium D to the β and γ radiations from 800 mg. of radium contained in a thin-walled glass tube. In the experiments with uranium, a layer of uranium oxide 2 mm. thick and 1 sq. cm. in area was used and the radium tube was placed on it and left in position for six weeks. The β radiation of the uranium X in equilibrium with the uranium was measured before and after the exposure, but no change as great as 0·1 per cent. in the β activity could be detected. Now from the quantity of uranium used the number of atoms breaking up per second was 15,000 and the measurements showed therefore that the disintegration was not accelerated by more than 15 atoms per second. Hevesy calculated that about 10^{10} γ ray quanta were absorbed in one second by the uranium atoms. Thus not more

* Jorissen and Vollgraf, Zeit. f. phys. Chem. 89, 151, 1915.
† Walter, Zeit. f. Phys. 39, 337, 1926.
‡ Hevesy, Nature, 110, 216, 1922.

than 15 in 10^{10} of the absorbed γ ray quanta were effective in inducing a change in the rate of disintegration of uranium.

In the experiments with radium D, 1·2 gm. of radio-lead chloride from Joachimstal was used and exposed in a similar way to the radiations from the radium tube for a period of 51 days. The β radiation of the preparation, which was due to radium E, was measured before and after exposure. Any change in activity was less than 0·2 per cent.

These experiments show very clearly that any effect of the γ radiations on the disintegration of a nucleus must be very small. Since the preparations were at the same time exposed to the β radiation from the radium tube, Hevesy's results show also that the β particles are unable to produce any detectable change in the rate of disintegration of these nuclei. A similar result was obtained when the preparations were exposed to intense bombardment by X rays.

§ 76 a. **Scattering of α particles by uranium.** It has been pointed out that a special interest is attached to the examination of the scattering of α particles by uranium. It seems on general grounds safe to conclude that an α particle of the same energy as that emitted by a nucleus should, if fired directly at this nucleus, penetrate its structure, while α particles of greater energy should do so even when not shot directly at the nucleus. For an experiment of this kind only uranium and thorium of all the radioactive elements are suitable. The range of the α particle emitted by uranium I, which is by far the main constituent of uranium, is 2·7 cm., so that on the above argument the α particles of radium C′ would be able to penetrate with ease into the structure of the uranium nucleus. The scattering of these α particles by uranium should therefore show divergences from the normal scattering which should be particularly marked at large angles corresponding to a considerable penetration into the nuclear structure.

This question was examined by Rutherford and Chadwick* in the series of experiments already described in § 45. The apparatus has been shown in Fig. 52. A thin film of uranium was deposited by cathode sputtering on a plate of graphite 1 sq. cm. in area. The thickness of the film corresponded to a weight of uranium of 1 mg. per sq. cm. or an air equivalent of about 2 mm., but the amount of uranium present was too small to enable the thickness to be determined with any

* Rutherford and Chadwick, *Phil. Mag.* **50**, 889, 1925.

accuracy. The scattering of α particles by this film was examined at a fixed mean angle of 135°. The source of α particles was radium (B + C), and beams of different ranges were obtained by placing suitable absorbing screens over the source. For each set of α particles the scattering by the uranium film was compared with that by a thin gold foil. Experiments, described in § 45, had shown that the scattering by gold was normal over the whole range under examination. Within the error of experiment the scattering was parallel to that of gold, and there was thus no indication of a departure from the usual laws. In these experiments the distance of approach of the α particle to the uranium nucleus was varied from 4×10^{-12} cm. to 8×10^{-12} cm. Special attention was paid to the scattering of particles of 4 to 2·5 cm. range, corresponding to distances of approach between 6 and 8×10^{-12} cm., for the reasons previously discussed. On account of the straggling of the α particle beam it was difficult to make accurate comparisons for low values of the range, but any marked divergence from the normal scattering would have been detected. Further, the absolute magnitude of the scattering of α particles of 6·7 cm. range was consistent with a nuclear charge of 92e, but again the error of measurement

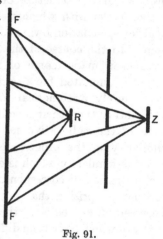

Fig. 91.

was large, about 20 per cent. in the number of scattered particles or 10 per cent. in the nuclear charge, owing to the difficulty of finding the thickness of the uranium film.

In some as yet unpublished experiments by Chadwick the scattering of uranium has been re-examined, using the annular ring method. The arrangement is shown in Fig. 91. The uranium film, deposited as before on a sheet of graphite, was placed at F; the source of α particles, radium (B + C), was at R, and the scattered particles were counted on the zinc sulphide screen Z. The average angle of scattering was 155°. As in the above experiments, beams of α particles of different velocities were used, but here attention was mainly directed to the scattering of the faster particles. It was found that the number of particles scattered to the screen Z was proportional to the fourth power of the velocity of the incident α particles within the experimental error of about 5 per cent. It follows therefore that the force between the uranium nucleus and the α particle cannot diverge appreciably from Coulomb's law at the distances examined in these collisions, from $3 \cdot 9 \times 10^{-12}$ cm. to $5 \cdot 5 \times 10^{-12}$ cm.

In both experiments the ranges of the particles scattered from uranium were examined, but no evidence was obtained that any particles were present which had ranges differing from those to be expected on the normal collision relations.

§ 77. **Theories of the structure of the radioactive nuclei.** The problem of the structure of nuclei and in particular of the radioactive nuclei with which we are here concerned offers a very wide field for speculation owing to the lack of adequate control by experiment. In the course of the last few years innumerable suggestions have been made, some of which have been useful in connecting hitherto unrelated facts or in pointing out regularities, some have been purely arithmetical formulations, and some have had little value other than their novelty. A glance at the table of transformations of the radioactive substances is sufficient to reveal a striking similarity in the progress of the disintegration through the three series, a similarity which is even more remarkable when the rates of disintegration of corresponding elements are compared. It is therefore not surprising that a number of interesting relationships have been established between the characteristics of the radioactive atoms, many of which have been different formulations of the same regularity. While it may be possible later to explain some of these suggested rules and hypotheses in terms of the structure of the radioactive nuclei, only one, the Geiger-Nuttall relation, has so far proved of real interest in this connection. This relation has already been described in § 35.

In this survey we shall direct attention to those hypotheses of the structure of the radioactive nuclei which have attempted to give a quantitative rather than a qualitative description of the facts of radioactivity. It has already been stated that a general explanation of the α ray disintegration of a nucleus has now been given. The earlier hypotheses are mentioned partly for their historical interest, but mainly to show the difficulties which were encountered in this problem on the laws of classical mechanics and to bring out the beautiful simplicity of the solution offered by the new quantum mechanics.

One of the earliest theories of the radioactive nucleus was put forward by Lindemann* to account for the exponential law of transformation of the radioactive elements and to give a derivation of the Geiger-Nuttall relation. He supposed that a nucleus contains particles in movement and that it becomes unstable when N independent particles, which rotate or oscillate with a mean frequency ν, all pass through some unknown critical position within a short interval of time τ. When this condition is fulfilled, the nucleus disintegrates and emits an α particle of energy $E = h\nu$. Each particle passes through the critical region ν times per second, so that the probability of its being there within the time τ is $\tau\nu$. Since the particles are supposed to move independently, the probability of N particles traversing the critical position within the time τ is $(\tau\nu)^N$. In a very large number X of similar nuclei the number which become unstable and disintegrate in the time dt is

$$dX = - X (\tau\nu)^N dt,$$

whence
$$X = X_0 e^{-(\tau\nu)^N t}.$$

The radioactive constant λ is thus equal to $(\tau\nu)^N = (\tau/h)^N . E^N$. The range R of the α particle is connected with its energy E by the relation $R = kE^{\frac{1}{2}}$, and therefore we can write

$$\log \lambda = N . \log (\tau/hk^{\frac{3}{2}}) + \tfrac{2}{3}N \log R.$$

This is of the same form as the Geiger-Nuttall relation

$$\log \lambda = A + B \log R.$$

Since the experimental value of B is about 53, it follows on comparing the constants that N is about 80.

Lindemann made the further assumption that the short time τ is to be regarded as the time taken by a strain to traverse the nucleus and he calculated the order of magnitude of the velocity of propagation of an elastic wave in the nucleus in terms of its mass, charge, and its unknown radius. This enabled him to express the constant A in the relation in terms of known quantities and the radius of the nucleus. By comparison with the experimental value of A he deduced a value for the radius of the radium nucleus of about 4×10^{-13} cm.

This theory provided an explanation of the probability nature of the transformation law by making the disintegration depend on a large number N of independent factors. Since the weights of the

* Lindemann, *Phil. Mag.* **30**, 560, 1915.

radioactive atoms are about 220 to 230 and the nuclear charges 80 to 90, the theory gives a reasonable value to N consistent with the general idea that the moving particles whose position determines the stability of the nucleus are positively charged particles. The value obtained for the radius of a nucleus is small but of the order of magnitude expected. The objections to this theory are however many and obvious, and in spite of its success in the above points it remained an artificial conception which was never developed further.

A theory of the structure of radioactive nuclei was suggested by Rutherford* in 1927 to explain the origin of the α particles. This theory endeavoured to reconcile the conflicting conclusions about the dimensions of the uranium nucleus derived from the energy of the emitted α particle and from the scattering experiments of Rutherford and Chadwick. The latter authors, in order to account for their results, suggested that the uranium nucleus might consist of a central core surrounded by satellites which might be either neutral doublets or a system of positive and negative charges with a very small net charge. Rutherford assumed that the nucleus of a radioactive atom consists of a very small core of radius less than 1×10^{-12} cm. surrounded at a distance by a number of neutral satellites describing quantum orbits in the field of the central nucleus. Some at least of these satellites are neutral α particles in which the neutralising electrons are more closely bound than in the free helium atom but not so closely as in the α particle. A neutral α satellite is regarded as stable only in very strong electric fields. The neutral α satellite is held in equilibrium by attractive forces due to the distortion or polarisation of the satellite in the field of the central nucleus. He showed that the force due to polarisation is of the right order of magnitude to hold the satellite in quantised orbits of radii between $1 \cdot 5 \times 10^{-12}$ cm. to 6×10^{-12} cm. The quantised orbit is essential to provide for the well-established fact that the α particles are liberated from a radioactive nucleus at a definite velocity characteristic of the element. The act of disintegration consists on this theory in the escape of a neutral α satellite from its orbit; when the satellite reaches a certain distance from the nucleus where the electric field has a certain critical value, the neutralising electrons break off and return to the core and the satellite finally escapes as a normal α particle. The energy of escape of the α particle will depend on the charge on the nucleus and on the quantum number of its

* Rutherford, *Phil. Mag.* **4**, 580, 1927.

orbit. The calculation of the energy is somewhat involved and will not be given here. In order to compare the results of the calculation with the experimental values three constants must first be fixed in addition to the value of the quantum number. These constants were fixed to ensure the best agreement with the α ray bodies in the uranium-radium series. The calculations then gave with the same constants very fair agreement with the observed energies of emission of the α particles of the thorium and actinium series, by substituting the known value of the nuclear charge and by suitable choice of the quantum number. The ranges of the long range α particles were also calculated with reasonable success. It has already been stated that the theory gives suitable radii to the orbits of the satellites. The dimensions of the satellite were also calculated and shown to agree with the apparent size of the α particle deduced from scattering experiments. The theory was also applied to considerations of the energy levels in the nucleus and the emission of γ rays.

This theory gives no clue to the exponential law of transformation of the radioactive elements or to the Geiger-Nuttall relation, but is concerned solely with the energy relations of the nuclear changes. Apart from certain difficulties in connection with the assumptions involved, the theory is open to the objection that it does not explain why, in experiments where α particles of high energy are shot at the uranium nucleus, no evidence is obtained either of disintegration of the nucleus or of any departure from Coulomb forces. It might be possible to show that on this theory the departure from Coulomb forces would not be very marked, but the penetration of an α particle into the satellite system should produce a disintegration.

A theory somewhat similar in general outline was put forward at the same time by Enskog*. He assumed that the α particle is held in equilibrium in the nucleus under the combined action of the Coulomb force of repulsion and an attractive magnetic force. He considered first the case in which the α particle and the small residual nucleus are at rest. If the particles have magnetic moments μ_1, μ_2 respectively and are placed so that their moments are in the same direction and in line, the force on the potential of the α particle will be a maximum at a distance $r_0 = \sqrt{\dfrac{6\mu_1\mu_2}{Ze.E}}$ and the maximum value will be $V(r_0) = \tfrac{2}{3}\dfrac{Ze.E}{r_0}$. Enskog then assumed that the α particle rotates around the nucleus in the attractive field within the critical distance r_0. It has therefore

* Enskog, *Zeit. f. Phys.* **45**, 852, 1927; **52**, 203, 1928.

a potential energy less than $V(r_0)$. In some way it happens occasionally that the α particle acquires sufficient kinetic energy to enable it to pass over the potential barrier. Cases in which it receives more than what is exactly necessary to escape are supposed to occur much less frequently. After the escape the total kinetic energy of the α particle and the recoiling nucleus corresponds to the potential energy $V(r_0)$. Putting $\mu_1 . \mu_2 = y . \mu^2$, where μ is the moment of the Bohr magneton, Enskog obtained a fair agreement with the known velocities of the α particles of the uranium-radium series by giving values to y such as $\frac{1}{3}$, $\frac{1}{2}$, 1, $\frac{4}{3}$, etc. The value of r_0 ranged from 2 to 4×10^{-12} cm. This model of the nucleus was then developed by assuming the α particle to rotate round the nucleus in a quantum orbit. The magnetic moments of the α particle and the residual nucleus were now supposed to be anti-parallel and in a direction perpendicular to the plane of the orbit. He took the moment of the α particle to be one Bohr magneton and that of the residual nucleus to be $2y$, so as to give the same value for $V(r_0)$ as before. He then calculated the effects of the magnetic force on the moving charge and of the electric force on the moving magnet. With the assumption of quantum numbers varying from -1 to $+5$, he was able to improve in this way the agreement between his calculations and the experimental values of the energies of the α particles emitted in the three radioactive series. The agreement was good in the case of most elements, but perhaps not better than could be expected when there are two constants which can be adjusted to suit each particular case.

This theory is open to most of the objections which have been raised in the previous discussions, and to the further objection that it postulates a considerable magnetic moment for the α particle, whereas, as we have seen on p. 276, the evidence from the band spectrum of the helium molecule shows that the helium nucleus has no magnetic moment whatever.

§ 77 a. **Application of the quantum mechanics.** A simple explanation of the radioactive disintegration and a solution of the apparent conflict of evidence between the radioactive data and the scattering results was offered simultaneously by Gamow* and by Gurney and Condon† on the basis of the new quantum mechanics. In this explanation it is assumed that there is an attractive force between the

---

* Gamow, *Zeit. f. Phys.* **51**, 204, 1928.
† Gurney and Condon, *Nature*, **122**, 439, 1928.

α particle and the nucleus which varies rapidly with the distance. The potential energy of an α particle in the field of, say, the uranium nucleus varies with the distance as shown by the curve of Fig. 92. Divergences from the Coulomb law begin to be appreciable at a distance r', say, while the region investigated by the scattering experiments is not closer than r''. At distances smaller than r_0 the force of attraction is greater than the Coulomb force and within this

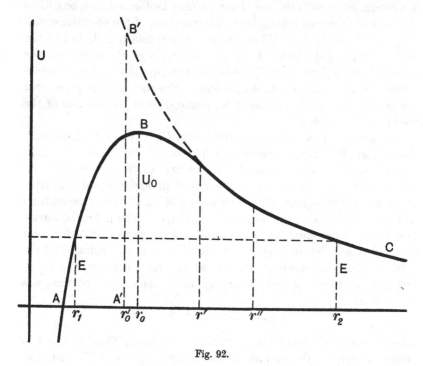

Fig. 92.

region the α particle will circulate in a quantum orbit. Its motion is not stable for its energy is positive, and any disturbance should cause the α particle to fly away. In order to get away the α particle must cross the potential wall of height U_0. For example, the value of U_0 for the nucleus of uranium I must be greater than the energy of the α particle of radium C', $7 \cdot 66 \times 10^6$ electron-volts, represented by the potential at r'' in the figure, while the energy of the α particle emitted by uranium I is $4 \cdot 07 \times 10^6$ electron-volts, represented by the potential at r_1. In order to escape, this α particle must cross the region between r_1 and r_2 in which its kinetic energy is negative. On classical mechanics this is not possible, but the difficulty no longer exists when the problem is considered from the point of view of the wave mechanics. On the wave mechanics a particle always possesses a finite chance of passing from one region to another of the same energy even when the two regions are separated by a potential wall of finite magnitude.

The problem of the emission of an α particle from the interior of a nucleus is analogous to the escape of an electron from a metal. A metallic body with its free electrons may be likened to a box filled with standing waves arising from the reflection of the electron waves at the walls of the box. When the waves are totally reflected at the walls, a wave disturbance is propagated through the walls into the surrounding medium, exactly as in the corresponding optical phenomenon. The square of the amplitude of the disturbance represents the probability of the escape of an electron from the interior of the metal into free space.

In the present case the potential hump U_0 is the wall of the box. The α particle inside the nucleus will have a finite chance of escape which will be greater the greater its energy E. Thus there will be a relation between the energy of the α particle and the disintegration constant of the nucleus. The exact nature of this relation depends on what assumptions are made about the shape of the potential curve. For a first approximation Gamow assumed a very simple form for the field round the nucleus in order to allow a simple solution of the wave mechanical equations. The relation found between the transformation constant λ and the energy of the emitted α particle was of the form

$$\log \lambda = \text{const.} + b \cdot E,$$

where b was a constant for all radioactive nuclei. This corresponds approximately to the empirical relation of Geiger and Nuttall, and the calculated value of b agreed well with the experimental value. The theory indicated that the relation between $\log \lambda$ and E is not strictly linear but that as E increases b must decrease, i.e. $\log \lambda$ increases less rapidly than E. This agrees with the measurements of Jacobsen on the period of radium C', which is much greater than the value calculated from the Geiger-Nuttall relation.

In a later paper, Gamow and Houtermans* made a closer approximation in the calculation of the chance of disintegration of a nucleus on this theory. It was assumed that the potential curve $A'B'C$

---

* Gamow and Houtermans, *Zeit. f. Phys.* **52**, 453, 1928.

(Fig. 92) would provide a good approximation to the true curve and would not seriously change the value of the disintegration constant. The potential curve is then that given by Coulomb's law up to a distance $r_0{}'$, when it falls very rapidly to a constant value less than E, the energy of the α particle, in the nucleus. The solution of the wave equations for this model gave the value of λ in terms of the velocity of the α particle and the charge contained within the orbit of the α particle, which was assumed to be two units less than the atomic number of the nucleus. Taking the known values of λ and velocities of the α particles for the radioactive emanations, the radii r_0 of these nuclei were deduced. The results are given in the table.

	r_0	Λ
Radon	$7 \cdot 35 \times 10^{-13}$ cm.	$6 \cdot 15 \times 10^{-13}$ cm.
Thoron	$7 \cdot 25$	$5 \cdot 74$
Actinon	$6 \cdot 63$	$5 \cdot 50$

The values of r_0 are in each case close to the values of Λ, the de Broglie wave-length of the particles, which are given in the last column of the table.

For the purpose of investigating how far the theory was able to give a quantitative agreement with the observed disintegration constants, it was assumed that the value of r_0 for a radioactive element was the same as that given above for the emanation of the same series. The disintegration constants of all the α ray elements were calculated from the nuclear charge and the velocity, v, of the α particle. The results for the uranium-radium series are shown in Fig. 93, where $\log \lambda$ is plotted against $\log v$.

The comparison of theory and experiment shows a general agreement which is very satisfactory in view of the approximations made in the calculations. The deviations shown by the bodies at the extremes of the series may perhaps be attributed to a progressive change in the radius of the nucleus such that the radius decreases as the disintegration proceeds. The jumps in the Geiger-Nuttall curve, at radioactinium for example, are to be ascribed to the change in the nuclear charge, which jumps by two units. The jump is smaller in the calculated than in the experimental values. This may be due partly to the change in radius, which has not been allowed for, but partly also to the assumption which has been made that the charge of the nucleus within the orbit of the α particle was two units less than the nuclear charge. This may not be quite true, for it is possible that some at least of the electrons which are emitted in the β ray disintegrations are further from the centre than the α particles and should therefore not be included in the charge of the residual nucleus.

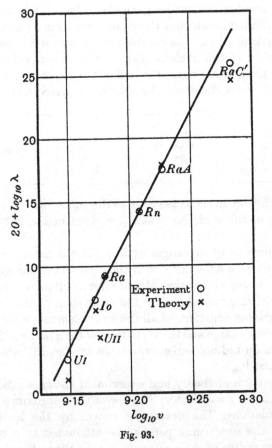

Fig. 93.

We may conclude that this application of the quantum mechanics is not only able to solve one of the chief difficulties in any picture of the radioactive nucleus but that it gives a satisfactory and quantitative explanation of the Geiger-Nuttall relation. While the main ideas and results are not contradicted by any known experimental facts, the theory is naturally still in a tentative form. It will be of great interest to see whether it can be developed further to account for the energy of emission of all the radiations in the radioactive transformations and to give information on the detailed structure of the nuclei. It may be noted here that the theory demands that the potential barrier around the radioactive nuclei must correspond to about 20 million electron-volts. It is not possible to obtain direct experimental evidence on this point without the use of much swifter α particles than are available today.

§ 78. **Evidence from β and γ transformations.** We have so far dealt almost exclusively with the problem of explaining the disintegration of a radioactive nucleus which results in the emission of an α particle. We shall now consider very briefly what evidence about the structure of the nuclei can be derived from the study of the β and more particularly the γ transformations.

It will be seen later that the γ radiations are emitted after the actual disintegration of the nucleus, and can be considered as due to its reorganisation. They constitute the spectrum of the nucleus and as such should provide valuable evidence about its structure. The range of wave-lengths in the γ radiations is less extended than one might expect from the great difference in the energies of the disintegration particles expelled by radioactive nuclei. The α particle of radium C' has an energy of seven million electron-volts, while the β particles from radium D have energies of the order of thirty thousand volts. Now while the highest frequency γ ray of which we have definite evidence has an energy corresponding to nearly three million volts, the greater number have energies in the region of two hundred to six hundred thousand volts. It would appear, if the analogy is permissible, that the γ radiations are the optical spectra of the nuclei, a view that is supported by the essential similarity of the radiations from different radioactive bodies. We are thus led to look for the origin of the γ rays either in some satellite in the outer regions of the nucleus or in some general property of the whole nucleus, such as rotation.

There are two lines of evidence which suggest that the γ rays cannot be emitted by the electrons in the nucleus. The first of these is a theoretical argument advanced by Kuhn. Combining the classical wave theory of radiation and the quantum view that emission is due to transition between two stationary states, Kuhn[*] deduces that the half width ν' of a γ ray line of frequency ν should

* Kuhn, *Zeit. f. Phys.* 43, 56, 1927.

have approximately the following value due to radiation damping alone:

$$\nu' = \frac{g_0}{g_1} \cdot \frac{4\pi e^2 \nu^2}{mc^3} \cdot f.$$

In this equation g_1 and g_0 are the statistical weights of the two states and f a constant which characterises the strength of the line. For a first approximation we may put both g_0/g_1 and f equal to unity. Taking the values of e and m appropriate to the electron, we obtain for a γ ray of frequency corresponding to a wave-length of 20 XU or an energy of 600,000 volts a half width, expressed in volts, of 10,000 volts. This represents a degree of inhomogeneity of the γ ray of 1 in 60. Now the β ray lines observed in the natural spectrum have been found to follow closely the theoretical form calculated for homogeneous particles, and the small degree of inhomogeneity which is found, rather less than 1 in 1000, may be partly instrumental. The γ rays which by their conversion in the atom give rise to these β ray lines cannot therefore be heterogeneous to an extent of more than 1 part in 1000. Even if allowance be made for the approximations for g_0/g_1 and f, it appears unlikely that the γ rays can be emitted by electrons. This argument thus suggests that the system responsible for the γ ray emission has a small value of e/m; the system may be a proton, an α particle, or the nucleus as a whole.

The second argument against the electron as the emitter of the γ radiation follows from the experimental evidence that the disintegration electrons are not expelled from the nucleus with a definite velocity but with some one velocity within a certain range, this range only being characteristic of the radioactive body. The question as to how this occurs will be discussed later; it appears that on almost any theory the electrons, if free in the nucleus, must be associated with an indefiniteness in energy that is quite incompatible with an emission by them of homogeneous γ rays.

A promising line of advance is to attempt to construct a system of nuclear levels, which will account for the γ rays. There is indubitable evidence* that level systems must exist, for in every spectrum investigated a great number of combination frequencies are found, that is, three γ rays are found to satisfy the relation $\nu_1 + \nu_2 = \nu_3$. Level systems have been proposed for radium B, radium C†, thorium

* Ellis, *Proc. Roy. Soc.* A, **101**, 1, 1922.
† Ellis and Skinner, *Proc. Roy. Soc.* A, **105**, 185, 1924.

C″, and mesothorium 2* which give a satisfactory account of the γ rays, but the difficulty arises that the accuracy of measurement of the frequencies of the γ rays is not sufficient to determine the level system uniquely. Usually one finds three or four possible alternatives, and some general principle is required to enable the final decision to be made. A knowledge of the absolute intensities of the γ rays will be of great value in this connection. The absolute intensity is the probability that a quantum of the γ ray will be emitted at any specified disintegration. Since at each disintegration there can be only one excitation of the level system, we obtain a criterion whether any given level system is possible or not. For example, it is clear that the sum of the intensities of the γ rays represented by transitions from one level must never exceed unity. We can go further than this and obtain a more exact criterion. If we define as the cross-sectional intensity the sum of the intensities of all γ rays whose transitions pass over the interval between two neighbouring levels, whether they begin or end on these levels or not, then this cross-sectional intensity must never be greater than unity. This is simply an expression of the physical assumption that there is only one excitation at each disintegration. The methods of measuring the intensities of γ rays are described in § 115, but they have not as yet been carried to a sufficient accuracy to be decisive for this purpose.

We have finally to consider the rôle of the electrons in the nucleus and the problem presented by the continuous distribution of energy among the expelled electrons. Since the energy of disintegration varies, it would appear that at some stage all the nuclei of a particular species cannot have the same total energy. This is often considered to be incompatible with the definiteness of the subsequent α particle disintegrations, so that it has been questioned whether it is correct to expect the conservation of energy to apply to each nucleus individually. This point of view has been particularly developed by G. P. Thomson†, who considers that an identical wave-packet, representing the electron, is emitted by each nucleus. This wave-packet is analysed by the experimental method of observation in a manner somewhat similar to the mode of action of a grating, but in any particular case the motion of the electron corresponds to one out of the infinite number of Fourier components. It is however by no means certain that the primary assumption is correct that all nuclei

---

* Black, *Proc. Roy. Soc.* A, **106**, 639, 1924.
† G. P. Thomson, *Phil. Mag.* **7**, 405, 1929.

of one species must have exactly the same total energy if they are subsequently to show the same behaviour. The experiments on the artificial disintegration of aluminium have suggested that nuclei of the same species can exist with different total energies, and we have no means of estimating to what extent, if any, small variations in this energy will affect subsequent α and γ ray emissions. It is therefore a possible standpoint to suppose that the continuous spectra show the actual energies of expulsion of the individual electrons and we must search for some mechanism that will cause this to happen.

This introduces the question of the rôle played by the electrons in the nucleus. There are many facts which suggest, although they do not prove, that electrons do not exist in the free state in the nucleus but are in a state of close combination with α particles. The resulting particles are usually called α' particles and are considered to be an α particle with two electrons in such close combination that the whole is of nuclear dimensions. Harkins* and Meitner† called attention to the possibilities of this idea and it has been used subsequently by Rutherford‡. It gives expression to the well-known fact that in the radioactive series β ray disintegrations are associated in pairs. Further, the results of Aston have shown that the majority of atomic species have an even number of electrons in the nucleus, and that, if the abundance of the elements is taken into account, it is true of the overwhelming majority of existing nuclei. It is certainly more satisfactory to imagine the nucleus to be built of particles all of comparable mass and similar charge, existing inside a potential barrier of the type already referred to; owing to the large masses of the particles, it is possible for their de Broglie wave-lengths to be of the order of the dimensions of the nucleus without supposing them to have unduly high energies.

* Harkins, *Journ. Amer. Chem. Soc.* 42, 1956, 1920.
† Meitner, *Zeit. f. Phys.* 4, 146, 1921.
‡ Rutherford, *Phil. Mag.* 4, 580, 1927.

CHAPTER XII

β RAY AND γ RAY SPECTRA

§ 79. Introduction. The β rays from radium and other radioactive bodies were early shown to consist of a stream of electrons projected with a wide range of speed, the swifter ones with a velocity close to that of light. Apart from a study of the apparent absorption and reflection of the β rays from different radioactive substances in traversing matter, further progress was comparatively slow, and it was not until recent years that a clear idea was obtained of the origin of the β rays and of their connection with radioactive transformations.

Compared with the complex emission of β rays, a study of the homogeneous emission of α particles and their absorption by matter presented a relatively simple and direct problem, and in a surprisingly short time the main facts of the material nature of the α rays and their origin were established and the results interpreted in terms of the transformation theory.

Radioactivity was then at once recognised to be an actual disintegration of the nucleus, which in the α ray case consisted in the emission of a helium nucleus and in the β ray case of an electron from the nucleus. These disintegration particles had to be distinguished from any subsidiary electrons that might be detached from the outer electronic levels during the disintegration. It will be seen later that this separation has been rendered possible by the recognition of the important part played by γ rays in the disintegration.

The early experiments on the nature of the radioactive radiations had shown that, in addition to the corpuscular α and β rays, there was also a third type, the γ rays, whose properties showed them to be electromagnetic radiations of high frequency. The emission of γ rays seemed to be the usual accompaniment of a β disintegration and was thought to be due to a reorganisation after the disturbance caused by the departure of the disintegration electron. Careful investigations by the absorption method of the γ rays emitted suggested that some at least were similar to the characteristic X rays discovered by Barkla and Moseley, indicating an action of the nuclear γ rays on the outer electronic structure. Since it was already known that an interaction of this type would involve the emission of high-speed electrons, it seemed possible that part of the electronic emission of a β ray body might be due to this cause.

The important work of von Baeyer, Hahn and Meitner had already shown the presence of definite groups of electrons with characteristic speeds in the emission of most β ray bodies. A new orientation of our ideas came from the experiments of Rutherford and Robinson, who investigated the electronic emission occurring when the γ rays passed through ordinary matter. They found that the characteristic groups of electrons again appeared, showing that these groups were a secondary effect and could not form the disintegration electrons. They also noticed that the velocity of these groups changed slightly from metal to metal and suggested that this effect would lead to a clear understanding of their origin. At this stage the War intervened and the experiments had to be broken off.

In 1921 further experiments were made on this subject which disclosed not only the origin of these β ray groups, but also gave a method of determining the frequencies of the γ rays. It appears that in a β ray disintegration there are only two main phenomena, the emission of the actual disintegration electrons, and the emission of nuclear γ rays. There are, however, many secondary effects, of which the chief is the relatively frequent conversion of these γ rays in the outer electronic structure and consequent emission of high-speed photo-electrons. These electrons form the characteristic groups already mentioned, and have proved subsequently to be of great importance for the information they yield about the disintegration. This is by no means the only secondary effect, for following this there is a succession of X ray and electronic emissions from the outer levels. When all these effects are superimposed on the results of the original disintegration, which by itself is sufficiently complicated, the complexity of the whole emission can easily be realised.

It is simplest to illustrate the general behaviour by reference to a particular case, such as radium B. This body is an isotope of lead, with atomic weight 214, and atomic number 82. It is formed from the preceding body radium A by the ejection of an α particle, and has an average life of about 39 minutes. During this period it behaves like an ordinary lead atom until for some cause of which we are ignorant the disintegration commences. The whole disintegration takes an extremely short time, how short is not known, but it is important to consider the sequence of the various events since this has a bearing on the interpretation to be placed on the experimental results. The first event is the actual disintegration, which is the ejection from the nucleus of a high-speed electron. This electron will be called the disintegration electron and the ejection is the fundamental phenomenon, since it leads to a change in the atomic number and the formation of a fresh atom. In the case of radium B, before the emission of the disintegration electron, the atomic number is 82 and the atom is isotopic with lead, after the ejection the atomic number is 83 and the atom is isotopic with bismuth. The new atom is called radium C, but it is more convenient to reserve this term for the relatively stable atom existing after all the secondary effects resulting from the disintegration have taken place. The next phenomenon after the disintegration electron has been ejected is the emission of γ rays, which is considered to be a direct result of the departure of this electron. There is evidence which indicates that in the normal radioactive atom the constituents of the nucleus are arranged in stable quantum states. The departure of the disintegration electron leaves the nucleus in an excited state and there will be a transition from one stationary state to others of lower energy with the emission of radiation. These radiations, the γ rays, represent the characteristic frequencies of the nucleus and, owing to the large forces in the nucleus, have usually very short wave-lengths. At the moment we need only note that the emission of the γ rays is the second main phenomenon in the disintegration and completes the part played by the nucleus. There are many other secondary effects but they all take place in the outer electronic system, and although they give valuable information about the disintegration they are not directly concerned in it. The greater proportion of the γ rays which are emitted from the nucleus escapes from the atom, and forms the γ radiation which is such a noticeable accompaniment of the β ray disintegration, but there is a fraction of the γ rays absorbed in the outer electronic structure of the same atom which emits them*.

The result of this absorption of the γ rays is the emission of high-speed electrons whose energy depends both on the frequency of the γ ray and the stationary state from which the electron comes. If ν

---

* It is not yet settled whether the γ rays are actually emitted from the nucleus and then reabsorbed in the electronic structure, or whether there is some kind of coupling between the nucleus and the outer electrons so that the extra energy of the nucleus appears at once as a high-speed electron. For convenience the former point of view is used in this preliminary account.

denotes the frequency of one γ ray, then, according as it is converted in the K, L_{I}, L_{II}, L_{III}, M or N levels, there will be an electron of energy

$$ h\nu - K, \quad h\nu - L_{\mathrm{I}}, \quad h\nu - L_{\mathrm{II}}, \quad ..., $$

where $K, L_{\mathrm{I}}, L_{\mathrm{II}}, ...$ denote the absorption energies of the levels, that is, the energies necessary to remove an electron from the level to a position at rest outside the atom.

It can now be seen that if the radioactive substance emits γ rays of several frequencies, then, since each one of these frequencies can give rise to as many different speed electrons as there are stationary states in the outer electronic structure, the result will be a whole series of groups of electrons of discrete and characteristic energies. These groups can be separated out by a magnetic field, and when recorded by a photographic plate give a picture not unlike an ordinary line spectrum. For this reason these electrons are usually referred to as the magnetic line spectrum or simply the β ray spectrum.

Supposing this internal conversion of the γ rays to occur in about one-tenth of the atoms which disintegrate, then this fraction of the atoms will now be capable of emitting the ordinary characteristic X rays. For instance, an atom in which a γ ray from the nucleus was converted in the K level will now be in a position to emit some line of the K X ray spectrum, say K_a, by the transition of an electron from the L level to the vacant place in the K level. Hence the next contribution to the total emission during disintegration is the entire X ray spectrum. Again, there is also the phenomenon of internal conversion of these X rays, giving a secondary magnetic line spectrum of considerably lower energies of which a typical example is the K_a radiation being converted in the L level, giving a group of energy

$$ h\nu_{K_a} - L. $$

Neither the X rays nor the secondary magnetic spectrum resulting from them show any trace of the particular γ ray causing their excitation, but are characterised only by the atomic number of the atom from which they come.

The process of disintegration involves a change from a normal radium B atom to a normal radium C atom and the start of this process is the emission of the disintegration electron. Once this has happened, all that follows is a gradual settling down into a normal radium C atom. It will be noticed that all the subsequent emissions are from an excited radium C nucleus and the electronic system of a radium C atom. To this extent then we must consider that these radiations show us the structure of the radium C nucleus and not the radium B nucleus, but since experimentally they can be obtained only from a radium B source and not from a radium C source (which gives the change 83–84) it is convenient to refer to them as belonging to the radium B disintegration. For this reason the γ rays and primary magnetic spectrum will always be given in the tables under the heading of the atom whose disintegration causes the emission. It may be noted at this point that several lines of evidence indicate clearly that these secondary processes, and in particular the emission of the γ rays, occur after the disintegration, i.e. after the expulsion of the primary disintegration particle from the nucleus.

§ 80. β ray spectra. The discovery of the homogeneous groups of electrons emitted by β ray bodies is due to the work of von Baeyer, Hahn and Meitner*.

In the years before 1910 a great deal of work had been carried out on the absorption of the various β radiations in matter. It had been found that in most cases the absorption curve in aluminium, for example, was exponential or could be expressed as a superposition of two or more exponential curves, and a considerable amount of discussion had centred round the point whether this result was evidence of the homogeneity of the rays. To test this point directly von Baeyer, Hahn and Me tner investigated the velocities of the electrons by determining their magnetic deflection. The apparatus they used was similar to that employed for determining the velocities of α rays and is shown in Fig. 7, p. 42. The source of radiation was a fine wire which had a thin layer of the radioactive body deposited on it. A weak magnetic field was applied perpendicular to the plane of the diagram and the thin sheet of rays from the source passed by the slit was bent in a circle whose radius was determined by the speed of the particles and the strength of the magnetic field. If all the electrons emitted from the source had the same velocity, then there would be only one line recorded on the photographic plate, whose displacement from the central line obtained with no magnetic field would give the velocity. Actually the results were more complicated, as can be seen from Plate XI, fig. 1, which is a reproduction of a photograph obtained by von Baeyer, Hahn and Meitner using

* von Baeyer and Hahn, *Phys. Zeit.* **11**, 488, 1910. von Baeyer, Hahn and Meitner, *Phys. Zeit.* **12**, 273, 378, 1911; **13**, 264, 1912.

mesothorium 2 as a source. Several well-marked lines can be seen of about the width to be expected from the dimensions of the source and the slit if homogeneous groups of β rays were emitted.

It is a curious fact and one of some interest that while these homogeneous groups of β rays discovered by von Baeyer, Hahn and Meitner have proved of the greatest importance they were not the disintegration electrons at all. The disintegration electrons are spread out in a continuous spectrum and although they are present in considerable excess the sharp lines due to the groups are far more prominent.

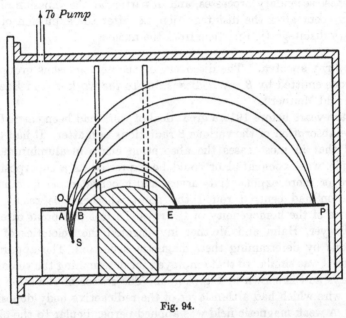

Fig. 94.

After the initial discovery it was clearly a matter of importance to obtain a better method of investigation of these groups, one that gave more detail, since it can be seen from the plate that this experimental arrangement could not hope to provide great accuracy. The so-called focussing method was introduced by Danysz in 1913 and developed by Rutherford and Robinson. It proved an immense improvement on the earlier method and has subsequently been used very widely.

The method is illustrated in Fig. 94. The source of the β radiation is placed at S. Usually the radioactive material is deposited on a fine wire, which is a few tenths of a millimetre in diameter and

5 to 10 mm. long arranged perpendicularly to the plane of the diagram. Above the source is a relatively wide slit AB and in a continuation of the plane of the slit is a photographic plate EP. The whole is contained in a box which can be evacuated and arranged between the poles of an electromagnet so that the magnetic field is perpendicular to the plane of the diagram and uniform over the path of the rays. Particles of the same velocity will describe circles of the same radius in the uniform magnetic field, and it can be seen from the diagram that even with a relatively wide slit the circles converge to a focus on the plate. The circle which passes through O, the foot of the perpendicular from S on the plane ABP, just completes a semicircle and attains the farthest distance along the plate from S, all other circles, whether passing through AB to the right or left of O, striking the plate nearer to the source. There is thus, no matter how wide the slit AB, a definite limit to the line on the side opposite the source, and further, owing to a fortunate property of the geometry of the circle, most of the rays are concentrated into this edge. The high-velocity side of the line is thus sharply defined and the momentum of the particles can be determined with considerable accuracy by the measurement of the position of this sharp edge. β particles of other velocities will also converge on to the photographic plate and give sharp-edged lines in a similar way, the lines due to particles of higher velocity being farther from the slit. The only conditions which it is necessary to observe in the disposition of the apparatus is that the slit AB should be in the continuation of the plane of the photographic plate and symmetrically disposed about the foot O of the perpendicular from S. The radius of curvature ρ of the path of the particles giving a particular trace is then given in terms of the lengths g and x by $4\rho^2 = g^2 + x^2$, where $g = OS$ and x is the distance between O and the high-velocity edge of the line. The momentum of the particles is $H\rho \cdot e_m$, where H is the value of the magnetic field and e_m the elementary charge in electromagnetic units. It is usual to use instead of the momentum the quantity $H\rho$, and it should be noticed that this is what is directly determined by experiment. The corresponding energy E in ergs can be calculated from the value of $H\rho$ by means of the relations

$$H\rho = \frac{m_0 c}{e_m} \frac{\beta}{\sqrt{1-\beta^2}},$$

$$E = m_0 c^2 \left(\frac{1}{\sqrt{1-\beta^2}} - 1 \right),$$

where β is the velocity in terms of the velocity of light c, and m_0 is the rest-mass of the electron. Again it is usual not to express the energy in ergs, but in terms of the potential difference V, measured in volts, through which the electron would have to fall in order to acquire this energy, so that, if e_s is the elementary charge in electrostatic units,

$$V = \frac{300E}{e_s}.$$

The great advantages of this method will be appreciated by comparing fig. 1 on Plate XI with figs. 2, 3, and 4, which are reproductions of photographs of portions of various spectra taken by means of this focussing method. In these photographs the velocities increase from the left to the right, and the sharp right-hand edges of the lines to which measurements are taken can be seen. The separation of the groups is far greater than in the other method, and at the same time owing to the focussing action photographs can be obtained with considerably shorter exposures. Another important advantage is that it is possible to protect the photographic plate from the direct action of the γ rays by interposing a block of lead.

The first experiments of Rutherford and Robinson* by this method showed its possibilities. They analysed the β ray spectrum of radium B and radium C, and were able to detect and measure accurately many more groups than had been found by previous observers. Subsequently the β ray spectra of a great many bodies have been measured by different workers. The results as a whole are very complex, most β ray bodies and some α ray bodies emit a whole series of groups, and it is to be expected that still more would be found with finer methods of observation. In a later section of this chapter will be found tables of the β ray spectra which have been investigated. In many cases these have been measured by several different observers and the tables contain the results of the latest work.

The interpretation of these complex spectra is treated in the next section, and at this stage only the experimental side of their measurement will be considered.

The simplicity of the apparatus has prevented any great divergences in design. The chief difficulty in detecting faint lines lies in the general background which is always present on the plates. A background due to the continuous spectrum formed by the disin-

* Rutherford and Robinson, *Phil. Mag.* **26**, 717, 1913.

PLATE XI

Fig. 1. Method of direct deviation.

Fig. 2. Part of RaAc and AcX β ray spectrum
$H\rho$ 800 to 2000 (Hahn and Meitner).

(a) AcX $H\rho$ 1502.

Fig. 3. Part of RaB β ray spectrum.

(a) $H\rho$ 661. (b) $H\rho$ 769.

Fig. 4. Part of RaB β ray spectrum.

(a) $H\rho$ 1410. (b) $H\rho$ 1677. (c) $H\rho$ 1938.

Fig. 5. Part of β ray spectrum excited in platinum by the γ rays
of RaB corresponding to 4.

[The magnetic field used for 5 was 3·5 per cent less than that for 4.]

tegration electrons clearly cannot be avoided, but, unless considerable care is taken, there will be in addition a general fog due to direct γ radiation and to electrons liberated from the walls by the γ rays. This can be diminished by disposing suitable screens in the paths of the rays and by making as much of the apparatus as possible of aluminium. Since the intensity of the lines decreases approximately linearly with increasing dimensions whilst such stray effects will decrease more rapidly it is an advantage to have the apparatus as large as possible, but a limit to this is set in practice by the difficulty of obtaining a homogeneous magnetic field over a large area and by the increase in the time of exposure.

The effect of slight inhomogeneities in the magnetic field has been investigated by Hartree*. If ϕ represents the angle turned through from the initial direction of propagation and the value of the magnetic field at this point is $H + \delta H$, then Hartree showed that the same point of impact on the photographic plate would be obtained in a uniform field of value $H + \int \frac{1}{2} \sin \phi \, \delta H \, d\phi$, where the integral extends from beginning to end of the path. Hence the "weighting function" for variations of H from normal over equal arcs in different parts of the trajectory is $\frac{1}{2} \sin \phi$, and is a maximum half-way along the trajectory and a minimum at the source and plate.

It is unfortunate that when we come to make use of the measurements of β ray spectra relative values of the energies of the groups, which can easily be obtained, are of little use and it is always the absolute values that are required. There are two factors involved in the absolute determination of $H\rho$, one a length depending on the geometry of the apparatus, and the other the value of the magnetic field. The magnetic field is the factor which so far has limited the accuracy, both due to the difficulty of ensuring uniformity and of measuring the field. The only definite attempt to attain a high accuracy has been made by Ellis and Skinner†, for the lines of the radium B and radium C spectra, for which they claim to have determined the values of $H\rho$ with an error less than one part in five hundred. A confirmation is desirable since many other workers have used these values as a method of calibrating their magnetic fields when measuring other spectra. The probable uncertainty of the $H\rho$ values in the other β ray spectra can be estimated as about one part in three hundred, although the relative values will be more accurate.

* Hartree, *Proc. Camb. Phil. Soc.* 21, 746, 1923.

† Ellis and Skinner, *Proc. Roy. Soc.* A. 105, 60, 1924.

The error in the energies will in this region of velocity be about 1·5 times the error in $H\rho$, and since it is from these energies that the wave-lengths of the γ rays are deduced the need for fresh work will be realised.

Wooster* has calculated the theoretical distribution of intensity throughout a line for the case of cylindrical sources taking into

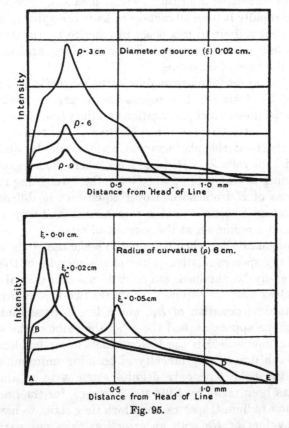

Fig. 95.

account the oblique rays. Some curves illustrating his results are shown in Fig. 95. The unsymmetrical character of the lines can be seen, the sharp high-velocity edge being on the left of the diagrams. The maximum occurs at a distance behind the high-velocity foot equal to the diameter of the source and for most purposes the remainder or tail of the line is of little importance. This tail can be

* Wooster, *Proc. Roy. Soc.* A, **114**, 729, 1927.

cut off at any desired point by adjusting the width of the slit. The distance from the head of the line to the tail is given approximately by $2\rho^2\theta^2/x$, where 2θ is the linear angle subtended at the source by the slit AB. There is a definite advantage in keeping the line as narrow as possible and just wider than the diameter of the source, for in this way the intensity of the continuous background is diminished while the intensity of the line is unaffected. For an accurate adjustment, where the diameter of the source is not negligible in relation to the width of the slit, it should be noticed that the slit must be arranged symmetrically about the foot of the perpendicular from the edge of the source nearer the photographic plate.

A helical method of focussing* has been used by Tricker† for investigating the effect of an electric field on the velocity of β rays. While with this method he was able to test accurately the relativity expressions for energy and momentum, it does not appear to be so suitable as the above focussing method for investigating β ray spectra.

§ 81. **Theories of the origin of β ray spectra.** Most of the early theories of the origin of β ray spectra assumed the groups to be formed by the disintegration electrons coming from the nucleus, and as such have only historical interest‡. They had, however, an effect on the development of the subject by suggesting an intimate connection between the β ray groups and the γ ray emission, and in this way stimulated investigation of the γ rays.

A complete change in the point of view was introduced by the experiments of Rutherford, Robinson and Rawlinson§. It was well known that when γ rays passed through matter they gave rise to high-speed electrons, but little was understood about the details of the process, although the penetrating power of the secondary electrons was known to be about the same as that of the primary β rays‖. The object of these experiments was to find the connection, if any, between these secondary electrons and the primary β rays from the radioactive substance. The method used was very simple and involved only the ordinary magnetic deflection apparatus. A small glass tube about 10 mm. long and 0·5 mm. diameter was filled with radon and thin foils of various metals wrapped round it. The

* Busch, *Phys. Zeit.* **23**, 438, 1922.
† Tricker, *Proc. Roy. Soc.* A, **109**, 384, 1925.
‡ Rutherford, *Phil. Mag.* **24**, 453, 893, 1912; **28**, 205, 1914.
§ Rutherford, Robinson and Rawlinson, *Phil. Mag.* **28**, 281, 1914.
‖ Biggs, *Phil. Mag.* **31**, 430, 1916.

whole formed a source of secondary β rays, the γ rays from the
radium B and radium C formed inside the glass tube liberating electrons from the metal sheath, and when the source was placed at S
(see Fig. 94) the magnetic spectrum of this electronic emission could
be obtained in the usual way. The glass and foil were of sufficient
thickness to stop many of the primary electrons and to straggle widely
even the most penetrating groups of β rays from the radioactive substance to such an extent that all traces of these lines disappeared. It
was found that well-marked homogeneous groups of electrons again
appeared which had inevitably to be associated with the action of
the γ rays on the metal sheath.

This result demonstrated in the simplest possible way that the
groups of electrons forming the β ray spectra of radioactive bodies,
which had for so long been regarded as the disintegration electrons,
must originate in a secondary process. It seemed probable that this
was a conversion of the γ rays in the radioactive atom, since to a
first approximation the groups emitted by radium B were similar to
those obtained when the γ rays went through lead, which had recently
been found to be isotopic with radium B*.

Fortunately at this moment, when the β ray groups were found
to be secondary in origin and not to be the primary disintegration
electrons, these latter were identified with another part of the emission
whose presence had just been discovered. It was found† when the
magnetic spectrum of radium B was analysed, either by counting
the electrons, or by an ionisation method, that the groups which
were so prominent on the photographs only formed a small fraction
of the total emission. The main portion was a continuous spectrum
of β rays which could be identified as the disintegration electrons, and
on this was superimposed the line spectrum formed by the groups.

These different lines of evidence were collected together to give
a general account of the disintegration by Rutherford‡. While our
present point of view is in many respects different, this theory has
certain features of interest. Rutherford supposed that the fundamental phenomenon was the emission of the disintegration electron
from the nucleus, and that by collision with the outer electrons it
lost varying amounts of energy, and as a statistical effect the continuous spectrum was formed. The escape of the β particle was also

* Rutherford and Andrade, *Phil. Mag.* 27, 854; 28, 263, 1914.
† Chadwick, *Verh. d. D. Phys. Ges.* 16, 383, 1914.
‡ Rutherford, *Phil. Mag.* 28, 305, 1914.

supposed to set certain regions of the atom in vibration causing the emission of characteristic γ rays, which partly escaped from the atom to form the ordinary γ radiation, and partly were converted in the same radioactive atom into the β ray form. The conversion of these characteristic γ rays gave rise to the groups of electrons forming the lines in the magnetic spectra and also caused the emission of the ordinary K and L X rays. The conversion of the γ rays was supposed to be controlled by some kind of quantum laws, and in a later paper Rutherford* showed that the actual energies of the β ray groups must be given approximately by the relation $E = h\nu$, ν being the frequency of the γ ray.

While this general point of view sufficed to show the secondary nature of the β ray lines and their connection with the γ rays, it was yet impossible to proceed to a quantitative analysis of the β ray spectra without more knowledge of the mode of conversion of the γ rays. At that time it was not known in what way, if any, the energy of the quantum was divided between the ejected electron and the remainder of the atom. The method followed by Ellis† to investigate this point was to measure the energies of the groups of electrons liberated from different metals by the γ rays in order to determine how these energies varied with the atomic number.

The most prominent feature of the natural β ray spectrum of radium B is three strong lines. Sets of three lines corresponding to these were found in the excited photographs, but owing to the retardation of the electrons coming from below the surface of the metal foil, the excited lines were broader and more diffuse than the natural lines. Fig. 1, Plate XII, is a reproduction of some of the original plates obtained in this experiment. The photographs show certain of the excited groups from tungsten, platinum, lead and uranium. The same magnetic field was used for each of the four photographs and the groups of lower energy are on the left.

The four photographs show a great similarity; on each there are two lines at the same distance apart, corresponding approximately to two of the strong natural lines already mentioned, but it will be noticed that as the atomic number increases the group of two lines shifts bodily to the left, that is to lower energies. The fact that the two lines change their position by the same amount when the atomic number changes suggested strongly that there were two γ rays acting,

* Rutherford, *Phil. Mag.* **34**, 153, 1917.
† Ellis, *Proc. Roy. Soc.* A, **99**, 261, 1921; **101**, 1, 1922.

each being responsible for one line, and that the energies observed could all be included in a statement as follows.

If E = energy of the ejected electron,

    W_1, W_2 represent energies characteristic of the γ rays and independent of the atom acted upon,

    W_a represents the work of extraction characteristic of the atom,

then for any one element we have two lines for which

$$E_1 = W_1 - W_a, \quad E_2 = W_2 - W_a,$$

and for another element W_a will be different, W_1 and W_2 remaining the same.

On this view the difference in energy between corresponding lines from two different elements would be simply $W_a'' - W_a'$, that is the difference in the work of extraction. This difference was found to be relatively large, for instance 9000 volts between the lines from platinum and lead. Such a large difference could only occur if the electron were to come from deep down in the electronic system, and the K level appeared to be the most probable origin. This point was easy to test by adding to the measured energy of each line the K absorption energy of the element from which it came, this latter quantity being obtained from X ray data. Within the limits of experimental error the same values for the characteristic γ ray energies W_1, W_2, W_3 were obtained from each element. This in itself constituted a proof of the validity of the Einstein photoelectric equation for these high frequencies and justified the identification of these energies with the $h\nu$'s of the three γ rays. Recently Thibaud * has made a very complete examination of these excited spectra and has been able to measure by this method γ rays of over a million volts energy. His results are very important in another connection and are described later, but it may be noted here that both he and subsequently Frilley have been able to measure the wave-lengths of γ rays by diffraction from a crystal and found agreement to 1 part in 250 with the value deduced from β ray measurements.

These experiments could be viewed from two standpoints. In the first place, quite apart from the natural β ray spectra, they showed the existence of monochromatic γ rays and gave reliable values for their wave-lengths. On the other hand, and this is the point with which we are immediately concerned, they suggested a simple explanation of the complexity of the natural β ray spectra.

    * Thibaud, *Thèse*, Paris, 1925.

The three strongest lines in the natural radium B spectrum fell exactly into the series of the three lines obtained in the excited spectra of tungsten, platinum, lead and uranium, being almost identical with those of lead. They were therefore considered to originate in the same way, except that, owing to the small amount of radioactive material present, the conversion in the K level must be internal, that is in the same atom that emitted the γ ray. The natural β ray lines, coming from the minute layer of radioactive material on the source, are much sharper and easier to see than the excited lines, so that it was plausible to expect also the presence of weaker lines due to the quantum conversion in the L, M and N levels. An example will show the application of this view. The energy of one of the strong lines which by these experiments is known to come from the K level is $2 \cdot 638 \times 10^5$ volts, so that, adding the K absorption energy $0 \cdot 900 \times 10^5$ volts, we have for the energy of the quantum $3 \cdot 538 \times 10^5$ volts. The three L levels L_I, L_{II}, L_{III} have absorption energies $0 \cdot 163$, $0 \cdot 157$ and $0 \cdot 134 \times 10^5$ volts, so that electronic groups from these should have energies $3 \cdot 375$, $3 \cdot 381$ and $3 \cdot 404 \times 10^5$ volts. Actually there is a strong line with energy $3 \cdot 379 \times 10^5$ volts which may be taken to be due to conversion in either L_I, L_{II} or both, but the group from L_{III} seems definitely to be absent.

There are five M levels whose absorption energies lie between $0 \cdot 040$ and $0 \cdot 026 \times 10^{-5}$ volts. The difference is small and groups from the M level should have energies $3 \cdot 498$ to $3 \cdot 512 \times 10^5$ volts. Again there is a measured line of energy $3 \cdot 502 \times 10^5$ volts.

There is even some evidence for a group from the N levels. The absorption energies lie between $0 \cdot 009$ and $0 \cdot 001 \times 10^{-5}$ volts, which would give groups $3 \cdot 529$ to $3 \cdot 537$ volts, and in the spectrum there is a faint line, which was difficult to measure, of energy $3 \cdot 536 \times 10^5$ volts which doubtless originates in this way.

It will be seen that one γ ray may give rise to two, three or more β ray lines, and the complex β ray spectrum can be ascribed to a small number of γ rays. Examples of the analysis of the different β ray spectra will be found in § 85 and will serve to illustrate this point sufficiently. It may be noted that no evidence has ever been obtained which suggests that the energy of a quantum may be divided between two ejected electrons or that two or more quanta may co-operate in ejecting one electron.

The most important result of this interpretation of the β ray spectra was that it gave at once a reliable method of determining the wave-lengths by the inverse of the process described above. If in the natural β ray spectrum there was a group of two or more lines whose energy differences were the same as those of the K, L, M states, then it was reasonable to infer the existence of one γ ray giving rise to these several lines, and to deduce the $h\nu$ value. It is in this way that the greater part of our knowledge of the wave-lengths of the γ rays of the different radioactive bodies has been obtained.

Shortly afterwards Meitner[*] independently reached the same results. The starting-point of her experiments was the important fact discovered by Hahn and Meitner[†], that certain α ray bodies emitted homogeneous groups of electrons forming β ray spectra analogous in every respect to those given by typical β ray bodies. Careful experiments showed that these β rays could not have originated in some subsequent β ray transformation and must therefore have come from the α ray body. This proved definitely that there was some process by which electrons of definite energies could be liberated from the outer electronic structure during radioactive disintegration and suggested at once that the groups found from β ray bodies were of similar origin. Meitner's experiments to test this point were similar to those already described and consisted in showing that the γ rays of thorium B passing through lead, which is isotopic with thorium B, liberated groups of electrons of similar energies to some of those in the natural β ray spectrum of thorium B. From this she inferred that the entire natural β ray spectrum was due to an internal photo-electric effect of the γ rays on the different electronic levels of the atom and supported her conclusions by examples taken from the β ray spectra of thorium B, radium D, radiothorium and radium.

It may be noted that in her first paper Meitner drew attention to the fact that it was important to know whether the emission of the γ rays preceded or followed the disintegration, because since the atomic number and so the strength of binding of the extra-nuclear electrons is changed after the disintegration, the energy of the β ray groups would be also affected. Meitner adopted the view, which was subsequently proved to be correct, that the γ rays were emitted after the disintegration.

§ 82. **The moment of emission of the γ rays.** It is a matter of both practical and theoretical importance to determine whether the

[*] Meitner, *Zeit. f. Phys.* 9, 131, 145, 1922.
[†] Hahn and Meitner, *Zeit. f. Phys.* 2, 60, 1920.

emission of the γ rays precedes or follows the departure of the actual disintegration electron. We have seen that it is possible to deduce the wave-lengths of the γ rays from the energies of the groups in the β ray spectra. In its simplest form this method consists in identifying a certain β ray line of energy E_β as being due to conversion of a γ ray in a particular level, say K or L, and then adding to E_β the absorption energy of this level. Now taking the case of radium B we find that the absorption energy of the K state is 87,400 volts before the disintegration, when the atomic number is 82, and is 89,900 volts afterwards, so that a perceptible error would be made in the deduction of the wave-length of the γ ray if the wrong choice were made. Without multiplying examples we can say that any detailed analysis of the β ray spectra is impossible until this point has been settled, and the theoretical importance scarcely needs emphasis.

There are two main methods by which this problem has been attacked. The first consists of an accurate analysis of the energies of the groups in the natural β ray spectra. Suppose both from the relative intensities and approximate energies that we can identify two groups of energies E_1 and E_2 as arising from conversion of the same frequency γ ray in the K and L states. Then $E_1 - E_2$ must equal the $K - L$ absorption differences of the radioactive atom at the moment the γ ray was emitted and may be compared with the X ray data corresponding to the two possible atomic numbers. This method was applied by Black* to his measurements on the β ray bodies of the thorium series. Owing to the multiplicity of sub-levels it was not possible to arrive at a clear-cut decision. For example $K - L_{III}$ for atomic number N differs little from $K - L_I$ for atomic number $N + 1$, but Black concluded that the balance of evidence was in favour of the γ ray emission following the disintegration.

Meitner† measured the β ray spectra from radioactinium and actinium X. These are both α ray bodies and in this case it is possible to obtain evidence directly from the β ray spectra in the manner already discussed. When the α particles are emitted the atomic number changes by two units, and it is now a question of deciding whether the measured energy difference between two β ray groups agrees with the $K - L$ absorption of one or other of two elements differing by two in atomic number, and not by one as in the β ray case. This is just sufficient to enable a decision to be reached. The

* Black, *Proc. Roy. Soc.* A, **106**, 632, 1924.
† Meitner, *Zeit. f. Phys.* **34**, 807, 1925.

results of Meitner's measurements for the γ rays of radioactinium are shown in the next table. For each group of β ray lines due to the conversion of one γ ray she calculated the $h\nu$ by adding on the appropriate absorption energies, and this was done twice, using in turn the absorption energies corresponding to the two possible atomic numbers. The table shows the maximum percentage divergence between the values obtained for $h\nu$ for each of the γ rays on these two assumptions.

Number of γ ray	1	2	3	4	5	6	7	8	9	10
Percentage divergence if γ ray follows disintegration	3·2	1·0	0·6	0·85	1·0	0·5	1·1	0·2	0·9	0·6
The same if γ ray precedes disintegration	5·4	2·6	1·2	2·8	1·5	3·1	3·8	2·0	2·3	2·1

It can be seen that the figures in the second row are in every case smaller than the corresponding ones in the third, and similar results were obtained for actinium X. These experiments showed definitely that in these cases the γ ray was emitted after the disintegration.

The method used by Ellis and Wooster* was based on the same principle but its sensitivity was greater, so that the difference due to a change of one in the atomic number, as in the β ray changes, could be detected with certainty.

Experiments were made with the γ rays of radium B and radium C, and photographs were taken simultaneously and on the same photographic plate of the natural β ray spectrum and β ray spectrum excited in platinum. The source consisted of a small glass tube containing radon and surrounded by a thick platinum sheath. The γ rays from the radium B and radium C liberated electrons from the platinum, those from the outside surface having energies $h\nu - K_{Pt}$, $h\nu - L_{Pt}$, ... At the same time a weak source of radium (B + C) was deposited on the outside of the platinum, and this emitted electrons with energies $h\nu - K_{RaB}$, $h\nu - L_{RaB}$, ... It will be seen that the difference of energy of corresponding groups is $K_{RaB} - K_{Pt}$.

The results of the measurements of this separation between corresponding groups for the three main γ rays of radium B and one of radium C, expressed not in energy but in units of $H\rho$, are given in the

* Ellis and Wooster, *Proc. Camb. Phil. Soc.* **22**, 844, 1926.

next table, and in the third and fourth columns are given the values of the same quantity calculated on the two possible values for the atomic number of the radioactive atom.

In the case of radium B this is either $K_{82} - K_{78}$, or $K_{83} - K_{78}$, quantities which are known accurately from X ray data and differ by 25 per cent. The chief advantage of the experimental method is that, except for quantities which are known more accurately than 1 per cent., the difference between the K absorption energies is determined simply by the separation of the lines. This could be measured with an error of less than 5 per cent.

γ ray tested gives natural β ray line	$\delta\, H\rho$ obs.	$\delta\, H\rho$ calculated if emission	
		precedes disintegration	follows disintegration
$H\rho\, 1410_{\text{Ra B}}$	$\begin{cases} 60 \\ 61 \end{cases}$	49	62
$H\rho\, 1677_{\text{Ra B}}$	$\begin{cases} 56 \\ 56 \\ 58 \end{cases}$	44	57
$H\rho\, 1938_{\text{Ra B}}$	$\begin{cases} 52 \\ 48 \end{cases}$	42	53
$H\rho\, 2980_{\text{Ra C}}$	57	46	56

The second method by which this problem was attacked consisted in the measurement of the secondary X ray phenomena which accompany the disintegration. We have seen how some of the γ rays emitted from the nucleus are converted in the K, L, M, ... states leading to the ejection of an electron. The atom is then capable of emitting some line in its X ray spectrum, and if the wave-length were determined accurately the atomic number could be deduced. Rutherford and Wooster* made a careful investigation by the crystal method of the α_1 and β_1 lines of the L X ray spectrum emitted by radium B in the course of its disintegration and found the wave-lengths agreed excellently with the accepted values for atomic number 83. Since the radium B disintegration gives a change of atomic number from 82 to 83, this result pointed clearly to the γ ray emission occurring after the atomic number had changed.

* Rutherford and Wooster, *Proc. Camb. Phil. Soc.* **22**, 834, 1925.

Another method of investigating the same problem was used by Black*. The L X ray spectrum measured by Rutherford and Wooster is partially internally converted in the M and N levels, giving rise to a corpuscular line spectrum, and this spectrum Black photographed by the magnetic focussing method in the usual way. His measurements agreed excellently both with the energies and the intensities of the groups calculated from an atom of atomic number 83. In addition Black investigated the corresponding β ray spectrum from thorium B. This is an isotope of radium B, and also undergoes a β ray change. If also in this case the γ rays are emitted after the disintegration, one would expect to obtain a corpuscular L spectrum from this substance exactly the same as from radium B. This Black was able to show to be the case, although owing to experimental difficulties the measurements were not so accurate as with radium B.

It is extremely satisfactory that this important problem has been settled so conclusively. It will have been noticed that five bodies have been investigated, of which two showed the α ray type of disintegration, and three the β ray type. In every case the γ rays were found to be emitted after the disintegration, and we can safely extend this conclusion to all radioactive bodies.

A very interesting point arises when we consider the first method in closer detail. In effect the energies of the electronic levels are found directly after the emission of the γ ray from the nucleus, and a close inspection of the results shows that the energies appear already to have taken up the final values corresponding to the second atomic number. If we consider the disintegration particle leaving the nucleus, we see that all the electrons have to change their states to those corresponding to the new atomic number. At present we have no means of estimating how long this will take, but should at any time this be possible then we should have at once a lower limit to the interval between the disintegration and the emission of the γ rays. Since the departure of the disintegration particle can be considered to be the cause of the γ rays in that it puts the nucleus into an excited state, we should have here a method of estimating the minimum life of an excited nucleus.

§ 83. General survey of the γ ray and β ray spectra of radioactive bodies. There are eighteen known radioactive bodies which show the β ray type of disintegration, and of these ten emit γ rays,

* Black, *Proc. Camb. Phil. Soc.* 22, 838, 1925.

while of the twenty-three α ray bodies five emit γ rays. It happens that the radioactive bodies which are mostly available for experiment are of the β ray type that give γ rays and of the α ray type which do not. As a result the view has generally been expressed that γ rays are the typical accompaniment of a β disintegration but are the exception in the α disintegration. There is considerable truth in this general statement about the occurrence of γ rays, but it must not be over-emphasised to the extent of regarding as exceptions those β ray bodies emitting no γ rays or those α ray bodies which do emit them. We may regard the emission of the γ rays as due to a reorganisation of the nucleus after the departure of the disintegration particle, and, to use a loose method of expression, the absence of γ rays is an indication that the new nucleus can stand the condition of strain set up by the departure of the disintegration particle without giving way to further reorganisations.

If the γ ray and β ray emission of the different bodies is compared on the basis of total energy emitted little order can be found, but there are general theoretical grounds for believing that this is only the result of the method of classification and that either a body (α or β type) emits no γ rays or it emits about one or two quanta per disintegration. Since the energy of the quanta can vary over a wide range, and the production of the photo-electric groups of β rays is far less efficient for high frequencies, we find large apparent differences in the "strength" of the radiations.

It will be noted from the examples which follow that no distinction can be drawn between the γ ray emission of α and β ray bodies on the score of complexity; radium D and radium form examples of extreme simplicity, radium C and radioactinium both emit a veritable spectrum of γ rays. It does appear, however, that the range of wavelengths capable of being emitted by β ray bodies is greater than in the case of α ray bodies.

There are four methods of investigating the γ rays of radioactive bodies: by absorption, by crystal reflection, by the excited corpuscular spectrum method (that is by the photo-electrons liberated from ordinary matter by the γ rays), and by the natural β ray spectrum.

The absorption method is, as would be expected, of little use for finding the wave-lengths of the lines constituting the γ ray spectrum. The crystal method is limited in its application, for it in general requires intense sources of γ rays and its resolving power is small for very short waves. It has up till now only yielded interesting results for the longer wave-length γ rays of radium B and radium C, both of which bodies can be obtained in relatively large quantities.

The excited corpuscular spectrum is direct, simple and certain, but it suffers from some serious disadvantages. It may be recalled that the principle of the method is to place the source of γ rays in contact with a thin sheet of a heavy metal, or inside a tube of the metal, and then to determine the energies of the groups of electrons liberated by the conversion of the γ rays with the usual focussing apparatus. Provided the metal is of sufficient thickness to render undetectable the groups of the natural β ray spectrum, any groups that are observed are indubitable evidence of γ rays. There is no uncertainty in the interpretation, since conversion in the K level is four to five times as probable as conversion in an L level. Should further proof be required, it is only necessary to change the radiator and observe the shift of the groups. In addition the relative intensities of the γ rays may be estimated, since, although the exact dependence on wave-length λ of the photo-electric absorption coefficient is not known, it appears probable that it varies according to a law intermediate between λ^2 and λ^3.

Unfortunately the method has also some undesirable features. Large sources or long exposures are necessary to obtain even the stronger groups and this is at once a great restriction on the method. The large exposures and the nature of the experiment always give photographs with a dense background on which it is impossible to see the fainter lines. Finally, and this is perhaps the most serious disadvantage, the lines are not sharp and accurate measurement of their position is difficult. These points will be appreciated by comparing Figs. 4 and 5 of Plate XI. The former shows the strong groups in the natural β ray spectrum of radium B, while the latter shows the corresponding groups liberated from platinum.

The excited spectrum method therefore, while invaluable for determining the intensities, is chiefly important for showing the validity of the more accurate method of deducing the γ rays from the natural β ray spectrum.

The great advantages of the method using the natural β ray spectra are that they are easily photographed even with weak sources, the background is sufficiently small to allow weak lines to be seen, and the lines are sharp and can be measured accurately. The method is admittedly somewhat indirect, but it has been confirmed so thoroughly by the excited spectrum method for the strong γ rays of several bodies that it may be relied upon with some confidence.

The analysis of the β ray spectra is helped by certain intensity rules. A line from the K level is several times stronger than lines from the L level, which in turn are stronger than those from the M level. It appears also to be a general rule, for the internal conversion of radiation of considerably greater frequency than the absorption limit, that the most firmly bound subgroup absorbs most strongly. This is similar to the interesting observations of Robinson* on the relative absorption of X rays in the different subgroups.

It is usually found that, while most of the lines in a β ray spectrum can be associated in pairs or more with γ rays in this way, there are some weak lines left over. The presumption in this case is that they are due to conversion in the K level of such weak γ rays that the corresponding L line escapes detection. This is a legitimate assumption, but it must be remembered that the existence of such γ rays is not assured. The K and L X ray spectra are also emitted by γ ray bodies as a result of the ejection of K and L electrons. In most spectra there are β ray lines in the region 60,000 to 90,000 volts, some of which are doubtless connected with these X rays. It will be seen that in certain spectra it has been possible to analyse such lines in detail.

The chief disadvantage of the method lies in the fact that no certain information can be obtained about the intensities of the γ rays. Very little is known about the probability of internal conversion and the factors which determine it. On the whole it appears to decrease with decreasing wave-length, but it is not certain that the relative intensities of two neighbouring γ rays is always similar to the relative intensities of the corresponding β ray lines† (see § 117). In the tables that follow, the intensities of the β ray lines, except when otherwise stated, represent visual estimates of the blackness of the lines on the photographic plate.

§ 84. β ray spectra from β ray bodies. *The natural β ray spectrum of uranium X_1 and uranium X_2.* This has been measured by von Baeyer, Hahn and Meitner‡, and recently by Meitner§, whose values

---

* Robinson, *Proc. Roy. Soc.* A, **104**, 455, 1923; Robinson and Cassie, *Proc. Roy. Soc.* A, **113**, 282, 1926.

† This was suggested by Smekal, *Ann. d. Phys.* **81**, 399, 1926.

‡ von Baeyer, Hahn and Meitner, *Phys. Zeit.* **14**, 873, 1913.

§ Meitner, *Zeit. f. Phys.* **17**, 54, 1923.

for the lines are given in the following table. Hahn and Meitner*
showed that these were to be attributed to uranium X_1. The close
agreement of the γ ray with the K_{α_2} X ray line, the energy of which
is 0.917×10^5 volts, suggests that this is not a nuclear γ ray at all, but
the method of excitation is not obvious since no harder nuclear γ rays
have been detected from uranium X_1.

By the method of direct deviation these authors have also detected
bands at $H\rho$ 1163 and 2450. These are possibly the disintegration
electrons of uranium X_1 and X_2 respectively. In addition a group,
appearing to consist of unresolved lines, was found at $H\rho$ 5800. Since
uranium X_2 emits some nuclear γ rays†, it is also possible that this
represents its true β ray spectrum.

The natural β ray spectrum of uranium X_1

Number of line	Intensity (visual estimate)	$H\rho$	Energy in volts $\times 10^{-5}$
1	m.	927	0·710
2	m.	1028	0·862
3	w.	1057	0·907

Analysis of the β ray spectrum of uranium X_1

Number of line	Intensity	Origin	Energy of β ray line + absorption energy in volts $\times 10^{-5}$	Energy of γ ray in volts $\times 10^{-5}$
1	m.	$L_{\rm I}$	0·710+0·211	0·921
2	m.	$M_{\rm I}$	0·862+0·053	0·915
3	w.	$N_{\rm I}$	0·907+0·013	0·921

γ ray of uranium X_1

Energy in volts	λ in X U‡
0.919×10^5	134·2

‡ 1 X U is 10^{-11} cm.

* Hahn and Meitner, *Zeit. f. Phys.* 17, 157, 1923.
† Meitner, *Zeit. f. Phys.* 17, 54, 1923.

The natural β ray spectra of radium B and radium C. These two spectra have been investigated many times. Measurements have been made by von Baeyer, Hahn and Meitner*, by Danysz†, and by Rutherford and Robinson‡. Recently Ellis and Skinner§ made a measurement of the absolute energies of the main lines of radium B, and using these values as a basis reinvestigated the entire spectra of both bodies‖. Their measurements are given in the following tables, including a few changes found in later work¶. The relative values of the $H\rho$'s of two lines differing by not more than a factor of 2 are probably correct to 1 part in 1000, the absolute accuracy in $H\rho$ may be taken as 1 in 500. The calculation of the energies is for convenience carried to more figures than is justified by this accuracy. The intensities are based on the photometric measurements of Ellis and Wooster**, and Ellis and Aston††. The values for radium B and radium C are on the same scale and refer to equal numbers of atoms disintegrating.

The γ rays of these bodies have also been investigated by Frilley‡‡ using the crystal method, and by Thibaud§§ and Ellis and Aston†† using the excited spectrum method (see § 83). Neither of these methods can give as accurate results as those obtained from the natural β ray spectrum, but the interpretation is entirely free from ambiguity. It is thus possible to obtain a reliable value for the wavelength of a γ ray from a single β ray line provided its existence is rendered certain by either of the other two experimental methods. The application of this point of view will be seen from the tables showing the γ rays of radium B and radium C.

It has not been possible as yet to determine how the latter γ rays are to be divided between the three possible bodies radium C, radium C′, and radium C″. The small number of atoms following the C″ branch renders it unlikely that any of the groups come from the α ray

* von Baeyer, Hahn and Meitner, *Phys. Zeit.* 12, 1099, 1911.

† Danysz, *Le Radium*, 10, 4, 1913; *Ann. de physique*, 30, 241, 1913.

‡ Rutherford and Robinson, *Phil. Mag.* 26, 717, 1913.

§ Ellis and Skinner, *Proc. Roy. Soc.* A, 105, 60, 1924.

‖ Ellis and Skinner, *Proc. Roy. Soc.* A, 105, 165, 1924; Ellis, *Proc. Camb. Phil. Soc.* 22, 369, 1924.

¶ Ellis and Aston, *Proc. Roy. Soc.* A, 119, 645, 1928.

** Ellis and Wooster, *Proc. Roy. Soc.* A, 114, 276, 1927.

†† Ellis and Aston, *Proc. Roy. Soc.* A, 129, 180, 1930.

‡‡ Frilley, *Thèse*, Paris, 1928.

§§ Thibaud, *Thèse*, Paris, 1925.

The natural β ray spectrum of radium B

Number of line	Measured intensity	$H\rho$	Energy in volts $\times 10^{-5}$	Number of line	Measured intensity	$H\rho$	Energy in volts $\times 10^{-5}$
1	17	660·9	0·3725	17	80	1410	1·529
2	5	667·0	0·3792	18	3·9	1496	1·697
3	1	687·0	0·4016	19	2·1	1576	1·858
4	11	768·8	0·4983	20	91	1677	2·067
5	8	793·1	0·5288	21	10	1774	2·275
6	4	799·1	0·5365	22	2·5	1832	2·402
7	2	833·0	0·5806	23	0·5	*1850	2·442
8	5	838·0	0·5872	24	100	1938	2·638
9	2	855·4	0·6106	25	12	2015	2·813
10	5	860·9	0·6172	26	2·1	†2064	2·926
11	1·5	877·8	0·6412	27	1·5	†2110	3·033
12	1·5	*896·0	0·6667	28	16	2256	3·379
13	3	926·2	0·7094	29	8	2307	3·502
14	3	949·2	0·7426	30	1·5	2321	3·536
15	2	1155	1·068	31	1·5	2433	3·809
16	1	1209	1·160	32	1·5	2480	3·925

* The values for these lines are less accurate.
† Possibly radium C lines.

Analysis of the β ray spectrum of radium B

Number of line	Intensity	Origin	Energy of β ray line + absorption energy in volts $\times 10^{-5}$	Energy of γ ray in volts $\times 10^{-5}$
1	17	L_{I}	0·3725 + 0·1634	0·536
2	5	L_{II}	0·3792 + 0·1569	0·536
3	1	L_{III}	0·4016 + 0·1339	0·535
4	11	M_{I}	0·4983 + 0·0400	0·538
5	8	N_{I}	0·5288 + 0·0096	0·538
6	4	O	0·5365 + 0·0014	0·538
17	80	K	1·529 + 0·899	2·428
21	10	L_{I}	2·275 + 0·163	2·438
22	2·5	M_{I}	2·402 + 0·040	2·442
18	3·9	K	1·697 + 0·899	2·596
23	0·5	L_{I}	2·442 + 0·163	2·605
20	91	K	2·067 + 0·899	2·966
25	12	L_{I}	2·813 + 0·163	2·976
26	2·1	M_{I}	2·926 + 0·040	2·966
24	100	K	2·638 + 0·899	3·537
28	17	L_{I}	3·379 + 0·163	3·542
29	8	M_{I}	3·502 + 0·040	3·542
30	1·5	N_{I}	3·536 + 0·010	3·546

disintegration of radium C or from the β ray body radium C″. There is at present no direct evidence to decide between the β ray disintegration of radium C or the a ray body radium C′, but the numerical agreement obtained in the analysis of the spectrum is better if radium C is chosen.

Six faint lines in the radium β spectrum are not accounted for, viz. nos. 15, 16, 19, 27, 31 and 32. They are possibly due to faint γ rays converted in the K ring whose energies can be obtained by adding 0.899×10^5 volts.

The group of lines 7–14 can be mostly accounted for by internal conversion of the K X radiation excited by the γ rays, as shown in the next table. It is interesting that the K_{a_2} radiation appears to account for the lines, although in the emission from an X ray tube it is weaker than K_{a_1}. A similar behaviour is found in the spectra of uranium X_1 and radiothorium.

Number of line	Energy of β ray line in volts $\times 10^{-5}$	Intensity	Calculated energies	
7	0·581	1	$K_{a_2}-L_{\mathrm{I}}$	0·579
8	0·587	3	$K_{a_2}-L_{\mathrm{II}}$	0·585
9	0·611	1	$K_{a_2}-L_{\mathrm{III}}$	0·608 $(K_{a_1}-L_{\mathrm{II}}$ 0·608)
10	0·617	3	....	....
11	0·641	1	....	$(K_{a_1}-L_{\mathrm{III}}$ 0·631)
12	0·667	1		....
13	0·709	2	$K_{a_2}-M_{\mathrm{mean}}$	0·709
14	0·743	2	$K_{a_2}-N_{\mathrm{mean}}$	0·737

γ rays of radium B

Number of γ ray	Energy in volts $\times 10^{-5}$	λ in X U	Number of β ray lines due to γ ray	Measured by crystal	Measured by excited spectrum
				volts $\times 10^{-5}$	
1	0·536	230·3	6	0·533	—
2	2·43(3)	50·8	3	2·40	2·43
3	2·60(0)	47·5	2	—	2·97
4	2·97(0)	41·6	3	2·94	2·97
5	3·54(0)	34·9	4	3·53	3·53
6	4·71	26·2	1	4·75	4·66

The natural β ray spectrum of radium C

Number of line	Measured intensity	$H\rho$	Energy in volts × 10⁻⁵	Number of line	Measured intensity	$H\rho$	Energy in volts × 10⁻⁵
1	v.w.	*703	0·420	33	0·4	5136	11·134
2	v.w.	*848	0·601	34	0·2	5178	11·256
3	v.w.	*871	0·632	35	0·68	5281	11·550
4	v.w.	*896	0·667	36	0·16	5428	11·970
5	v.w.	*944	0·725	37	0·16	5552	12·327
6	v.w.	*964	0·764	38	0·17	5775	12·970
7	0·1	*1379	1·470	39	4·76	5904	13·340
8	0·1	1438	1·582	40	0·2	5948	13·467
9	1·0	1557	1·819	41	0·2	6030	13·704
10	0·2	1594	1·894	42	0·70	6161	14·082
11	0·2	*1834	2·406	43	0·26	6212	14·234
12	0·1	*1912	2·580	44	0·07	6350	14·629
13	0·5	2085	2·975	45	0·05	6523	15·131
14	0·1	*2156	3·142	46	0·05	6656	15·518
15	0·3	*2256	3·379	47	0·05	6800	15·937
16	0·1	*2390	3·704	48	0·13	6932	16·321
17	0·1	2550	4·100	49	0·13	6998	16·515
18	0·2	2720	4·529	50	0·79	7109	16·838
19	0·1	2840	4·846	51	0·06	7240	17·222
20	0·1	2890	4·965	52	0·18	7380	17·631
21	7·6	2980	5·199	53	0·04	7530	18·069
22	0·5	3145	5·632	54	0·04	7690	18·537
23	0·5	3203	5·785	55	0·04	7974	19·370
24	1·7	3271	5·966	56	0·04	8090	19·710
25	0·3	3307	6·062	57	0·04	8313	20·365
26	0·3	3326	6·113	58	0·18	8617	21·259
27	0·6	3584	6·807	59	0·04	8885	22·047
28	0·4	3824	7·462	60	0·04	9165	22·874
29	0·77	4196	8·489	61	0·04	9425	23·641
30	0·5	4404	9·070	62	0·04	9655	24·320
31	2·42	4866	10·370	63	0·04	10020	25·390
32	0·2	4991	10·747				

* The values for these lines are less accurate. There is evidence that radium C emits several groups of higher energy, see § 87.

It is probable by analogy with the radium B spectrum that line number 1 is due to a γ ray converted in the L_I level, and numbers 2–6 have an origin similar to numbers 7–14 of radium B.

There are probably many fainter γ rays giving the weak lines not accounted for.

Analysis of the β ray spectrum of radium C

Number of line	Intensity	Origin	Energy of β ray line + absorption energy in volts $\times 10^{-5}$	Energy of γ ray in volts $\times 10^{-5}$
9	1·0	K	1·819+0·926	2·745
12	0·1	L_I	2·580+0·169	2·749
11	0·2	K	2·406+0·926	3·332
14	0·1	L_I	3·142+0·169	3·311
13	0·5	K	2·975+0·926	3·901
16	0·1	L_I	3·704+0·169	3·873
15	0·3	K	3·379+0·926	4·305
17	0·1	L_I	4·100+0·169	4·269
21	7·6	K	5·199+0·926	6·125
24	1·7	L_I	5·966+0·169	6·135
25	0·3	M_I	6·062+0·042	6·104
26	0·3	N_I	6·113+0·011	6·124
31	2·42	K	10·370+0·926	11·296
33	0·4	L_I	11·134+0·169	11·303
34	0·2	M_I	11·256+0·042	11·298
35	0·68	K	11·550+0·926	12·476
37	0·16	L_I	12·327+0·169	12·496
39	4·76	K	13·340+0·926	14·266
42	0·70	L_I	14·082+0·169	14·251
43	0·26	M_I	14·234+0·042	14·276
50	0·79	K	16·838+0·926	17·764
52	0·18	L_I	17·631+0·169	17·800
58	0·18	K	21·259+0·926	22·185
59	0·04	L_I	22·047+0·169	22·216

γ rays of radium C

Number of γ ray	Energy in volts $\times 10^{-5}$	λ in X U	Number of β ray lines due to γ ray	Measured by crystal	Measured by excited spectrum
				volts $\times 10^{-5}$	
1	0·589	209·5	1	0·588	—
2	2·75	44·9	2	—	—
3	3·32	37·2	2	—	—
4	3·89	31·7	2	—	—
5	4·29	28·8	2	4·20	4·26
6	5·03	24·5	1	5·15	5·06
7	6·12	20·2	4	6·17	6·12
8	7·73	16·0	1	7·70	7·70
9	9·41	13·1	1	—	9·38
10	11·30	10·92	3	—	11·29
11	12·48	9·89	2	—	12·44
12	13·90	8·80	1	—	13·90
13	14·26	8·66	3	—	—
14	17·78	6·94	2	—	17·78
15	22·19	5·57	2	—	22·20

The natural β ray spectrum of radium D. The β ray spectrum of radium D is very simple, consisting of a group of lines all attributable to the action of one γ ray. It has been measured by Danysz*, Ellis†, Meitner‡, and by Black§. There is general agreement about the results which are given in the following tables.

The natural β ray spectrum of radium D

Number of line	Intensity (visual estimate)	$H\rho$	Energy in volts $\times 10^{-5}$
1	50	600	0·309
2	2	606	0·315
3	0·5	628	0·338
4	20	714	0·433
5	10	738	0·461

Analysis of the β ray spectrum of radium D

Number of line	Intensity	Origin	Energy of β ray line + absorption energy in volts $\times 10^{-5}$	Energy of γ ray in volts $\times 10^{-5}$
1	50	L_I	0·309 + 0·163	0·472
2	2	L_{II}	0·315 + 0·157	0·472
3	0·5	L_{III}	0·338 + 0·134	0·472
4	20	M_I	0·433 + 0·040	0·473
5	10	N_I	0·461 + 0·010	0·471

γ ray of radium D

Energy in volts	λ in X U
$0·472 \times 10^5$	261

The natural β ray spectrum of mesothorium 2. The examination of this spectrum was first made by von Baeyer, Hahn and Meitner‖, by the method of direct deviation. It was later studied in great detail by Black¶, using the focussing method, and his measurements are given

* Danysz, *Le Radium*, **9**, 1, 1911.
† Ellis, *Proc. Camb. Phil. Soc.* **21**, 125, 1922.
‡ Meitner, *Zeit. f. Phys.* **11**, 35, 1922.
§ Black, *Proc. Roy. Soc.* A, **109**, 166, 1925.
‖ von Baeyer, Hahn and Meitner, *Phys. Zeit.* **13**, 264, 1912.
¶ Black, *Proc. Roy. Soc.* A, **106**, 632, 1924.

The natural β ray spectrum of mesothorium 2

Number of line	Intensity (visual estimate)	$H\rho$	Energy in volts $\times 10^{-5}$	Number of line	Intensity (visual estimate)	$H\rho$	Energy in volts $\times 10^{-5}$
1	100	668	0·381	17	20	1469	1·644
2	85	700	0·416	18	6	1538	1·782
3	65	796	0·532	19	16	1692	2·099
4	45	822	0·566	20	8	1782	2·291
5	6	842	0·593	21	6	2094	2·99
6	4	870	0·631	22	2	2173	3·18
7	16	907	0·682	23	8	2317	3·52
8	50	953	0·749	24	4	2679	4·42
9	35	982	0·791	25	2	2738	4·58
10	16	1077	0·929	26	6	4035	8·04
11	35	1170	1·093	27	3	4238	8·61
12	25	1191	1·129	28	2	4371	8·97
13	22	1257	1·245	29	2	4555	9·49
14	6	1276	1·279	30	1	6468	14·97
15	4	1308	1·337	31	1	6604	15·37
16	18	1345	1·406				

Analysis of the β ray spectrum of mesothorium 2

Number of line	Intensity	Origin	Energy of β ray line + absorption energy in volts $\times 10^{-5}$	Energy of γ ray in volts $\times 10^{-5}$
1	100	L_I	0·381+0·204	0·585
2	85	L_{III}	0·416+0·162	0·578
3	65	M_I	0·532+0·052	0·584
4	45	N_I	0·566+0·013	0·579
5	6	L_I	0·593+0·204	0·797
6	4	L_{III}	0·631+0·162	0·793

The M and N lines would be exactly masked by the intense lines 8 and 9.

11	35	L_I	1·093+0·204	1·297
12	25	L_{III}	1·129+0·162	1·291
13	22	M_I	1·245+0·052	1·297
14	6	N_I	1·279+0·013	1·292
8	50	K	0·749+1·092	1·841
17	20	L_I	1·644+0·204	1·848
18	6	M_I	1·782+0·052	1·834
16	18	K	1·406+1·092	2·498
20	8	L_I	2·291+0·204	2·495
19	16	K	2·099+1·092	3·19
21	6	L_I	2·99 +0·204	3·19
22	2	N_I	3·18 +0·01	3·19
23	8	K	3·52 +1·09	4·61
24	4	L_I	4·42 +0·20	4·62
25	2	M_I	4·58 +0·05	4·63
26	6	K	8·04 +1·09	9·13
28	2	L_I	8·97 +0·20	9·17
27	3	K	8·61 +1·09	9·70
29	2	L_I	9·49 +0·20	9·69

in the above tables. The values of the $H\rho$'s were obtained by comparison with the strong lines of radium (B + C) and the error should not exceed 1 part in 250. Yovanovitch and d'Espine* have also measured this spectrum, and their results are in good agreement with Black's. Since they used the less accurate direct deviation method, Black's measurements are to be preferred. Mesothorium 2 also appears to emit some β rays of very high energy. An account of these will be found in § 87.

Thibaud† has investigated the γ ray emission of this body by the excited spectrum method and identified those γ rays marked by a star in the following table. The γ rays 7 and 8 could only be found in this way, since the natural β ray groups due to them coincide almost exactly with those due to γ rays 5 and 6.

γ rays of mesothorium 2

Number of γ ray	Energy in volts × 10⁻⁵	λ in X U	Number of γ ray	Energy in volts × 10⁻⁵	λ in X U
1	0·581	212	7*	3·38	36·5
2	0·795	155	8*	4·08	30·2
3	1·29(4)	95·4	9*	4·62	26·7
4*	1·84(1)	67·0	10*	9·15	13·5
5	2·49(7)	49·4	11*	9·70	12·7
6	3·19	38·7			

The natural β ray spectrum of thorium B and thorium C". The β ray spectra of these bodies have been investigated by von Baeyer, Hahn and Meitner‡, by Meitner§, by Ellis‖, by Yovanovitch and d'Espine¶, and by Black**. Thibaud† has also detected a number of the γ rays by the excited spectrum method, using lead as a radiator. These γ rays are marked with a star in the tables. The γ rays numbers 1 and 2 of thorium B, and number 11 of thorium C", are based in each case on only one β ray line, but their existence is shown by the occurrence of relatively strong lines in the excited β ray spectrum. The analysis

* Yovanovitch and d'Espine, *C.R.* **178**, 1811; **179**, 1162; 1924.
† Thibaud, *Thèse*, Paris, 1925.
‡ von Baeyer, Hahn and Meitner, *Phys. Zeit.* **12**, 1099, 1911.
§ Meitner, *Zeit. f. Phys.* **9**, 131, 1922.
‖ Ellis, *Proc. Camb. Phil. Soc.* **21**, 121, 1922; *Proc. Roy. Soc.* A, **101**, 1, 1922.
¶ Yovanovitch and d'Espine, *C.R.* **180**, 202, 1925.
** Black, *Proc. Roy. Soc.* A, **109**, 166, 1925.

The natural β ray spectrum of thorium B

Number of line	Intensity (visual estimate)	$H\rho$	Energy in volts $\times 10^{-5}$	Number of line	Intensity (visual estimate)	$H\rho$	Energy in volts $\times 10^{-5}$
1	10	835	0·583	10	4	1452	1·610
2	4	856	0·611	11	30	1701	2·118
3	2	926	0·709	12	80	1764	2·254
4	2	946	0·738	13	30	1820	2·376
5	4	1020	0·849	14	6	1831	2·399
6	20	1118	1·006	15	1	1926	2·605
7	1	1185	1·119	16	3	2037	2·854
8	30	1352	1·419	17	1	2095	2·998
9	200	1398	1·506				

Analysis of the β ray spectrum of thorium B

Number of line	Intensity	Origin	Energy of β ray line + absorption energy in volts $\times 10^{-5}$	Energy of γ ray in volts $\times 10^{-5}$
9	200	K	1·506 + 0·900	2·406
12	80	L_I	2·254 + 0·163	2·417
13	30	M_I	2·376 + 0·040	2·416
14	6	N_I	2·399 + 0·010	2·409
11	30	K	2·118 + 0·900	3·018
16	3	L_I	2·854 + 0·163	3·016
17	1	M_I	2·998 + 0·040	3·038

γ rays of thorium B

Number of γ ray	Energy in volts $\times 10^{-5}$	λ in X U	Number of γ ray	Energy in volts $\times 10^{-5}$	λ in X U
1*	1·91	64·6	3*	2·41(2)	51·2
2*	2·32	53·2	4*	3·02	40·9

The natural β ray spectrum of thorium C''

Number ne	Intensity	$H\rho$	Energy in volts $\times 10^{-5}$	Number of line	Intensity	$H\rho$	Energy in volts $\times 10^{-5}$
1	v.s.	541	0·252	22	f.	1817	2·369
2	s.	548	0·259	23	m.	1852	2·446
3	m.	568	0·278	24	m.f.	1916	2·583
4	v.s.	658	0·369	25	m.s.	1939	2·640
5	m.	668	0·380	26	m.s.	1990	2·756
6	s.	684	0·398	27	f.	2312	3·515
7	m.	689	0·404	28	f.	2475	3·913
8	f.	830	0·577	29	v.s.	2622	4·281
9	f.	960	0·758	30	v.s.	2913	5·025
10	f.	1056	0·906	31	m.f.	2961	5·150
11	m.	1157	1·071	32	m.f.	3057	5·400
12	m.s.	1249	1·231	33	m.s.	3182	5·729
13	f.	1278	1·283	34	m.	3432	6·397
14	m.s.	1373	1·458	35	m.f.	3650	6·990
15	f.	1421	1·550	36	f.	3910	7·70
16	m.s.	1478	1·661	37	m.s.	3960	7·836
17	m.	1501	1·706	38	f.	4040	8·057
18	v.s.	1604	1·915	39	f.	4310	8·81
19	m.f.	1623	1·954	40	s.	10080	25·58
20	s.	1665	2·042	41	m.	10340	26·35
21	m.f.	1723	2·165	42	f.	10380	26·46

of the lines 29 to 33 depends essentially on Thibaud's results. In the excited spectrum he finds the line corresponding to number 30 stronger than that corresponding to number 29. We therefore infer a γ ray of $5·90 \times 10^5$ volts giving line 30 by K conversion which happens to coincide with the L conversion of the γ ray $5·16 \times 10^5$ volts.

The β ray groups obtained with a source of thorium C are attributed to thorium C″, since thorium C emits γ rays, if at all, only in very small amount*.

Analysis of the thorium C″ β ray spectrum

Number of line	Intensity	Origin	Energy of β ray line + absorption energy in volts × 10^{-5}	Energy of γ ray in volts × 10^{-5}
1	v.s.	L_I	0·252+0·158	0·410
2	s.	L_{II}	0·259+0·152	0·411
3	m.	L_{III}	0·278+0·133	0·411
4	v.s.	M_I	0·369+0·038	0·407
5	m.	M_V	0·380+0·025	0·406
6	s.	N_I	0·398+0·009	0·407
7	m.	N_I or O	0·404+0·001	0·405
8	f.	K	0·577+0·875	1·452
13	f.	L_I	1·283+0·158	1·441
12	m.s.	K	1·231+0·875	2·106
19	m.f.	L_I	1·954+0·158	2·112
14	m.s.	K	1·458+0·875	2·333
21	m.f.	L_I	2·165+0·158	2·323
16	m.s.	K	1·661+0·875	2·536
22	f.	L_I	2·369+0·158	2·527
17	m.	K	1·706+0·875	2·581
23	m.	L_I	2·446+0·158	2·604
18	v.s.	K	1·915+0·875	2·790
25	m.s.	L_I	2·640+0·158	2·798
20	s.	K	2·042+0·875	2·917
26	m.s.	L_I	2·756+0·158	2·914
29	v.s.	K	4·281+0·875	5·156
30	v.s.	L_I	5·025+0·158	5·183
31	m.f.	M_1	5·150+0·038	5·188
30	v.s.	K	5·025+0·875	5·900
33	m.s.	L_I	5·729+0·158	5·887
35	m.f.	K	6·990+0·875	7·86
36	f.	L_I	7·70 +0·158	7·86
40	s.	K	25·58 +0·875	26·46
41	m.	L_I	26·35 +0·158	26·51
42	f.	M_I	26·46 +0·034	26·49

* Marsden and Darwin, *Proc. Roy. Soc.* A, **87**, 17, 1912; Meitner and Hahn, *Phys. Zeit.* **14**, 873, 1913; Rutherford and Richardson, *Phil. Mag.* **26**, 937, 1913.

γ rays of thorium C″

Number of γ ray	Energy in volts ×10⁻⁵	λ in X U	Number of γ ray	Energy in volts ×10⁻⁵	λ in X U
1	0·408	303	8	2·915	42·4
2	1·45	85·2	9*	5·17	23·9
3	2·11	58·5	10*	5·90	20·9
4	2·33	53·0	11*	7·27	17·0
5	2·53	48·8	12*	7·86	15·7
6	2·59	47·7	13*	26·49	4·66
7*	2·794	44·2			

The β ray spectrum of actinium $(B + C)$. This has been measured by Hahn and Meitner‡. The experimental difficulties were considerable because of the short period (36 minutes) and the small amount of active material available. It is probable that only the strongest lines were detected and since there were indications of many weak lines we can conclude that these spectra would be just as rich in lines as those of the other B and C bodies. Only one line, the first in the table, could be assigned with certainty to actinium B.

The natural β rays of actinium $(B + C)$

Number of line	Intensity (visual estimate)	Hρ	Energy in volts ×10⁻⁵	Number of line	Intensity (visual estimate)	Hρ	Energy in volts ×10⁻⁵
1	20	734	0·456	6	50	2418	3·772
2	100	1942	2·647	7	30	2472	3·906
3	35	2184	3·208	8	20	2670	4·402
4	35	2263	3·396	9	15	2772	4·662
5	30	2314	3·519				

Analysis of the β ray spectrum of actinium C″

Number of line	Intensity	Origin	Energy of β ray line + absorption energy in volts ×10⁻⁵	Energy of γ ray in volts ×10⁻⁵
2	100	K	2·647+0·874	3·521
4	35	L_I	3·396+0·158	3·554
5	30	M_I	3·519+0·038	3·557
6	50	K	3·772+0·874	4·646
8	20	L_I	4·402+0·158	4·560
7	30	K	3·906+0·874	4·780
9	15	L_I	4·662+0·158	4·820

‡ Hahn and Meitner, *Zeit. f. Phys.* **34**, 795, 1925.

γ rays of actinium C"

Number of γ ray	Intensity	Energy in volts ×10^{-5}	λ in X U
1	v.s.	3·54	34·9
2	s.	4·60	26·8
3	m.	4·80	25·7

§ 85. β ray spectra from α ray bodies. *The natural β ray spectrum of radium.* Hahn and Meitner* investigated this spectrum in 1909 by the direct deviation method and found two distinct groups. Later† they carried through a more accurate investigation by means of the focussing method. This spectrum has a special interest since radium is an α ray body. It is technically in some respects the most difficult to measure of all the spectra owing to the danger of contaminating the whole apparatus and laboratory with the emanation.

The absolute values of $H\rho$ were obtained by comparison with the radium B lines.

The natural β ray spectrum of radium

Number of line	Intensity (visual estimate)	$H\rho$	Energy in volts ×10^{-5}
1	s.	1037	0·876
2	m.s.	1508	1·720
3	w.	1575	1·856

Analysis of the radium β ray spectrum

Number of line	Intensity	Origin	Energy of β ray line + absorption energy in volts ×10^{-5}	Energy of γ ray in volts ×10^{-5}
1	s.	K	0·876+0·981	1·86
2	m.s.	L_{I}	1·720+0·180	1·90
3	w.	M_{I}	1·856+0·045	1·90

γ ray of radium

Energy in volts	λ in X U
1·89 ×10^{5}	65·2

---

* Hahn and Meitner, *Phys. Zeit.* **10**, 741, 1909.
† Hahn and Meitner, *Zeit. f. Phys.* **26**, 161, 1924.

The natural β ray spectrum of protactinium, radioactinium, and actinium X. These spectra are interesting as they are fully as complex as any from the typical β ray bodies. The identification of these spectra and the proof that the β rays are of secondary origin is the work of Hahn and Grosse[*] for protactinium, and Hahn and Meitner[†] for radioactinium and actinium X.

The results given here are the measurements of Meitner[‡] on protactinium, and Hahn and Meitner[§] on radioactinium and actinium X. In a few cases the figures given for the energies of the groups are slightly different from those appearing in the original paper. These have been altered with the consent of the authors to correspond to the values of the constants and system of approximations used in this book.

The natural β ray spectrum of protactinium

Number of line	Intensity (visual estimate)	$H\rho$	Energy in volts $\times 10^{-5}$	Number of line	Intensity (visual estimate)	$H\rho$	Energy in volts $\times 10^{-5}$
1	60	956	0·753	7	30	1855	2·450
2	40	980	0·788	8	30	1909	2·573
3	40	1055	0·905	9	60	1985	2·746
4	30	1077	0·939	10	30	2039	2·869
5	100	1595	1·896	11	40	2103	3·016
6	70	1736	2·194	12	20	2173	3·182

Analysis of the β ray spectrum of protactinium

Number of line	Intensity	Origin	Energy of β ray line + absorption energy in volts $\times 10^{-5}$	Energy of γ ray in volts $\times 10^{-5}$
1	60	L_I	0·753+0·198	0·951
2	40	L_{III}	0·788+0·158	0·946
3	40	M_I	0·905+0·050	0·950
5	100	K	1·896+1·064	2·960
9	60	L_I	2·746+0·198	2·944
10	30	M_I	2·869+0·050	2·919
6	70	K	2·194+1·064	3·258
11	40	L_I	3·016+0·198	3·214
12	20	M_I	3·182+0·050	3·232

[*] Hahn and Grosse, *Zeit. f. Phys.* **48**, 1, 1928.
[†] Hahn and Meitner, *Phys. Zeit.* **9**, 697, 1908; *Zeit. f. Phys.* **2**, 60, 1920.
[‡] Meitner, *Zeit. f. Phys.* **50**, 15, 1928.
[§] Hahn and Meitner, *Zeit. f. Phys.* **34**, 795, 807, 1925.

γ rays of protactinium

Number of γ ray	Energy in volts × 10⁻⁵	λ in X U
1	0·949	130
2	2·94	41·9
3	3·23	38·2

The natural β ray spectrum of radioactinium

Number of line	Intensity (visual estimate)	Hρ	Energy in volts × 10⁻⁵	Number of line	Intensity (visual estimate)	Hρ	Energy in volts × 10⁻⁵
*1	20	378	0·125	26	40	996·5	0·813
*2	30	407	0·144	27	20	1010	0·834
*3	20	429	0·160	28	50	1075	0·936
4	50	534	0·246	29	30	1093	0·965
*5	20	543	0·255	30	30	1108	0·990
*6	15	551	0·262	31	20	1132	1·029
*7	10	561	0·271	32	30	1159	1·074
*8	25	571	0·281	33	10	1178	1·106
*9	15	580	0·290	34	10	1195	1·135
*10	30	590	0·299	35	80	1291	1·305
*11	20	596	0·305	36	30	1367	1·445
*12	15	611	0·320	37	60	1396	1·501
13	15	624	0·333	38	30	1525	1·753
14	40	630	0·340	39	60	1546	1·796
15	10	652	0·363	40	20	1597	1·899
16	10	675	0·388	41	50	1634	1·976
17	90	707·5	0·425	42	20	1663	2·036
18	100	732	0·454	43	20	1703	2·121
19	20	759	0·486	44	20	1745	2·211
20	70	822·5	0·567	45	10	1773	2·271
21	50	846	0·598	46	40	1808	2·348
22	20	876	0·639	47	30	1872	2·488
23	20	912	0·689	48	20	1930	2·618
24	20	951	0·745	49	20	2010	2·800
25	15	982	0·791				

* Difficult to measure accurately.

Analysis of the β ray spectrum of radioactinium

Number of line	Intensity	Origin	Energy of β ray line + absorption energy in volts $\times 10^{-5}$	Energy of γ ray in volts $\times 10^{-5}$
1	20	L_I	$0 \cdot 125 + 0 \cdot 192$	$0 \cdot 317$
3	20	L_III	$0 \cdot 160 + 0 \cdot 154$	$0 \cdot 314$
6	15	M_I	$0 \cdot 262 + 0 \cdot 048$	$0 \cdot 310$
7	10	M_II	$0 \cdot 271 + 0 \cdot 044$	$0 \cdot 315$
9	15	M_V	$0 \cdot 290 + 0 \cdot 031$	$0 \cdot 321$
10	30	N_I	$0 \cdot 299 + 0 \cdot 012$	$0 \cdot 311$
11	20	N_VI	$0 \cdot 305 + 0 \cdot 003$	$0 \cdot 308$
12	15	...	$0 \cdot 320$	$0 \cdot 320$
4	50	L_I	$0 \cdot 246 + 0 \cdot 192$	$0 \cdot 438$
5	20	L_II	$0 \cdot 255 + 0 \cdot 185$	$0 \cdot 440$
8	25	L_III	$0 \cdot 281 + 0 \cdot 154$	$0 \cdot 435$
16	10	M_I	$0 \cdot 388 + 0 \cdot 048$	$0 \cdot 436$
14	40	L_I	$0 \cdot 340 + 0 \cdot 192$	$0 \cdot 532$
19	20	M_I	$0 \cdot 486 + 0 \cdot 048$	$0 \cdot 534$
17	90	L_I	$0 \cdot 425 + 0 \cdot 192$	$0 \cdot 617$
20	70	M_I	$0 \cdot 567 + 0 \cdot 048$	$0 \cdot 615$
21	50	N_I	$0 \cdot 598 + 0 \cdot 012$	$0 \cdot 610$
26	40	L_I	$0 \cdot 813 + 0 \cdot 192$	$1 \cdot 005$
29	30	M_I	$0 \cdot 965 + 0 \cdot 048$	$1 \cdot 013$
30	30	N_I	$0 \cdot 990 + 0 \cdot 012$	$1 \cdot 002$
18	100	K	$0 \cdot 454 + 1 \cdot 035$	$1 \cdot 489$
35	80	L_I	$1 \cdot 305 + 0 \cdot 192$	$1 \cdot 497$
36	30	M_I	$1 \cdot 445 + 0 \cdot 048$	$1 \cdot 493$
28	50	K	$0 \cdot 936 + 1 \cdot 035$	$1 \cdot 971$
38	30	L_I	$1 \cdot 753 + 0 \cdot 192$	$1 \cdot 945$
40	20	M_I	$1 \cdot 899 + 0 \cdot 048$	$1 \cdot 947$
37	60	K	$1 \cdot 501 + 1 \cdot 035$	$2 \cdot 536$
46	40	L_I	$2 \cdot 348 + 0 \cdot 192$	$2 \cdot 540$
47	30	M_I	$2 \cdot 488 + 0 \cdot 048$	$2 \cdot 536$
39	60	K	$1 \cdot 796 + 1 \cdot 035$	$2 \cdot 831$
48	20	L_I	$2 \cdot 618 + 0 \cdot 192$	$2 \cdot 810$
41	50	K	$1 \cdot 976 + 1 \cdot 035$	$3 \cdot 011$
49	20	L_I	$2 \cdot 800 + 0 \cdot 192$	$2 \cdot 992$

γ rays of radioactinium

Number of γ ray	Energy in volts $\times 10^{-5}$	λ in X U	Number of γ ray	Energy in volts $\times 10^{-5}$	λ in X U
1	$0 \cdot 315$	392	6	$1 \cdot 493$	$82 \cdot 6$
2	$0 \cdot 437$	282	7	$1 \cdot 954$	$63 \cdot 2$
3	$0 \cdot 533$	232	8	$2 \cdot 53(7)$	$48 \cdot 7$
4	$0 \cdot 614$	201	9	$2 \cdot 82(0)$	$43 \cdot 7$
5	$1 \cdot 007$	122	10	$3 \cdot 00$	$41 \cdot 2$

The natural β ray spectrum of actinium X

Number of line	Intensity (visual estimate)	Hρ	Energy in volts × 10⁻⁵	Number of line	Intensity (visual estimate)	Hρ	Energy in volts × 10⁻⁵
1	20	524	0·237	12	50	1321	1·361
2	80	733	0·455	13	25	†1335	1·387
3	10	756	0·483	14	15	1380	1·471
4	100	816·5	0·559	15	15	1402	1·513
5	40	845	0·597	16	100	1502	1·708
6	15	*900	0·672	17	25	1527	1·759
7	10	*983	0·793	18	15	*1547	1·800
8	10	*1000	0·819	19	30	1753	2·230
9	10	1140	1·043	20	30	1817	2·369
10	10	1191	1·129	21	30	1880	2·508
11	50	1265	1·259				

* Difficult to measure accurately.   † Double line.

Analysis of the β ray spectrum of actinium X

Number of line	Intensity	Origin	Energy of β ray line + absorption energy in volts × 10⁻⁵	Energy of γ ray in volts × 10⁻⁵
2	80	K	0·455 + 0·980	1·435
11	50	L_I	1·259 + 0·180	1·439
13	25	M_I	1·387 + 0·045	1·432
4	100	K	0·559 + 0·980	1·539
12	50	L_I	1·361 + 0·180	1·541
14	15	M_I	1·471 + 0·045	1·516
15	15	N_I	1·513 + 0·011	1·524
5	40	K	0·597 + 0·980	1·577
13	25	L_I	1·387 + 0·180	1·567
15	15	M_I	1·513 + 0·045	1·558
9	10	K	1·043 + 0·980	2·023
18	15	L_I	1·800 + 0·180	1·980
16	100	K	1·708 + 0·980	2·688
21	30	L_I	2·508 + 0·180	2·688

γ rays of actinium X

Number of γ ray	Energy in volts × 10⁻⁵	λ in X U
1	1·435	86·0
2	1·53	80·6
3	1·57	78·6
4	2·00	61·7
5	2·69	45·9

The natural β ray spectrum of radiothorium. The β ray spectrum of this α ray body was first detected by von Baeyer, Hahn and Meitner*. It has recently been carefully investigated by Meitner†, whose measurements are given in the table. There are two points of exceptional interest in this spectrum. In the first place the γ rays which are deduced from the β ray spectrum agree so well with the K_a X ray lines of the resulting product that little doubt can exist that this is their origin. Yet since no other γ rays are found in sufficient intensity we are led to conclude that these X rays are excited by the α rays in their passage through the atom. The intensity of the β ray lines suggests that this happens in the majority of the disintegrations. The other point of interest is that the β ray groups due to the K_{a_2} line are stronger than those due to K_{a_1}, whereas in emission under normal circumstances the latter line is twice the intensity of the former.

The natural β ray spectrum of radiothorium

Number of line	Intensity	$H\rho$	Energy in volts × 10⁻⁵
1	w.	806	0·547
2	w.	827	0·567
3	s.	891	0·656
4	m.s.	911	0·688
5	m.s.	988	0·801
6	m.	1010	0·834

Analysis of the β ray spectrum of radiothorium

Number of line	Intensity	Origin	Energy of β ray line + absorption energy in volts × 10⁻⁵	Energy of γ ray in volts × 10⁻⁵
3	s.	$L_{\rm I}$	0·656 + 0·192	0·848
5	m.s.	$M_{\rm I}$	0·801 + 0·048	0·849
4	m.s.	$L_{\rm I}$	0·688 + 0·192	0·880
6	m.	$M_{\rm I}$	0·834 + 0·048	0·882

γ rays of radiothorium

Energy in volts × 10⁻⁵	λ in X U
0·848	145
0·881	140

* von Baeyer, Hahn and Meitner, *Phys. Zeit.* **16**, 6, 1915.
† Meitner, *Zeit. f. Phys.* **52**, 637, 645, 1928.

§ 86. Measurement of γ rays by crystal method. Rutherford and Andrade* were the first to show that it was possible to use the crystal method for investigating the radiations emitted by radioactive bodies and subsequently Thibaud† and Frilley‡ have made valuable contributions.

There are two points of interest in this work. In the first place it is most important to detect and measure the γ rays by this direct method and obtain a confirmation of the accuracy of the photo-electric law for these high frequencies, since this is the basis of all the deductions of γ ray frequencies from the β ray spectra. In the second place this method is particularly suitable for investigating the K and L X ray spectra which are excited in the radioactive atom during the disintegration.

The experimental arrangements follow the usual X ray technique, the source being a fine tube containing the radioactive material and placed close to the collimating slit. However, the experiments are difficult, owing to the extremely small glancing angles required for these frequencies, and owing to the necessity of protecting the plate from general scattered β and γ rays. Rutherford and Andrade avoided the necessity of adjusting the small glancing angle by a very ingenious arrangement of the slit and crystal, but the tendency of the later work has been to follow standard practice.

Remarkable progress in detecting radiations of high frequency has been made by Frilley. Recently he has been able to measure a γ ray of 770,000 volts, that is a wave-length of about 16 X U. When it is realised that this radiation has a glancing angle of less than 10 minutes and is relatively feebly scattered by the crystal, the difficulty of the experiments can be appreciated. The apparatus consisted of three carriages mounted on a rigid steel bench 1·50 metres long, holding the source, the crystal, and the plate. In front of the source was arranged a lead slit 45 cm. long and whose width varied from 0·3 mm. at one end to about 0·6 mm. at the other. Between the two faces a potential difference of a few hundred volts could be applied to deviate the β radiation. The crystal was of rock-salt, 3×9 cm. in size and 0·6 to 0·8 mm. thick, and was mounted on a carrier capable of adjustments in all directions. Two photographic plates were used, placed face to face with an intensifying screen of calcium tungstate between them, and they were protected from

* Rutherford and Andrade, *Phil. Mag.* **27**, 854; **28**, 262, 1914.
† Thibaud, *Thèse*, Paris, 1925.   ‡ Frilley, *Thèse*, Paris, 1928.

stray scattered radiations by screens of lead. The distance between the source and the plates was varied between 60 and 160 cm. The crystal was rotated slowly during the exposure, following the usual technique, and while the strongest lines in the following table could be obtained with 100 mg. and a rotation of 2° in 24 to 48 hours, the weaker lines required 200 to 300 mg. with a rotation of only 30' in 24 hours. The error is estimated at less than 1 X U.

Reproductions of two photographs taken by Frilley are shown in Plate XII, figs. 4 and 5. The numbers refer to the following table. Photograph 4 shows the strong lines 6, 7 and 8 of radium B remarkably clearly, and the lines 1 and 2 are interesting because they are the shortest wave-lengths yet measured by the crystal method. Photograph 5 shows the K lines 16, 17 and 18 and also very clearly the line 22 which is responsible by its photo-electric effect for the entire group of lines shown in Plate XI, fig. 3.

Frilley's measurements on the wave-lengths emitted by radium B and radium C are shown in the first of the next two tables, and this is followed by Rutherford and Andrade's measurements on the softer radiations from the same bodies. The latter have been corrected for

Measurements by the crystal method of the γ and X radiations emitted by radium B and radium C

Number	λ in X U	Energy in volts $\times 10^{-5}$	Intensity	γ rays from β ray spectra in volts $\times 10^{-5}$
1	16	7·70	m.	7·73 Ra C
2	20	6·17	m.s.	6·12 Ra C
3	24	5·15	v.w.	5·03 Ra C
4	26	4·75	v.w.	4·71 Ra B
5	29·5	4·20	w.	4·29 Ra C
6	35	3·53	m.s.	3·54 Ra B
7	42	2·94	m.s.	2·97 Ra B
8	51·5	2·40	m.s.	2·43 Ra B
9	65	1·90	w.	—
10	119	1·04	v.w.	—
11	135	0·914	w.	—
12	140	0·884	s.	—
13	144	0·858	s.	—
14	147	0·840	v.w.	—
15	149	0·828	v.w.	—
16	155	0·798	m.	$\rbrace K_{a_1}$ and K_{a_2}
17	161	0·767	v.s.	from at. nos.
18	166	0·744	s.	83 and 84
19	170	0·725	w.	—
20	190	0·650	w.	—
21	210	0·588	w.	0·59 Ra C
22	232	0·533	m.s.	0·536 Ra B

a small systematic error brought to light by the later measurements of Rutherford and Wooster*.

γ and X rays from Ra (B + C)		L series X rays	
λ in X U	Intensity	At. No. 83	At. No. 84
768	m.	761 γ_4	768 γ_2
784	m.	787 γ_6	785 γ_1
812	m.	811 γ_1	—
827	m.	—	—
888	w.	—	—
924	m.	—	919 β_1
952	s.	950 β_1	—
985	m.	975 β_4	—
998	m.	992 β_6	—
1022	w.	—	1024 η
1041	w.	—	—
1066	w.	1057 η	—
1105	m.	—	1110 α_1
1139	s.	1141 α_1	1123 α_2
1160	m.	1153 α_3	—
1181	w.	—	—
1226	w.	—	—
1246	w.	—	—
1274	w.	—	—
1307	m.	—	—
1323	m.	—	—

If we consider first the shorter wave-lengths at the beginning of the table, we see that Frilley's values agree excellently with the results deduced from the β ray spectra. The four γ rays of radium B, and the one γ ray of radium C (nos. 4 to 8), appear from the β ray spectra to be of comparable intensity and are the only strong radiations in the region investigated. Previous to Frilley's work, Thibaud, using a similar apparatus, had detected from thorium B a radiation of energy $2\cdot36 \times 10^5$ volts, while the β ray spectrum had indicated that the only strong γ ray in this region was one of energy $2\cdot41 \times 10^5$ volts.

These experiments therefore constitute a most important and convincing proof of the validity of the Einstein photo-electric equation for these high frequencies, and of the explanation given of the β ray spectra.

While it is quite understandable that the strong γ rays, numbers 2, 6, 7 and 8, should be detected by this method, it is unexpected that so many other γ rays should be found which give relatively weak lines in the natural spectrum. Such marked apparent differences in

* Rutherford and Wooster, *Proc. Camb. Phil. Soc.* **22**, 834, 1925.

PLATE XII

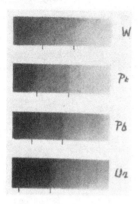

Fig. 1. β ray spectrum excited in different metals by the γ rays of RaB.

Fig. 2. 5·72 mg. per cm.²

Fig. 3. 2·65 mg. per cm.²

Retardation of β rays by mica (White and Millington).

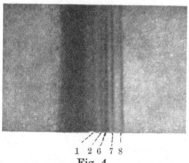

1	$\lambda - 16$ XU
2	20 XU
6	35 XU
7	42 XU
8	51·5 XU

1 2 6 7 8
Fig. 4.

12	140 XU
13	144 XU
16	155 XU
17	161 XU
18	166 XU
19	170 XU
22	232 XU

12,13 16 17 18 19   22
Fig. 5.

Measurement of the wave-length of γ rays by reflection from a crystal (Frilley).

the relative intensities of the γ rays observed by the two methods are of great interest and importance. As we shall see later, these observations throw light on the variation of the internal absorption of γ rays of different frequencies in their escape from the radioactive atom.

Turning to the other lines in the table, it will be seen that Frilley detected the K spectra and Rutherford and Andrade the L spectra corresponding to atomic numbers 83 and 84. As has already been pointed out, this is precisely what we should expect from a source of radium B and radium C if the γ rays are emitted after the disintegration. The completeness with which the spectra have been found and the agreement of the intensities with those of the normal X ray spectra are noteworthy and indicate that the atom had completely settled down to the states corresponding to the new atomic number by the time these radiations were emitted.

§ 87. β **rays of very high energy.** Radium C, thorium C and mesothorium 2 are exceptional in emitting β rays of very high energy. From radium C Danysz[*] found two groups with $H\rho$ 10,850 and 18,900 (energies 3 to 6 million volts), and later Rutherford and Robinson[†] stated that there seemed to be many faint lines between $H\rho$ 10,000 and $H\rho$ 20,000. Ellis[‡] measured one line of $H\rho$ 10,020 but was unable to find any of higher energy. A careful search was made by White (unpublished) to find the lines whose presence was suspected by Rutherford and Robinson, and he concluded that if present they were much weaker than the line $H\rho$ 10,020, which is itself very weak.

Black[§], and Yovanovitch and d'Espine[||], found three lines of $H\rho$ 10,080, 10,340, 10,380 in the spectrum of thorium C which appeared to be due to the action of a γ ray of energy 2·65 million volts on the K, L and M states.

These results taken together would suggest a limitation of the β ray spectra to energies of under 3 million volts. Recently, however, Yovanovitch and d'Espine (*loc. cit.*) have published an account of experiments carried out by the method of direct deviation in which they found β rays of far greater energies. In this type of apparatus the displacement of a ray is inversely proportional to $H\rho$,

[*] Danysz, *Le Radium*, **10**, 4, 1913.
[†] Rutherford and Robinson, *Phil. Mag.* **26**, 717, 1913.
[‡] Ellis, *Proc. Camb. Phil. Soc.* **32**, 369, 1924.
[§] Black, *Proc. Roy. Soc.* A, **109**, 168, 1925.
[||] Yovanovitch and d'Espine, *J. de Physique*, **8**, 276, 1927.

and the dispersion near the central spot is very small. A diffuse band can give the appearance of a line, and can have apparently a quite sharp upper limit. The same band investigated by the focussing method, where the displacement is directly proportional to $H\rho$, would be reproduced more nearly in its true form and would probably not be detected. It seems likely therefore, as Yovanovitch and d'Espine indicate, that these very high energy β rays are not homogeneous groups but wide bands. They are shown as such in the table which follows.

β rays of $H\rho$ greater than 10,000

$H\rho$	Intensity	Energy in volts	
Mesothorium 2			
10,000	w.	$2 \cdot 6 \quad \times 10^6$	
16,700	v.w.	$4 \cdot 2$	Bands (Yovanovitch and
21,000	v.w.	$5 \cdot 9$	d'Espine)
28,000	w.	$8 \cdot 0$	
Thorium C			
10,080	m.s.	$2 \cdot 551 \times 10^6$	Normal β ray lines
10,340	m.	$2 \cdot 631$	(Black)
10,380	w.	$2 \cdot 65$	
18,000	w.	$4 \cdot 9$	Bands (Yovanovitch and
*40,000	v.w.	$11 \cdot 0$	d'Espine)
Radium C			
10,020	w.	$2 \cdot 5 \quad \times 10^6$	β ray line (Ellis)
10,700	w.	$2 \cdot 74$	Head of band (Yovanovitch and d'Espine)
15,000 to 27,000	w.	$4 \cdot 0$ to $7 \cdot 6$	Band (Yovanovitch and d'Espine)

* Extremely weak, existence doubtful.

These bands are difficult to detect and were only found after considerable improvement in the technique of the experiment. The apparatus used is shown in Fig. 96, and was built of aluminium except for the slit D which was of lead. Very fine slits, as small as $0 \cdot 1$ mm., were used, and the distance between the source S and slit (about 5 cm.) was much greater than that between the slit and the photographic plate P (about 1 cm.). The photographic plate was in certain experiments covered with a thin sheet of aluminium $0 \cdot 01$ to $0 \cdot 05$ mm. thick, which, while presenting no obstacle to the very fast β rays, greatly diminished the effect of slower scattered electrons. This type of apparatus cannot give very accurate results, for the very reason that it shows up bands as lines and the results are to be considered only as indicating the general positions.

The origin of these bands presents a problem of considerable interest. If they are due to the internal conversion of γ rays, then, since at these high frequencies the probability of conversion is most likely very small, we should have to assume the presence of an amount of high-frequency γ rays that is not easily compatible with other experimental results. We are thus led to the interesting speculation that they may be disintegration electrons and come from the nucleus. The appearance of bands would indicate a similar distribution of energy as that in the continuous spectra of the main disintegration electrons. These β rays would therefore be analogous to the long-range α particles and represent an abnormal and infrequent mode of disintegration. Attempts to determine the number of these electrons have been made in the following way, by Cave* and by Cave and Gott. Cave placed a small thin-walled radon

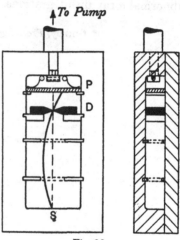

Fig. 96.

tube at the centre of the circular polepieces of a large Weiss electromagnet. With a given field only β particles with energies above a certain value could escape into the region outside the polepieces, where they were detected by an electroscope. The results showed that on the average less than one of these high-energy β particles was emitted in every five hundred disintegrations. In the experiments of Cave and Gott an absorption method was used in which a radon source was surrounded by successive layers of thin lead foil and the unabsorbed β particles were measured by a Faraday cylinder which surrounded both the source and foils. The interpretation was complicated by the secondary β rays expelled by the γ rays, but the results nevertheless showed definitely that the fast β rays were very few in number and certainly less than one in one thousand disintegrations. We can form a rough estimate of the relative numbers of these electrons

* Cave, *Proc. Camb. Phil. Soc.* 25, 222, 1929.

as follows. Gurney* found that about five electrons were emitted in the radium B line $H\rho$ 1938 in every hundred disintegrations, so that it appears from the intensities given in § 84 that about twenty electrons are emitted in the radium C line $H\rho$ 10,020 in every 10^6 disintegrations. We know very little about the photographic action of such very high speed electrons, and the estimates of intensity are extremely rough, but we can take the same figure as giving the order of magnitude of the number of electrons in these high-energy bands. It is interesting to note that this is the same order of magnitude as is found for the abnormal α ray disintegrations.

* Gurney, *Proc. Roy. Soc.* A, **109**, 540, 1925.

CHAPTER XIII

THE DISINTEGRATION ELECTRONS

§ 88. The greater portion of the β ray emission of a radioactive body is formed by the disintegration electrons. Useful qualitative information can be obtained by investigation of the total emission by methods such as measuring the absorption in aluminium, but while these suffice to show the great difference in penetrating power of the β radiation from different bodies they are not suitable for a detailed analysis.

The chief difficulty in investigating the disintegration electrons lies in distinguishing them from those forming the β ray spectra and any other electrons ejected from the outside electronic structure of the atom by subsidiary processes.

The two lines of research which have yielded the most conclusive evidence have been the determination of the total number of electrons emitted by a known quantity of radioactive material, and further the investigation of the distribution of energy among the emitted electrons. The importance of the first type of experiment lies in the fact that since we know there must be one electron emitted from the nucleus of each disintegrating atom, any excess of electrons above this number must be due to secondary processes such as conversion of the γ rays, collisions and so on. We obtain in this way direct and valuable evidence on the extent to which such processes occur.

The distribution of energy among the disintegration electrons is particularly interesting, since it has brought to light a behaviour quite unlike that of the α ray bodies. Instead of a β ray body emitting electrons all of one speed from the nucleus, they appear to be distributed continuously over a wide range of velocity. This phenomenon of the continuous spectrum of β rays is one of the fundamental differences between the α and β ray types of disintegration.

Since the β particles carry a negative charge, the number emitted from a radioactive substance can be estimated either by measuring the negative charge communicated to a body in which the β particles are stopped or by the gain of positive charge of a conducting body containing the radioactive material. The earliest experiments of this kind were made by M. and Mme Curie, using an electrometer method. The escape of electrons from a radioactive body can be strikingly illustrated by an apparatus devised by Strutt*, which is shown in Fig. 97. To a sealed tube AA containing the radium was attached at one end a pair of thin gold leaves CC in metallic connection with the radium, which was insulated by means of a quartz rod B. The inner surface of the tube was coated with tinfoil EE connected to earth. The glass surface of AA was made conducting by a thin coating of phosphoric acid. The air in the outer tube was exhausted as completely as possible by means of a mercury pump, in order to reduce the ionisation in the gas, and consequently the loss of any charge gained by the gold leaves. After some hours the gold leaves were observed to diverge to their full extent, in-dicating that they had acquired a large charge. This could only have occurred by the emission from the radium of charged particles capable of travers-ing the glass walls AA. The divergence of the leaves could not therefore be due to the emission of α particles and so showed in a striking manner the existence of the β rays.

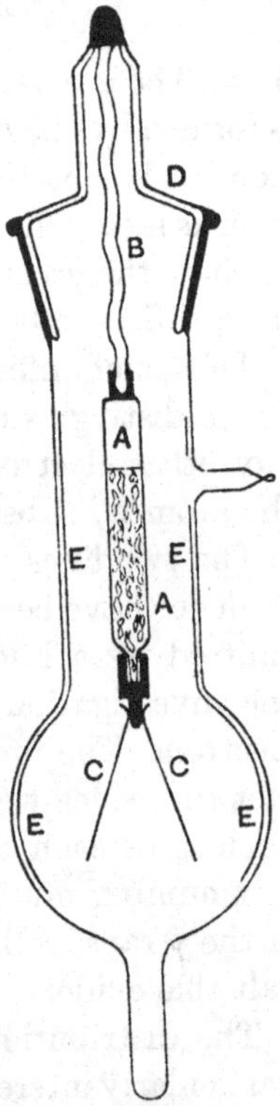

If the tube is filled with 30 mg. of radium bromide, the leaves diverge to their full extent in the course of about a minute. If it is arranged that the gold leaf, at a certain angle of divergence, comes in contact with a piece of metal connected to earth, the apparatus can be made to work auto-matically. The leaves diverge, touch the metal, and at once collapse, and this periodic movement of the leaves should continue, if not indefinitely,

Fig. 97.

at any rate as long as the radium lasts. For this reason this apparatus has been referred to as a "radium clock."

The potential to which an insulated body in a vacuum may be raised by the emission of β particles has been carefully examined by Moseley†. A silver-coated glass sphere, thick enough to stop the α particles, but thin enough to allow many of the β particles to traverse it, was filled with about 100 millicuries of radon. The sphere was insulated by a long quartz rod inside a glass envelope which was exhausted to the limit possible by the use of charcoal and liquid air.

* Strutt, *Phil. Mag.* 6, 588, 1903.
† Moseley, *Proc. Roy. Soc.* A, **88**, 471, 1913.

The potential of the sphere, which was measured by the attraction on a neighbouring conductor, was found to rise in the course of a few days to about 150,000 volts and then subsequently to fall as the radon decayed. Since from other evidence we know that β rays of energy up to two million volts are expelled from the product of radon radium C, the maximum potential was no doubt due to a balance between the loss of negative charge due to the escape of the β particles and the gain of negative charge due to electrons liberated from the outer glass sphere by the action of the primary β and γ rays. But for this action the potential would rise to a value sufficient to prevent the escape of the β particles.

§ 89. The number of β particles emitted by radium B and radium C. Early experiments on this subject were made by Wien*, Rutherford†, and Makower‡, but while they demonstrated clearly that the number of emitted particles was approximately the same as the number of disintegrating atoms, the results are not suitable for the detailed consideration suggested in the preceding section.

The problem was attacked again by Moseley§ in 1912 with improved methods. The principle is simple and consists in measuring the charge carried by the electrons emitted from a known amount of radioactive material, which on division by the product of the elementary charge and the number of atoms disintegrating per unit time gives at once the average number of β particles emitted at each disintegration.

The active material was placed inside the paper cylinder P (see Fig. 98) which was covered on the outside by aluminium foil. In the first series of experiments the source was a thin-walled tube containing radon and the active deposit, and the total thickness to be traversed could be varied between the limits equivalent to 0·037 mm. and 0·117 mm. aluminium by putting on more india paper. The apparatus was highly exhausted and the charge was collected by the thick brass cylinder B, and measured by an electrometer in absolute units by a compensation method.

One of the difficulties in this type of experiment is in avoiding errors due to secondary electrons liberated by the primary β and γ rays. Moseley found that the β particles in their passage through

* Wien, *Phys. Zeit.* **4**, 624, 1903.
† Rutherford, *Phil. Mag.* **10**, 193, 1905.
‡ Makower, *Phil. Mag.* **17**, 171, 1909.
§ Moseley, *Proc. Roy. Soc.* A, **87**, 230, 1912.

metals produced slow-speed electrons, δ rays, the number of which
was, however, very small compared with that given by the same
number of α particles. A special investigation was made with the
purpose of finding the most suitable method of correcting for this
secondary radiation, which originates both from the central electrode
P and the brass cylinder B.

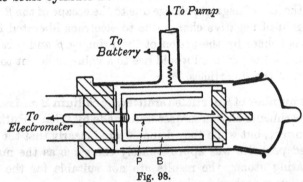

Fig. 98.

In order to prevent the α particles from the source reaching the
collecting electrode B, the source must always be covered. This
covering will also partially absorb the β rays, and the only way of
determining this absorption is to carry out experiments with different
thicknesses of material covering the source and to attempt to extrapolate back to zero thickness.

Total absorption in equivalent mg./cm.² of aluminium	Number of β particles for each pair of disintegrating atoms of RaB and RaC	Number of β particles from each disintegrating atom of	
		RaB	RaC
10	1·64	—	—
12·7	1·51	—	—
13·2	—	0·634	0·888
15·4	1·44	—	—
17·5	—	0·547	0·854
18·1	1·38	—	—
20·8	1·29	—	—
24·8	—	0·410	0·812
26·2	1·19	—	—
31·2	1·09	—	—

Moseley's results for the number of electrons penetrating different
thicknesses are shown in the above table. The radioactive source,
as has been already mentioned, was in this case a thin-walled glass
tube containing radon, constituting as far as β particles are concerned

a source of equal numbers of disintegrating atoms of radium B and radium C.

The figures in this table differ slightly from those in Moseley's paper, as they have been corrected for the later values of the radium standard, the elementary charge, and the number of atoms disintegrating per second in one curie. The last two columns show how

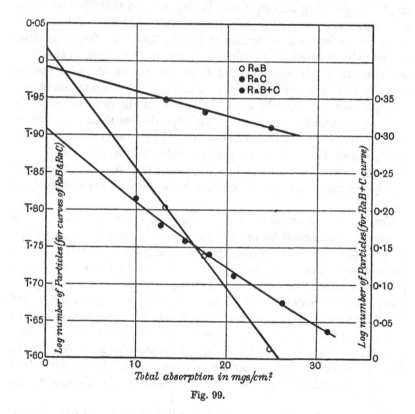

Fig. 99.

these particles are divided between radium B and radium C. This was determined in the usual way by preparing a source of radium B by recoil and observing the change with time of the number of electrons emitted as the radium C was gradually formed.

These results are plotted logarithmically in Fig. 99, from which it will be seen that to obtain the total number of electrons emitted a large uncertain extrapolation to zero thickness has to be made. Carrying this out as shown in the diagram, we find 1·04 electrons for radium B, 0·98 for radium C, and 2·04 from the curve for radium (B + C). Now while the measurements given in the table are probably correct to 2 per cent., the extrapolated result, which is what we really require, may be in error to 5 or 10 per cent.

Since a similar extrapolation has so frequently to be carried out in this type of measurement, it is of interest to consider shortly what are the underlying assumptions. It will be seen that Moseley's first measurement was of the β particles transmitted through 10 mg./cm.^{2}. Since we know from Schonland's* results that very few β particles are actually stopped in the first one-fifth of the range, it is safe to say that all β particles emitted from the source of speed greater than $H\rho$ 1800 are included in the value 1·64. The correctness of this extrapolation can thus be seen to depend on the occurrence of the appropriate number of electrons with speeds less than $H\rho$ 1800, or more accurately on the form of the low velocity portion of the distribution curve. Electrons of even the slowest energies are not neglected, they are only tacitly assumed to be present in a definite amount. It will be realised that the result may be in error either positively or negatively.

Very shortly afterwards, Danysz and Duane† published an account of the charges carried by the α rays and β rays emitted from a thin-walled glass bulb containing radon. The main difference in the method was that a magnetic field was used to bend away the δ particles and also to separate the charge carried by the α rays and β rays. In this way the ratio of the charges carried by the α rays and the β rays could be determined, and it was not necessary to assume a figure for the number of atoms disintegrating per second in 1 gram of radium. Danysz and Duane found that the charge carried by the β rays was 0·63 of that carried by the α rays, but this figure involved a rather uncertain correction for reflection of the β rays at the source.

Wertenstein‡ used an apparatus similar in principle but adapted to work with a source of active deposit on a small cylinder of thin aluminium leaf of thickness 0·003 mm. By observing the decay it was possible to separate the effect due to radium B and radium C, and Wertenstein found 1·04 β particles for each atom of radium C disintegrating and 1·35 for radium B, making 2·4 in all.

* Schonland, *Proc. Roy. Soc.* A, **104**, 235, 1923; **108**, 187, 1925.

† Danysz and Duane, *Le Radium*, **9**, 417, 1912.

‡ Wertenstein, *C.R. Soc. Sci. de Varsovie*, **9**, 929, 1916.

Quite recently measurements by an entirely different method have been made by Gurney*, following up the preliminary measurements of Curtiss†. The apparatus is shown in Fig. 100. The chief difference from the preceding methods was that the β rays were separated out into a spectrum by a magnetic field and the successive portions measured by a Faraday cylinder. The total emission was then obtained by integration.

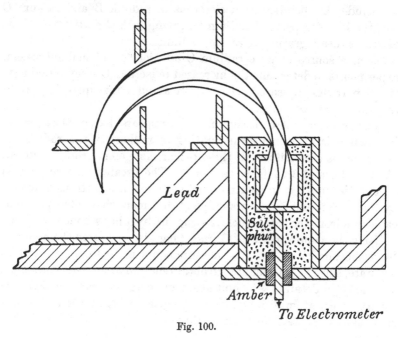

Lead

Sul-phur

Amber

To Electrometer

Fig. 100.

Using a small glass tube filled with radon, Gurney found that the amount of radium B and radium C in equilibrium with 1 mg. of radium emitted together $8 \cdot 45 \times 10^7$ β particles per second with a velocity greater than 500 $H\rho$. The stopping power of the walls was $3 \cdot 5$ mg./cm.². He was able to measure directly the absorption in the walls of the tube by a separate experiment, using a bare source on brass. The brass source was not used in the main experiments owing to the difficulty of estimating the amount reflected, but it could be found in this special experiment by comparison with the radon tube results for higher velocities. The above value gives $2 \cdot 28$ electrons

* Gurney, *Proc. Roy. Soc.* A, **109**, 540, 1925.

† Curtiss, *Proc. Camb. Phil. Soc.* **22**, 597, 1925.

from each pair of atoms disintegrating, in good agreement with previous values. Gurney's experiments are probably accurate to about 3 per cent. It was an important advantage of the method that he could determine directly a lower limit to the number of β particles in the line spectrum. He found that at least 0.57×10^7 β particles per "milligram" of radium B and radium C originated in this way, nearly all of which came from radium B. This leaves 7.88×10^7 β particles to include the disintegration electrons of radium B and radium C and the line spectrum of radium C, giving 2.13 β particles per pair of atoms disintegrating as an upper limit.

Using a source of pure radium C on nickel, and making special experiments to determine the increased reflection, Gurney found that the β particles appeared to come in about equal quantities from radium B and radium C.

Considering all these measurements, it appears that, on the average, for each pair of atoms disintegrating of radium B and radium C, 2.3 electrons are emitted; of these rather more come from radium B, about 1.25, and 1.05 from radium C. The greater amount of this excess over unity is to be attributed to the photo-electrons liberated by the γ rays, and it appears unlikely that more than a few per cent. can come from the extra-nuclear structure by other secondary effects.

Using the same apparatus, Gurney[*] has also measured the emission from thorium (B + C). His measurements in this case are not so accurate, since he was not able to obtain such large sources. His results are shown in detail in § 92 and again indicate rather more than one β particle per atom disintegrating, the general explanation being the same as that for radium B and radium C.

§ 90. The number of electrons emitted by radium E.
Many investigations have been made on the number of β particles emitted by radium E. Since this body emits no γ rays, there can be in this case no photo-electrons superimposed on the disintegration electrons. Any excess over one per atom can only be interpreted as due to secondary electrons ejected by collision with the primary electrons.

Moseley, using the same apparatus as in his experiments with radium B and radium C, found only 0.57 electrons from each disintegrating atom. This was a very curious result and could only have been explained on the assumption that many of the electrons were emitted with very small energies. This seemed improbable, and the measure-

[*] Gurney, *Proc. Roy. Soc.* A, **112**, 380, 1926.

ment has been repeated by Emeléus and by Riehl, who find that the number of electrons emitted from each disintegrating atom is close to one. It is difficult to account for the low value found by Moseley.

Emeléus* adopted an experimental arrangement shown in Fig. 101. The radioactive material is placed at S and only the particles emitted within a small solid angle are measured. Instead of determining the number of these particles by measuring their charge, a direct count of them was made by a counter at C. The discharges from the point due to the entrance of an ionising particle are shown by the string electrometer G earthed through the high resistance R. The brass box D could be evacuated and the minimum absorption due to the foils covering the opening of the counter and the end of

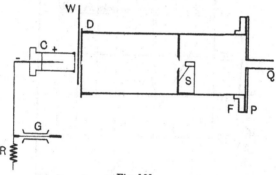

Fig. 101.

the box amounted to 2·5 cm. air equivalent. The radioactive source consisted of pieces of a 3-year old glass emanation tube, virtually radium (D + E + F) in equilibrium. It was not necessary to know the amount of material, since counts were first taken of the combined β and α rays and then the β rays were bent away by a magnetic field. The β rays of radium D cannot penetrate even the smallest absorptions used, so that the ratio of β to α counts gives at once the number of β rays from each disintegrating atom of radium E. Emeléus' results for the number of β particles when different thicknesses of absorbing material were placed in the path are shown in Fig. 102 plotted against the absorption expressed in equivalent centimetres of air. The extrapolation to zero thickness is shown by the curve, giving a result of 1·43 β particles for every α particle. Applying a correction of 23 per cent. for reflection of the β particles at the glass of the source, Emeléus

* Emeléus, *Proc. Camb. Phil. Soc.* **22**, 400, 1924.

reduced this to 1·1 β particles from each disintegrating atom. Before discussing this result in detail, an account will be given of a similar experiment recently carried out by Riehl*. The principle of the experiment is exactly the same as Eméléus', so that a description of the apparatus is not necessary. The experiment is noteworthy for the great

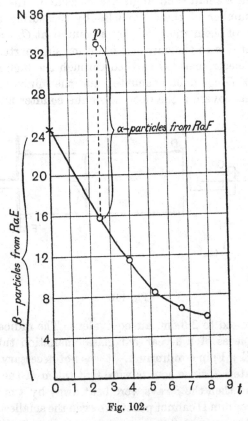

Fig. 102.

care taken to find the conditions under which all the β particles were counted, and for the extremely small initial absorption used. By special arrangement of the apparatus this was reduced to the equivalent of 0·98 cm. of air. The source was either a platinum or aluminium point which had been exposed for a long time to radium emanation, giving radium D and radium E in equilibrium and an amount of radium F which could be calculated from the periods.

* Riehl, Zeit. f. Phys. 46, 478, 1928.

A weak magnetic field was used to deflect the β rays of radium D, and subsequently a stronger one to separate the β rays of radium E from the α rays of radium F. Knowing the amount of radium F present, it was possible from the latter count to deduce the number of α particles that would be emitted from the equilibrium amount of radium F. Measurements were also taken with an absorption of 2·04 cm. of air to enable the extrapolation to zero thickness to be made. The result of the extrapolation was 1·33 β particles to every α particle, which again has to be corrected for reflection at the source. If now we consider the actual experimental results of these two investigations we find fair agreement. Emeléus, extrapolating through 2·5 cm. of air with the source on glass, found 1·43 times as many β particles as α particles, while Riehl, extrapolating through only 0·98 cm. of air with about the same reflection due to the source being on aluminium, found 1·33 times as many β particles as α particles. While then there is little doubt about the experimental results, considerable divergencies of opinion have been shown as to their interpretation, the difficulty lying in the allowance to be made for reflection. Schonland* used homogeneous rays generated artificially in an ordinary discharge tube and found the reflection not to vary with the energy of the rays at least up to 100,000 volts. For aluminium he found 13 per cent. to be reflected, and if we extrapolate this result to the much higher energies of the β rays of radium E we obtain from Riehl's measurement 1·18 β particles to be emitted per atom. Riehl also carried out experiments with the radium E deposited on platinum when he found 1·56 times as many β particles as α particles from the equilibrium amount of radium F. Now for gold Schonland found 50 per cent. to be reflected and this taken to apply to platinum gives 1·04 β particles per disintegration. Emeléus' measurements, when corrected for the reflection at glass, will give a result of about the same magnitude. It should be noted that the conditions under which Schonland measured the reflection are different to those occurring in these experiments. Schonland determined the fraction of a directed beam which is scattered through more than 90°, whereas in the present case the primary beam is incident on the material at all angles and we require the fraction so scattered as to leave it normally. Chalmers has calculated that in this second case the reflection should be 2·32 times greater, so that a 30 per cent. correction

* Schonland, *Proc. Roy. Soc.* A, **104**, 235, 1923 and **108**, 187, 1925.

should be applied to Riehl's result, and the figure of 23 per cent. assumed by Emeléus is probably not far from the correct value. For these reasons no great accuracy can be attached to these values and in addition, quite apart from the statistical error in counting and general experimental error (in both cases 3 to 4 per cent.), there is the uncertainty about the validity of the extrapolation.

The final conclusion that may be drawn from these experiments is that the number of β particles emitted from radium E is not much more than one per disintegration and may be written as $1\cdot1 \pm 0\cdot1$.

An entirely different method of investigating this problem has been used by Kikuchi*. A fine silk fibre activated with a preparation of radium (D + E + F) in equilibrium was stretched across the cloud chamber of a Wilson condensation apparatus and the tracks of the rays issuing from this line source obtained in the usual manner. Four types of tracks were observed, of which three originated from the fibre. These were α ray tracks; long high-speed β ray tracks; and short, dense and tortuous low velocity β ray tracks. The fourth type originated in the gas of the chamber and corresponded to low velocity electrons. These four classes were assigned to radium F, radium E, radium D and the photo-electrons from the γ rays of radium D respectively. A statistical study was made of the relative frequency of occurrence of these four types of tracks. Taking into account the fact that α tracks can be formed during a longer interval in the expansion than β tracks, he concluded that his measurement showed that to within the statistical error of 10 per cent. there is one β ray from each radium E disintegration. Thus this experiment confirms the other results without, however, attaining any greater accuracy. It is, however, especially noteworthy for one feature which will now be considered. If the number of β rays is greater than one per disintegration, then a few atoms must emit two. These would appear in the photographs as branched tracks, both starting from the same spot on the silk fibre. Kikuchi paid great attention to this point and states that he found no branched β ray tracks, either of the long or short type. Including certain other plates, he examined carefully in all three hundred radium E tracks and found no sign of two β rays starting at the same point. Tracks of β rays as slow as 200 or 300 $H\rho$ should have been observable, though with such slow β rays there is a possibility of their

* Kikuchi, *Jap. Journ. of Physics*, 4, 143, 1927.

being absorbed in the silk fibre or in an adhering water film. Allowing
for this and the statistical error we can state that the number of β rays
from radium E is very unlikely to be more than a few per cent.,
say three or four, greater than the number of atoms disintegrating,
and, what is most important, the extra electrons, if present, are
very slow.

We summarise all these experiments on radium E by taking
$1\cdot06 \pm 0\cdot06$ as the number of β particles emitted per disintegration
by radium E.

§ 91. The number of electrons emitted by radium D. It has been
already mentioned in the discussion of the previous experiment that
Kikuchi found a number of short tracks which he attributed to
radium D. The lengths of these tracks showed that at least some of
them were formed by the photoelectrons liberated by the internal
conversion of the γ ray of radium D. It is clear, however, that if
the remainder were formed by disintegration electrons, frequent
pairing of these long tracks should have been observed. Kikuchi
found no cases of pairing and concluded that all the observed β tracks
were due to photoelectrons. On this view the disintegration electrons
must be emitted with considerably smaller energies than the photo-
electrons of about 40,000 volts, and escape detection owing to the
shortness of the tracks they form. Petrova* has made a systematic
statistical study of the occurrence of tracks of different lengths. By
working at reduced pressure he was able to measure tracks formed
by electrons of small energies. He concluded that the upper limit
to the continuous spectrum was about 10,000 volts. The interpretation of these measurements is difficult and it is possible to analyse
the measurements to show a continuous spectrum extending to
25,000 volts. It is a matter of considerable interest that disintegration electrons can be emitted with such small energies.

Kikuchi estimated the number of photoelectrons to be 0·9 per
disintegrating atom, but recent work by Gray and O'Leary† has
shown that a more probable figure is 0·5.

We may conclude tentatively that there are about 1·5 electrons
emitted at each disintegration, this figure consisting of one very slow
disintegration electron and on the average 0·5 electron due to the
conversion of the γ ray.

* Petrova, *Zeit. f. Phys.* **55**, 628, 1929.
† Gray and O'Leary, *Nature*, **123**, 568, 1929.

§ 92. **Collected results on the number of electrons emitted by**
β **ray bodies.** The following table contains a summary of the determinations described in the preceding sections. The second column
shows what is considered to be the best value for the average number
of electrons emitted from each disintegrating atom. The excess of
these numbers over unity must be ascribed to photo-electrons and
electrons arising by subsidiary processes such as collisions. This is
shown in the third column. The fourth column contains the number
of the photo-electrons when it is known, and the fifth, by subtraction,
the number of the subsidiary electrons.

Average number of electrons emitted at each disintegration

	Total number of electrons	Photo-electrons + subsidiary electrons	Photo-electrons	Subsidiary electrons
Radium B + Radium C	2·3 ±0·06	0·30 ±0·06	0·27 ±0·03	0·03 ±0·03
Radium D	1·5 ±0·1	0·5 ±0·1	—	—
Radium E	1·06 ±0·06	0·06 ±0·06	not detected	0·06 ±0·06
Thorium B	1·45 ±0·14	0·45 ±0·14	>0·3	<0·15
Thorium C + C″	1·28 ±0·13	0·28 ±0·13	—	—

Thorium C and thorium C″ are grouped together in this table, since it
was impossible in Gurney's experiments to separate the effects of these
two bodies. The photo-electrons, however, are known to come entirely
from thorium C″. The number of photo-electrons from thorium B
and thorium C″ is not known with any accuracy. However, in the
case of thorium B, Gurney found the very strong line $H\rho$ 1398 to
give 0·25 electron per disintegration. This line is due to conversion
of the γ ray in the K level, and Ellis and Wooster found the number
originating in the $L + M$ levels was about 20 per cent. of these. Since
there are other γ rays, the figure 0·3 electron per disintegration is
an under-estimate. For thorium C″ we can only note that the observed
line spectrum is intense enough to account approximately for the
observed number of electrons.

The main conclusion that may be drawn is that there are at most
only a few per cent. of subsidiary electrons. This point is of importance
in considering the cause of the velocity distribution of the disintegration electrons.

§ 93. The distribution of energy among the disintegration electrons.

It is a curious fact that although the β ray bodies were the first to be discovered it is only recently that the actual disintegration electrons have been identified. The early experiments on the β radiation, which of course consists mainly of the disintegration electrons, were carried out by the absorption method, and it appeared possible to resolve the emission into a series of components each absorbed according to an exponential law. When a little later von Baeyer, Hahn and Meitner discovered the existence of the homogeneous groups by the photographic method, it was an obvious step to identify these with the components of the absorption curves. On the other hand the experiments of Schmidt, and more definitely those of W. Wilson and J. A. Gray, showed that a homogeneous beam of β particles was not absorbed exponentially, and thereby rendered difficult the above interpretation of the absorption and photographic experiments. It was also shown by the photographic method that the homogeneous groups were appreciably reduced in velocity by passage through matter, a fact which was not easy to reconcile with exponential absorption. Wilson at once suggested that the correct explanation of the exponential absorption of the total β radiation from a radioactive body, was that the β rays formed a continuous velocity spectrum and that exponential absorption was a proof of heterogeneity not of homogeneity. The series of contradictions which had by then been arrived at was resolved by Chadwick's* discovery of the continuous spectrum, when it was seen that the total emission consisted mainly of a continuous distribution, capable on Wilson's views of explaining the absorption experiments, on which were superimposed the homogeneous groups. Chadwick showed that the prominence of these groups in the photographs was due chiefly to the ease with which the eye neglects background on a plate.

The apparatus Chadwick used is shown in Fig. 103, and it is interesting since it was the first application of the focussing method to electrical measurements. The radioactive source, a fine glass tube filled with radon, is fixed at Q inside the evacuated box V, and the β particles, passing out through the slit, are sorted out into a velocity spectrum by a magnetic field perpendicular to the plane of the paper. By varying the magnetic field successive portions of the spectrum entered the counter T and the amount could be determined

---

* Chadwick, *Verh. d. D. Phys. Ges.* **16**, 383, 1914.

by the frequency of the throws of the string electrometer attached to
the counter. A movable screen is shown at B by which the β particles
could be cut off entirely and the inevitable stray effect and the effect

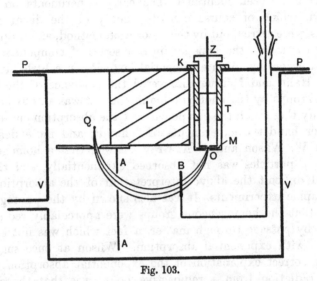

Fig. 103.

of those γ rays passing the lead block L could be determined. Chadwick also used an ionisation chamber in place of the counter and the
results of the two measurements are shown in Fig. 104.

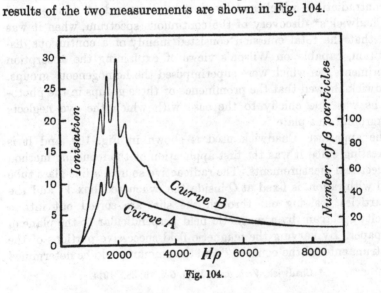

Fig. 104.

The two curves differ since the counter measures the number of β particles (curve A), whereas the ionisation chamber measures the ionising action (curve B) which varies with the velocity. These curves show clearly that the line spectrum is only a relatively small effect superimposed on a continuous spectrum.

This important fact that the β ray emission fell naturally into two parts was explained by Rutherford, who pointed out that the groups were certainly secondary in origin and due to the conversion of the γ rays, so that since the disintegration electrons had to be found somewhere it only remained to identify them with the continuous spectrum. In 1922 Meitner pointed out that a nucleus, presumably quantised, ought not to emit electrons of varying energy, and that it was quite possible that the inhomogeneity was introduced after the actual expulsion from the nucleus. Several experiments* were carried out to test this standpoint, and the general conclusion† was that it was difficult to find any evidence of the introduction of this inhomogeneity. As an example we may consider the possibility that the electron emits continuous γ rays in its passage out through the intense electric fields of the atom. Some such effect as this probably does occur, but experiment shows that it happens to such a small extent as to be negligible in the present discussion. Radium E is a β ray body showing a normal continuous spectrum formed by its disintegration electrons, but there are no homogeneous groups superimposed and therefore no monochromatic γ rays. Actually there is a small amount of relatively hard γ radiation which might arise in the above way. The amount of γ ray energy emitted has been measured by G. H. Aston‡ and he found that on the average only 10,000 volts is emitted per disintegration in the form of γ rays, probably in the form of one quantum of relatively hard radiation emitted on the average once in every thirty disintegrations. When it is realised that the observed spread of the radium E β ray spectrum, if produced in this way, would involve an average emission of γ ray energy equal to 600,000 volts per disintegration, it is clear that there is no possibility of explaining the continuous spectrum by this process.

It would be a long task to analyse the various ways in which the heterogeneity might be introduced if sufficient freedom in the choice

---

* Chadwick and Ellis, *Proc. Camb. Phil. Soc.* **21**, 274, 1922; Pohlmeyer, *Zeit. f. Phys.* **28**, 216, 1924.

† Ellis and Wooster, *Proc. Camb. Phil. Soc.* **22**, 859, 1925.

‡ Aston, *Proc. Camb. Phil. Soc.* **22**, 935, 1927.

of hypotheses were allowed, but all such hypotheses would have as their one object to make the total energy the same for each disintegration. If the continuous spectrum extends from an upper limit of energy $E_{max.}$, then this energy of disintegration could not be less than $E_{max.}$. If on the contrary the disintegration electrons were actually ejected from the nucleus with various energies, then the average energy of disintegration should correspond to the mean energy of the continuous spectrum. A final decision between the two views would thus be reached by measuring the total energy of disintegration of a β ray body whose continuous spectrum was known. Ellis and Wooster* have recently carried this out for radium E, which is particularly suitable, since it emits no γ rays which might complicate the interpretation. It can also be obtained free from α particle bodies whose energy of disintegration would be large compared to that of the β particle body. The average energy of disintegration can be measured by finding the heat produced when a known number of atoms disintegrate inside a calorimeter so thick that no β rays can

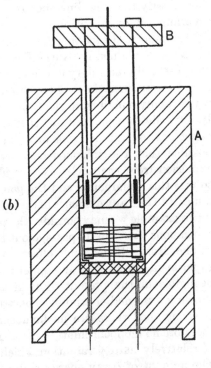

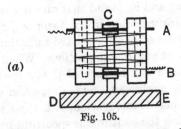

Fig. 105.

escape. This has then to be compared with the data obtainable from the continuous spectrum.

The continuous spectrum of radium E is shown in Fig. 109. It will be noticed that the maximum energy is 1,050,000 volts, whilst the average energy calculated from this curve is 390,000 volts.

* Ellis and Wooster, *Proc. Roy. Soc.* A, **117**, 109, 1927

Owing to experimental difficulties this latter figure may be in error by 15 per cent. A measurement of the heating effect therefore provides a clear distinction between the two hypotheses, one predicting a value of 390,000 volts per disintegration, the other 1,050,000, or 2·6 times as great.

The radium E was deposited on either a short platinum or nickel wire. This was enclosed in a very thin brass case and could be easily lowered into or removed from the calorimeter (Fig. 105 (a)), which consisted of a lead tube of rather more than 1 mm. thickness so that all the β rays were absorbed. To avoid external disturbances the apparatus was made symmetrical, a second identical calorimeter being constructed into which a dummy non-activated wire was lowered. The whole was enclosed in a massive copper block, as can be seen from Fig. 105 (b). The steady temperature difference set up between the two calorimeters, when the rate of heat supply equalled the heat loss, was measured by a system of thermocouples attached to a sensitive galvanometer. The heating effect given by the radium E source was followed over several days. The heating at any time is due partly to what remains of the radium E, and partly to the radium F that is grown. Knowing the periods of the two bodies their two effects could be separated. Since at any moment the ratio of the number of radium E atoms disintegrating to the number of radium F atoms disintegrating is known, it was possible to deduce the ratio of the energy given out by a radium E disintegration to that given out by a radium F disintegration, the latter being known from the energy of the α particle.

The result of these experiments gave 350,000 volts ± 40,000 volts, in striking agreement with the mean energy of the emitted β particles as shown by the continuous spectrum, that is 390,000 volts ± 60,000 volts, and entirely incompatible with the value 1,050,000 volts predicted by the alternative theory that the energy of disintegration is always the same*. Further evidence on this important question can be obtained from some experiments of Moseley and Robinson[†] and of Gurney[‡]. The former (see p. 497) found that the β rays from the radium B and radium C in equilibrium with one gram of radium formed $9·65 \times 10^{14}$ pairs of ions per second in air. Taking a round

* This experiment has been repeated, with several improvements, by Meitner and Orthmann (*Zeit. f. Phys.* **60**, 143, 1930). They obtain 337,000 ± 20,000, in excellent agreement with the above result.

† Moseley and Robinson, *Phil. Mag.* **28**, 327, 1914.

‡ Gurney, *Proc. Roy. Soc.* A, **109**, 540, 1925.

figure of 40 volts to form an ion, this yields 10^6 volts as the sum of the average energies of the radium B and radium C β particles. From Gurney's analysis of the còntinuous spectra of these bodies we find the sum of the mean energies to be about 10^6 volts, while the sum of the energies corresponding to the upper limits of the two curves is $3\cdot7 \times 10^6$ volts. We may safely generalise this result obtained for radium E and radium (B + C) to all β ray bodies and must conclude that in a β ray disintegration the nucleus can break up with emission of an amount of energy that varies within wide limits.

No satisfactory interpretation has yet been given of this curious result so much in contrast with the high degree of homogeneity and definiteness shown by the α ray disintegrations.

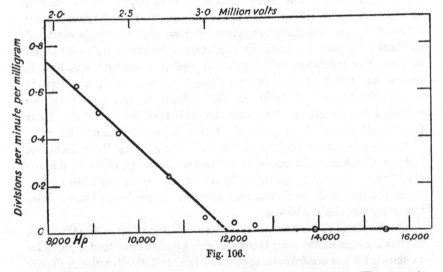

Fig. 106.

§ 94. **The continuous spectra of different β ray bodies.** The experiments that have been already carried out on this subject were designed more with a view to settling fundamental points than to determining the distribution curves with great accuracy. However, the results already obtained enable a preliminary survey of the field to be made and suffice to direct attention to some special points.

The bodies whose continuous spectra have been analysed by a magnetic field are radium B, radium C, thorium B, thorium (C + C″), and radium E. All the curves show a fairly definite upper limit, a point that has been emphasised by Madgwick*, and at present this

* Madgwick, *Proc. Camb. Phil. Soc.* **23**, 982, 1927.

is the most easily determined characteristic constant. Gurney's experiments have already been described on p. 391, and his measurements on the end of the radium C curve are shown in Fig. 106. The ordinates represent the actual electrometer currents at different magnetic fields, and it can be seen how there appears to be a definite

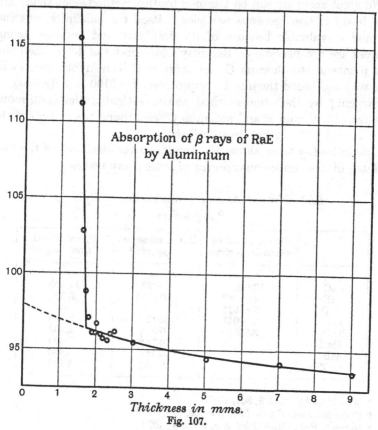

Absorption of β rays of RaE
by Aluminium

Thickness in mms.
Fig. 107.

end at $H\rho$ 12,000. This end point can also be detected by simple absorption experiments with aluminium and an estimate made of the velocity from the limiting range. Careful measurements on the absorption of the β rays of radium C in aluminium were made by Schmidt*, and although the kink is difficult to detect in the presence of the intense γ radiation there is no doubt that the β ray effect ends at an absorption of about 1·35 gm./cm.² corresponding to $H\rho$ 11,400.

* Schmidt, *Ann. d. Phys.* **21**, 632, 1906.

The best illustration of the method is obtained with radium E, where there are no γ rays to disturb the measurements. Schmidt's* results with this body are shown in Fig. 107 and a sharp break in the curve can be seen at 1·7 mm. corresponding to $H\rho$ 5300, in fair agreement with the magnetic analysis, which gives $H\rho$ 4920.

No great accuracy can be attained by this method since the range of a beam of homogeneous particles is itself too indefinite, but the method is valuable because of its simplicity and because strong sources are not necessary. Recently Chalmers† has determined the end points of the thorium C and thorium C″ continuous spectra in this way and found them to be respectively $H\rho$ 9100 and $H\rho$ 7800.

Sargent‡ by the same method has investigated the continuous spectra of actinium B and actinium C″ and found the end points to be 0·24 and 0·62 gm./cm.² respectively.

The following table shows the results so far obtained on the end points§ of the continuous spectra of some β ray bodies.

End points of the continuous spectra of different β ray bodies

	$H\rho$ measured by magnetic spectrum	Range measured in gm./cm.²	$H\rho$ calculated from range
RaB	3,500‖	—	—
RaC	12,000‖	1·35¶	11,400
RaE	4,920**	0·475††	5,300
MsTh₂	6,800‡‡	—	—
ThB	2,300§§	0·12†	2,400
ThC	8,860§§	0·98†	9,100
ThC″	—	0·79†	7,800
AcB	—	0·24‡	3,430
AcC″	—	0·62‡	6,140

* Schmidt, *Phys. Zeit.* **8**, 361, 1907.

† Chalmers, *Proc. Camb. Phil. Soc.* **25**, 331, 1929.

‡ Sargent, *Proc. Camb. Phil. Soc.* **25**, 514, 1929.

§ Terroux has found recently that there is a slight tail to the RaE spectrum extending well beyond the apparent end point. The matter is still under investigation and there is as yet no information whether it is a general feature of all continuous spectra or whether this tail represents the continuous spectrum of an abnormal mode of disintegration of RaE.

‖ Gurney, *Proc. Roy. Soc.* A, **109**, 541, 1925.

¶ Schmidt, *Ann. d. Physik.* **21**, 632, 1906.

** Madgwick, *Proc. Camb. Phil. Soc.* **23**, 982, 1927.

†† Schmidt, *Phys. Zeit.* **8**, 361, 1907.

‡‡ Yovanovitch and d'Espine, *Journ. de Phys.* **8**, 280, 1927.

§§ Gurney, *Proc. Roy. Soc.* A, **112**, 380, 1926.

The experimental evidence about the low energy portion of the curve is unsatisfactory and contradictory. We might anticipate one of three things, the distribution curve against energy following one of the typical curves shown in Fig. 108. Curve 1 has a definite lower limit which would presumably be different for different bodies, curve 2 descends continuously to zero energy, while curve 3 shows the presence of a large number of electrons of very low energy. It is scarcely necessary to emphasise the importance of obtaining information on these points, since either curve

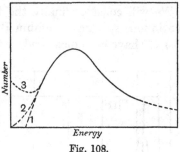

Fig. 108.

would at least provide a basis on which to commence a theory of the continuous spectrum.

The distinction between these three curves is largely lost in the experimental curves obtained by Gurney's method showing the number of electrons entering the Faraday cylinder in different magnetic fields, since not only has each ordinate of the experimental curve to be divided by the value of the magnetic field to give the distribution with $H\rho$, but again by the velocity to give the distribution with energy. The result is that each of the experimental curves corresponding to the three curves of Fig. 108 will descend sharply towards the origin and very accurate measurements would be necessary to distinguish between them. Unfortunately it is just in this region that the measurements become uncertain owing to the difficulty of applying the proper correction for reflection. To investigate low speed electrons it is essential to use a bare source, and trouble is experienced not only with the δ rays liberated by the a rays of the other radioactive bodies which are usually present in the source but also with the reflected electrons. These may consist not only of electrons reflected without sensible loss of speed but also of electrons coming from the higher energy portions of the spectrum and which have lost energy in the process of reflection. Bearing these points in mind, we must conclude that at present we have no definite information about the distribution curves below about 100,000 volts.

This is a matter which urgently needs investigation, and it would appear that the most feasible method would be by means of the expansion chamber method with a source spread out along a very fine fibre as in Kikuchi's experiments. An investigation of the frequency and mode of occurrence of the short tracks would at least provide valuable evidence on this point.

We will consider finally the form of the curves as a whole. The continuous spectra of radium B, radium C, thorium B and thorium (C + C″) have been measured by Gurney (*loc. cit.*), using the apparatus

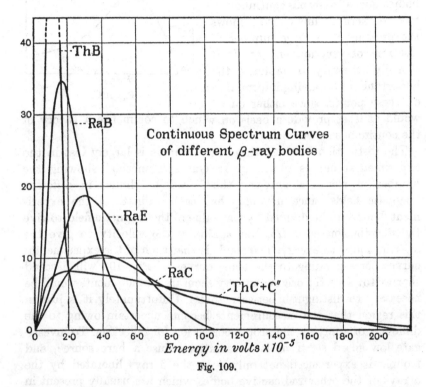

Fig. 109.

described on p. 391. These curves are shown in Fig. 109, in which is also included the continuous spectrum of radium E, after the measurements of Madgwick by the ionisation method. It seems most reasonable to take the distribution against energy, and the curves have been adjusted to have the same area. As has been already emphasised, the lower energy portions of the curve are very uncertain, and in this diagram they have been drawn to pass through the origin merely in order to complete the curves on some common basis.

Here again caution must be exercised in handling the experimental evidence, since owing to the small effects to be measured and the ever-recurring trouble of reflection we cannot be certain that the curves are as accurate as the mere reproducibility of the experiment would suggest. However, there does appear to be a great similarity between the different curves with the exception of the curve for thorium (C + C″) which shows an extended flat maximum. It must be remembered that this curve is a combination of the thorium C and thorium C″ curves, and since we have already seen that they have different end points their maxima will be in different positions and the superposition of the two curves could easily produce the observed result.

We have already referred briefly in chapter XI to the problem of how we are to explain the occurrence of the continuous spectrum, and it was suggested there that the simplest view was that the separate nuclei of a radioactive substance did disintegrate in the β ray case with different amounts of energy, and that the reason why the individual nuclei did not subsequently show different behaviour due to their different energy content was to be sought for in the detailed structure of the nucleus. Although it is not possible to propose any theory to account quantitatively for this phenomenon, it is of interest to develop this point of view a little further.

We have seen that it is unlikely that the electrons exist in the free state in the nucleus and that many facts suggest that they are bound in pairs to an α particle, forming what has been called an α′ particle. The nucleus may thus be considered as built up of α particles, protons, and α′ particles, all in definite quantum states. Owing to the large masses of these particles there is no difficulty in associating them with plausible energies but yet keeping their de Broglie wave-lengths of the order of the dimensions of the nucleus.

The usual development of this view is that the β ray changes are initiated by an α particle disintegration involving the departure of the α particle of one of the α′ particles. This, however, would again present the difficulty of having two free electrons in the nucleus. It appears more reasonable to consider the order to be reversed and that an α′ particle breaks up by first losing one of its electrons. The period of a β ray body on this view would be the mean life of an α′ particle, the moment the electron is set free it is incapable of remaining in the nucleus and escapes at once. The singly charged particle left behind is again unstable and after a certain interval will set free the remaining electron, leaving a normal α particle in the nucleus.

Certain regularities in the periods of the radioactive bodies may be mentioned in support of this view. When we consider the great range in periods of transformation of the α ray bodies, by a factor of 10^{14} in the actinium series, 10^{24} in the uranium series, and 10^{27} in the thorium series, it is striking to find only a range of 10^6 in the β ray bodies of the thorium and actinium series and 10^7 in the uranium series*. This difference becomes more easily understandable if the β ray period is in all cases the mean life of an α' particle.

On this view the energy of the emitted electrons would be the difference in the energies of the stationary states occupied by the α' particle before and after the departure of one of its electrons less the energy required to set free the electron from its state of close binding. It is this last quantity which might be considered to vary and to account for the spread of the continuous spectrum, and it would only show itself in the nucleus by a slightly varying mass of the α' particles. Except for this feature all the nuclei will be identical and in the same definite stationary states.

* We are not considering here the disintegration of potassium and rubidium which are β ray bodies of extremely long period.

CHAPTER XIV

THE PASSAGE OF β PARTICLES THROUGH MATTER

§ 95. The study of the passage of β particles through matter encounters serious difficulties, both experimental and theoretical, which in most cases can be traced back to the ease with which these particles are scattered under the conditions in which experiments can be carried out. This scattering together with the rate of loss of energy constitute the two fundamental phenomena with which we are concerned, but unfortunately it is difficult to study the loss of energy without being seriously inconvenienced by the scattering.

In this respect β particles are far less amenable to investigation than α particles, which in the great majority of cases pursue straight paths until near the end of their range.

A convenient method of giving a general survey of the behaviour of β particles in traversing matter is to compare them with α particles. From this comparison we shall see how to differentiate between those experiments which can be easily analysed to give information about the mode of interaction of a β particle with an atom, and those experiments which, while valuable from a practical standpoint, yet prove on detailed consideration to be essentially complicated.

In the first place there is a considerable difference in penetrating power. While a few centimetres of air suffices to stop α particles, almost a hundred times as much is required to stop β particles. If in this preliminary comparison we neglect the slightly longer paths followed by the β particles due to their being scattered, we see that in a given distance the average number of collisions of a definite type with the atomic electrons will be the same for both sets of particles. Mainly because of their lower velocities, but partly due to their higher charge, the α particles will transfer several hundred times as much energy except at the close collisions. Although their initial energies are greater than those of the β particles, they will have been seriously reduced in energy or even completely stopped before the β particles have lost more than a few per cent. of their energies.

Another important point to consider is the straggling, a term used to denote the inhomogeneity introduced into a beam initially homogeneous by passage through matter.

If we consider the passage of a charged particle through a thin
sheet of some substance of thickness Δx we may divide the collisions
into groups according to the value of the impact parameter p, so that
there are on the average A_r collisions with p between p_r and p_{r+1},
resulting each in a loss of energy Q_r. Now provided this grouping
can be so arranged that while all the A's are large yet the variation of
Q for the different collisions within a group is small, it is then legitimate to apply probability considerations and it is possible to deduce
the probability $W(\Delta T)\,dT$ of a total loss of energy between ΔT and
$\Delta T + dT$

$$W(\Delta T)\,dT = (2\pi P\Delta x)^{-\frac{1}{2}}\,\epsilon^{-\frac{(\Delta T - \Delta_0 T)^2}{2P\Delta x}}\,dT,$$

where
$$\Delta_0 T = \Sigma A_r Q_r$$

and
$$P\Delta x = \Sigma \frac{(A_r Q_r)^2}{A_r} = \Sigma A_r Q_r^2.$$

As a more comprehensible expression for the straggling we may take
the energy spread between the two values for which the exponential
takes the value of one-half, and divide it by the mean remaining
energy, viz.

$$\frac{\sqrt{5\cdot54\,\Sigma A_r Q_r^2}}{T - \Sigma A_r Q_r},$$

which indicates clearly the dependence of straggling on the number
of collisions (the A's) and the general order of magnitude of the energy
transference (the Q's).

Now while, as we have already mentioned, an average β particle
with energy 200,000 to 400,000 volts has much less energy than
an average α particle with 5,000,000 to 7,000,000 volts, yet at similar
types of collision the α particle will, except in the closest collisions,
lose far more energy. If we consider two experiments, one with
α particles and one with β particles, in both of which there is the
same percentage loss of the initial energy, that is $\Sigma A_r Q_r/T$ is the
same in both cases, we can see in a general way from the above
expression that the relative straggling of the two types of particles
will be of the same order of magnitude.

If therefore we had only to consider the rate of loss of energy and
the true straggling, we should find β particles almost as convenient
to work with as α particles and merely more penetrating. The great
distinction between them arises not from the causes already mentioned

but from the scattering, which is far greater for β than for α particles. Scattering, if it merely meant removal of particles from the beam, would not be serious, but when the scattering is so large that particles scattered out of the beam are frequently scattered back into it again, it becomes impossible to define the average distance traversed by the beam. Particles which have followed a zig-zag path will have traversed a considerably longer distance than those going straight through. This point can be appreciated from photographs taken by the expansion chamber method, one of which is shown in Plate IV, fig. 2. This shows a few centimetres of the path of a β particle, and the difficulty of accounting for its loss of energy and ionisation by considering it to have followed a straight path across the chamber will readily be appreciated. Unfortunately this is precisely what has to be done, since in any experiment in which beams of β particles traverse thin foils or a few centimetres of gas, the only length which is measurable is that of the thickness of the foil or gas layer, and it is only to this that the experimental results can be referred.

It may be noted that from a practical standpoint this shows itself as an increased straggling, and as a result there is a smaller range of experiments which are capable of simple interpretation than for α particles. Only measurements carried out with small thicknesses of matter justify accurate analysis, although from a purely practical standpoint measurements of the apparent total range of β particles are of some interest. The ionisation produced by β particles is in the same way rendered difficult to investigate, and in this connection a further complication introduced by the large scattering may be considered. We refer here to the so-called reflection of β particles. If a beam of β particles impinges on a sheet of aluminium about one-fifth will be apparently diffusely reflected, and with heavy elements such as gold the fraction rises as high as one-half. This reflection is of course nothing other than multiple scattering, and, owing to the complication of the phenomenon, it yields little additional information about the fundamental problem of the collision of a β particle with an atom. Its presence, however, is obvious in any experiment and almost impossible to avoid. The "reflected electrons" will have all velocities and will add their effects to whatever is being investigated. The extent to which this may disturb the measurements can be seen from the magnitude of the effect and has already been referred to in connection with the determination of the number of β particles emitted in a radioactive disintegration.

§ 96. **The absorption of β particles.** The first experiments on the absorption of homogeneous beams were made by W. Wilson* in 1909. By means of a magnetic field he separated out approximately homogeneous beams of β particles and measured the absorption curves in aluminium by an ionisation method. To a first approximation the curves he obtained could be described as showing a linear relation

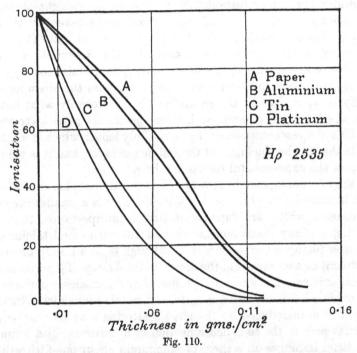

Fig. 110.

between the ionisation and the thickness of material traversed, a result in striking contrast both with the previously obtained exponential absorption curves and with the Bragg absorption curve shown by α particles.

A general explanation of these differences was indicated by Wilson along the following lines. The typical exponential absorption curves for β particles had been obtained with the total β radiation from various radioactive bodies. At that time it was always assumed that these radiations were homogeneous in velocity from analogy with the α rays, but Wilson pointed out that the disagreement with his experiments disappeared if the total β radiation from the radioactive

* Wilson, *Proc. Roy. Soc.* A, **82**, 612, 1909.

substances was heterogeneous in velocity, a result subsequently proved to be correct.

Subsequently similar experiments have been carried out by Varder*, Madgwick† and Eddy‡. The general results may be illustrated by curves taken from Varder's paper shown in Fig. 110. This refers to β particles of $H\rho$ 2535 and shows the transmission through paper, aluminium, tin and platinum. There is a progressive change in shape of the absorption curves which may be ascribed to the relative effects of scattering in the different materials. For a substance like paper, which contains only elements of low atomic weight, the effect of scattering is less important than for aluminium, so that a greater

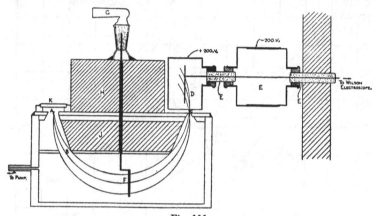

Fig. 111.

fraction of β rays will penetrate a given thickness of matter, whereas the opposite is the case for tin and platinum, and the curves for these metals are convex to the origin.

The form of the absorption curve appears therefore to have small importance, since being mainly controlled by the scattering it is very dependent on the design of the apparatus and the varying extent to which scattered rays are measured. A typical apparatus† used for this type of work is shown in Fig. 111. The radioactive source A can be easily manipulated by removing the plate K, and the shutter F enables the stray radiation to be measured. At D, just above the opening C of the evacuated box, is placed the measuring instrument,

* Varder, *Phil. Mag.* **29**, 726, 1915.
† Madgwick, *Proc. Camb. Phil. Soc.* **23**, 970, 1927.
‡ Eddy, *Proc. Camb. Phil. Soc.* **25**, 50, 1929.

either ionisation chamber or counter. In the diagram an ionisation chamber D is shown and also an arrangement E for compensating the greater portion of the γ ray effect. It is usual to have the line AC near the top edge of the magnetic field, since this is convenient and does not introduce any great error in the focussing. Eddy* showed that different forms of absorption curves were obtained when different angular fractions of the transmitted beam were included, and that with aluminium as absorber it was possible to reproduce an entire

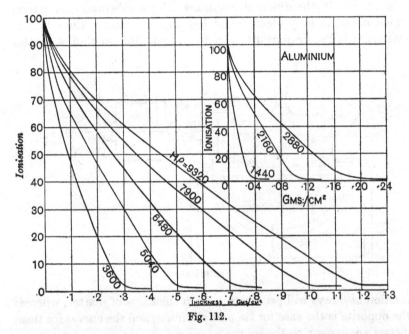

Fig. 112.

range of curves similar to those shown in Fig. 110 merely by varying the angular fraction of the transmitted beam which is measured. The curves convex to the origin are obtained with small angular width.

A further complication in this type of measurement has been the use of ionisation chambers of such shape and size that they do not respond equally to particles entering in different directions. This tends to emphasise the effect of scattering and increases the convexity of the curves towards the origin. Eddy* was able to show that using a counter and including a large fraction of the transmitted

* Eddy, *Proc. Camb. Phil. Soc.* **25**, 50, 1929.

beam the absorption curves became initially concave to the origin and showed a marked flattening approaching the form of an α ray absorption curve. This initial flattening was still noticeable even in the case of absorption by heavy elements.

An interesting feature of these curves is the end point representing the greatest thickness that can be penetrated, that is, the range of the β particles. This is, as one would expect, independent of the method of measurement, although it is more difficult to fix with curves convex to the origin. Examples of absorption curves in aluminium of different speed β particles are shown in Fig. 112, which is taken from Madgwick's paper. The end point cannot be determined very definitely and it is difficult to be sure whether the small tail is real or is only a stray effect. It is found, however, that if the main portion of the curve is extrapolated to cut the axis, a reproducible figure is obtained which may be called the effective range.

The effective range increases rapidly with the speed of the particle, being 5000 times greater for β particles of $H\rho$ 12,000 than for those of $H\rho$ 350.

The most accurate measurements of ranges have been made in aluminium, and a summary of the results obtained by Varder and Madgwick is included in the table on page 422.

This phenomenon has been very thoroughly investigated by Schonland*, using cathode rays of $H\rho$ 340 to 1000. The conditions are more favourable with cathode rays, since the intense beams available enable the currents to a Faraday cylinder to be measured with a galvanometer, and further there is no trouble with γ rays. This made it possible to follow in detail the fate of the entire beam incident on the foil and to find the fractions transmitted, absorbed and reflected. The essential parts of the apparatus he used are shown in Fig. 113.

The beam of electrons from a cathode ray tube entered the water-cooled solenoid SS through the hole H. A is an evacuated box, in which the rays, originally at right angles to the plane of the paper, are bent upwards through a right angle, to pass through the opening B into the measuring apparatus. F is a search coil which could be manipulated into a correct position for measuring the field. The divergent beam of electrons passed through a set of slits designed to cut down scattered rays and fall upon the foil W. By suitably insulating the different parts of the apparatus, it was possible to measure i_t, the portion penetrating the foil, i_r, the portion reflected, and

---

* Schonland, *Proc. Roy. Soc.* A, **104**, 235, 1923; **108**, 187, 1925.

i_a, the portion stopped in the foil. The emission of secondary rays from the foil was prevented by maintaining the grids G at -200 volts with reference to the foil, and care was taken to correct for the small fraction of the primary beam stopped by the first grid.

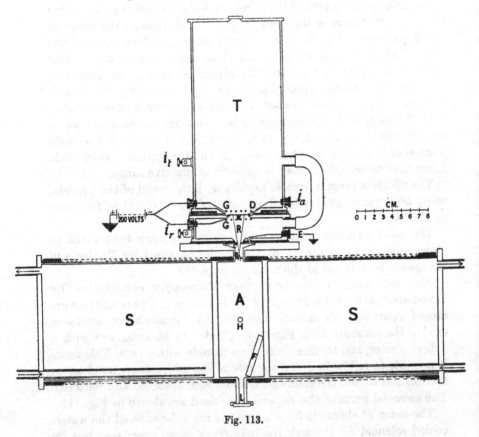

Fig. 113.

In this type of experiment it is easier to keep the foil unchanged and to vary the speed of the rays, giving curves of different form to those already considered. The results of an experiment with a gold foil of thickness $1\cdot87 \times 10^{-5}$ cm. are shown in Fig. 114, where the observations have been plotted against $1/(H\rho)^4$, which in this region is approximately equal to $1/(\text{velocity})^4$.

Considering first the fraction i_r/I scattered back, it will be seen that at sufficiently small velocities this is constant and amounts to $0\cdot50$. This constant value continues as the velocity of the rays is increased (the curve being read from right to left) until a critical value of the velocity is reached, when a decrease begins, becoming more rapid at high velocities. This means that were we to consider electrons of given velocity and investigate the amount reflected as the thickness of foil was increased, we should find a steadily increasing reflection until a thickness of about half the range was reached, when further increase in thickness would produce no change in the reflection since electrons reaching the deeper layers even if turned back would not be able to get out.

The variation of the fraction absorbed can be seen both from this figure and from Fig. 115. The different curves in this figure refer to

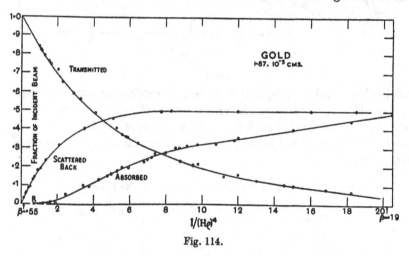

Fig. 114.

different thicknesses of foil at W in Fig. 113. With sufficiently thick foils or sufficiently slow electrons no electrons are transmitted, and as a constant fraction of the beam is found to be reflected, a constant fraction is absorbed. As the velocity of the rays is increased (the curves being read from right to left) the true absorption begins to diminish. It is to be noted that the curves do not pass through the origin, indicating that rays possessing a velocity higher than that corresponding to the point R (Fig. 114) do not suffer any appreciable absorption. The transmitted fraction represents what is left of the beam after reflection and absorption have played their part.

In Fig. 115 we can estimate the speed of particles having a range equal to that of the foil employed by producing the linear and the horizontal portions to meet at a point like P. This is found to be

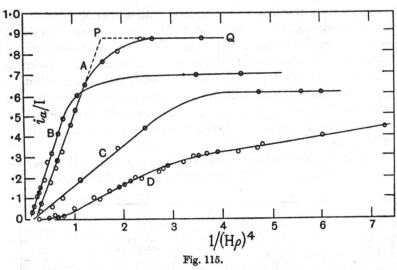

Fig. 115.

applicable to all metals, and Schonland's results for four absorbers and various velocities are shown in the next table and plotted in Fig. 116, from which the interesting point appears that the range measured in gm./cm.² is, to a first approximation, independent of the material.

Range-velocity in various metals for various thicknesses

Element	R. Thickness in gm./cm.²	Range-velocity $H\rho$
Aluminium	$2 \cdot 56 \times 10^{-4}$	346
	3·20	380
	4·80	421
	6·40	453
	9·60	500
	14·0	550
Copper	$3 \cdot 99 \times 10^{-4}$	379
	7·98	471
Silver	$4 \cdot 05 \times 10^{-4}$	372
	6·09	460
Gold	$6 \cdot 01 \times 10^{-4}$	446
	9·01	483

The amount of reflection was measured directly by Schonland for these four elements and he found respectively 13 per cent., 29 per cent., 39 per cent. and 50 per cent. to be reflected and to be independent of the speed. We have already mentioned (p. 395) that the percentage reflected depends on the experimental conditions. In this case the incident beam is confined to one direction, and any electrons emerging on the incidence side are counted as reflected electrons. Chalmers'* work suggests that when the electrons are

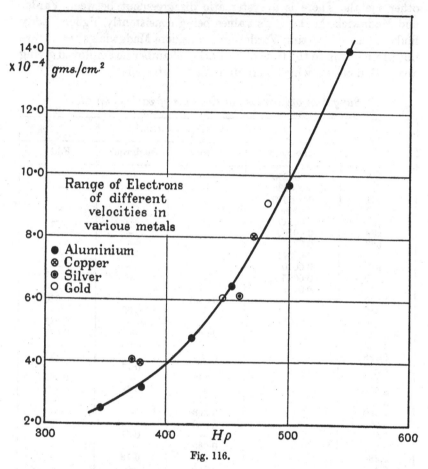

Fig. 116.

incident on the material at all angles, and reflection normal to the surface only is considered, then the percentage reflection will be about twice the above values.

The experiments of Wilson, Varder, Madgwick, Eddy and Schonland that have been described provide the evidence for the existence of an effective range, and give values over an extended range of

* Chalmers, *Phil. Mag.* (in publication).

velocities. Some of these are shown in the next table. The measurements with β rays refer to the range in aluminium and this was also
chiefly used by Schonland, but his results just quoted suggest that
the same effective range measured in gm./cm.^{2} would be found in
other metals. There is a systematic disagreement between Varder
and Madgwick, Madgwick's values being consistently higher. Eddy
finds values supporting Varder and considers Madgwick's results are
uncertain owing to the type of ionisation chamber he employed giving
absorption curves which were difficult to extrapolate.

Ranges in aluminium of electrons of various speeds

$H\rho$	Range in gm./cm.^{2}			
	Schonland	Varder	Madgwick	Eddy
342	0·000250	—	—	—
372	0·000336	—	—	—
401	0·000375	—	—	—
465	0·000664	—	—	—
527	0·00117	—	—	—
583	0·00169	—	—	—
627	0·00232	—	—	—
694	0·00341	—	—	—
916	0·0070	—	—	—
1,010	0·0095	—	—	—
1,040	0·0108	—	—	—
1,380	—	(0·018)	—	0·020
1,440	—	—	0·03	—
1,680	—	—	—	0·045
1,930	—	0·064	—	0·063
2,160	—	—	0·09	—
2,535	—	0·124	—	0·127
2,880	—	—	0·17	—
3,170	—	0·189	—	0·190
3,600	—	—	0·29	—
3,790	—	0·279	—	—
4,400	—	0·360	—	—
5,026	—	0·440	—	—
5,040	—	—	0·51	—
6,230	—	0·580	—	—
6,480	—	—	0·72	—
7,490	—	0·785	—	—
7,900	—	—	1·0	—
8,590	—	0·925	—	—
9,320	—	—	1·2	—
11,370	—	1·36	—	—

Another method of investigating the range of β particles which has
a greater theoretical interest is to measure the total length of the
track in an expansion chamber. This method of investigation has

from technical reasons so far been confined to slow β rays of less than 80,000 volts. The electrons are conveniently generated by photo-electric effect of X rays on the gas in the chamber. The most comprehensive results have been obtained by E. J. Williams and were described on p. 143.

§ 97. **Reduction in velocity by the electrical method.** The first experiments by the electrical method were carried out by W. Wilson*. They are important because they provided the first definite proof that β particles were slowed down in passing through matter. The source was a thin glass bulb containing radon, and the β particles from the products radium B and radium C were separated out into a velocity spectrum by a magnetic field. After traversing a quarter of a circle in this field, a small portion of the rays was allowed to pass through a small hole into a second independent magnetic field which bent them through another quarter of a circle into an electroscope where they were measured. This method does not give a very homogeneous beam and it was found that, when the first magnetic field was kept constant, β particles entered the electroscope for a considerable range of values of the second magnetic field. However, there was one value of this field where the ionising effect was a maximum, and if now the hole through which the particles entered the second field was covered with a thin sheet of aluminium, it was found that a smaller value of the second magnetic field was required in order to obtain the maximum effect in the electroscope. This showed clearly that passage through the aluminium diminished the velocity of the β particles, since they were now more easily bent in the second magnetic field. To obtain measurable effects Wilson had to use thick sheets of aluminium, which renders the interpretation of the results rather difficult, but he was able to show definitely that the amount of retardation in passing through a given screen increased rapidly with decreasing velocity of the particle.

Recently Madgwick † has carried out experiments on this subject by a greatly improved method. His apparatus has already been shown in Fig. 111. The source A was a small plate coated with thorium active deposit, and by varying the magnetic field successive portions of the spectrum could be allowed to enter the ionisation chamber D. With an apparatus of this type, which uses the focussing method, it is

* Wilson, *Proc. Roy. Soc.* A, **84**, 141, 1910.
† Madgwick, *Proc. Camb. Phil. Soc.* **23**, 970, 1927.

possible to isolate the stronger of the homogeneous lines emitted by the source. The top curve of Fig. 117 shows the strong line $H\rho$ ca. 1400, with the background (continuous spectrum) for a small range on either side. If now absorbing screens are placed over the source A it will be seen that the line moves bodily to smaller $H\rho$'s and at the

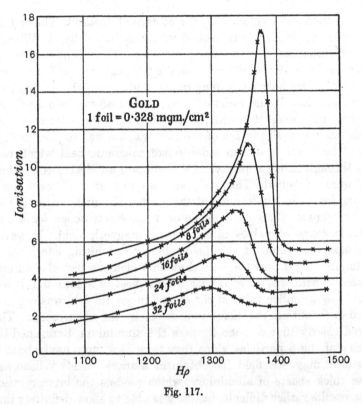

Fig. 117.

same time becomes broader, until finally it is almost lost in the background. The great advantage of using a β ray line by which to observe the reduction in velocity can be seen by inspection of these curves. There are clearly two points on which attention must be fixed, firstly the average loss of velocity, represented approximately by the shift back of the peak of the line, and secondly the amount of inhomogeneity introduced, which is represented by the broadening. From the theoretical standpoint the amount of heterogeneity introduced, or more accurately the velocity distribution of an initially homogeneous beam, is the fundamental quantity to which all calculations would refer, and were it possible to find it this would be the most interesting deduction to make from these experiments. However, owing to the size of source and slit, the curve of the unretarded line is so wide that little accurate information can be deduced about the form of the distribution function, although the amount can be

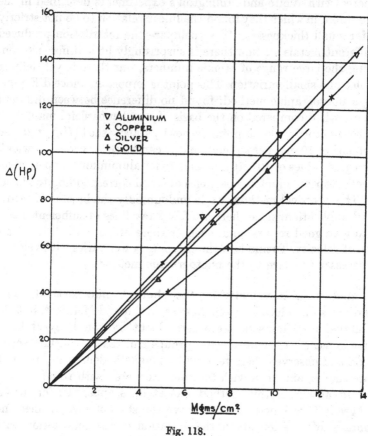

Fig. 118.

gauged from the fact that the high-velocity foot of the curve only moves back at about half the rate that the peak does. For instance a gold sheet of 8 mg./cm.² retards the peak by 60 $H\rho$ but the high-velocity foot by only about 30 $H\rho$. The initial homogeneous beam of β particles of velocity $H\rho$ ca. 1400 has now become a diffuse beam consisting of particles between $H\rho$ 1350 and, say, $H\rho$ 1280, with a maximum at $H\rho$ 1320.

It is possible to measure the distance the peak moves back with fair accuracy, and the results for screens of gold, silver, copper and aluminium are shown in Fig. 118, where the change in $H\rho$ is plotted against the weight per sq. cm. of the screen. The relation appears to be linear, at least up to a retardation of 10 per cent. This may be compared with White and Millington's experiments described in the next section, in which they found the linear relation to be not strictly true for small thicknesses. If we compare the retardation produced by different metals we find there is surprisingly little difference considering the large range of atomic numbers, but there is yet definite evidence of a small variation. This point is important, since d'Espine, using a photographic method, found no difference between different elements when compared on the basis of equal superficial mass.

It is usual to give the stopping power by stating the $\delta(H\rho)$ produced by screens of 10 mg. per sq. cm., and it can be seen that Madgwick's results give values of 107, 101, 98 and 78 for aluminium, copper, silver and gold respectively, for β particles of speed corresponding to $H\rho$ ca. 1400. The electrical method, while undoubtedly the best, is of rather limited application, since there are few β ray lines of sufficient intensity to give good results, and, such as there are, are all in the same region of velocity. To find how the stopping power varies with velocity it is necessary to turn to the photographic method.

§ 98. Reduction in velocity by the photographic method. This method was introduced by von Baeyer, Hahn and Meitner*, but the first actual measurements were carried out by von Baeyer† later. He made use of the homogeneous groups emitted by thorium active deposit and observed their velocity by magnetic deflection first with a bare source, and then with the source covered with thin foils. The great advantage of this method was at once apparent, for in one photograph it was possible to observe the shift of several lines and to obtain information about the variation of the retardation with velocity.

Von Baeyer used the method of direct deflection and the shift of the lines was only of the order of one-third of a millimetre, but he noted that the reduction in velocity appeared to depend only on the superficial mass and not on the nature of the absorbing material, and further he showed that the faster groups were far less retarded than

* von Baeyer, Hahn and Meitner, *Phys. Zeit.* **12**, 275, 1911.
† von Baeyer, *Phys. Zeit.* **13**, 485, 1912.

the slower ones. At the same time Danysz[*] had been making measurements on this problem, but he used the superior focussing method and obtained values over a range from $H\rho$ 1410–5840.

He confirmed von Baeyer's conclusions and determined the variation with velocity more accurately. A great improvement in technique first suggested by Danysz was introduced by Rawlinson[†]. A fine activated wire was covered for half its length by a sheet of the metal under investigation and this source placed in an ordinary focussing apparatus. Both halves of the source will give a spectrum consisting of lines extending right across the plate, but the groups coming from the covered half of the source will be retarded relatively to those from the uncovered half. Except for a very small correction, the loss in velocity ($\Delta H\rho$) is given by measuring the distance between the un-retarded and retarded lines and multiplying by half the value of the magnetic field. It is thus measured directly and not found as the differ-ence between two large $H\rho$'s measured in two distinct experiments. Rawlinson was the first to direct attention to the necessity for taking measurements to a definite portion of the displaced line, since it was obvious from the photograph that the displaced line was considerably broadened. As far as was possible by visual estimation, he measured to. the maximum of the retarded line, and his results are therefore directly comparable with Bohr's calculations (see § 99). The most important new result that he obtained was that for equal superficial masses the heavier elements showed a smaller stopping power than the light elements, and this is supported by Madgwick's observations. The variation with atomic number is not large except for slow β particles but appears to be quite definite.

Variation of stopping power with atomic number

Stopping power $= \Delta (H\rho)$ in traversing foil 10 mg./cm.^{2}

Origin of group	$H\rho$	Mica	Aluminium	Copper	Silver	Tin	Gold	
ThB	1398	—	107	101	98	—	78	Madgwick
RaB	1410	138	—	—	—	89	—⎫	
	1677	101	—	—	—	67	—⎪	
	1938	78	—	—	—	57	—⎬ Rawlinson	
RaC	4866	47	—	—	—	38	32⎪	
	5281	49	—	—	—	38	—⎪	
	5904	43	—	—	—	32	33⎭	

* Danysz, *Journ. de physique*, 3, 949, 1913.
† Rawlinson, *Phil. Mag.* 30, 627, 1915.

We have already pointed out that d'Espine* observed no difference between the stopping powers of aluminium, copper, silver and gold, but in this case the foils were wrapped round the source. This method may introduce errors, since it is more difficult to estimate the average distance traversed in the foil. The results of Rawlinson and of Madgwick seem to be the most reliable and appear to be confirmed by the recent work of White and Millington†. It may be pointed out that little progress was likely to be made until far more attention was paid to the technique of the experiments. It was essential that the lines on the photographic plates should be photometered so that the distribution of intensity in the lines could be found and that proper precautions should be taken to ensure that the β particles passed through the foil at approximately the same angle. Another point which had been neglected in the past was the effect of the diameter of the source. It had been always assumed that the distance between the maxima of the unretarded and retarded lines measured the average loss of energy, but while this is approximately true for an infinitely narrow source, it is by no means necessarily true for sources whose diameter is comparable with the shift to be measured. This is rather a complex matter, depending on the exact form of the β ray lines given by a focussing apparatus, and also on the velocity distribution of the particles after passing the foil, and is discussed by White and Millington.

It was these considerations which formed the starting point of a very thorough reinvestigation of the problem by White and Millington†. They used the strong β ray groups of radium (B + C), employing the focussing method and mica as stopping material. The mica sheet was placed immediately over the source and was inclined at a small angle so that the majority of the rays traversed it normally. They estimate that the paths of the different rays in the screen did not differ by more than 2 per cent. Two holes were pierced in the mica, leaving a portion of the source bare, and the amount of this uncovered portion was adjusted so that the traces of the unretarded and retarded beams were approximately of the same density. Reproductions of two of their photographs are shown in Plate XII, figs. 2 and 3. These show the three strong lines of radium B, which reading left to right are $H\rho$ 1410, 1677 and 1938, and were taken with screens

---

* d'Espine, *C.R.* **182**, 458, 1926.
† White and Millington, *Proc. Roy. Soc.* A, **120**, 701, 1928.

of mica 5·72 mg./cm.² and 2·65 mg./cm.² respectively. The magnetic fields were slightly different in the two cases, so the lines have not the same position. The two sets of lines can easily be seen, the sharp narrow ones are the unretarded lines and immediately on the left of each is the retarded line. By comparing these two photographs one can see how the retardation increases with the thickness of the screen, and decreases with increasing speed of the group, and further how the straggling, shown by the broadening of the trace, is greater with the thicker screen. These photographs were then photometered and the results corrected for the characteristic curve of the photographic plate to give the true distribution of the number

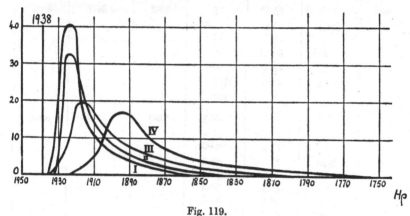

Fig. 119.

of particles along the plate. Since the unstraggled trace had a finite breadth, the straggled trace did not show that curve which would have been obtained with an ideally fine unstraggled line. A numerical way of effecting this correction was worked out, for the details of which the original paper should be consulted.

The results for β particles of speed $H\rho$ 1938 are shown in Fig. 119, where the successive curves labelled I to IV show the distribution with velocity of an initially homogeneous beam after traversing screens of mica 2·25, 2·65, 3·95 and 5·72 mg. per sq. cm. It is at once apparent how much more definite and complete is the information yielded by this type of analysis. We will first consider the most probable retardation as shown by the peak of these curves, since this corresponds most nearly to what the previous observers had measured, and is further a definite feature on which theory can be compared with experiment. White and Millington's results for the peak displacement are given in the next table, which includes measurements on some higher-velocity lines.

Line ($H\rho$)	Thickness of screen in mg./cm.² $=10\sigma$	$\delta(H\rho)$	$I=\delta(H\rho)/\sigma$	Average I	β^3	$I\beta^3$
1410	2·25	23·6*	105	105·5	0·2615	27·6
	2·65	25·0*	94·3			
		28·8	107			
	3·95	41·6*	105			
		46·2	117			
1677	2·25	16·3*	72·4	89·2	0·3478	30·9
	2·65	19·4*	73·6			
		21·1	79·4			
	3·95	33·6*	85·0			
	5·72	58·5*	99·5			
	7·92	85·5	108			
	11·1	117	105			
1938	2·25	16·4*	94·3	78·0	0·4265	33·3
	2·65	16·4*	61·4			
		17·5	65·9			
	3·95	23·8*	60·3			
	5·72	45·5*	79·9			
		46·2	80·8			
		49·8	87·1			
	7·92	63·8	80·5			
	11·1	87·2	78·5			
		98·0	86·1			
		92·3	83·1			
2256	5·72	36·3	63·4	67·5	0·5112	34·5
		40·0	70·0			
	11·1	75·9	68·4			
		75·7	68·1			
2980	5·72	31·8	55·6	55·8	0·6566	36·6
		32·1	56·1			
		37·7	65·8			
	11·1	56·6	51·0			
		56·1*	50·6			
4866	14·2	62·4†	44·0	44·0	0·8426	37·0
5904	14·2	62·0*	43·6	43·6	0·8880	38·8

* These were measured from the photometer curves.
† Only one doubtful reading for this line.

The unstarred values represent figures deduced directly from other photographs which were not subjected to the detailed analysis referred to above. These additional values should, however, also be rather more dependable than any obtained previously, since the authors had the advantage of the experience of the complete analysis of the main plates. The stopping power I is put equal to $\delta(H\rho)/\sigma$, where σ is the thickness of the foil in centigrams/cm.[2] White and Millington considered that these figures showed that I was not constant, as had been found by previous observers, but was slightly less for small thicknesses. If the starred $\delta(H\rho)$ values in the table are plotted against the thickness σ, it will be found that on the whole there is a distinct tendency for the curves to be convex to the axis of σ, but the effect is not large. An effect of this kind is predicted by Bohr's theory and is due to the term $\log_e \sigma$ in equation (4), § 100 (infra). It appears, however, that the rate of variation actually found is more rapid than that predicted by the equation and that as regards this matter the theory needs amendment.

It is important to consider the variation of the stopping power I with speed of the particles. Bohr's theory suggests that, if β is the velocity of the particles in terms of the velocity of light, $I\beta^3$ should be nearly constant. The exact equation will be found in § 100, p. 440, and a detailed comparison with theory is made later. At the moment we will draw attention only to the important point that the term β^3 accounts for the greater part of the variation of the retardation with velocity.

The experimental evidence about the straggling of the β particles has already been illustrated by means of the curves of Fig. 119. These have been drawn so that the area under the various curves is the same, that is, as if no particles had been lost from the beam.

The curves all show the same general features; a well-defined peak of approximately Gaussian form, which merges on the side of greater loss of velocity into a tail which increases rapidly both in height and length relatively to the peak, as the thickness of straggling material increases or the initial velocity decreases.

It is important to examine whether all the curves contained in the set of curves can be summarised into a single curve.

If the tangents at the two points of inflection of each of these curves are produced to cut the axis of abscissae in F and B (as in Fig. 120) and if O is the abscissa corresponding to no loss of velocity, P the abscissa corresponding to the maximum of the curve, then the quantities OF, OP and OB, which we may call the "foot," "peak" and "back" displacements respectively, will be independent of the scale of ordinates chosen.

It is found that as the three points F, P, B move back with increasing thickness the ratios of their mutual distances apart remain approximately constant; in other words, there is not only a motion of the curve as a whole but also a proportional spreading.

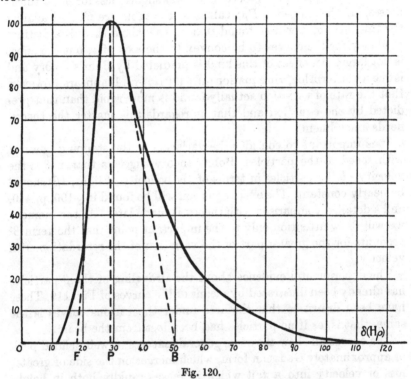

Fig. 120.

Now OP the peak displacement has already been shown to be roughly proportional to the thickness σ, and inversely as the cube of the velocity β^3, so to a first approximation if we write l, m and n for OF, OP and OB, we have

$$l = l_0 \frac{\sigma}{\beta^3}, \quad m = m_0 \frac{\sigma}{\beta^3}, \quad n = n_0 \frac{\sigma}{\beta^3},$$

where l_0, m_0, n_0 are constants.

Corresponding to this general increase of abscissae proportionally to σ/β^3, there must be a fall of ordinates proportional to β^3/σ, if the area of the curve is to remain constant. With these suppositions, if

$$y = g(x)$$

is the equation to one of the curves for which β^3/σ is unity, x being written for $\delta(H\rho)$, the equation for the general curve is

$$y = \phi(x) = \frac{\beta^3}{\sigma} g\left(\frac{\beta^3 x}{\sigma}\right).$$

From the curves similar to those in Fig. 119 for $H\rho$ 1410, 1677, and 1938, the most probable shape of the curve $y = g(x)$ was found; the values 18, 29, and 50 were obtained for l_0, m_0, and n_0 respectively, σ being measured, as is usual, in centigrams per sq. cm. The curve in Fig. 120 represents $y = g(x)$ and in the next table are given the values of $g(x)$ for a series of values of x, when the maximum at $x = 29$ is adjusted to be 100, from which the curve $y = \phi(x)$ can be drawn for any values of β and σ.

Values of $g(x)$. Max. at $x=29$ adjusted$=100$.

x	$g(x)$	x	$g(x)$	x	$g(x)$	x	$g(x)$
12	0	40	62·0	68	13·0	96	1·3
14	4·5	42	55·5	70	11·7	98	1·0
16	10·5	44	49·5	72	10·5	100	0·8
18	17·8	46	44·0	74	9·4	102	0·6
20	27·5	48	39·0	76	8·4	104	0·4
22	43·8	50	34·5	78	7·4	106	0·3
24	66·3	52	30·5	80	6·5	108	0·2
26	89·3	54	27·0	82	5·6	110	0·1
28	99·5	56	24·0	84	4·8	112	0·0
30	99·8	58	21·5	86	4·1	—	—
32	96·8	60	19·5	88	3·4	—	—
34	86·3	62	17·7	90	2·8	—	—
36	77·3	64	16·0	92	2·2	—	—
38	69·3	66	14·4	94	1·7	—	—

We must now consider the errors introduced by the assumption that the retardations, i.e. l, m, and n, are strictly proportional to the thickness. Since the experimental values for the foot and back displacements are less reliable than the corresponding peak displacements, we can allow for this effect to the order of accuracy of the experiment by putting in the expression for $\phi(x)$ not the actual value σ of the thickness, but an effective value σ' adjusted to give the true peak displacement. It is found that within the limits of experimental error the effective thickness is independent of the initial velocity over the ranges of thickness and velocity considered. The next table gives the values of σ' for a series of values of σ from which with the modified equation $y = \dfrac{\beta^3}{\sigma'} \, g \left(\dfrac{\beta^3}{\sigma'} x \right)$ the straggling curve for any velocity and thickness in the given range can be computed.

Effective thickness σ' corresponding to a thickness of foil σ in centigrams per sq. cm.

σ	σ'
0·1	0·08
0·2	0·17
0·3	0·28
0·4	0·41
0·5	0·56
0·6	0·73

Millington has also carried out experiments with gold, and finds that for the same thickness the peak shift is smaller than for mica, but the actual spreading is greater. The variation of $I\beta^3$ with the thickness σ is more pronounced but the law of proportional spreading still holds, the values for l_0, m_0 and n_0 being respectively 10, 21 and 60.

§ 99. **Theories of the loss of energy of β particles in passing through matter.** The theory of the loss of energy of β particles traversing matter is fundamentally the same as that for α particles which has been considered in an earlier chapter. However, in the case of β particles considerable limitations are imposed upon theoretical calculation and experimental observation by the excessive scattering which these particles suffer, and the formulae derived for α particles are not applicable. The velocities of β particles closely approach that of light, and relativity effects, which are negligible for α particles, must also be taken into account.

Calculations on this problem were made by Bohr* in 1915 which went so deeply to the root of the matter that it is most convenient to refer to later work by pointing out the modifications introduced into the original theory. A short account will therefore be given of the special application of Bohr's theory to β rays.

If m and e denote the mass and charge of an electron and β the velocity of the β particle expressed in terms of the velocity of light,

---

* Bohr, *Phil. Mag.* 25, 10, 1913; 30, 581, 1915.

then as in chapter v it follows under the assumption of inverse square forces and neglecting relativity corrections that the energy given to an electron initially free and at rest is

$$Q = \frac{2e^4}{mV^2} \cdot \frac{1}{p^2 + a^2} \qquad \ldots\ldots(1),$$

where $a = 2e^2/mV^2$, and p is the perpendicular distance from the electron to the path of the β particle before collision.

In passing through a thickness Δx of matter containing n atoms per c.c., each with Z electrons, the average number dA of encounters of the β particles with electrons for which p lies between p and $p + dp$ is

$$dA = 2\pi nZ \Delta x p . dp \qquad \ldots\ldots(2),$$

so that provided we could neglect the effect of the binding forces on the electrons we should have an average loss of energy

$$\Delta T = \frac{4\pi e^4 nZ \Delta x}{mV^2} \int \frac{p . dp}{p^2 + a^2} \qquad \ldots\ldots(3).$$

Now while for close collisions the energy transfer is so great and so rapid that it is quite correct to neglect the intra-atomic forces, this cannot be the case for distant collisions. It is just some such limitation to the upper limit that is required to keep the integral finite. This point has been treated in chapter v, but for the sake of completeness Bohr's argument will be repeated here. The electrons in the atom are supposed to be bound in the atoms by restoring forces varying inversely as the displacement so that every electron has a natural period of vibration τ. The influence of these atomic forces on the energy acquired by an electron from the β particle is negligible if the time of collision, which is of the order of p/V, is small compared with τ, and the energy lost is therefore given by (1). Now if the β particle passes the electron at a distance so great that the force it exerts on the electron is of the same order as the atomic restoring forces and if the time of collision is of the order of the natural period τ, the transference of energy to the electron is considerably restricted and is much less than the value expressed by (1). By considering this process in detail, Bohr showed that the loss of energy was the same as would be obtained from the expression (3) (which is only valid for free electrons) if an upper limit of $p_r = \frac{1 \cdot 123}{2\pi} \frac{V}{\nu}$ is inserted, ν being the frequency corresponding to τ. A separate limit must in general be taken for each electron in the atom which will give rise to a summation sign in the final formula.

The lower limit for p is of course zero, corresponding to a head on collision, and inserting this in (3) and integrating from $p = 0$ to $p = p_r$ we should obtain the result already given in the section on α rays. Now this value does represent the mean loss of energy per β particle and should ideally be measurable, but in practice there is no way of doing so. The very violent collisions, represented by p very small, will be very few in number but will each involve a large loss of energy, and will give rise to particles losing far more energy than the average. The object of the calculation must be to find the form of the velocity distribution curve and the position of the maximum since this is determinable by experiment.

In the first section of this chapter a general formula was given for the probability $W(\Delta T) dT$ of an energy loss between ΔT and $\Delta T + dT$. This formula was

$$W(\Delta T)\, dT = (2\pi P\Delta x)^{-\frac{1}{2}}\, \epsilon^{-\frac{(\Delta T - \Delta_0 T)^2}{2P\Delta x}}\, dT \qquad \ldots\ldots(4)$$

and $P\Delta x$ was given by $\Sigma A_r Q_r{}^2$, A_r being the number of collisions with the impact parameter p between p_r and p_{r+1} and Q_r the corresponding energy loss.

The validity of this probability argument depends entirely on all A_r's being sufficiently large, while the variation of Q_r corresponding to the range p_r to p_{r+1} must be small. These conditions will not be satisfied by very close collisions, and we will write p_e for that impact parameter inside which we cannot go without seriously invalidating the probability arguments. The analytical expression for this criterion is that $dA \big/ \dfrac{dQ}{Q}$ is not small compared with unity, which from the previous equations is identical with $\pi n Z\Delta x\,(p_e{}^2 + a^2)$ not being small compared with unity.

Considering only the collisions which satisfy this criterion, we can write

$$P\Delta x = \int_{p_e}^{p_r} Q^2 dA, \quad P\Delta x = \frac{4\pi n Z e^8 \Delta x}{m^2 V^4}\cdot\frac{1}{p_e{}^2} \qquad \ldots\ldots(5),$$

neglecting a in relation to p_e and $1/p_r$ in relation to $1/p_e$. This value of $P\Delta x$ determines the distribution function, and the energy loss corresponding to the peak is

$$\Delta_0 T = \frac{4\pi e^4 n \Delta x}{m V^2} \overset{2}{\underset{1}{\Sigma}} \log\frac{p_r}{p_e} \qquad \ldots\ldots(6).$$

The average number of collisions per β particle which have not been taken into consideration is

$$A_0 = \pi n Z \Delta x\ (p_e{}^2) \qquad \ldots\ldots(7),$$

which clearly we wish to be as small as possible. On the other hand the validity of the probability argument depends on $\pi n Z \Delta x\ (p_e{}^2 + a^2)$ being not small compared to unity. It is thus necessary to effect a compromise, which Bohr shows is most conveniently arrived at by putting A_0, that is $\pi n Z \Delta x\ p_e{}^2$, equal to unity. With this value, p_e is large compared to a, and it will be noticed that on the average we are neglecting one close collision for each β particle. Inserting this value of p_e in (5), we obtain from (4) and (6) formulae for the distribution of energy loss for all collisions excepting on the average one close collision per β particle. The neglect of these collisions will not make a great deal of difference to the distribution. For example ϵ^{-1} of the particles will not make this extra collision, and their energy losses will be given accurately by these formulae. The remainder must lose Q_e or more where Q_e is the value from (1) obtained by putting $p = p_e$, and the distribution curve for these particles (4) will have to be moved back by an amount of the order of Q_e, which for the usual thicknesses (Δx) of foil used is of the order of one-tenth of $\Delta_0 T$. Now it must be noted that those collisions for which $Q \gtreqless Q_e$ will involve large angle deflections and many of the particles making one of these collisions will thereby escape measurement in experiments either like White and Millington's or in range experiments. It will thus be seen that while the value of $\Delta_0 T$ given by (6) is undoubtedly slightly under-estimated, the error cannot be very large. The formula for $\Delta_0 T$, the average energy loss, will be more accurate than that for $W (\Delta T)$, the distribution, which may be seriously in error for energy losses more than the average.

The value for the most probable loss of energy obtained by the above considerations is

$$\Delta_0 T = \frac{2\pi e^4 n \Delta x}{m V^2} \sum_1^z \log \frac{(1\cdot 123)^2 V^2 n Z \Delta x}{4\pi \nu^2} \qquad \ldots\ldots(8).$$

For high-speed β particles it is necessary to include the relativity corrections. These only make a slight alteration, leading to two additional terms in the logarithm, which at low speeds nearly cancel

$$\Delta_0 T = \frac{2\pi e^4 n \Delta x}{m V^2} \sum_1^z \left[\log \frac{(1\cdot 123)^2 V^2 n Z \Delta x}{4\pi \nu^2} - \log (1 - \beta^2) - \beta^2 \right]$$

$$\ldots\ldots(9).$$

From this expression by integration it is possible to obtain the average range R of β particles of given speed

$$R = \frac{m^2 c^4}{2\pi e^4 nS} \left[(1 - \beta^2)^{\frac{1}{2}} + (1 - \beta^2)^{-\frac{1}{2}} - 2 \right],$$

where $\quad S = \sum_1^s \log \left[\frac{(1\cdot123)^2 V^2 nZR}{4\pi \nu^2} - \log(1 - \beta^2) - \beta^2 \right] \quad \ldots(10).$

The expression in equation (9) differs from the corresponding expression for α rays by the occurrence of Δx inside the logarithm, and the rate of loss of energy corresponding to the peak of the distribution curve thus depends slightly on the thickness. The reason for the occurrence of this Δx is of course that the lower limit to the integral p_s has been taken as a function of Δx according to equation (7). From a physical standpoint we may put it as follows. For any given thickness of matter there are certain close collisions which for purposes of calculation it is desirable at first to leave out. It is subsequently shown that since these collisions remove the particles concerned either from observation or from the neighbourhood of the maximum of the distribution curve, they will have little effect on the position of the maximum. It is legitimate to omit these collisions in the calculation of the position of the maximum. Now the criterion of the collisions which are omitted is their infrequency of occurrence, so that for very small thicknesses we must omit collisions with larger impact parameters than for larger thicknesses. The result is that the stopping power of the material measured by the retardation of the maximum will become greater for larger thicknesses since more of the actually occurring collisions contribute to it.

It will be noticed that the essential point in which the theoretical considerations on the passage of β rays through matter differ from those for α rays is in the consideration of the close collisions. The treatment of the more distant collisions is identical in the two cases. All that has been said in chapter v about the modifications of Bohr's outlook introduced by Henderson[*] to fit the original quantum theory, and by Gaunt[†] to fit the modern wave theory, also apply here. This is also true of E. J. Williams'[‡] and Thomas'[§] refinements to the theory to take into account the effect of the velocities of the electrons which are essentially concerned with the more distant collisions.

[*] Henderson, *Phil. Mag.* 44, 680, 1922.
[†] Gaunt, *Proc. Camb. Phil. Soc.* 23, 732, 1927.
[‡] E. J. Williams, *Proc. Manch. Lit. and Phil. Soc.* 71, No. 4, 23, 1927.
[§] Thomas, *Proc. Camb. Phil. Soc.* 23, 713, 1927.

It will be remembered from chapter v that the general effect of the modern theories is to change slightly the picture of the mode of transfer of energy in distant collisions without seriously modifying the formulae for the mean energy transfer. In view of the rather low accuracy of the results so far obtained for β rays, it appears quite sufficient to use Bohr's original formulae for comparison with experiment. There is one point, however, which may be noted. On Bohr's theory there is a slight transference of energy for all collisions, although the energy so transferred is exceedingly small and far less than any ionisation or resonance energy. The modern quantum view of such collisions is that in the aggregate approximately the same amount of energy is transferred as on the classical theory, but that this happens by a few cases of transference of large amounts of energy of the order of the ionisation energy of the atom. In a collision of a certain impact parameter p there is a probability q of transferring approximately the ionisation energy Q, and a probability $1-q$ of there being no loss of energy. The statistical result is much the same as on Bohr's theory, since qQ is approximately equal to the classical energy transfer corresponding to the parameter p.

§ 100. **Comparison of theory with experiment.** If in equation (9) of the preceding paragraph we change from loss of energy ΔT to loss of momentum $\Delta H\rho$ and further from the thickness Δx of the sheet to its superficial mass σ measured in centigrams, we obtain after rearrangement

$$\beta^3 \frac{\Delta (H\rho)}{\sigma} = I\beta^3 = \frac{2\pi e^3 N}{100\, mc^2} \cdot \frac{Z}{A} \left[\log_e \frac{(1\cdot123)^2 c^2 N}{100} \frac{1}{4\pi} - \log_e \frac{A}{Z} \right.$$
$$\left. + \log_e \sigma - \log_e \frac{1-\beta^2}{\beta^2} - \beta^2 - \frac{2}{Z} \overset{z}{\underset{1}{\Sigma}} \log_e \nu \right] \quad \ldots\ldots(1),$$

where Z is atomic number, A atomic weight and N Avogadro's constant. The quantity which is unknown and which expresses the essential stopping power of the material is $\frac{2}{Z} \overset{z}{\underset{1}{\Sigma}} \log_e \nu$ which we shall henceforth denote by $2C$.

If we insert the values of the constants, we obtain

$$I\beta^3 = 5\cdot10\, \frac{Z}{A} \left[96\cdot1 - \log_e \frac{A}{Z} + \log_e \sigma - \log_e \frac{1-\beta^2}{\beta^2} - \beta^2 - 2C \right] \quad \ldots(2),$$

which for aluminium becomes

$$I_{Al}\,\beta^3 = 2\cdot45\left[95\cdot4 + \log_e\sigma - \log_e\frac{1-\beta^2}{\beta^2} - \beta^2 - 2C_{Al}\right]\ \ \ldots\ldots(3).$$

We may obtain the expression for mica, which is

$$H_2\,(\text{K or Na})\,Al_3\,Si_3\,O_{12},$$

by noticing that the expression $-\log_e A/Z + \log_e \sigma$, that is $\log_e \sigma Z/A$, is proportional to the number of electrons in unit area of the foil. Inserting the values of the constants gives

$$I_{mica}\,\beta^3 = 2\cdot53\left[95\cdot4 + \log_e\sigma - \log_e\frac{1-\beta^2}{\beta^2} - \beta^2 - 2C_{mica}\right]\ \ \ldots\ldots(4),$$

an expression almost identical with that for aluminium.

We can test this equation by White and Millington's results by seeing whether the variation with velocity is given correctly. As already pointed out, this is mainly due to the term β^3 on the left of the equation. The results are shown in the next table, the measured values of the stopping power I being in the second column. The calculation has been carried out on the assumption that the values refer to the thicknesses σ as shown in the fourth column. This is a rough method of taking into account the fact that the lower the speed, the smaller is the thickness over which measurements can be taken.

The fifth column contains the values of C_{mica} calculated from the equation, and these should be constant.

$H\rho$ of β particle	I	$I_{mica}\beta^3$	σ	C_{mica}
1410	105·5	27·6	0·265	40·8
1677	89·2	30·9	0·4	40·6
1938	78·0	33·3	0·8	40·7
2256	67·5	34·5	1·0	40·6
2980	55·8	36·6	1·0	40·4
4866	44·0	37·0	1·4	40·7
5904	43·6	38·8	1·4	40·6

It is clear that the variation of the stopping power, I, with velocity is very satisfactorily given by this formula.

In judging the extent of the agreement it must be remembered that the quantity C_{mica} is not very sensitive to errors in I, in fact 6 per cent. change in I only produces a 1 per cent. change in C_{mica}.

We may attempt to estimate how the characteristic stopping constants vary from body to body from the measured stopping powers quoted in § 98, p. 427, by applying equation (2).

Material	$H\rho$ of β particles	$I_{\text{meas.}}$	C
Aluminium	1398	107	41·1
Copper	1398	101	41·1
Silver	1398	98	41·0
Tin	1410	89	41·3
Gold	1398	78	41·7

The similarity in the values of C is noteworthy. It appears that the observed variation in the stopping power I is mainly accounted for by the variation in the factor Z/A, what remains has little effect on the constant C owing to the insensitivity of the equation. These results cannot be compared directly with White and Millington's results for mica, owing to differences in the technique of the experiment.

The method of comparison which has been used has, in effect, only tested the ability of Bohr's formula to account for the variation of the stopping power with velocity. To determine whether it can give the absolute magnitudes we might attempt to correlate the characteristic constants C with $\Sigma \log \nu$, ν being some frequencies characteristic of the atom, but we should at once encounter difficulties in the choice of these frequencies. E. J. Williams* has developed another method of treating this problem which avoids this difficulty. Referring to equations (1) and (2), p. 435, it will be found by elimination of p, that the chance $\phi(Q)\,dQ$ of a loss of energy between Q and $Q + dQ$ at one collision in passing through a thickness Δx is

$$\phi(Q)\,dQ = \frac{2\pi e^4}{mV^2}\, nZ\Delta x\, \frac{dQ}{Q^2} \qquad \ldots\ldots(5).$$

We may test the correctness of our theories of the interchange of energy between the β particles and the electrons in the atom by investigating the validity of this simple equation.

The first point investigated by Williams was the form of the straggling curve for β particles. He showed that excellent agreement could be obtained on Bohr's theory with the results of White and Millington. For example the width at half maximum should on theory vary as $(\Delta x)^{n-1}/V^2$ if n is the exponent of the law of force, while experimental results showed that the width at half maximum varied as $(\Delta x)^{1\cdot0\pm0\cdot1}/V^{1\cdot8\pm0\cdot5}$. The relevant collisions in these experiments were of a distance of approach of the order of 10^{-11} cm. and it appears that in this region the inverse square law is valid. That part of the formula involving $1/V^2Q^2$ is thus strongly confirmed. However, the

* E. J. Williams, *Proc. Roy. Soc.* A, **125**, 420, 1929.

absolute magnitude of the straggling is about 2·4 times the theoretical amount, and even allowing for the effect of scattering in the foil is still certainly twice too large. We can express this result by multiplying the right-hand side of equation (5) by two.

The same result is obtained by considering in detail the variation of the stopping power with thickness of foil traversed. It has already been mentioned that White and Millington's experiments showed that the stopping power $I = \delta\,(H\rho)/\sigma$ increased more rapidly with σ than was predicted by Bohr's theory. Williams showed that agreement of theory with experiment could be obtained by insertion of the 2 as above.

It is interesting to note that a similar result was deduced from the consideration of α particle straggling. It appears that in both the cases of α particles and β particles, collisions involving a given energy loss of about the magnitude considered occur twice as frequently as calculated on the classical theory.

The other feature which has been measured by experiment and calculated by theory is the range of the β particles. An expression for the range was quoted in equation (10), § 99. If as usual we express the range in grams per sq. cm. (G), we obtain the following expression

$$G = \frac{m^2c^4}{2\pi e^4 N \left(\dfrac{Z}{A}\right)} \cdot \frac{(1-\beta^2)^{\frac{1}{2}} + (1-\beta^2)^{-\frac{1}{2}} - 2}{\log_e \dfrac{(1\cdot123)^2}{4\pi} c^2 N \dfrac{Z}{A} + \log_e G - \log_e \dfrac{1-\beta^2}{\beta^2} - \beta^2 - 2C}$$
$$\dots\dots(6),$$

the symbols being as before.

Inserting the values of the constants, we obtain

$$G = 3\cdot34\,\frac{A}{Z} \cdot \frac{(1-\beta^2)^{\frac{1}{2}} + (1-\beta^2)^{-\frac{1}{2}} - 2}{100\cdot8 + \log_e G - \log_e \dfrac{A}{Z} - \log_e \dfrac{1-\beta^2}{\beta^2} - \beta^2 - 2C}$$
$$\dots\dots(7).$$

We can compare this expression with the measured ranges in aluminium quoted in § 96 in the following way. Inserting the values of the atomic weight and atomic number for aluminium, we get after rearrangement

$$50\cdot05 - C_{\text{Al}} = 3\cdot48\,\frac{F\,(\beta)}{G} + \tfrac{1}{2}\phi\,(\beta) - \tfrac{1}{2}\log_e G \quad \dots\dots(8),$$

where
$$F\,(\beta) = (1-\beta^2)^{\frac{1}{2}} + (1-\beta^2)^{-\frac{1}{2}} - 2$$
$$\phi\,(\beta) = \log_e \frac{1-\beta^2}{\beta^2} + \beta^2.$$

The next table shows the extent to which this equation fits the experimental facts. The first two columns show the measured ranges of β particles of different speeds. The next three columns show the values of the three quantities occurring on the right-hand side of the equation, and the last column shows their sum which should be constant.

Col. I	Col. II	Col. III	Col. IV	Col. V	Col. VI
$H\rho$	Measured range G gm./cm.^{2}	$\dfrac{3\cdot48F(\beta)}{G}$	$\frac{1}{2}\phi(\beta)$	$\frac{1}{2}\log_e G$	$50\cdot05 - C_{\text{Al}}$
342	$2\cdot50 \times 10^{-4}$	5·61	1·50	$-4\cdot14$	11·2
372	3·36	5·71	1·47	$-4\cdot00$	11·2
401	3·75	7·25	1·45	$-3\cdot94$	12·6
465	6·64	6·70	1·35	$-3\cdot66$	11·7
527	$1\cdot17 \times 10^{-3}$	6·33	1·20	$-3\cdot38$	10·9
583	1·69	6·45	1·15	$-3\cdot19$	10·8
627	2·32	6·10	1·10	$-3\cdot03$	10·2
694	3·41	6·22	1·02	$-2\cdot84$	10·1
916	7·0	8·10	0·75	$-2\cdot48$	11·3
1,010	9·5	8·50	0·66	$-2\cdot33$	11·5
1,040	$1\cdot08 \times 10^{-2}$	8·29	0·65	$-2\cdot26$	11·2
1,380	1·80	11·6	0·39	$-2\cdot01$	14·0
1,930	6·40	9·41	0·17	$-1\cdot37$	10·9
2,535	$1\cdot24 \times 10^{-1}$	9·95	$-0\cdot05$	$-1\cdot04$	10·9
3,790	2·79	10·9	$-0\cdot36$	$-0\cdot63$	11·2
5,026	4·40	11·6	$-0\cdot61$	$-0\cdot41$	11·4
7,490	7·85	12·0	$-1\cdot00$	$-0\cdot12$	11·1
11,370	13·6	13·4	$-1\cdot41$	0·15	11·8

In the first place we may note that the variation of the range by a factor of 5000 is almost completely accounted for by the velocity term $(1 - \beta^2)^{\frac{1}{2}} + (1 - \beta^2)^{-\frac{1}{2}} - 2$ (see col. III) and the variation that remains is accounted for approximately by the remaining two terms.

The success of this equation in accounting for the experimental results is striking, and suggests the possibility of determining C_{Al}, the characteristic constant of aluminium. It can be seen that a value of about 39 would be obtained which at first sight appears to be in fair agreement with the value 41 deduced from the stopping power experiments. Actually this apparent agreement is largely illusory. If C_{Al} is put equal to 41, the figures in column VI will become 9·05, instead of about 11, and the corresponding range will be changed by 25 per cent.

The reason for this discrepancy lies mainly in the definition of the experimental range, since as can be seen from the curves in Fig. 112

certain of the β particles do penetrate farther. The range of these latter β particles would correspond more nearly with the values calculated from the equation with $C_{Al} = 41$, and the final lack of agreement could be put down to the fact that even these exceptional particles will not have pursued straight paths and it is the total path which is calculated.

A method of calculating the range from experimental data has been used by Sargent*. From the measurements described in § 98 it is known how the loss of momentum $\delta (H\rho)$ in passing through a thin foil depends on the momentum $H\rho$ of the β particle. By a direct numerical calculation it is possible to find the thickness which will reduce the momentum to zero. The ranges obtained in this way are, as would be expected, considerably greater than the experimental ranges.

In § 97 it was shown that Schonland's measurements with fairly slow electrons gave the same ranges (measured in gm./cm.^{2}) in different metals, whereas from equation (7) it can be ascertained there will be a slight variation mainly due to the term A/Z. The general success of the formulae deduced by Bohr suggests that they will also be accurate in this respect, and we must attribute the experimental result to the fact that the measured range approximates to the true range in varying degrees for the different metals. In this connection it may be mentioned that W. H. Bragg† showed that, on certain assumptions, the relative ionisation produced by γ rays in closed vessels of the same shape but of different materials should be proportional to the total true range of the secondary β particles in the material. A dependence of the range on A/Z of the type to be expected was shown by these measurements. Gray and Sargent‡ have used an ionisation method which depends chiefly on measuring the amount of β radiation reflected from different materials and obtained a similar result.

§ 101. The ionisation due to β particles. It is usual to discuss this phenomenon with relation to the quantity k, the ionisation produced per β particle per cm. of path at N.T.P. in the gas considered. This specific ionisation in different gases is roughly proportional to the number of electrons in the molecule, so that for example the ionisation

---

* Sargent, *Trans. Roy. Soc. Canada*, 22, 179, 1928.
† W. H. Bragg, *Phil. Mag.* 20, 385, 1910.
‡ J. A. Gray and Sargent, *Trans. Roy. Soc. Canada*, 21, 173, 1927.

per cm. path in hydrogen is only one-eighth that in oxygen, and one thirty-seventh that in carbon tetrachloride. The departures from this proportionality are of the order of 15 per cent. The relative ionisation in two gases is according to C. G. Barkla and A. J. Philpot* independent of the velocity of the β particle.

The specific ionisation is zero for electrons of energies less than the critical potential of the gas, but once this is passed it increases rapidly, reaching a maximum which was estimated by G. Anslow† to occur close to 1000 volts. From this point on it decreases rapidly, being 1140 pairs of ions per cm. at 4000 volts‡ in air and dropping down to the order of 50 for very high-speed β rays. The order of magnitude and the type of variation to be expected in the specific ionisation in the β ray region was first ascertained by Geiger and Kovarik§, who, using the complex radiations from various β ray bodies, obtained values ranging from 130 to 70 pairs of ions per cm. path in air.

The ionisation produced by a β particle in traversing a gas is partly due to the ejection by the β particle of electrons from the atoms of the gas and partly to secondary ionisation produced by the secondary electrons. The former is known as the primary ionisation. The majority of the secondary electrons are so slow that their range in air at N.T.P. is only a fraction of a millimetre. The ionisation due to these secondary electrons, amounting to rather more than the primary ionisation, is thus concentrated along the track of the β particle. Photographs of the track of a β particle taken by the expansion method show that occasionally the β particle makes a sufficiently direct collision with an atomic electron to be able to transfer a considerable fraction of its energy. The ejected electron will in this case produce far more than the average secondary ionisation, and may travel a considerable distance, several centimetres, from the path of the β particle. This type of collision is so rare, however, that we may consider without sensible error the ionisation of a β particle to be due to primary ionisation giving rise to electrons with energy of the order of 50 volts which in their turn produce a few ions.

We will first consider W. Wilson's‖ measurements on the total ionisation due to β particles of different speeds. His apparatus is

* Barkla and Philpot, *Phil. Mag.* 25, 832, 1913.
† Anslow, *Science*, N.S. 60, 432, 1924.
‡ Glasson, *Phil. Mag.* 22, 647, 1911.
§ Geiger and Kovarik, *Phil. Mag.* 22, 604, 1911.
‖ W. Wilson, *Proc. Roy. Soc.* A, 85, 240, 1911.

shown in Fig. 121. The source A was a thin glass bulb filled with radon and the β rays coming from the products radium B and radium C were sorted out into a velocity spectrum by means of a magnetic field at right angles to the plane of the diagram. It will be seen that the rays are only bent through 90° which does not give good resolution, but since the ionisation varies only slowly with $H\rho$, this defect was not serious. Wilson only measured the relative ionisation and for that purpose he first determined the number of particles passing the slit D at any setting of the field by measuring the rate of communication of charge to the Faraday cylinder E, on the top of which was mounted a small gold-leaf system. When the number entering the cylinder had been found for a sufficient variation of $H\rho$, an ionisation chamber "a" of exactly similar shape was substituted for the Faraday cylinder and the ionisation found for each of the values of the magnetic field previously used. The ionisation chamber contained air at a low pressure so that only the ionisation corresponding to a small drop in energy was determined. While the greater portion of the β particle energy was communicated to the walls, a certain amount of reflection occurred and for this a correction had to be applied. This was determined by means of the apparatus shown at b which was

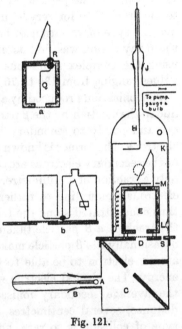

Fig. 121.

placed with the aperture covering the hole D, after the chamber O had been entirely removed. The ionisation chamber shown in b contained air at atmospheric pressure and was made of thin aluminium leaf fixed over a cage of fine wire. The ionisation in this was first measured and then a vessel similar to E, but without a base, was placed over it, and the increase in ionisation was taken to measure the correction for reflection.

The results obtained by Wilson are shown in the next table. Wilson only obtained the ionisation in arbitrary units but his results have been reduced to absolute units, that is pairs of ions per cm.

path at atmospheric pressure, by comparison with S. Bloch's* measurement. In this latter experiment the heterogeneous rays furnished by the total radiation of uranium X were used as a source of β rays but owing to the slow rate of variation of the ionisation with $H\rho$ this is not as serious a drawback as might at first be anticipated. It will still, however, be realised that we cannot expect these figures to give more than the approximate absolute values.

Number of pairs of ions formed by β rays of different velocity in traversing 1 cm. of air at N.T.P.

$H\rho$	Number of pairs of ions (k)	$k\beta^2$
850	214	43
1150	155	47
1390	127	49
1650	94	45
1890	84	46
2160	74	45
2440	68	46
2750	61	44
3300	56	44
3900	55	46
4800	53	47
5430	50	45
5910	47	43
6350	46	43

It will be seen that the specific ionisation k is closely proportional to the inverse square of the velocity. Now in these experiments the length of the chamber E was 4 cm., its diameter 3·3 cm., and the pressure of air varying between 1 mm. and 50 mm. As we have seen in the preceding sections, the loss of momentum in passing through a small mass of material is to a first approximation proportional to the inverse cube of the velocity, so that the energy lost is approximately as the inverse square of the velocity. The constant values of the $k\beta^2$ shown in the third column of the table may therefore be taken to show that for all velocities the ionisation accounts approximately for the same fraction of the total energy loss. In view of the uncertainty of the absolute values, the question must be left open whether all the energy loss can be accounted for by ionisation.

C. T. R. Wilson† showed that photographs of the tracks of β particles in an expansion chamber provided a convenient means of

---

* Bloch, *Ann. d. Phys.* **38**, 559, 1912.
† C. T. R. Wilson, *Proc. Roy. Soc.* A, **104**, 192, 1922.

investigating the primary ionisation. He found that the track of a
β particle could be resolved into a series of separate droplets. Each
of these must be interpreted as showing a primary ionisation, since
the secondary ionisation is produced by slow electrons in the immediate neighbourhood of the primary ion. For example, for β particles
of speed $0·32c$ he found 96 droplets per cm., but only about 20 for
the fast β particles of radium C.

This method has been used recently by E. J. Williams and Terroux*
in a thorough investigation of the primary ionisation. They photographed the tracks given by β particles of radium E and arranged
that a strong magnetic field could be applied parallel to the axis of
the chamber at the moment of expansion. The speed of each particle
could be estimated from the curvature of the track. Their results for
the number of primary ions formed per cm. by particles of different
speeds are shown in the next table.

$H\rho$	$\beta = v/c$	Number of primary ions per cm. at N.T.P. in	
		Oxygen	Hydrogen
875	0·454	—	18·3
1100	0·538	43·0	12·6
1500	0·660	34·0	8·9
1850	0·738	28·4	7·6
2690	0·845	27·0	7·1
3170	0·880	—	6·4
4100	0·920	25·2	6·1
5180	0·950	—	5·1
7000	0·972	22·2	—

The point of greatest general significance is that for large values
of $H\rho$ the ionisation tends to become constant.

The electrons in the gas traversed by the β particles are situated
in definite energy levels and have a definite ionisation potential.
The primary ionisation thus depends on the frequency of the collisions
in which the β particle gives more energy to an atomic electron than
its ionisation potential. If $\phi(Q)\,dQ$ is defined as the frequency of
the collisions in which a β particle loses energy between Q and
$Q + dQ$, then the primary ionisation is given by

$$I = \int_{Q-J} \phi(Q)\,dQ \qquad \qquad ......(1),$$

where J is the ionisation potential.

* E. J. Williams and Terroux, *Proc. Roy. Soc.* A, **126**, 289, 1930.

For a β particle traversing an atmosphere of free electrons initially at rest the value of $\phi(Q)$ according to the classical theory is

$$\phi(Q) = \frac{2\pi e^4 n}{m V^2} \cdot \frac{1}{Q^2} \quad\dots\dots\dots\dots\dots(2),$$

where n is the number of electrons per unit volume, V the velocity of the β particle, and e and m the electronic charge and mass respectively.

The primary ionisation is therefore, according to this formula,

$$I = \frac{2\pi e^4 n}{m V^2} \left[\frac{1}{J} - \frac{1}{T}\right] \quad\dots\dots\dots\dots\dots(3),$$

T being the energy of the β particle.

In all actual cases J/T is negligible, so that

$$I = \frac{2\pi e^4 n}{m V^2} \cdot \frac{1}{J} \quad\dots\dots\dots\dots\dots(4).$$

This formula for the primary ionisation was first given by J. J. Thomson. It depends on the assumption that in ionising collisions the electron behaves as if it were free. In ionising collisions the maximum time of collision, measured by p/V where p is the impact parameter, is only of the order of 1/100,000 of the natural period of the outer electrons. The considerations advanced by Bohr (see § 99) show that under these conditions the above assumption is valid.

Thomas[*] and E. J. Williams[†] have calculated the correction to be applied to (2) to take into account the velocity of the atomic electrons in their orbits. If ϵ is the mean kinetic energy of the atomic electrons, then formula (4) becomes

$$I = \frac{2\pi e^4 n}{m V^2} \left(1 + \frac{2\epsilon}{3J}\right) \cdot \frac{1}{J} \quad\dots\dots\dots\dots(5).$$

It might also be anticipated for the high speed β particles considered that relativity effects would be appreciable. However, calculation shows that for ionising collisions the classical energy transfer and the value of $\phi(Q)$ remain unaltered.

It will be noticed that the dependence of the ionisation I on the velocity V is given by these formulae as

$$I \propto 1/V^2.$$

[*] Thomas, *Proc. Camb. Phil. Soc.* **23**, 713, 1927.
[†] E. J. Williams, *Nature*, **119**, 489, 1927; *Manchester Memoirs*, **71**, 25, 1926-7.

The results of Williams and Terroux however can be expressed within the limits of experimental error by

$$I = 5 \cdot 2 \Big/ (V/c)^{1 \cdot 5 \pm 0 \cdot 2} \qquad \text{for hydrogen,}$$

$$I = 22 \Big/ (V/c)^{1 \cdot 1 \pm 0 \cdot 2} \qquad \text{for oxygen.}$$

The disagreement of theory with experiment is also shown by comparison of the absolute values with those measured. It is not possible to evaluate (5) with great exactness owing to uncertainty in the values of the ionisation potentials in the molecules, but there appears to be no doubt that for very fast particles, where $V \sim c$ and the disagreement in the variation with velocity is of less account, the theory gives values two or three times too small.

It may be repeated that the collisions to which this result refers are those in which the energy loss is of the order of 50 volts. In connection with the straggling of β particles where the important energy transfers are of the order of 1000 volts, it was found (§ 100) that again classical theory predicted a probability of energy transfer about half that found, although in this case the dependence on velocity appeared to be correct. Williams and Terroux have also investigated the probability of energy transfers of amount 5000 to 10,000 volts by observing the frequency of branch tracks. Here also the actual probability of transfer of energy is greater than that calculated on the classical theory and the dependence on velocity is also in disagreement. Since we have seen that to within about 20 per cent. the total loss of energy is given correctly by the classical theory, it is clear that there must be a divergence from theory in the opposite direction for non-ionising collisions. It appears that the energy actually lost in non-ionising collisions is less than one-seventh of that calculated on classical theory.

CHAPTER XV

THE SCATTERING AND ABSORPTION OF γ RAYS

§ 102. The discovery of the penetrating radiations known as the γ rays was made by observations of the ionisation in an electroscope, and an idea of the penetrating power was obtained by placing a screen of absorbing material between the source and the electroscope. In this way it was found that these radiations were far more penetrating than the β rays, a centimetre of lead being required in some cases to reduce the ionising effect to one-half. It was observed that the γ rays in passing through matter gave rise to swift β rays, and the ionisation in the electroscope was ascribed to β rays liberated from the walls and the gas.

The chief source of γ radiation in the early experiments was radium (B + C), and a number of experiments were made to test the law of absorption by the simple method outlined above. While it was found that with considerable thicknesses the absorption curve approximated closely to a simple exponential law, yet with small thicknesses a greater absorbability was noticed. This led naturally to the deduction that a large part of the radiation was homogeneous in character and that superimposed on this were softer radiations which were more rapidly absorbed.

Systematic observations were made by Soddy and Russell of the absorption of different metals which showed that for the lighter elements the absorption depended only on the mass per square centimetre of the absorber and did not vary with the atomic weight. For the heavier elements such as lead the absorption was twice as great as would have been expected from the results with the light elements. It is now known that the interpretation of this result depends on the recognition of the existence of two distinct absorbing processes, one, the photoelectric, varying more rapidly with atomic number than the other, the scattering process.

However, even at this early stage it was inferred that the γ rays were a type of very penetrating X rays and it was anticipated that the absorption measured in ordinary ways should consist of two parts, scattering and true absorption. The early experiments of Florance and J. A. Gray and the later work of Compton brought out the surprising result that the quality of the scattered radiation varied markedly with the angle of scattering. In this respect the γ rays appeared to show different properties from the X rays where the work of Barkla had shown that the scattered radiations were similar in quality to the primary beam. It seemed clear that there was some degradation of the radiation in the process of scattering, but it was difficult at that time to draw definite conclusions since there was no information, except from absorption results, about the homogeneity of the primary γ rays.

In subsequent measurements attempts were made to obtain an idea of the relative magnitude of the scattering coefficient and true absorption coefficients. Progress became much more rapid as definite information was obtained about the frequency of the γ rays, since it was then possible to compare the absorption with that to be expected by extrapolation of the relations found for X rays. The later developments of this subject have depended on methods which differentiate between absorption due to scattering and that due to the photoelectric effect, and in this field the Compton laws of scattering have been applied with considerable success. While the essential processes involved in the absorption of γ rays are thought to be understood, there is still considerable doubt as to the validity of the various formulae that have been proposed to account for their dependence on frequency. Since the γ radiations of radioactive bodies are in general complex, consisting of radiations spread over a wide range of frequencies, it has proved difficult to arrive at precise laws for the absorption of high frequency radiation. We shall see, however, that it has been shown that with increasing frequency there is a gradual change from the well-known laws applicable to ordinary X rays.

In the early days of radioactivity a marked difference in the penetrating power of the γ radiation of different bodies was noticed, and a series of measurements were made by Rutherford and Richardson to find in a general way, by analysis of the absorption curves, the different types of radiation present. It was observed in this way that in addition to the typical γ rays, there were always some softer radiations, some of which could be identified with the characteristic X rays. Further information on this point was obtained by observing the quality and amount of the radiations excited in different materials when exposed to swift β particles. In most cases not only were soft radiations observed, but also radiations of a more penetrating type.

We shall first describe these experiments on the comparison of the γ rays of different radioactive bodies, since quite simple experimental methods give a general idea of the quality of the radiation present. Later we shall deal with attempts that have been made to correlate the coefficients with the frequency of the radiation under observation. This is still in a transition stage but sufficient evidence has been obtained to draw deductions of value.

§ 102 a. The formal and usual method of describing the absorption of a beam of radiation is by means of one or more "absorption coefficients." Since we shall be continually referring to such coefficients, it is convenient to set out the definitions and terminology that will be employed.

Consider a parallel beam of homogeneous radiation of frequency ν, and denote by E the flux of energy. In passing through a small thickness Δx cm. of matter the fraction of the energy that is removed per second per unit area of the beam will be

$$\Delta E/E = -\mu \cdot \Delta x \qquad \ldots\ldots(1).$$

If the quality of the beam is not changed in its passage through matter, this equation may be integrated to give

$$E = E_0 e^{-\mu x} \qquad \ldots\ldots(2).$$

The quantity μ is termed an absorption coefficient and has dimensions length $^{-1}$. If on physical grounds there is reason to divide up the energy lost by the beam into several parts, $\Delta E_1, \Delta E_2, \ldots$, each of which appears in a distinct form, e.g. photo-electrons, scattered radiation, etc., we may write

$$\Delta E_1 = -\mu_1 \Delta x E, \qquad \Delta E_2 = -\mu_2 \Delta x E \qquad \ldots\ldots(3),$$

so that

$$E = E_0 e^{-(\mu_1 + \mu_2 + \ldots)x} \qquad \ldots\ldots(4),$$

and

$$\mu = \mu_1 + \mu_2 + \ldots \qquad \ldots\ldots(5).$$

The separate coefficients $\mu_1, \mu_2, \ldots$ would now be referred to as the photo-electric absorption coefficient, the scattering coefficient, etc.

If ν is the frequency of the radiation, it can be seen by dividing equations (1), (2), (3), and (4) by $h\nu$ that the same absorption coefficients will describe the change in the flux of quanta of the beam.

If the beam is complex, consisting of radiations of several frequencies, it is still possible to define mean absorption coefficients by the relations (1) and (3) and equation (5) will still be valid. It will not however be correct to integrate expressions (1) or (3) to obtain equations (2) or (4), since the constitution of the beam will alter in passing through a finite thickness owing to the more rapid weakening of the longer wave-length components. In this case the mean absorption coefficient will be a function of the thickness of absorber traversed. It will also be clear that the same mean absorption coefficients will not express the diminution in the flux of energy and in the flux of quanta.

We shall have occasion later to refer to a complication that occurs in practice even with a homogeneous beam of radiation. Owing to the change of wave-length that can accompany the process of scattering, an initially homogeneous beam does not remain homogeneous. Even with a strictly parallel beam two scattering processes can lead to the admixture of longer wave-length radiation. However, from a formal standpoint this latter radiation is distinct from the primary radiation, although it may not be possible to maintain the distinction experimentally, and the definitions and properties of the absorption coefficients can be maintained.

If instead of using the thickness x of material traversed we use the superficial mass m, measured in gm./cm.^{2}, it will be seen that equation (1) becomes

$$\Delta E = - \mu . \frac{\Delta m}{\rho} . E \quad \dots\dots\dots\dots\dots\dots(6),$$

where ρ is the density of the material.

The quantity μ/ρ is termed the mass absorption coefficient. Its use is to be preferred to that of μ since the same mass absorption coefficient will describe the absorption in the gaseous, liquid and solid states of a substance, the absorption of γ rays being an atomic phenomenon. In the same way if P_A and P_e be respectively the number of atoms and electrons in a portion of the absorber 1 cm.^{2} cross-section, and m gm./cm.^{2} thick, we can define absorption coefficients per atom $_A\mu^*$ and per electron $_e\mu^*$ by the relations

$$\Delta E = - {}_A\mu . P_A . E, \qquad \Delta E = - {}_e\mu . P_e . E \quad \dots\dots(7).$$

It follows by comparing (6) and (7) that

$$_A\mu = {}_e\mu . Z = \frac{\mu}{\rho} . \frac{A}{N},$$

where Z, A, and N are respectively the atomic number, atomic weight and Avogadro's number.

* The subscripts A and e are written in front of the symbol to leave room for a subscript after the symbol to distinguish between scattering absorption and pure scattering (see § 104).

§ 103. **The investigation of the** γ **rays of different radioactive bodies by the absorption method.** The absorption method has proved a simple and rapid way of obtaining a general idea of the character of the γ ray emission of a radioactive body. It can seldom give very definite information and it is unnecessary to push the refinement of the experiment to any great lengths. In most cases all that has been done is to place the source close to a simple electroscope and to observe the change in the ionisation current when different thicknesses of lead or aluminium are interposed. It is generally found that after the absorber has reached a certain thickness the curve becomes closely exponential. This part of the curve is extrapolated back to zero thickness according to the same law and is subtracted from the original readings. The new absorption curve of the softer fraction of the radiation is then treated in the same way. In this way it is possible to analyse the experimental curve into a series of exponentials whose individual coefficients then serve to describe the original complex radiation. Too much weight must not be laid on the exact numerical values of the coefficients obtained from measurements with such simple apparatus, since it is found that the absorption curve depends slightly on the size and disposition of the apparatus. The method of analysis is clearly not unique, nor is it possible from the physical standpoint to give an accurate description in terms of three or four absorption coefficients of a radiation containing many components of which the frequencies are spread over an octave or more. The method, however, serves to classify the radiation into hard, medium, or soft γ rays and X rays and to give a rough idea of their relative amounts.

An inspection of the next table will show the results that have been obtained.

The column headed X rays shows which characteristic X rays have been detected. The values of the absorption coefficients obtained for these radiations have not been found to show a close agreement with X ray data, but this is hardly to be expected from the method of analysis and from the different experimental arrangements that have been used. In addition there are frequently nuclear γ rays of about the same penetrating power which will affect the values. The two columns headed (μ/ρ) Al and (μ/ρ) Pb show the mass absorption coefficients of those components which may be reasonably attributed to γ rays of nuclear origin. It is hardly to be anticipated that the aluminium and lead absorption curves will be analysed into corresponding com-

ponents, since while the absorption in lead is partly photo-electric
and partly scattering, that in aluminium is almost entirely scattering.

*Analysis by the absorption method of the γ ray emission
of different radioactive bodies*

	X rays	(μ/ρ) Al	(μ/ρ) Pb
UX_1 (a)	L and K	—	—
UX_2 (a)	—		0·20
		0·052	0·06
Io (b)	M, L and K	—	—
Ra (c)	L	130	—
RaB (d)	L and K	—	0·41
			0·13
RaC (d)	—	0·085	0·132
		0·0424	0·047
RaD (e)	L	0·37	—
RaE (e)	—	0·092	0·43
RaF (c, j)	—	215	—
RdAc (c, f)	L and K	0·07	—
AcB (f)	L and K	44	—
AcC'' (f)	—	0·073	—
MsTh$_2$ (f)	L and K		0·25
		0·043	0·062
ThB (f)	L and K	59	—
ThC'' (f, g)	—	0·036	0·041
ThC, ThX (h) and Rn (i)	—	—	—

The L radiation from radioactive bodies has (μ/ρ) Al of order 14·0,
and K radiation (μ/ρ) Al of order 0·17.

(a) Hahn and Meitner, *Zeit. f. Physik*, **7**, 157, 1923. Soddy and Russell, *Phil. Mag.* **18**, 620, 1909. Richardson, *Phil. Mag.* **27**, 22, 1914.

(b) Chadwick and Russell, *Proc. Roy. Soc.* A, **88**, 217, 1913.
(The amount is very small, see § 111.)

(c) Russell and Chadwick, *Phil. Mag.* **27**, 112, 1914.
(The intensity of the radiation from radium is only about 1 per cent. of that of the γ radiation of radium in equilibrium with its short-lived products.)

(d) Rutherford and Richardson, *Phil. Mag.* **25**, 723, 1913. Richardson, *Proc. Roy. Soc.* A, **91**, 396, 1915. Kohlrausch, *Wien. Ber.* **126**, 2a, 683, 1913; *Jahrb. d. Rad.* **15**, 64, 1918. Enderle, *Wien. Ber.* **131**, 2a, 589, 1922. Moseley and Makower, *Phil. Mag.* **23**, 302, 1912. Fajans and Makower, *Phil. Mag.* **23**, 292, 1912.

(e) Rutherford and Richardson, *Phil. Mag.* **26**, 324, 1913. Curie and Fournier, *C.R.* **176**, 1301, 1923. G. H. Aston, *Proc. Camb. Phil. Soc.* **23**, 935, 1927.
(Radium E emits a very weak γ radiation giving less than 2 per cent. the ionising effect of the γ rays from radium D. Part may have been excited in the support.)

(f) Rutherford and Richardson, *Phil. Mag.* **26**, 937, 1913.

(g) Soddy and Russell, *Phil. Mag.* **21**, 130, 1911.

(h) Hahn and Meitner, *Phys. Zeit.* **14**, 873, 1913.
(γ rays are possibly present in very small amount.)

(i) Slater, *Phil. Mag.* **42**, 904, 1921.
(No measurable amount of radiation.)

(j) Recent experiments have suggested that this may not be γ radiation but H particles liberated from the source by a ray impact.

The use and applicability of this method of investigation can be judged from this table and from the notes appended to the references. The three bodies which emit the most penetrating radiations are mesothorium 2, radium C and thorium C″, while that from actinium C″ is considerably softer. We shall see in § 108 that the penetrating radiations of radium C and thorium C″ have been of great use in studying the absorption and scattering of high frequency radiation.

Particular interest is attached to the identification of the characteristic X rays, since this was the first suggestion of an interaction of the main γ rays on the electronic system of the disintegrating atom.

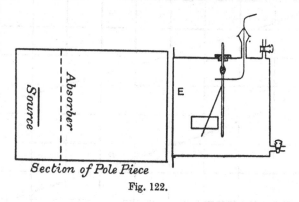

Section of Pole Piece

Fig. 122.

These X rays are very easily absorbed in comparison with the γ rays, and are not greatly different in penetrating power from the β rays. To detect their presence a slightly more elaborate arrangement is necessary than that already indicated.

The experimental method of Rutherford and Richardson to whom many of the above results are due will illustrate this. A typical apparatus used by them is shown in Fig. 122. The source is placed between the poles of an electromagnet to enable the β rays to be deflected and is about 10 cm. from the electroscope. This was of aluminium (10 × 10 × 10 cm.) with one face E made of a thin sheet of mica equivalent in α ray stopping power to about 2 cm. of air. It was thus possible to take the first reading with an absorption equivalent to only a few centimetres of air. A typical absorption curve in aluminium with the electroscope filled with air is shown by A, Fig. 123. In this curve the logarithm of the ionisation is plotted against the thickness of the absorber. This curve was obtained from a tube filled with radon and is due to the γ rays of radium B and radium C. While the effect of the penetrating γ rays of radium C is large compared with that due to the softer types of radiation, the presence of these latter can be seen from the initial rapid drop of about 10 per cent. If the electroscope is filled with a mixture of methyl iodide and hydrogen, the curve B is obtained showing how the initial drop is increased, due to the relatively greater absorption of the soft components of the radiation in the methyl iodide. Under these conditions it is possible to investigate this radiation in some detail and to separate it into several components.

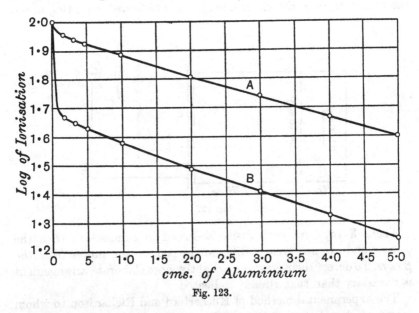

Fig. 123.

An interesting application of the different absorbabilities of the γ radiations from radium and thorium preparations has been made by Bothe* to decide in what proportions a preparation contains radium, mesothorium or radiothorium. This is a problem that occurs in practice, since on the one hand the raw material from which mesothorium is obtained always contains a certain amount of its isotope radium, while on the other hand radium is not in all cases isolated from minerals free from thorium. The principle of the method is to compare through different thicknesses the absorption in lead of the γ rays of the preparation with those of a standard radium source

* Bothe, *Zeit. f. Phys.* **24**, 10, 1924.

which is known to be pure. The mass absorption coefficients in lead of the hard radiations from the three bodies are approximately

Ra (RaC)	MsTh (MsTh$_2$)	RdTh (ThC")
0·047	0·062	0·041 cm.$^{-1}$

so that, if after 5 to 10 cm. of lead have been interposed, the apparent radium equivalent of the preparation is found to fall, the presence of mesothorium can be inferred with certainty. If, on the other hand, the radium equivalent is found to rise, it is clear that the mesothorium in the preparation is sufficiently old to have grown radiothorium. Bothe describes a special experimental arrangement which is suitable for these large thicknesses of lead and gives curves from which the relative proportions of the three bodies may be calculated from the observed variation in the radium equivalent. Yovanovitch* has worked out a calorimetric method for determining the proportion of mesothorium in radium preparations, based on the measurement of the growth of heat due to the formation of radiothorium.

§ 104. **The laws of absorption and scattering of high-frequency radiation.** The early experiments on the absorption of the γ rays of radium C showed that after passing through about two centimetres of lead the radiation was absorbed exponentially. This point was investigated specially by Soddy and Russell† and in great detail by Russell‡, who found the absorption exponential up to thicknesses of over 20 cm. of lead or mercury. This was thought to indicate an extreme homogeneity of the radiations, a conclusion that later proved difficult to reconcile with the complexity of the spectral distribution found by investigation of the β ray spectrum.

In 1904 the important observation was made by Eve § that secondary γ rays were emitted by any body or radiator through which the γ rays of radium C passed, the secondary rays appearing to be less penetrating than the primary. Investigations on the nature of these secondary rays were subsequently carried out by Kleeman ‖, Madsen ¶ and Florance**, the latter giving the most definite information. Florance obtained the important result that while the secondary rays

* Yovanovitch, *J. de Physique et le Radium,* **9**, 297, 192?.
† Soddy and Russell, *Phil. Mag.* **18**, 620, 1909.
‡ Russell, *Proc. Roy. Soc.* A, **88**, 75, 1913.
§ Eve, *Phil. Mag.* **8**, 669, 1904.
‖ Kleeman, *Phil. Mag.* **15**, 638, 1908.
¶ Madsen, *Phil. Mag.* **17**, 423, 1909.
** Florance, *Phil. Mag.* **20**, 921, 1910.

appeared to be independent of the nature of the radiator, they showed a progressively greater absorbability as the angle between their direction and that of the primary beam increased. Florance considered these secondary rays to be scattered primary rays, and as these latter were heterogeneous he explained his results by assuming that the softer rays were scattered to a relatively greater extent through large angles. However, J. A. Gray* showed experimentally that as the intensity of the primary rays was diminished by lead screens the softer scattered rays were not cut down as quickly as would be expected. For example with scattering at 110° from carbon he found the scattered rays were reduced to about 12 per cent. of their initial value by a sheet of lead 3 mm. thick placed between the scatterer and the electroscope. However, if the lead were placed between the source and the scatterer it required over 3 cm. of lead to do this. Gray concluded that when homogeneous γ rays are scattered, there is a change of quality, i.e. wave-length, and this change of quality is greater the greater the angle of scattering.

It is clear that we have here an example of the Compton effect which much later has been carefully studied in the case of X rays. This phenomenon may be most shortly described if we employ the language of the light quantum hypothesis. A quantum $h\nu_0$ in passing through matter may "collide" with an electron, and if we consider the quantum to have momentum $h\nu_0/c$ it is clear that energy will be imparted to the electron. This energy comes from the radiation with the result that the scattered quantum is of lower frequency ν'. The equations expressing the conservation of energy and momentum (see Fig. 124) are

$$h\nu_0 = mc^2 \left(\frac{1}{\sqrt{1-\beta^2}} - 1 \right) + h\nu',$$

$$0 = \frac{mc\beta}{\sqrt{1-\beta^2}} \sin \phi + \frac{h\nu'}{c} \sin \theta,$$

$$\frac{h\nu_0}{c} = \frac{mc\beta}{\sqrt{1-\beta^2}} \cos \phi + \frac{h\nu'}{c} \cos \theta.$$

These equations suffice to determine the drop in frequency $\nu_0 - \nu'$, the velocity $c\beta$ and angle ϕ of recoil of the electron for each angle of scattering of the radiation. The diagrams in Fig. 124 illustrate the relations between the energies of the scattered quantum and the recoil electrons (shown by the lengths of the radius vector) for

* J. A. Gray, *Phil. Mag.* **26**, 611, 1913.

various angles of scattering and for two different frequencies. A convenient formula is obtained by considering the increase in wave-length for different angles of scattering:

$$\lambda' - \lambda_0 = \frac{h\,(1 - \cos\theta)}{mc} = 24 \cdot 2\,(1 - \cos\theta)\ \mathrm{X\,U}.$$

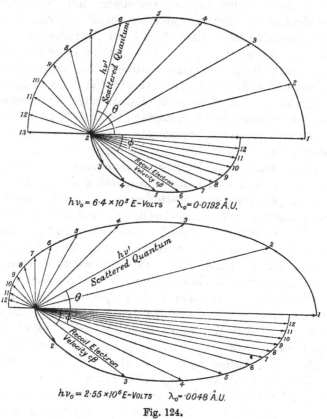

$h\nu_0 = 6 \cdot 4 \times 10^5\ E\text{-}V\!olts \qquad \lambda_0 = 0 \cdot 0192\ \text{Å.U.}$

$h\nu_0 = 2 \cdot 55 \times 10^6\ E\text{-}V\!olts \qquad \lambda_0 = \cdot 0048\ \text{Å.U.}$

Fig. 124.

We should also anticipate that there would be scattering without change of wave-length such as was envisaged by the classical theory. This would occur in collisions of the quantum with tightly bound electrons when the impact of the quantum was insufficient to supply the ionisation energy. This has been investigated with X rays, when, under suitable conditions, it is found that the scattered radiation is partly of the primary frequency and partly of a lower one as given by the above equations. The unmodified scattering will become progressively less important as the frequency increases and it probably plays only a small part in the scattering of γ rays except at very small angles. An experiment of Compton's* gives some evidence on this point. He used the method, initiated by J. A. Gray, of comparing the diminution of the scattered radiation produced by an absorber when it is placed either between the source and the scatterer or between the scatterer and the ionisation chamber. From his results Compton concluded that extremely little of the radiation of radium (B + C) when filtered through several centimetres of lead was unchanged in wave-length on being scattered. The scattering in these experiments was at 90° but it is probable that at smaller angles there is a greater proportion of unmodified scattering. We have seen in § 86 that Frilley was able to photograph the line 7.73×10^5 volts by reflection from a crystal by presumably unmodified scattering where the scattering was through less than one degree.

The phenomenon of the change of wave-length on scattering is particularly relevant with γ rays since, owing to the short wave-length of the primary γ rays, scattering through quite a small angle produces a marked change in the frequency. That it occurs is beyond doubt and is perhaps best, though indirectly, shown by Skobelzyn's experiments. These will be referred to later (p. 504), but it may be noted here that he passed the γ rays of radium C through an expansion chamber and obtained tracks of high-speed electrons. From other evidence we know that the photo-electric absorption (i.e. the complete conversion of the quantum) in air is negligible, compared with the scattering, so we are forced to associate these high-speed electrons with the latter process. On present-day views this implies a lower frequency of the scattered radiation in order that $h (\nu_0 - \nu')$ shall provide the energy of the recoil electrons.

Hoffmann† has attempted to determine the change of wave-length directly by observing the absorbability of the γ rays from radium (B + C) filtered through 1 cm. of lead after scattering from carbon at various angles. The absorption coefficient μ in lead increased from 1·02 for scattering at 15° to 11·8 for scattering at 150°. A detailed comparison with theory is rendered difficult by our lack of exact knowledge of the absorption laws, and by the fact that the heterogeneity of the radiation introduces complications about the relative scattering of different wave-lengths at a given angle. However, we

---

* Compton, *Phil. Mag.* **41**, 749, 1921.
† Hoffmann, *Zeit. f. Phys.* **36**, 251, 1926.

can see that qualitatively the change of wave-length is demonstrated by the experiments.

Many attempts have been made to deduce a theoretical formula for the intensity of scattering. Of these the most frequently used are those of Compton*, of Dirac† and of Klein and Nishina‡. They all agree in postulating that all the electrons in whatsoever atom scatter equally and independently, so that it is sufficient to give a scattering coefficient per electron. The atomic scattering coefficient for any atom will then be Z times this where Z is the atomic number. In describing this process we have not only to distinguish between the mass absorption coefficients μ/ρ, σ/ρ, etc., and the corresponding coefficients calculated per electron or per atom, but also between the two parts of the total scattering coefficient which describe respectively the deflection of radiation and the conversion of energy into the electronic form by the Compton process. These latter we shall write as σ_s and σ_a so that $\sigma = \sigma_s + \sigma_a$, and we shall use a subscript before the symbol such as $_e\sigma$, $_A\sigma$ or $_A\mu$ to denote that the corresponding coefficient has been calculated per electron or per atom. The Compton formula for the energy removed per electron by the scattering process from the primary beam is

$$_e\sigma = \frac{\pi e^4}{m^2 c^4} \frac{8}{3} \frac{1}{1+2\alpha}, \text{ where } \alpha = \frac{h\nu}{mc^2}.$$

This was obtained before the advent of the new mechanics and may be compared with the Dirac formula based on this theory:

$$_e\sigma = \frac{\pi e^4}{m^2 c^4} 2 \left\{ \frac{1+\alpha}{\alpha^2} \left[\frac{2(1+\alpha)}{1+2\alpha} - \frac{1}{\alpha} \log_e (1+2\alpha) \right] \right\}.$$

Recently Klein and Nishina have modified this formula by considering Dirac's relativistic interpretation of the quantum mechanics. They obtain

$$_e\sigma = \frac{\pi e^4}{m^2 c^4} 2 \left\{ \frac{1+\alpha}{\alpha^2} \left[\frac{2(1+\alpha)}{1+2\alpha} - \frac{1}{\alpha} \log_e (1+2\alpha) \right] \right.$$
$$\left. + \frac{1}{2\alpha} \log_e (1+2\alpha) - \frac{1+3\alpha}{(1+2\alpha)^2} \right\}.$$

Each of these formulae converges for small values of α (long wave-lengths) to the classical value $_e\sigma_0 = \frac{\pi e^4}{m^2 c^4} \frac{8}{3}$, but the Klein-Nishina formula differs considerably from each of the other two in the γ ray region.

* Compton, *Phys. Rev.* **21**, 483, 1923.
† Dirac, *Proc. Roy. Soc.* A, **111**, 405, 1927.
‡ Klein and Nishina, *Zeit. f. Phys.* **52**, 853, 1928.

Since the Klein-Nishina formula is to be preferred on theoretical grounds to the two others, and as we shall see later agrees on the whole well with experiment, we shall simplify the discussion by retaining it alone.

The amount of energy I scattered per unit solid angle per electron in a direction θ is given in terms of the incident energy I_0 by Klein-Nishina as

$$I = I_0 \frac{e^4}{2m^2c^4} \frac{1 + \cos^2 \theta}{\{1 + \alpha(1 - \cos \theta)\}^3} \left\{1 + \frac{\alpha^2(1 - \cos \theta)^2}{(1 + \cos^2 \theta)(1 + \alpha[1 - \cos \theta])}\right\}.$$

Whence by integration it is possible to obtain the value of the true scattering coefficient per electron, that is the coefficient describing how much of the energy of the incident beam is deflected as radiation:

$$_e\sigma_s = \frac{\pi e^4}{m^2c^4} \left[\frac{1}{\alpha^3} \log_e(1 + 2\alpha) + \frac{2(1 + \alpha)(2\alpha^2 - 2\alpha - 1)}{\alpha^2(1 + 2\alpha)^2} + \frac{8\alpha^2}{3(1 + 2\alpha)^3}\right].$$

The scattering absorption coefficient $_e\sigma_a$, describing how much of the energy of the incident beam is converted into electronic energy, is obtained by difference, since $_e\sigma = {_e\sigma_s} + {_e\sigma_a}$.

Fig. 125 shows the variation of these quantities with wave-length according to the Compton and Klein-Nishina formulae. The Dirac formula does not give results greatly different from that of Compton.

In addition to this method of conversion of the γ ray energy there is also the photo-electric conversion in which the whole energy of the quantum is given to the atom. The energy relations during this process have been considered in detail in chapter XII. The amount of the absorption is described by a coefficient which from a physical standpoint should be given as the absorption per atom $_A\tau$. The photo-electric absorption increases rapidly with the strength of binding of the absorbing electron and is not the same for the different electrons in the atom. The K electrons are the most efficient, absorbing four to five times as much as the L electrons. In the X ray region the formula $_A\tau = CZ^4\lambda^n$ for the atomic absorption photo-electric coefficient is found to fit the observations with reasonable accuracy, where n is about 3 for medium X rays, but there is evidence that the change of τ with wave-length becomes less rapid for the short wave-length γ rays. No satisfactory formula for $_A\tau$ has yet been proposed on theoretical grounds.

We see therefore that the total atomic absorption coefficient $_A\mu$ may be written $$_A\mu = {_A\tau} + [{_e\sigma_a} + {_e\sigma_s}]Z,$$

where, to recapitulate, the first two terms on the right-hand side represent conversion of radiation energy into corpuscular form by the photo-electric effect and the Compton effect respectively, and the last term represents the deflection of the radiation, this deflected radiation consisting almost entirely of modified radiation of progressively lower frequency as the angle of deflection increases.

As explained in § 102 a, the atomic coefficients may be deduced from the experimentally measured coefficients by the relation $_A\mu = \dfrac{\mu}{\rho}\dfrac{A}{N}$, A being the atomic weight, N Avogadro's number, and ρ the density of the material.

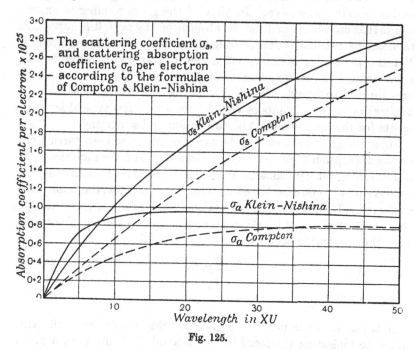

Fig. 125.

The general method which has been followed in attempting to find out experimentally how these absorption coefficients depend on the wave-length has been to compare the absorption coefficients of a given radiation in two elements of widely different atomic numbers, such as aluminium and lead. The photo-electric absorption in aluminium is negligibly small, so that if we have determined the two atomic absorption coefficients $_A\mu_{Al}$ and $_A\mu_{Pb}$ we may write

$$_A\mu_{Al} = (_e\sigma_a + {}_e\sigma_s)\,Z_{Al},$$
$$_A\mu_{Pb} = {}_A\tau_{Pb} + (_e\sigma_a + {}_e\sigma_s)\,Z_{Pb}.$$

The measurement with aluminium gives directly the scattering absorption per electron, while the atomic photo-electric absorption coefficient in lead is given by

$$\left[{}_A\mu_{Pb} - {}_A\mu_{Al}\left(\frac{Z_{Pb}}{Z_{Al}}\right)\right].$$

Further important subjects of investigation are the amount and distribution of the scattered radiation and of the recoil electrons.

§ 105. **The experimental methods for measuring the absorption and scattering of γ rays.** In view of the great number of experiments that have been made on the absorption of γ rays, it is impossible to describe in detail all the different experimental arrangements. Yet the possibility of interpreting the measurements depends to a great degree on the appreciation of certain disturbing factors which are difficult to avoid. We shall describe these and illustrate the application of these points of view by treating a few typical arrangements for the measurements of absorption and scattering.

Since the great majority of the experiments have been carried out on the absorption of the γ rays of radium C, where the radiation even after filtering through considerable thicknesses of lead is still complex, we have first to consider the exact meaning of an observed absorption coefficient. If E_1, E_2, E_3, ... are the energies and μ_1, μ_2, μ_3, ... the absorption coefficients of the different components of the complex beam, then the absorption coefficient of the complex beam will be

$$\mu = \frac{\sum\limits_1^n \mu_r E_r}{\sum\limits_1^n E_r}.$$

It is not however possible to measure this absorption coefficient, since the ionisation produced in the measuring chamber for a given amount of incident energy depends on the wave-length. If we write $K_1, K_2, K_3, \ldots$ for what are called the ionisation functions of the different beams, expressing the relative ionisation produced in a specified ionisation chamber by equal amounts of energy, then the absorption coefficient that will be actually measured will be related to the true absorption coefficients by the relation

$$\bar{\mu} = \frac{\sum\limits_1^n K_r \mu_r E_r}{\sum\limits_1^n K_r E_r}.$$

The ionisation function depends not only on the wave-length, but on the material of the electroscope and the nature of the gas. We have already seen how Rutherford and Richardson were able to increase the relative effect of the soft γ rays in the complex beam from radium (B + C) by changing the gas in the electroscope from air to methyl iodide, while Chalmers* found that the γ rays of radium B gave about 13 per cent. of the total ionisation from a radium (B + C) source measured in an aluminium electroscope but 26 per cent. in a lead electroscope. The importance of the ionisation functions is thus clear, but it is a matter of some difficulty to form a quantitative estimate of their magnitude. L. H. Gray† has shown that the ionisation produced by γ rays in a small air cavity in a mass of metal is proportional to the energy lost by the γ rays in the same volume of metal. Thus in a small ionisation chamber having a volume of 1 to 2 c.c. K should be proportional to $(\tau + \sigma_a)$ for the wave-length considered. The conditions in a large ionisation vessel are of course somewhat different as one-third to one-half of the ionisation may be due to β particles generated in the gas. In the absence of more definite information it seems reasonable to write for an aluminium electroscope $K \propto {}_e\sigma_a$ and for a lead electroscope $K \propto {}_e\sigma_a + \tfrac{1}{2}{}_e\tau_a$, thus allowing for a photo-electric effect from the walls but not from the gas.

When it is desired to compare the energy content of two beams of different hardness (as in scattering experiments), J. A. Gray‡ has attempted to find an effective value of K for each beam experimentally. He used a wooden electroscope covered internally with a layer of graphite a few mm. thick and measured the absorption coefficient of the beam in graphite, placing the absorber in contact with the electroscope. The absorption coefficient found in this way was taken as the effective K of the beam. This absorption coefficient will be slightly greater than the effective σ_a of the beam.

The next problem that must be considered in interpreting the results of any absorption experiment is the extent to which scattered radiation may reach the ionisation chamber. Suppose a source S is placed at a certain distance from an ionisation vessel I, and measurements are made of the weakening of the beam caused by an absorber A, first when placed close to S, and then when placed close to I.

* Chalmers, *Phil. Mag.* **6**, 475, 1928.

† L. H. Gray, *Proc. Roy. Soc.* A, **122**, 647, 1929.

‡ J. A. Gray, *Trans. Roy. Soc. Canada*, **21** III, 163, 1927.

In the first case the greater part of the radiation scattered in A will miss the ionisation chamber I, and the decrease in ionisation will be a measure of $\tau + \sigma_a + \sigma_s$; in the second case most of the radiation scattered forward will still enter I, and the coefficient which is determined will more nearly approach $\tau + \sigma_a$. If this were all that occurred it would be possible to evaluate in each case the exact fraction of the scattered radiation entering the chamber and so finally to deduce separate values for $\tau + \sigma_a$ and for σ_s. Owing however to the change of wave-length on scattering, any admixture of scattered radiation introduces great complications. The constitution of the beam is not representative of the γ ray emission from the radioactive body but is influenced by the geometrical arrangement. Oba* found that under suitable conditions the absorption coefficient in aluminium of the γ rays of radium C increased with the thickness of the absorber. An experiment of Bastings† shows this directly. A source of radiothorium was placed 10·9 cm. from a simple electroscope, and the absorption coefficient of the rays determined in a lead absorber of thickness 1·625 cm. placed in contact with the ionisation chamber. Various filters of lead were placed directly in front of the source, and the following absorption coefficients were found:

Filter thickness	Absorption coefficient
2·4 cm.	0·417 cm.$^{-1}$
5·8 cm.	0·419 cm.$^{-1}$
9·2 cm.	0·425 cm.$^{-1}$

Under these conditions of filtering most of the ionisation is due to one high-frequency γ ray of thorium C″ of wave-length 4·7 X U. Any longer wave-length components if present would be relatively more reduced in intensity by the thicker filters, and a slight decrease of the absorption coefficient with the thicker filters would be anticipated. The increase that was observed can only be interpreted as showing the progressive building up of longer wave-length radiation in the beam by Compton scattering in the filter of the one high-frequency γ ray.

It will be clear from what has been said that it is not easy to obtain an absorption coefficient which can be referred definitely to certain monochromatic γ rays, and that great care must be taken about two points, the construction of the ionisation chamber to give determinable

* Oba, *Phil. Mag.* 27, 601, 1914.   † Bastings, *Phil. Mag.* 5, 785, 1928.

ionisation functions, and the arrangement of the filters, absorbers, stops, etc., so as to avoid contamination of the beam by degraded radiation.

§ 106. **Methods of investigating the absorption.** The arrangement used by Kohlrausch* is admirable in both the respects referred to at the end of the previous section, but involves the use of large sources. It is shown in Fig. 126. The source was placed in the centre of an iron sphere K of 15 cm. radius filled with mercury and the radiation canalised by a 12 cm. long tube of lead as indicated. MM represents an electro-magnet to deviate any β radiation liberated from the further defining slits Pb, Pb. The filter and absorber A were in most cases sufficiently far from the ionisation chamber, and the beam was so well

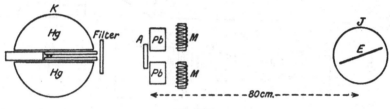

Fig. 126.

canalised that very little scattered radiation could have been measured. Under these conditions the observed absorption coefficient will refer to the primary radiation and will be a measure of the photo-electric absorption plus the total scattering absorption $\sigma_a + \sigma_s$. The ionisation chamber J was cylindrical and made of thin aluminium, the central electrode E being a plate placed at grazing angle to the beam. With this arrangement it is possible to make an approximate estimation of the ionisation function.

The apparatus used by Ahmad† was designed primarily to give a high accuracy with small sources. While Kohlrausch was able to measure absorption coefficients with an accuracy of about 5 per cent. using a source of about 100 millicuries, Ahmad claimed half per cent. with sources of about 5 millicuries. This accuracy and sensitivity was obtained partly by an ingenious balance arrangement, but partly by placing the source closer to the ionisation chamber, and using a wider beam, with consequent loss of definiteness in the experiment.

* Kohlrausch, *Wien. Ber.* **126**, 441, 683, 887, 1917.
† Ahmad, *Proc. Roy. Soc.* A, **105**, 507, 1924.

Fig. 127 illustrates the disposition of the apparatus which consisted of two identical sets of ionisation chambers, defining stops and sources, only one of which is shown in the diagram. The potentials on the ionisation chambers were of opposite sign, and the electrometer used as a null instrument to detect the equality of the ionisation in the two chambers. The radiation was filtered through 1 cm. of lead placed at P and the process of measuring the absorption coefficient of a given material was as follows. The currents in the two ionisation chambers with no absorbers present were first adjusted to exact equality by altering one of the slits D, the absorber to be investigated was introduced

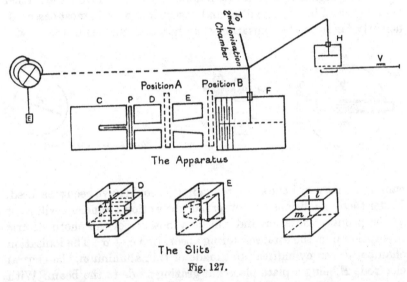

The Apparatus

The Slits

Fig. 127.

at A, and aluminium of such a thickness that the currents were again equal was placed at the corresponding position in the other beam. The material was then removed and aluminium inserted in its place of thickness again to balance the electrometer. Under these conditions it will be seen that the relative absorption coefficients of the material and of aluminium could be obtained to a very high accuracy. The absolute absorption of aluminium in the position A was carefully determined by the usual methods. In this experiment part of the scattered radiation reached the ionisation chamber, and to estimate the amount two positions A and B for the absorbers were used, at which the ionisation chamber subtended half angles of 20° and 80° respectively. The scattering correction was calculated from the difference in the absorption coefficients in the two positions. It is precisely this calculation which is difficult to carry out accurately. The scattered rays pass through the ionisation chamber obliquely, but on the other hand have had their wave-length and hence their ionisation functions altered by the scattering process. The ionisation chambers contained systems of iron plates arranged as indicated in the figure, with the result that the ionisation function varied with the obliquity of the ray. This type of ionisation chamber is not to be recommended, since in view of the considerable absorption in the iron plates it is very difficult to form an estimate of the ionisation functions. This objection is the more important in this particular experiment since the γ rays, being those of radium (B + C) and only filtered through 1 cm. of lead, consisted of components of widely different wave-length.

J. A. Gray* has used a simple method of reducing the harmful effects of a filter in producing degraded radiation while keeping the source close to the ionisation chamber. It consists simply in sub-dividing the filter into a number of small plates, separated from each other by as large a distance as is allowed by the dimensions of the apparatus. In this case a large part of the radiation scattered from one filter will hit the walls of the defining canal and can only enter the beam by two further scattering processes. With this arrangement Gray and Cave observed the mass absorption coefficient in lead (μ/ρ) of the γ rays of radium (B + C) to decrease from 0·0491 to 0·0472 when the filter thickness was increased from 3·28 cm. of lead to 5·44 cm. This decrease is about what would be expected from the intensity distribution in the spectrum, and if compared with Russell's results of the constancy of μ over a range of filter thicknesses of 2 to 22 cm. of mercury, brings out forcibly the advantages of this method of filtering.

Tarrant† has used an arrangement in which the absorber is placed midway between the source and the ionisation chamber, but the filter is placed in contact with the ionisation chamber. Since most of the radiation scattered by the filter still enters the ionisation chamber, it will be clear that the filtering is mainly due to the photo-electric effect. Tarrant has applied this method to an investigation of the absorption of the γ rays of thorium C″ where it is particularly suitable owing to the energy being concentrated mainly in two regions of

* J. A. Gray and Cave, *Trans. Roy. Soc. Canada*, **21**, 163, 1927.
† Tarrant, *Proc. Roy. Soc.* A, **128**, 345, 1930.

long and short wave-length. A further advantage of this arrangement is that to a first approximation the measurements are independent of the degraded radiation produced in the filter.

§ 107. **Methods of investigating the scattering.** The simplest method of investigating the scattering has already been referred to, that of comparing the absorption of a plate of material when it is placed at some distance from, and then close to the ionisation chamber. The integrated scattering through a certain range of angles is obtained by difference, but the method is inevitably inaccurate for the reasons already mentioned.

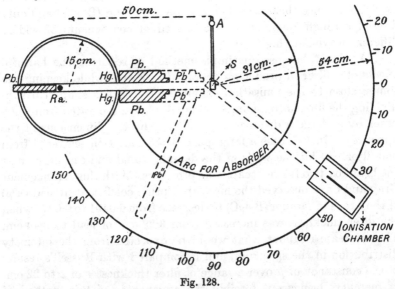

Fig. 128.

The direct measurement of the total scattered radiation by an ionisation vessel completely surrounding the scatterer is greatly to be preferred, but it meets with serious experimental difficulties.

Of these the greatest is in choosing the dimensions of the scatterer. If this is too small, the effect to be measured will be small in comparison with the inevitable effect of the primary rays, however the ionisation chamber is shielded by lead screens. If it is too large there will be a considerable correction for absorption of the scattered radiation in the scatterer. This latter error is especially marked for scattering through large angles and in fact is so serious that no reliable absolute measurement of σ_s has yet been made in this way. The relative values of the integrated scattering between 0° and 90° for different elements might however be determined.

More detailed information is obtained by measuring the radiation scattered at different angles. The apparatus used by Kohlrausch* is shown in Fig. 128, and it will be noticed that the penetrating power of the radiation scattered at each angle could be measured under comparable conditions by having an absorber rotatable around an inner circle.

An original and powerful method of investigation has been developed by Skobelzyn†. Using an expansion apparatus he determined the distribution with angle of the recoil electrons liberated by the scattering process by the γ rays of radium B and radium C. This is equivalent to measuring the angular distributions of the scattered radiations, since each recoil electron ejected at angle θ corresponds to a quantum of radiation scattered at an angle ϕ where θ and ϕ are connected by the simple energy and momentum equations.

§ 108. **Results of the measurement of absorption and scattering in light elements.** There is good reason to believe that the absorption in light elements is almost entirely due to scattering, the photo-electric absorption being negligible. By confining the attention first to this portion of the experimental results we are thus able to investigate the scattering laws, and then subsequently we may attribute the difference between the absorption of heavy elements and of light elements to the photo-electric effect.

All theories of scattering assume that for sufficiently short wavelength all electrons scatter equally, whatever their state of binding, and it is important to consider the experimental evidence on this point. If it is true, the absorption per electron of a beam however constituted should be the same for all light elements. The results of Kohlrausch's‡ and Ahmad's§ measurements using the γ rays of radium (B + C) with different degrees of filtering are shown in the next table:

| | Filter | Absorption per electron × 10²⁵ | | | | | |
		H	C	O	Mg	Al	S
Kohlrausch	> 3·5 cm. Pb	—	1·58	—	1·59	1·60	1·51
Ahmad	1·0 cm. Pb	1·900	1·916	1·907	—	1·908	1·942

* Kohlrausch, *Wien. Ber.* **128**, 853, 1919.   † Skobelzyn, *Nature*, **123**, 411, 1929.
‡ Kohlrausch, *Wien. Ber.* **126**, 441, 683, 887, 1917.
§ Ahmad, *Proc. Roy. Soc.* A, **105**, 507, 1924.

It will be seen that the measurements support the constancy of the scattering per electron, but that they do not agree sufficiently well among themselves to render this conclusion certain. The relative values for sulphur, for example, are remarkably different in the two series. However, to a first approximation the absorption per electron is constant and we shall proceed to give an account of the analysis of the experimental results with this as basis.

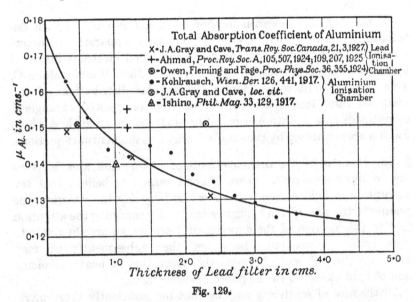

Fig. 129.

The values that have been obtained by different observers for the total absorption coefficient in aluminium, that is $\sigma_s + \sigma_a$, of the rays of radium (B + C) under different conditions of filtering, are shown in Fig. 129, the full curve being drawn through Kohlrausch's values. The only serious divergence is shown by the value of Owen, Fleming and Fage which is largely in excess of the measurements of Kohlrausch, and of Gray and Cave for the same degree of filtering. The other divergencies in the values of different observers probably have their origin in the different experimental arrangements used, in particular in the construction of the ionisation chambers, and in the relative positions chosen for the filter and absorber. Viewing the results as a whole one would anticipate that the curve drawn in Fig. 129 represents correctly the relative values of the absorption coefficient for different conditions of filtering, and that the absolute values as shown in the curve are more likely to be too great, possibly to the extent of 5 per cent., than too small.

A comparison of the theoretical values to be expected on the Klein-Nishina formula with these measurements can only be effected by assuming some particular spectral distribution. The experiments which will yield a distribution best suited for comparison are those of Skobelzyn*. An account of this work will be found on p. 504, and it will suffice here to note that, using the Wilson cloud method, he measured the number and velocity of all electrons ejected from air whose initial directions made an angle of less than 20° with the direction of the γ-ray beam. Using these data and employing the scattering laws which are under discussion, he was able to deduce the number of quanta in any frequency range.

If p_r is the relative number of quanta in a certain range of frequencies, and $h\nu_r$ is the mean energy of a quantum, then $E_r = p_r . h\nu_r$ is the total energy associated with the interval. The quantities p_r and ν_r were determined directly in Skobelzyn's experiments. The total absorption coefficient of the beam is then

$$\sigma = \frac{\overset{n}{\underset{1}{\Sigma}} E_r \sigma_r}{\overset{n}{\underset{1}{\Sigma}} E_r},$$

σ_r being the value calculated for the particular wave-length from the Klein-Nishina formula.

To effect a closer comparison with experiment we must take the ionisation function of the ionisation chamber into account, which we shall do by weighting each value by σ_a. Hence finally we obtain for the effective absorption coefficient

$$\sigma_{eff} = \frac{\overset{n}{\underset{1}{\Sigma}} (\sigma_a)_r E_r \sigma_r}{\overset{n}{\underset{1}{\Sigma}} (\sigma_a)_r E_r}.$$

Calculation yields the value 0·155 cm.$^{-1}$, while for this degree of filtering (3·5 mm. of lead and 2 mm. of glass) experiment gives a value rather less than 0·163 cm.$^{-1}$. This agreement can be considered provisionally as satisfactory.

For the rays filtered through considerable thicknesses of lead (about 4 cm.) it is clear that most of the energy of the beam will be

---

* Skobelzyn, *Zeit. f. Phys.* **43**, 354, 1927; **58**, 595, 1929.

of short wave-length. Reference to the table of intensities given later (see p. 509) and also to Skobelzyn's results shows that a fair estimate of the absorbability of the beam will be obtained by taking the coefficient for the strong line of 6·95 X U. The Klein-Nishina formula for this wave-length gives 0·123 cm.$^{-1}$ as compared with the experimental value 0·125 cm.$^{-1}$ found by Kohlrausch.

The γ rays of thorium C″ should provide a better opportunity for testing the absorption formula, since so much of the energy resides in the line 4·7 X U and there are no other strong lines in the vicinity. Unfortunately the experimental data are scanty. Rutherford and Richardson[*] found that all the soft rays were absorbed in 2 mm. of aluminium and that then the absorption coefficient remained constant up to 9 cm. of aluminium. They found $\sigma = 0·096$ cm.$^{-1}$, whereas the Klein-Nishina formula gives 0·097. This numerical agreement may be deceptive since the technique of γ ray absorption measurements was not fully developed at the time these measurements were made.

§ 108 a. **The absorption of energy in the scattering process.** As is seen from Fig. 125 the theoretical value of σ_a remains sensibly constant over the whole of the radium B spectrum and part of the radium C, and varies only slowly over the remainder of the radium C region. The calculated value of σ_a for radium (B + C) rays will therefore not be very sensitive to spectral distribution. Using intensities deduced from Skobelzyn's results, one finds $\sigma_a = 0·068$ cm.$^{-1}$. Attempts to measure σ_a as the apparent absorption coefficient of a light element when all the scattered radiation enters the ionisation chamber have been made by Ishino[†] and by Owen, Fleming and Fage[‡]. The method is not easy to carry out as there are many corrections to be applied. The chief of these is due to the reabsorption of the scattered radiation which is abnormal owing to its increase in wave-length. A more promising method of attack would appear to be on the lines of an experiment by L. H. Gray[§]. A source is situated at the centre of a mass of absorbing material, and the loss of energy per c.c. is measured at different points along a radius by finding the ionisation in a small air cavity in a large mass of aluminium. Since

$$\sigma_a = \frac{1}{E}\frac{dE}{dx},$$ it may be deduced at once from the graph of the

[*] Rutherford and Richardson, *Phil. Mag.* **26**, 939, 1913.
[†] Ishino, *Phil. Mag.* **33**, 129, 1917.
[‡] Owen, Fleming and Fage, *Proc. Phys. Soc. London*, **36**, 355, 1924.
[§] L. H. Gray, unpublished.

experimentally measured quantity dE/dx against x. Gray finds values decreasing from 0·068 cm.$^{-1}$ to 0·055 cm. within the first 5 cm. of absorber. The reason for this decrease is not clear and may be connected with the fact that in these preliminary experiments brick was used to replace part of the aluminium.

§ 108 b. **The angular distribution of the scattered γ rays and the recoil electrons.** A careful series of experiments has been carried out by Kohlrausch with the apparatus described on p. 472. His results for scattering from carbon are shown in Fig. 130, where the radius vector shows the intensity of radiation scattered at that

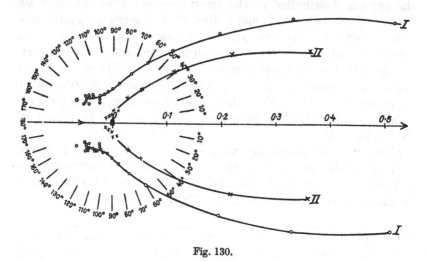

Fig. 130.

angle. It will be seen that the scattering is predominant in the forward direction, and further, by comparing curve I with no absorber with curve II with 3 mm. of lead in the path of the scattered rays, it will be noticed that the radiation is softer the greater the angle of scattering. The latter result is a consequence of the Compton change of wave-length. While all modern theories of scattering have been arranged to give the excess forward scattering, they differ considerably in their quantitative predictions. Experiments of this type should therefore provide a valuable test of the accuracy of different theories. At present there are no experimental results sufficiently dependable to justify a close analysis and it is unsafe to say more than that Kohlrausch's measurements probably support the Klein-

Nishina formula. J. A. Gray and Cave* took great care to estimate the ionisation function for the beam scattered at each angle. However, they only filtered their rays through 2·2 cm. of lead, and in such a way that the full scattered radiation would be mixed with the beam. It is difficult to form an estimate of the constitution of the beam, and if a certain latitude is allowed in the choice of the mean wavelength either the Compton, Dirac or the Klein-Nishina formula may be made to fit.

The angular distribution of the recoil electrons is merely another aspect of the above problem. We have already mentioned that Skobelzyn investigated by means of the expansion chamber method the angular distribution of the recoil electrons liberated from air by the γ rays of radium B and radium C. Two series of experiments were carried out in which the γ rays were filtered through 3·5 mm. and 1·0 cm. of lead respectively. If we consider the former series, we may use directly the intensity distribution found by Skobelzyn in his other experiments referred to on p. 475. The observed distribution of recoil electrons is the mean of the distributions given by the separate components, weighted in proportion to the number of quanta in that component.

Skobelzyn† measured the number of electrons emitted within various angular ranges from the γ ray beam. The distribution of 903 tracks is shown in the next table and is compared with that to be expected for 900 tracks from the Klein-Nishina formula.

Angular range	0°–10°	10°–20°	20°–40°	40°–60°	60°–80°	80°–90°
No. of tracks:						
Observed	117	149	242	215	161	19
Calculated	92	152	266	224	146	20

In view of the difficulty and statistical nature of the experiment we may consider this as strong support for the accuracy of the theoretical formulae.

If we view as a whole the various comparisons that have been made between theory and experiment, we may conclude that there is nothing in the accumulated experimental data which is definitely in contradiction to the Klein-Nishina formula. The order of disagree-

---

* J. A. Gray and Cave, *Trans. Roy. Soc. Canada*, 21, 163, 1927.
† Skobelzyn, *Nature*, March 16, 411, 1929.

ment of theory with experiment is 5 per cent. and is not systematic. The older theories of Compton and Dirac on the other hand gave systematic disagreement of the order of 20 to 40 per cent.

§ 109. The results of the measurement of absorption in heavy elements. If it is admitted that the scattering per electron is constant, and independent of the state of binding, we may obtain the absorption due to the photo-electric effect by subtracting from the absorption per electron $_e\mu'$ of a heavy element such as lead the mean value of the absorption per electron $_e\mu$ found for the lighter elements. The atomic coefficient $_A\tau$ is thus $Z\,(_e\mu' - _e\mu)$. It has been usual to use the X ray formula $_A\tau = CZ^4\lambda^3$ for expressing results which while not exact is convenient, and to obtain exact agreement in any one range by slightly altering the exponents of Z and λ. In the γ ray region $_A\tau$ appears to follow approximately the same law with respect to Z. We have referred to Ahmad's experiments in which he made an accurate measurement of the absorption coefficients of a series of elements from uranium to aluminium for the γ rays of radium (B + C) filtered through 1 cm. of lead. Provided $_A\tau$ is given by an expression of the form $_A\tau = f\,(Z)\,\phi\,(\lambda)$, the dependence on Z should be given by these measurements independently of the constitution of the beam. Ahmad obtained $_A\tau$ as above by subtracting the absorption in aluminium and found that approximately $_A\tau$ varied as Z^4. Some results of Kohlrausch's carried out under quite different conditions of filtering lead to the same conclusion, so there is good reason to believe in the substantial accuracy of a formula of the type $_A\tau = CZ^4\phi\,(\lambda)$. Further investigation on this point is desirable since there are certain discrepancies in the values of τ for the heaviest elements which need explaining.

The dependence on λ is not so simple and can scarcely be represented by λ^n. While in the X ray region n is approximately 3, we shall see that a smaller value for n must be taken if we wish to keep the same form for γ rays. J. A. Gray* has pointed out that the γ rays of radium (B + C) when filtered through 1·6 cm. of lead give recoil electrons sufficiently fast to penetrate nearly 3 mm. of aluminium, and requiring 0·5 mm. to reduce them to half-value. From this he deduced that the effective value of the wave-length of the beam was unlikely to be greater than 8 X U. The photo-electric absorption coefficient of the beam is deduced as usual by subtraction

* J. A. Gray, *Nature*, **115**, 13, 86, 1925.

of the aluminium absorption. This coefficient for 8 X U is then compared with the measured value for X rays of 100 X U, and it is found that in this range the average value of n cannot be much greater than 2·2*.

Fig. 131† shows the measurements of different observers of the absorbability in lead of the γ rays of radium (B + C) filtered through different thicknesses of lead. For any given filter thicknesses we may determine the effective wave-length by the Klein-Nishina formula, using the absorption in aluminium shown already in Fig. 129. The

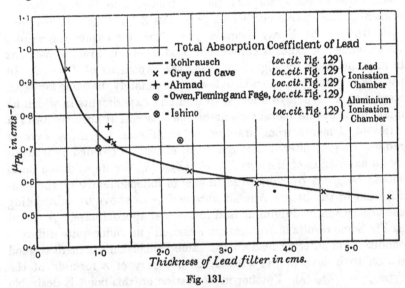

Fig. 131.

photo-electric absorption for this degree of filtering is then obtained by subtraction of the aluminium from the lead absorption. In this way it is found that $\tau \propto \lambda^{2\cdot2}$. The exponent of λ will have been over-estimated since the mean effective wave-length for photo-electric absorption is always greater than the effective wave-length for scattering.

While much weight cannot be laid on the values of the exponent n determined in this way, it seems clear that there is a departure from the X ray laws. Further evidence on this point is considered in chapter XVI, § 116.

---

* J. A. Gray and Cave, *Trans. Roy. Soc. Canada*, **21**, 163, 1927.

† The full curve attributed to Kohlrausch was obtained by graphical methods from the corresponding data of ionisation current and filter thickness given in the paper.

§ 110. The excitation of γ rays by β particles. The first experiments which gave definite evidence of the excitation of γ radiation by the impact of β particles on ordinary matter were those of J. A. Gray *. Using the β particles of radium E, which body is particularly suitable since it emits so few primary γ rays, he was able to detect γ radiation from several different materials. The amount of the γ radiation became greater as the atomic weight of the radiator increased. In a subsequent paper† he showed that the γ radiation in the forward direction was generally more intense and more penetrating than that in the backward direction, this difference being greater for elements of small atomic weight. Chadwick‡, using the β rays of radium C, confirmed and extended these results. Though more intense sources were available, it proved difficult to measure the secondary γ radiation in the presence of the intense primary γ radiation from radium C. However, by concentrating the β particles on to the radiator by means of a convergent magnetic field, and arranging the apparatus in such a way that a large block of lead could be interposed between the source and the ionisation chamber, it was possible to determine the relative amounts of radiation excited in different materials, and their approximate absorption coefficients. His results are shown in the next table.

Material	Relative amount of excited rays	$(\mu/\rho)_{\text{Fe}}$	
		0–1·3 cm.	1·3–2·6 cm.
Uranium	100	0·15	0·10
Lead	92	0·17	0·10
Tin	82	0·18	0·11
Zinc	79	0·19	0·12
Aluminium	75	0·23	0·13

It is clear from these figures that no considerable quantity of the characteristic X rays of the radiators can have been excited, or a far greater change in the absorption coefficient of the radiation would have been found between uranium and aluminium. It is probable that this radiation was emitted directly by the impinging β particles in the same manner as at an anticathode in an X ray tube. We should expect in this case that the radiation would be continuous in character and show a hardening with increasing thickness, such as was in fact found.

---

* J. A. Gray, *Proc. Roy. Soc.* A, **85**, 131, 1911.
† J. A. Gray, *Proc. Roy. Soc.* A, **86**, 513, 1912.
‡ Chadwick, *Phil. Mag.* **24**, 594, 1912.

§ 111. Excitation of X rays and γ rays by α particles. It is known that when α particles pass through matter they ionise the atoms, and ·while most of the electrons that are removed come from the outer regions of the atoms, in certain cases K and L electrons will be ejected. It is to be expected therefore that a small amount of the characteristic X radiations of the material will be excited.

The first experiments on this subject were made by Chadwick[*], who showed that a small but detectable amount of radiation was excited by the α rays of radium C when they impinged on matter. A detailed investigation of the radiation was not possible owing to the intense primary γ radiation from radium C. Chadwick and Russell[†] found that from a mixture of ionium and thorium oxides a weak radiation was emitted which appeared to be composed of the K, L and M radiations of an element of high atomic number. It was impossible to settle whether the radiations were excited in the ionium atoms or in the surrounding atoms of thorium, but the former conclusion is the more probable.

Another series of experiments led to more definite results. The same authors found that, when polonium was deposited on aluminium, an approximately homogeneous radiation with μ/ρ in aluminium of about 215 cm.$^{-1}$ was emitted (see p. 456, footnote j). When the active material was deposited on copper, in addition to this radiation, they found one with μ/ρ about 1300. They suggested that this latter was the L radiation of copper and was excited by the impact of the α particles.

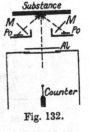

Fig. 132.

A very careful and detailed series of experiments have recently been carried out by Bothe and Fränz[‡].

The arrangement of the apparatus can be seen from Fig. 132. The α rays from two polonium sources Po fell on the substance S to be investigated, and any radiation emitted could be detected by a Geiger counter. The character of the radiation could be estimated from its absorbability, determined by inserting screens of aluminium before the counter, while the effect of α particles of different speed was determined by placing thin sheets of mica at M. The counter was filled with a mixture of 80 per cent. argon and 20 per cent. nitrogen,

---

* Chadwick, *Phil. Mag.* **25**, 193, 1913.
† Chadwick and Russell, *Proc. Roy. Soc.* A, **88**, 217, 1913; *Phil. Mag.* **27**, 112, 1914.
‡ Bothe and Fränz, *Zeit. f. Phys.* **52**, 466, 1928.

since with this filling the effect of any H particles liberated from S is small compared with the radiation effect. Fourteen elements were investigated, ranging from atomic numbers 12 to 83. From magnesium (12) to zinc (30) the absorption coefficients showed clearly that it was the K X radiation which gave the main effect. With zinc (30) the K radiation was already very weak and with selenium (34) a new soft radiation appeared which proved to be the L X rays. The amount of this also steadily decreased with increasing atomic number, until with bismuth it was the M radiation which gave the chief effect. These results are easily understandable in a general way, for the necessary preliminary to the emission of K or L radiation is the removal of a K or L electron, and this becomes increasingly less probable as the binding energy increases. The particular atomic number for which the L radiation is first found depends on the arrangement of the apparatus, since for atomic numbers below 34 it was so soft that it was absorbed before reaching the counter. No continuous radiation was found of the same order of hardness or of intensity as the characteristic radiation, and in this respect there is a distinct difference from the behaviour of β particles. The very much harder impact radiation, which as we shall see later was found by Slater, would scarcely have been detected in these experiments.

The dependence of the amount of excited radiation on the range of the α particles was investigated in detail for aluminium, and from these results, virtually by differentiation, was obtained a curve showing $z(R)$ against R, where $z(R)\,dR$ is the amount of radiation excited by an α particle between ranges R and $R + dR$. This curve is analogous to the ordinary ionisation-range curve for α particles but it shows only the ionisation due to the removal of the K electrons. With aluminium the curve appeared to be approaching a maximum value, although this was not quite reached with the available range 3·85 cm. It may be noted that K radiation was excited even with α particles of 1 cm. range, whereas according to the ordinary classical theory of collisions no α particles of range less than 1·6 cm. could give the K ionisation energy to the electron. With calcium, iron and tantalum the theoretical minimum range which should have been necessary for excitation was actually greater than the full range of the α particle. This shows a distinct departure from classical laws and is probably only explainable on modern quantum views. Born's* theory of collisions permits, in the case of infinite nuclear mass, the

* Born, *Zeit. f. Phys.* **38**, 803, 1926.

whole energy of the α particle to be transferred to the electron. Experiments leading to somewhat similar results have been performed on the efficiency of excitation of the K radiation by electrons of different speeds.

Slater* investigated in considerable detail the radiations excited by the impact on various materials of the α particles from radon. A glass tube lined with a foil of the material under investigation was, by a special arrangement, filled very rapidly with radon. Under these

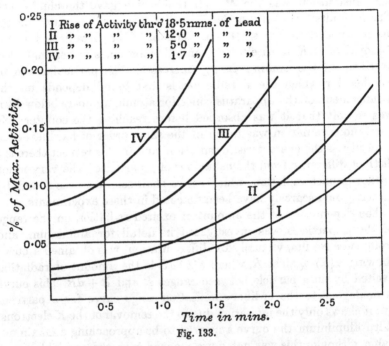

Fig. 133.

conditions the primary γ radiation due to radium B and radium C is very small for the first few minutes (see § 5). Great care was taken that no active deposit was introduced into the tube, which was then placed close to the base of a large electroscope. Measurements were commenced usually 30 seconds after the introduction of the radon. Typical curves of the rise of the γ ray activity are shown in Fig. 133. The increase from the initial value is due to the growth of the products radium B and radium C, but the extrapolation to zero time shows clearly that a small amount of γ radiation is associated with the radon.

* Slater, *Phil. Mag.* 42, 904, 1921.

The difference between the radiations at zero time obtained with a lead lining and a paper lining in the glass tube was taken as a measure of the radiation excited by the α particles of radon on the lead. In this way it was shown that a hard γ radiation was emitted, which differed but little in quality when the radiator was changed from one of high atomic number to one of medium atomic number. For example the mass coefficients (μ/ρ) in lead were 0·16 for a radiator of lead and 0·18 for one of tin. The intensity obtained was very small, and only a small fraction of the impinging α particles could have been effective. For the same absorption conditions the intensity is about 50 per cent. greater for the lead radiation than for the tin radiation. These results are what would be expected if this radiation was emitted during the close collision of an α particle with a nucleus of one of the atoms of the radiator. It is most important to decide whether this is continuous radiation and analogous to the continuous radiation emitted by high-speed electrons hitting the anticathode of an X ray tube, or whether it is due to a disturbance of the nucleus by the impact of the α particle. As is discussed in § 57, it is possible that the nucleus suffers a marked distortion or polarisation during a close collision with an α particle, and it is therefore possible that some radiation is emitted by the nucleus during the subsequent relaxation.

In addition to this hard radiation, Slater detected the K and L characteristic radiations of the radiators, which have already been discussed in relation to the experiments of Chadwick and Russell, and Bothe and Fränz.

§ 112. Penetrating radiation.

The study of the penetrating radiation had its origin in the observations of Elster and Geitel* and of C. T. R. Wilson† that there was always a small amount of residual ionisation in an electroscope. They found that there was a definite transport of charge to the insulated system even after all possible precautions had been taken to reduce electrical leakage over the insulators. The order of magnitude of this residual ionisation corresponded to the production of about twenty pairs of ions per second per cubic centimetre. McLennan and Burton‡ and Rutherford and Cooke§ found that the ionisation could be considerably reduced by

---

* Elster and Geitel, *Phys. Zeit.* 2, 116, 560, 590, 1900.

† C. T. R. Wilson, *Proc. Camb. Phil. Soc.* 11, 32, 1900 and *Proc. Roy. Soc.* A, 68, 151; 69, 277, 1901.

‡ McLennan and Burton, *Phys. Rev.* 16, 184, 1903.

§ Rutherford and Cooke, *Phys. Rev.* 16, 183, 1903.

surrounding the electroscope with a thick layer of some material, and they concluded that part of the ionisation must be due to an external penetrating radiation. That the greater part of this effect was due to radioactive material in the ground was shown by McLennan, Wulf, Gockel, Wright and others, who found that the ionisation was reduced when the apparatus was set up over a deep lake. Bergwitz and Gockel attempted to pursue this further by making observations in balloons. If the radiation came entirely from the earth the effect should decrease with increasing height, due both to the increased distance and to the absorption of the air. Hess* was the first to obtain definite results by this method, and showed that, while the ionisation decreased slightly up to a distance of 1000 metres, above 2000 metres it began again to increase, and at 5000 metres it was already two or three times the value found at ground-level. This indicated definitely the presence of radiation travelling downwards through the earth's atmosphere, and in view of the possible cosmic origin of the radiation gave a fresh interest and importance to the investigations. These experiments were continued by Kolhörster† with improved apparatus and he was able to extend the measurements to a height of 9000 metres. He observed that the absorption coefficient of the radiation increased with altitude up to 7000 metres and then fell sharply. Quite recently he has made other balloon ascents which have confirmed the maximum of the absorption coefficient between 6000 and 7000 metres, but so far no quantitative theoretical explanation has been offered.

Millikan and Bowen‡ have reported the results obtained from two flights of small pilot balloons carrying recording electroscopes which reached heights of 11·2 and 15·5 kilometres. The average discharge rate above the 5 kilometre level was about three times the rate at the surface. Hess§ and Kolhörster‖ have pointed out that this ratio was less than that to be estimated from their results.

A more fruitful method of investigation has been to measure the ionisation at the surface of a snow-fed lake¶ at a considerable altitude, say 5000 metres, and then by sinking the electroscope in the lake to obtain the absorption curve in water. An advantage of this method

---

* Hess, *Phys. Zeit.* 12, 988, 1911; 13, 1084, 1912.
† Kolhörster, *Phys. Zeit.* 14, 1066, 1153, 1913.
‡ Millikan and Bowen, *Phys. Rev.* 27, 351, 1926.
§ Hess, *Phys. Zeit.* 27, 159, 1926.
‖ Kolhörster, *Zeit. f. Phys.* 28, 404, 1926.
¶ Under these conditions the water is free from contamination by radioactive bodies.

is that by sinking the electroscope to a sufficient depth the true natural leak of the electroscope can be estimated. It is further possible to compare the measurements with those taken at sea-level by allowing for the water equivalent of the intervening mass of air. It is assumed that only scattering absorption is relevant, and that it is only the number of electrons which determines the absorption. On this assumption the normal atmosphere has an absorbing effect equivalent to 10·33 metres of water, while that portion of the atmosphere above 5000 metres is equivalent to 5·3 metres of water.

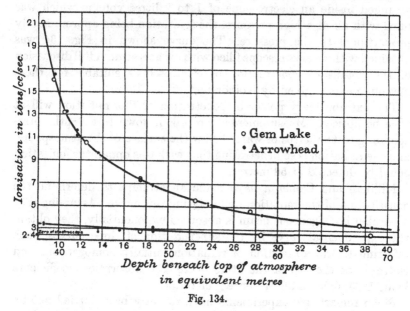

Fig. 134.

Following this general method, Millikan and Cameron* in a long series of experiments have been able to establish the results given in Fig. 134. This shows the absorption curve of the penetrating radiation in water obtained by combining the measurements obtained at two different localities, the value at 10·33 metres corresponding to that at sea-level. The ionisation is expressed in terms of the number of ions produced per c.c. per sec. in air at N.T.P. This unit is usually denoted by I. Millikan and Cameron find the ionisation at sea-level to be about 1·4I, while Kolhörster reports values nearer to 2I. Values falling on this curve were also obtained from obser-

---

* Millikan and Cameron, *Phys. Rev.* **28**, 851, 1926; **31**, 163, 921, 1928.

vations made in lakes situated both in the northern and southern hemisphere, so that it furnishes reliable experimental material from which we may attempt to determine the characteristics of the radiation. The possibility of carrying out measurements of this type has depended to a great extent on the development of suitable electroscopes. The general type used is due in principle to Wulf and consists of two sputtered quartz fibres, mounted on a suitable quartz support, whose mutual repulsion measured by their separation shows the charge on the system. In the experiments of Millikan and Cameron this was arranged inside an electroscope of 1 to 2 litres volume which was filled with air at a high pressure, the ionisation being approximately proportional to the pressure. The curve shown in Fig. 134 was obtained with an electroscope filled with air at 8 atm. A full discussion of the details of construction of electroscopes suitable for these measurements is given by Kolhörster*.

The extraordinary power of penetration of this radiation will at once be noticed. Measurements were taken down to a depth corresponding to 68 metres of water below the surface of the atmosphere, and the radiation is so penetrating that with this apparatus ionisation could be detected at 58 metres.

Millikan and Cameron have made an attempt to determine the constitution of the radiation by analysing this curve. Assuming that a parallel homogeneous beam is absorbed exponentially, they obtain the experimental curve by the superposition of three curves, each referring to the absorption of radiation initially homogeneous and isotropic at the top of the atmosphere, the absorption coefficients being 0·35, 0·08 and 0·04 metre^{-1} of water.

Some remarkable experiments have recently been carried out by Regener† in which he was able to detect ionisation at a depth of 231 metres of water below the surface of Lake Constance. He used a steel ionisation vessel of 39 litres capacity filled with carbon dioxide at 30 atmospheres pressure, and the position of the fibre of the electrometer was registered on a photographic plate at hourly intervals. His results are shown in the next table, and the great extension in our knowledge of the penetrating radiation provided by these experiments will be appreciated from the fact that even the second value at 78·6 metres is beyond the region explored by Millikan and Cameron.

At the time of writing the full results have not been worked out,

* Kolhörster, *Zeit. f. Instrumentkunde*, **44**, 333, 1924.

† Regener, *Naturwiss.* **11**, 183, 1929.

but it appears likely that a single absorption coefficient of value rather less than 0·018 metre⁻¹ will describe the absorption of the radiation at depths greater than 80 metres.

Depth below the surface in metres	Ionisation current (arbit. units)	Ionisation current less natural effect
32·4	4·33	3·55
78·6	1·65	0·87
105·2	1·31	0·53
153·5	1·00	0·22
173·6	0·93	0·15
186·3	0·89	0·106
230·8	0·83	0·051

The absorption of penetrating radiation has been studied in media other than air and water, notably by Büttner*, Steinke†, Hoffmann‡ and Myssowsky and Tuwim§. The experimental results of the latter authors are particularly clear-cut. After the first 9 cm. the absorption per electron in lead is the same as in water, but during the initial period the absorption is abnormally rapid. This abnormality of the transition region from a medium of low to one of high atomic number was referred by Hoffmann to the photoelectric absorption of the degraded radiation in the second medium, but it seems doubtful whether a quantitative explanation is possible along these lines.

The account we have given of the experiments on the penetrating radiation may have implied that this was electromagnetic radiation, a type of ultra γ radiation. We shall see that while this is perhaps the most natural assumption it is by no means rendered inevitable by the data, in fact there is one experiment which can be interpreted as direct proof of the corpuscular nature of this radiation. C. T. R. Wilson‖ first pointed out that β particles of sufficient energy could produce most of the phenomena observed in connection with the penetrating radiation, and there may well exist conditions in thunderstorms where β particles may acquire the necessary great energies of more than 10⁹ volts. Apart from the possibility how β particles of such energy could arise, it is a matter of great interest to see whether there is any evidence that they constitute the penetrating radiation.

* Büttner, *Zeit. f. Geophys.* **3**, 161, 1927.
† Steinke, *Zeit. f. Phys.* **42**, 570, 1927; **48**, 647, 1928.
‡ Hoffmann, *Phys. Zeit.* **26**, 669, 1925; *Ann. d. Phys.* **82**, 413, 1927.
§ Myssowsky and Tuwim, *Zeit. f. Phys.* **50**, 273, 1928.
‖ C. T. R. Wilson, *Proc. Camb. Phil. Soc.* **22**, 534, 1926.

Bothe and Kolhörster* have recently reported experiments which they interpret as favouring this latter view. They arranged two counters, of a special design due to Geiger, at a small distance apart inside a protecting shield of iron. A certain number of the kicks occurred simultaneously in the two counters and were certainly due to the same β particle traversing both counters. To determine whether these β particles were of secondary origin and liberated by a type of ultra γ ray radiation, or whether they themselves constituted the penetrating radiation, they placed a block of gold 4·1 cm. thick between the two counters and observed a 25 per cent. decrease in the number of simultaneous kicks.

Since it is known that the introduction of such a block of gold would lead to about the same reduction in the ionisation produced by penetrating radiation, Bothe and Kolhörster concluded provisionally that the primary penetrating radiation was in fact very high speed β particles. The number of these particles would be very small, about 1/100 per cm.² per sec., but their individual energies must be very high, probably greater than 10^9 volts.

It is extremely unlikely that the very penetrating radiation observed by Regener could be corpuscular in nature, but it is possible that it is true electromagnetic radiation excited in matter by the impact of the high-speed corpuscles which are responsible for the ionisation observed at smaller absorptions. A detailed examination of this view shows that the chief difficulty will be to account satisfactorily for the phenomena in the transition region.

At present it is not possible to form any definite idea as to the nature of these corpuscles, they might be either electrons or protons. At such high energies as 10^9, volts, electrons and protons have comparable masses and momenta and the chief remaining distinction is the sign of the charge.

While the experiment we have described is the only evidence in favour of the corpuscular nature of the radiation, it must be remembered that on the other hand there is no positive evidence for the alternative view. However, the assumption that the radiation is electromagnetic is so natural and has been so widely adopted that it is of interest to consider some of the consequences of this view.

Many attempts have been made to obtain an idea of the nature of the radiation by analysing the absorption curves. It is of course simple to split up the observed curve into a number of components

* Bothe and Kolhörster, *Zeit. f. Phys.* **56**, 751, 1929.

and then on the basis of some absorption formula to obtain their wave-lengths, but L. H. Gray* has pointed out the necessity for considering the effect of the degraded radiation of longer wave-length produced by the scattering process. Whenever a primary quantum of energy is scattered there is produced a recoil electron, the energy of which is transformed into "primary" ionisation, and a quantum of "degraded" γ ray energy, the whole of which is ultimately transformed into ionisation. It is evident that the degraded radiation will

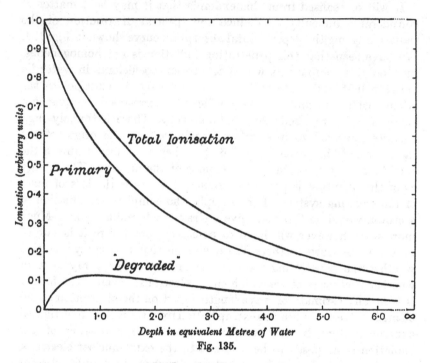

Fig. 135.

build up from the primary in a manner analogous to the growth of a short-lived radioactive substance from a decaying parent body. After equilibrium has been reached the total ionisation (primary + degraded) will fall off in a manner identical with the absorption of the primary radiation alone. Whether or not experimental absorption curves may be used as the bases of calculation of wave-lengths will therefore depend upon how quickly equilibrium is reached between degraded and primary ionisation.

* L. H. Gray, *Proc. Roy. Soc.* A, **122**, 647, 1929.

L. H. Gray* has calculated on the basis of the Klein-Nishina formula that the degraded radiation produced by the absorption of a primary radiation of wave-length 0·094 X U (corresponding to an absorption coefficient 0·22 metre⁻¹ water) will have reached 85 per cent. of its equilibrium value after traversing 2 metres of water. Equilibrium is of course reached more slowly in the case of shorter wave-lengths. The relative contributions of the primary and degraded ionisation to the total ionisation can be seen from Fig. 135.

It will be realised from this example that it may be a matter of considerable difficulty to deduce the spectral distribution of the radiation giving the experimental absorption curve shown in Fig. 134. We have seen that this penetrating radiation is not homogeneous, but contains components whose absorption coefficients in water lie between 0·35 and 0·018 metre⁻¹. If it is valid to extrapolate the Klein-Nishina formula, these coefficients correspond to quantum energies of 90 to about 2000 million volts. These extremely high energies have led to interesting speculations on the origin of the radiations. It has been suggested that they are emitted during the combination of nuclei in remote regions of the universe. The energy $h\nu$ of the quantum is put equal to mc^2, where m is the loss of mass of the emitting system. For example, the complete annihilation of a proton would on this view give a γ ray of 940 million volts. Much more work however will have to be carried out before it is safe to pursue these views further. We have seen that it is not yet settled whether the penetrating radiation is in fact of the γ ray type or whether it consists of very high speed electrons. Even if the former is true, any estimate of wave-length based on the absorption coefficients involves an enormous extrapolation from the region investigated experimentally. In the ordinary theories the scattering of the radiation is supposed to be confined to the extra-nuclear electrons, but if we are dealing with a quantum of energy of the order of some hundred million volts, it is not unlikely that the nuclear electrons may also be effective in scattering. So far, experiments have been mainly confined to measuring the ionisation produced in a sealed electroscope. Further experiments are required which will give definite indication of the energy of the swift electrons present in the atmosphere, for this will give valuable information on the minimum frequency of the radiation that could produce them, independently of the exact truth of our theories of absorption.

* L. H. Gray, loc. cit.

Some interesting observations on this point have been reported by Skobelzyn*. During the course of his experiments, in which he observed by means of an expansion chamber the recoil electrons produced by γ rays, he noticed certain tracks whose occurrence was quite unconnected with the presence of the radioactive source. Although a vertical magnetic field of about 1500 gauss was applied at the moment of expansion, these tracks were in most cases quite unaffected. This was due partly to the high energy of the particles and partly to the fact that most of the tracks made only small angles with the direction of the magnetic field, and it was only possible to set a lower limit to the momentum of the particles. Assuming the particles to be electrons, their energy was estimated at greater than 15 million volts. In all, 32 of these tracks were observed during 613 expansions, which corresponds to about 1 electron per minute crossing 1 cm.^{2} of a horizontal plane.

It is a subsidiary point of some interest that the density of ionisation along the tracks did not differ greatly from that produced by β particles of 1 or 2 million volts energy. This indicates that the ionisation must be mainly a function of the velocity and not of the energy, in agreement with the results described in § 101, and leads to an estimate of about 1 ion per c.c. per sec. for the total ionisation produced by these particles. Since, as we have already seen, the ionisation produced in electroscopes by penetrating radiation is about 1·4 ions per c.c. per sec., it seems clear that these β particles of high energy are the ultimate ionising particles.

Several attempts have been made to determine whether the radiation shows any directional effect which can be associated with the distribution of matter in the universe. Kolhörster has investigated this matter very carefully, the general principle of the experiments being to take a long series of measurements at one point, and to determine whether there are any systematic fluctuations which can be correlated with the sidereal time. It appears that there are fluctuations within the period of a few hours which are two or three times as great as the error of the experiment. Systematic observations have been made by Steinke†, Büttner and Feld‡, and Kolhörster and von Salis§. A critical examination of these measurements has

* Skobelzyn, Zeit. f. Phys. 54, 686, 1929.
† Steinke, Zeit. f. Phys. 42, 570, 1927.
‡ Büttner and Feld, Zeit. f. Phys. 45, 588, 1927.
§ Kolhörster and v. Salis, Ber. d. Preuss. Akad. d. Wiss. p. 92, 1927.

been made by Corlin*, who concludes that on the whole there is some
regularity. In all cases the intensity is relatively high at 15h and
low at 10h and there is possibly another maximum between 5h and 8h.
Since Hess and Matthias† did not observe these fluctuations, using
an electroscope covered with 7 cm. of iron, it was concluded provisionally that only the softer components of the penetrating radiation followed sidereal time. However, Regener‡ in his recent work
observed definite fluctuations in the penetrating radiation at a depth
of 78 metres below the surface of Lake Constance, and it is possible
that even the hardest components show the same phenomenon.

* Corlin, *Zeit. f. Phys.* **50**, 808, 1928.
† Hess and Matthias, *Ber. d. Akad. d. Wiss. in Wien*, **137**, 2 a, 327, 1928.
‡ Regener, *Naturwiss.* **11**, 183, 1929.

CHAPTER XVI

INTENSITY PROBLEMS CONNECTED WITH THE EMISSION OF γ RAYS

§ 113. Measurement of the total energy emitted in the form of γ rays. It was early shown by Rutherford and Barnes* that the γ rays could only account for a small fraction of the heating effect of radium and its products. Later Rutherford and Robinson†, in the course of their work on the heating effect of the α rays (see § 32), were able to estimate the energy emitted in the form of γ rays by radium B and radium C as about 7 per cent. of the total disintegration energy of radon in equilibrium with its products. The method employed was to measure first the heating effect due to the α and β rays, the walls of the calorimeter being sufficiently thin to allow practically all the γ rays to escape, and then to determine the increased heating effect when a certain fraction of the γ rays were absorbed by lead screens placed inside the calorimeter. No great accuracy was possible as the small heating effect of the γ rays was measured as the difference between two large effects.

Ellis and Wooster‡ devised a method of automatically compensating the large α ray and β ray effect, enabling that due to the γ rays to be measured directly. The calorimeter consisted of a hollow cylinder of four equal sectors, the two opposing ones A and B being respectively of lead and aluminium, while the intermediate ones were of insulating material. The difference of temperature between A and B was measured by a system of thermocouples. The radioactive source, a small glass tube filled with radon, was placed in a hollow copper rod arranged along the axis of the apparatus so that it did not touch any of the sectors. The α ray and β ray heating was automatically compensated by this arrangement, since although the copper rod became heated, as much heat was received from it by the aluminium sector as by the lead sector. These had exactly equal heat capacities and therefore the rise in temperature due to the absorption of the α and β rays was the same for both. On the other hand the greater absorption of the γ rays in the lead sector than in the

---

* Rutherford and Barnes, *Phil. Mag.* **9**, 621, 1905.
† Rutherford and Robinson, *Phil. Mag.* **25**, 312, 1913.
‡ Ellis and Wooster, *Phil. Mag.* **50**, 521, 1925.

aluminium gave rise to a temperature difference, and a corresponding deflection of the galvanometer connected to the thermocouple system The deduction of the total energy from the observed heating effect due to the partial absorption of the γ rays involved the use of the-mean energy absorption coefficients of the γ rays in lead and aluminium. These could not be measured directly but had to be calculated from scattering and absorption measurements. The final result was that the amount of radium B and radium C in equilibrium with 1 gram of radium emits 8·6 calories per hour in the form of γ rays. We shall see later that this figure is almost certainly too low and a more probable value is 9·4 calories per hour. While the use of later values of the energy absorption coefficients would lead to almost this figure, it appears unlikely that the heating method can ever give very dependable results in this case, since so much of the γ ray energy is associated with the higher frequency γ rays. Suppose we consider a high-frequency beam of γ rays to traverse a thickness x of lead and to give a heating effect Q in it. The fraction absorbed in the lead is deduced from ionisation measurements and can be written $(I_0 - I_x)/I_0$. The total energy would then have to be calculated from the expression $Q . I_0/(I_0 - I_x)$, which involves the ratio of the two small quantities, Q and $I_0 - I_x$, which have to be measured independently. Under these circumstances the possible error is necessarily large.

A more promising method is to measure the total number of ions formed by the γ rays and to multiply this by the average energy necessary to form an ion.

A simple method of finding the total ionisation produced in air by any radiation which is absorbed exponentially was suggested by Eve*. Let n be the number of pairs of ions produced per c.c. per second by the radiation at a distance r from the source, then the total ionisation N is given by

$$N = \int_0^\infty 4\pi r^2 n \, dr \qquad \ldots\ldots(1);$$

but since the radiation has been supposed to be absorbed according to an exponential law

$$n = K e^{-\mu r}/r^2,$$

where K is a constant. Hence

$$N = 4\pi K/\mu \qquad \ldots\ldots(2).$$

* Eve, *Phil. Mag.* 22, 551, 1911; 27, 394, 1914.

The total ionisation can therefore be calculated at once when we know the absorption coefficient of the radiation and the value of n at any one distance. Equation (2) was used by Eve to determine both the total ionisation due to the β rays and that due to the γ rays from radium (B + C). The measurement of n was made by finding the ionisation in a thin-walled ionisation chamber placed at a known distance from a source of radium.

Similar experiments to determine the total ionisation due to the β rays and γ rays of radium B and radium C were carried out by Moseley and Robinson*. Realising that the β radiation is not absorbed according to an accurate exponential law they used the exact equation (1); that is they made a number of observations of n at different distances from the source up to 3 metres and found the area of the experimental curve relating nr^2 and r. The remainder of the ionisation, amounting to 14 per cent., was calculated by extrapolation. The ionisation due to the γ rays in this region was small and was corrected for. In order to determine the total ionisation due to the γ rays, Moseley and Robinson were forced to assume an exponential absorption, and, selecting the value of the ionisation measured at 3 metres as being the most reliable and adopting a value of μ found by Chadwick†, inserted these values in equation (2). The result obtained was $N = 1\cdot22 \times 10^{15}$ pairs of ions per second per gram. This result is almost certainly too low since the appropriate absorption coefficient in this case is clearly σ_a, the scattering absorption coefficient. Chadwick's value $0\cdot000059$ cm.$^{-1}$ is probably intermediate between σ_a and $\sigma_a + \sigma_s$, and it would be better to take $\mu = \sigma_a = 0\cdot000032$ cm.$^{-1}$, leading to $N = 2\cdot25 \times 10^{15}$. However, little weight can be placed on this value since there are many disturbing factors in a measurement of this kind. For instance, the values of K (see equation (2)) found by Eve and by Moseley and Robinson do not agree, and both measurements were clearly influenced by scattering from the walls of the room, and from parts of the apparatus. The β particles, giving the ionisation which is measured, come from a distance of several metres, originating in the air between the source and ionisation vessel, and their amount can be considerably influenced by surrounding objects. A serious objection also is that it cannot be valid to assume an exponential law of absorption. The ionisation due to a source of monochromatic γ rays placed in a homogeneous medium

---

* Moseley and Robinson, *Phil. Mag.* **28**, 327, 1914.
† Chadwick, *Proc. Phys. Soc. London*, **24**, 152, 1912.

will only fall off according to an exponential law after the degraded radiation has built up to the equilibrium value (§ 112, p. 491). The only reliable method of determining the total ionisation is to follow the method used by Moseley and Robinson for the β ray ionisation, that is to measure the value of nr^2 at a sufficient number of different distances to enable the curve to be integrated. This is clearly not possible in air, where it would be necessary to take measurements at hundreds of metres away from the source. An ingenious method of overcoming these difficulties has been used by L. H. Gray*. He has shown on theoretical grounds that the energy lost per unit volume in a homogeneous medium completely surrounding the source is ρ times the energy lost per unit volume in an air cavity situated at the same spot, where ρ is the ratio of the distances, short compared with the range, in air and in the medium in which the ejected β particle loses the same amount of energy. He has shown further that the quantity ρ should be independent of the velocity of the β particle and approximately equal to the ratio of the densities of the medium and air. Knowing the energy required to produce a pair of ions in air, it is thus possible to deduce the energy lost per c.c. in a material such as aluminium from measurements of the ionisation in a small air cavity. The source, a radon tube, was placed at the centre of a large block of aluminium and by taking measurements up to a distance of 40 cm. from the source almost the complete curve of nr^2 against r was obtained. Preliminary experiments, using, on practical grounds, dense bricks to replace the greater portion of the aluminium, gave the result of $N = 2 \cdot 2 \times 10^{15}$ pairs of ions per gram per second. The measurements using a large homogeneous mass of aluminium have not yet been completed, but they indicate the slightly lower value of $N = 2 \cdot 13 \times 10^{15}$ pairs of ions per gm. Ra per sec.

Before the total energy of the γ rays can be obtained from this result it is necessary to know the average energy expended by a β particle in forming each pair of ions. No measurements of this quantity have been made with β rays as the experimental difficulties are great, but since the general evidence suggests that for high-speed electrons the energy lost in forming a pair of ions is constant, we can use the result obtained with artificial electron beams, or with electrons generated by X rays. The experimental information on this point is contradictory, although a great number of investigations have been made. Some of the results of these investigations are given in the

* L. H. Gray, unpublished.

table. In the experiments of Eisl the energy of the electrons was varied between 10,000 and 60,000 volts and the average energy expended in forming an ion pair was found to be constant in this region. He estimates his error to be not more than ± 0.5 volt. Although this value is lower than all the previous measurements except Buchmann's, it appears to be the most reliable.

Author	Volts/ion pair
Kuhlenkampf[1]	35
Crowther[2]	42·5
Lehmann and Osgood[3]	45
Buchmann[4]	31
Schmitz[5]	42
Eisl[6]	32·2

[1] Kuhlenkampf, *Ann. d. Physik*, **79**, 97, 1926.
[2] Crowther, *Phil. Mag.* **6**, 401, 1928.
[3] Lehmann and Osgood, *Proc. Roy. Soc.* A, **115**, 608, 1927.
[4] Buchmann, *Ann. d. Physik*, **87**, 509, 1928.
[5] Schmitz, *Phys. Zeit.* **29**, 846, 1928.
[6] Eisl, *Ann. d. Physik*, **3**, 379, 1929.

We shall therefore assume that the average loss of energy in forming a pair of ions is 32.2 ± 0.5 volts. It is of interest to recall (§ 17) that an upper limit to the average energy required to form a pair of ions by α particles and by the δ particles they eject is about 35 volts.

The energy emitted in the form of γ rays by the radium B and radium C in equilibrium with 1 gram of radium is then 9·4 cals. per hour. It is convenient to express this in terms of the mean γ ray energy emitted during the disintegration of one atom of radium B and one atom of radium C. This value is 1·86 million volts.

§ 113 a. The relative γ ray energies emitted by different bodies. It is important to consider how this energy emission of 9·4 cals. per gram per hour is to be divided between radium B and radium C. One of the simplest methods of determining this ratio is to measure with an electroscope the fraction of the ionisation due to each body. Since the γ rays of radium B are much more easily absorbed than those of radium C, it is sufficient to take measurements through increasing thicknesses of lead. After a certain thickness practically only the γ rays of radium C will reach the electroscope, and it is found that from this point they are absorbed according to a simple exponential law. The ionisation due to this radiation is then extrapolated back to zero thickness and subtracted from the total effect to give the ionisation due to the γ rays of radium B. To deduce the relative energies from the ionisations requires a knowledge of the relative ionisation functions of the electroscope. Chalmers* has paid special attention to this point and has attempted to calculate the ionisation functions for the aluminium electroscope he used. The ratio of the ionisations of radium B to radium C was found experimentally to be 0·137, from which he calculated the ratio of energies to be 0·10. If the ionisation function were taken as proportional to the scattering absorption coefficient (see p. 467), the ratio of the energies would be 0·12.

Moseley and Robinson† in their measurement of the number of ions formed by the γ rays of radium B and radium C attempted to separate the effects of the two bodies on the basis of Rutherford and Richardson's‡ absorption curves. They concluded that the ratio of the energies was 0·075. Skobelzyn has deduced a value 0·15 from his measurements of the relative intensities of the γ rays after passage through 3·5 mm. of lead. This is likely to be too high since he appears to have over-estimated the absorption of the radium B γ rays. It is difficult to be certain what is the best value, but it is probably not far from 0·12.

Radium B and radium C are the only bodies for which the energy of the γ rays has been investigated at all systematically, but rough estimates of the γ ray energy emitted by mesothorium 2, thorium B and thorium C″ can be made in the following way from ionisation measurements. McCoy and Henderson§, and McCoy and Cartledge‖ showed that the γ ray effects of the mesothorium 2 and thorium (B + C) in equilibrium with 1 gram of thorium were respectively equivalent to that of $0·52 \times 10^{-7}$ gm. and $0·956 \times 10^{-7}$ gm. of radium. The γ ray comparisons were made with a simple brass electroscope through 2 mm. of lead, and should provide a fair basis for estimating the relative energies from radium C, mesothorium 2, and thorium C″, since the γ rays from these three bodies do not differ greatly in penetrating power. It follows from these figures that per atom disintegrating mesothorium 2 emits 0·42 and thorium C″ 2·2 times the energy emitted by radium (B + C). In making this calculation it must be remembered that only 35 per cent. of the atoms follow the thorium C″ branch. An independent estimate of the energy of

* Chalmers, *Phil. Mag.* **6**, 745, 1928.
† Moseley and Robinson, *Phil. Mag.* **28**, 327, 1914.
‡ Rutherford and Richardson, *Phil. Mag.* **25**, 722, 1913.
§ McCoy and Henderson, *Am. Chem. Soc. J.* **40**, 1316, 1918.
‖ McCoy and Cartledge, *Am. Chem. Soc. J.* **41**, 50, 1919.

the thorium C″ γ rays may be made from some measurements of Shenstone and Schlundt* who compared the number of α particles emitted by sources of thorium C and radium C of equal γ ray activity. It appears that for equal γ ray activity measured through 6 mm. of lead the number of α rays of range 8·6 cm. emitted by thorium C is approximately 0·77 the number of α rays of range 7 cm. emitted by radium C. Remembering that the γ rays come from thorium C″, of which 35 atoms disintegrate for every 65 atoms of thorium C′ which emits the 8·6 cm. α particles, we find that a thorium C″ atom on the average has 2·4 times the γ ray activity of a radium C atom. This agrees well with the previous estimate. Gurney† in the course of similar measurements with thorium (B + C) sources found that the γ rays of thorium B were responsible for 25 per cent. of the initial ionisation. Since the γ rays of thorium B are of much longer wave-length than those of thorium C″, it is difficult to estimate the correction for the ionisation function. We may take as a provisional estimate that the energy of the γ ray emission from thorium B is less than one-twentieth of that from thorium C″.

The results of the measurements and estimates described in this section are shown in the following table.

γ ray energy emitted by different bodies

Body	Relative energy	Energy per atom in volts $\times 10^{-6}$
Ra (B + C)	1	1·86
RaB	0·11	0·20
RaC	0·89	1·66
MsTh 2	0·42	0·78
ThC″	2·3	4·3

It is a matter of great interest that thorium C″ emits so much γ ray energy per disintegration. The highest frequency γ ray that has been identified in the emission of thorium C″ is $2·65 \times 10^{6}$ volts. In order to account for the emission of 4·3 million volts it would be necessary for each disintegrating atom to emit almost the whole spectrum of γ rays shown on p. 371. Another possibility is that there are other γ rays of still higher frequency which account for the β rays of very high energy discussed in § 87. However, it is then

* Shenstone and Schlundt, *Phil. Mag.* **43**, 1039, 1922.
† Gurney, *Proc. Roy. Soc.* A, **112**, 380, 1926.

difficult to understand why radium C emits relatively so little γ ray energy, since it appears to emit about the same number of abnormally fast β rays as thorium C".

§ 114. The number of γ ray quanta emitted by radium B and radium C. Kovarik* has attempted to determine this quantity by measuring, in effect, the number of electrons ejected from a thin sheet of metal and dividing it by the absorption coefficient of the γ rays in the metal. A Geiger counter with a complete metal front was set up about 3 metres from a radium source of 1·3 mg. The effect in the counter was due partly to electrons ejected from the metal front and partly from the walls. The effect from the front alone was found by carrying out experiments with no front on the counter. If d represents the thickness of the metal front of the counter and dx a thin layer distant x from the incident side, then the number of electrons liberated in dx will be

$$N\mu_1 e^{-\mu_1 x} dx \qquad \ldots\ldots(1),$$

where N is the number of quanta incident on the plate and μ_1 is their absorption coefficient in the material of the plate. The number of these electrons which reach the counter was put equal to

$$N\mu_1 e^{-\mu_1 x} e^{-\mu_2 (d-x)} dx \qquad \ldots\ldots(2),$$

thereby assuming that all the liberated electrons were initially thrown forward, and that they were reduced in number according to an exponential law of absorption. The first assumption is certainly true for the recoil electrons, and a fair approximation for the photo-electrons. The second assumption however appears doubtful in the light of Eddy's work (see p. 416). However, integrating this last expression it is seen that

$$N = n \frac{\mu_2 - \mu_1}{\mu_1} \frac{1}{e^{-\mu_1 d} - e^{-\mu_2 d}},$$

where n is the number of electrons reaching the counter from the front of thickness d. The measurement consisted in the determination of the quantities n, μ_1 and μ_2. The number of counts n for a given thickness d was investigated carefully, effects due to scattered radiation, electrons from the walls of the counter, and natural discharge of the counter being corrected for. It appears likely however that Kovarik's value will be an under-estimate, since in the first place the counter does not respond to all of the faster electrons,

* Kovarik, *Phys. Rev.* 23, 559, 1924.

and further the "no front" correction will have been too large owing to the entry of electrons generated in the air, which could not enter when the front was on. Consideration will show that for μ_1, the absorption coefficient of the γ rays, the total absorption coefficient $\tau + \sigma_a + \sigma_s$ is required. This was found by keeping the same front on the counter, but placing absorbers of the same material at a distance of 90 cm. from the source. As in all measurements of absorption coefficients, it was necessary to determine the zero effect. This was done by placing lead plates 12 cm. square and aggregating 19 cm. in thickness in place of the absorber. It seems possible that this may have introduced some error by shielding off too much of the air in between the lead plates and the counter, which during the actual absorption measurements would scatter the radiation. The absorption coefficient of the β rays, μ_2, was found from the above expressions by finding how the count n depended on the thickness of the front. The results for the number of γ ray quanta emitted per second by 1 gram of radium in equilibrium with its products of short life is shown in the next table.

Metal used as front of counter	γ ray quanta emitted per second	
	I	II
Pb	$7 \cdot 33 \times 10^{10}$	$5 \cdot 56 \times 10^{10}$
Pt	6·97	5·82
Sn	7·29	5·78
Cu	7·60	5·04
Al	7·18	5·58
Mean values	7·26	5·56

Column I shows the results for the total γ rays when there was only 1 mm. of glass round the radium, column II when there was in addition a tube of brass which absorbed about 2 per cent. of the rays and a filter of lead 1·55 cm. thick. Approximately therefore column I shows the number of quanta from radium B and radium C, and column II the number of quanta from radium C alone.

Considering the difficulty of the measurement, the agreement between the values with different metals is excellent, and would appear to lend strong support to the assumptions that have been made. In spite of this agreement there are, as we have already mentioned, serious grounds for doubting the absolute accuracy of these figures. It would appear likely that the above values are too small. This determination of the total number of γ ray quanta emitted is of fundamental importance and it is unfortunate that the measurement is beset with so many difficulties.

§ 115. The intensities of the γ rays. An important series of experiments has been carried out on this subject by Skobelzyn* using the γ rays of radium (B + C). A narrow pencil of γ rays was allowed to pass horizontally through an expansion apparatus, and a magnetic field was applied perpendicularly to the plane of the chamber. The tracks of the recoil electrons ejected by the γ rays were thus curved and a measurement of the curvature enabled the velocity of each recoil electron to be ascertained. It follows by applying the energy and momentum equations to the scattering process that the energy E_β of an electron ejected at an angle ϕ to the γ ray beam is given by

$$E_\beta = h\nu \frac{2\alpha}{1 + 2\alpha + (1 + \alpha)^2 \tan^2 \phi} \qquad \ldots\ldots(1),$$

where as usual $\alpha = h\nu/m_0 c^2$.

If therefore in addition to observing the energy of the recoil electron the angle at which it is ejected is also measured, it is possible to calculate the frequency of the γ ray responsible for the scattering process. The number of tracks associated with each particular γ ray will thus give its relative intensity multiplied by a function involving the scattering absorption coefficient. The method is analogous to that of investigating the γ rays by the photo-electric effect, and, while as we shall see the sharpness of resolution of the different γ rays is far less, it possesses the great advantage that the correction function of the scattering absorption coefficient is practically independent of the frequency. It is especially applicable to high frequency γ rays where the photo-electric method is difficult to use.

Only recoil tracks making an angle of less than 20° with the direction of the γ ray beam were considered, and from the measured $H\rho$ of each track starting at an angle ϕ to the γ ray beam Skobelzyn first calculated, in effect by equation (1), the $H\rho_{\max}$ value it would have had if it had started at angle $\phi = 0$. The quantity $H\rho_{\max}$ is in the region considered practically a linear function of the frequency of the γ ray. A statistical study was then made of the frequency of occurrence of tracks of different $H\rho_{\max}$, and the results are shown in the next table. Column I shows the $H\rho_{\max}$ intervals that were chosen

---

* Skobelzyn, *Zeit. f. Phys.* **43**, 354, 1927; **58**, 595, 1929. These measurements are shown by a diagram in the paper. We are indebted to Dr Skobelzyn for permission to give the actual values.

and column II the numbers of tracks found to correspond to each interval. The γ rays in these experiments were filtered through 3·5 mm. of lead and 2 mm. of glass.

Col. I	Col. II	Col. I	Col. II	Col. III	Col. IV
$H\rho_{max}$	No. of tracks	$H\rho_{max}$	No. of tracks	$h\nu$ of γ ray in volts × 10⁻⁵	$H\rho_{max}$
925	9	4750	14	2·43	1223
1230	10	4950	7	2·97	1446
1330	5	5279	24	3·54	1672
1440	14	5500	20		
1558	14	5731	4	6·12	2638
1685	16	5970	11	7·70	3203
1820	6	6220	13	9·41	3801
1969	9	6480	30	11·30	4457
2128	14	6750	10	12·48	4864
2300	23	7034	5	13·90	5349
2488	35	7328	8	14·26	5473
2690	19	7635	6	17·78	6669
2909	16	7954	9	22·19	8144
3145	14	8300	1		
3400	10	8634	6		
3676	14	8995	1		
3876	11	9371	5		
4033	5	9763	0		
4202	6	10170	0		
4377	17	10590	2		
4560	10	10900			

We have already mentioned that the Klein-Nishina scattering formula, which appears to agree well with the facts (ch. xv), shows that the number of electrons ejected within 20° of the γ ray beam is practically independent of the frequency. We may therefore take the figures in column II to represent the relative number of quanta of the γ rays capable of giving recoil electrons within the limits shown in column I. Columns III and IV contain respectively the $h\nu$ of the main known γ rays of radium B and radium C with the calculated $H\rho_{max}$ of the electrons they could eject. In effecting a comparison it must be noticed that owing to the inevitable errors in measuring the $H\rho$'s of the recoil electrons the tracks due to a homogeneous γ ray will be spread over a range of about ± 5 per cent. Although the total emission consists largely, if not entirely, of homogeneous γ rays, it will be seen that the correct way to present these results is to give a normal distribution curve against $H\rho_{max}$. If this is done, a series of peaks are found which agree remarkably well in position with the

γ rays of columns III and IV. In fact taking into account that there are certainly many weaker monochromatic γ rays, which will be detected with equal efficiency by this method, we see that it is probable that the entire γ ray emission consists only of the monochromatic γ rays responsible for the natural β ray spectrum. The possibility must not be lost sight of, however, that there may be a certain amount of γ ray energy emitted in the form of a continuous spectrum. A point of great interest is that tracks of recoil electrons were found with $H\rho$ as great as 10,000. This indicates the emission of γ rays from radium C of about 3 million volts energy, that is of higher frequency than have yet been detected by any other method.

Another method of investigating the intensities of the γ rays has been employed by Ellis and Aston*. They utilised the photo-electric effect, and while only the intensities of the stronger γ rays were measurable the resolution of the method was sufficiently high to enable their effect to be separated both from that of the weaker γ rays and of any continuous spectrum.

A small platinum tube was filled with radon, and the magnetic spectrum of the electrons ejected from the platinum by the γ rays of radium B and radium C was recorded photographically with the usual semicircular focussing apparatus. The electronic groups due to the photo-electric effect of the different γ rays appeared as broadened lines superimposed on a general background as can be seen from Fig. 5 of Plate XI. By a photometric study of the density of the photographic plates it was possible to determine the relative intensities of the different lines.

It is clear that the intensity of the β ray groups will depend jointly on the intensity I of the γ ray (measured in relative number of quanta emitted) and on the value of the photo-electric absorption coefficient τ for that frequency. The calculation of this quantity $I\tau$ from the experimental results is greatly complicated by the straggling of the electrons from the lower layers of the platinum tube and by the non-isotropic emission of the photo-electrons. However, it was shown that a detailed study of the form of the photometer curves for the different groups enabled certain simplifying assumptions to be made and that to a good approximation the quantity $I\tau$ was proportional to the peak intensity of the line divided by $H\rho\beta^3$. Column II of the next table shows the relative values of $I\tau$ for such γ rays of radium B and radium C as were detectable by this method. Column III

* Ellis and Aston, *Proc. Roy. Soc.* A, **129**, 180, 1930.

contains relative values of the photo-electric absorption coefficient τ obtained in a manner to be described, and hence by division it is possible to obtain the intensities shown in column IV.

Col. I	Col. II	Col. III	Col. IV
$h\nu$ of γ ray in volts $\times$ 10^{-5}	$I\tau$ Relative values	τ Relative values	I Relative values
		Radium B	
2·43	67	9·00	7·4
2·97	89	5·35	16·7
3·54	100	3·45	29·0
		Radium C	
6·12	40	0·940	42·5
7·73	2·4	0·568	4·2
9·41	1·6	0·370	4·3
11·30	3·4	0·256	13·3
12·48	0·85	0·210	4·0
13·90	0·70	0·170	4·1
14·26	—	—	—
17·78	1·8	0·108	16·7
22·19	0·35	0·073	4·8

There is no direct experimental evidence about the variation of the photo-electric absorption coefficient with wave-length in the region under consideration. It seems clear that an extrapolation of the X ray result*, $\tau \propto \lambda^{2\cdot92}$, to these γ ray frequencies is not justifiable. It was pointed out in § 109 that it was impossible with this formula to account for the difference between the absorption coefficients of γ rays in lead and aluminium. An empirical rule such as

$$\log_{10}\tau = \text{cons.} + a \log_{10} \lambda + b (\log_{10}\lambda)^2$$

can be made to have the same slope and absolute value as that given by X ray measurements at some chosen wave-length such as 100 X U, and leaves one constant to be adjusted to give the best fit with the γ ray absorption results. A more elaborate empirical formula than this is not justified by the present state of our data. Working on this basis L. H. Gray† has concluded that the values $a = 1\cdot0$, $b = 0\cdot480$, where λ is measured in X U, lead to fair agreement with the experimental results on the absorption of γ rays. The relative values of τ used in column III of the last table were obtained in this way.

It will be noticed that no entry is made against the γ ray of energy $14\cdot26 \times 10^5$ volts. In the natural β ray spectrum of radium C there

* S. J. Allen, *Phys. Rev.* 27, 266, 1926.　　　　† L. H. Gray, unpublished.

are prominent lines whose energies correspond to the conversion of such a γ ray in the K, L and M states. Thibaud* was the first to observe, however, that no evidence could be obtained by the excited spectrum method that γ radiation of this frequency was actually emitted. Ellis and Aston paid special attention to this point and confirmed Thibaud's conclusion. It appears that if this γ ray is emitted its intensity is less than 2, but it is very difficult in this case to understand why the natural β ray lines are so prominent. We can, however, with certainty infer the existence of a nuclear quantum switch of energy $14\cdot26 \times 10^5$ volts, and we must assume there is some peculiarity of the levels involved, which, in distinction from the normal case, renders the emission of radiation extremely improbable or impossible, while the transference of the energy to the extra nuclear electrons is not so affected (see § 116).

§ 116. We have so far discussed only the relative intensities of the γ rays, but since the γ rays are emitted at the moment of disintegration it is clearly possible to define the absolute intensity of the γ ray as the probability that a quantum will be emitted at any specified disintegration. It is helpful however to go further, and following the method of Ellis and Wooster†, to define as p_r the probability that at any specified disintegration a quantum switch will occur which is associated with the liberation of energy $h\nu_r$. In a certain fraction of the cases α_r this results in the ejection of electrons from the K, L, M levels by "internal conversion," in the remainder a γ ray is emitted. The intensities we have been considering in the previous paragraph have thus been the relative values of $p_r (1 - \alpha_r)$. The total energy emitted on the average per disintegration from the atom in the form of monochromatic γ rays is thus

$$\sum_1^n p_r\, h\nu_r\, (1 - \alpha_r).$$

In § 113 we saw that on the average the combined energy emitted by one disintegration of radium B and one of radium C was $1\cdot86 \times 10^6$ volts. We may put this equal to the above expression if the summation is extended over all the γ rays of radium B and radium C. Utilising the relative values of the intensities shown on p. 507, we obtain the following table of absolute intensities. These values are upper limits since the figure for the total energy will include the

* Thibaud, *Thèse*, Paris, 1925.
† Ellis and Wooster, *Proc. Camb. Phil. Soc.* 23, 717, 1927.

weaker γ rays, and the continuous spectrum if there is any, while the summation can only be extended over the γ rays which are sufficiently strong to be measured separately.

Absolute intensities of the main γ rays of radium B and radium C

$h\nu$ of γ ray in volts $\times 10^{-5}$	$p(1 - a_r)$ Average number of quanta emitted per disintegration
Radium B	
2·43	0·115
2·97	0·258
3·54	0·450
Radium C	
6·12	0·658
7·73	0·065
9·41	0·067
11·30	0·206
12·48	0·063
13·90	0·064
14·26	—
17·78	0·258
22·19	0·074

§ 117. **Internal conversion of the γ rays.** Even before the exact origin of the γ ray spectra was understood, it was realised that the electrons forming the homogeneous groups must have originated in the disintegrating atoms, and later it appeared possible to interpret this phenomenon as an internal absorption of the γ ray. It was considered that the nucleus was left in an excited state after the emission of the disintegration particle, and the γ ray, which was subsequently emitted, sometimes escaped from the atom, but sometimes was absorbed in one of the electronic levels in a manner analogous to that occurring in the normal photo-electric effect.

While this view sufficed for a description of the phenomenon, it was pointed out* that there was no real justification for assuming the existence of the γ ray in the case of those disintegrations which resulted only in the emission of a photo-electron. It was maintained that it was more correct to state that the atom as a whole had two possibilities of getting rid of its excess energy; either by emitting a quantum of radiation, or by ejecting a high speed electron. Before discussing these two alternative standpoints, a short review will be given of the experimental evidence.

* Smekal, *Zeit. f. Phys.* **10**, 275, 1922; *Ann. d. Phys.* **81**, 399, 1926; Rosseland, *Zeit. f. Phys.* **14**, 173, 1923.

It is convenient first to define a coefficient of internal conversion as follows:

Let p_r be the probability of occurrence of a quantum switch of amount $h\nu_r$, then if the probability of the internal conversion of the energy in the K, L or ... electronic state of the atom be $_K\alpha_r$, $_L\alpha_r$, etc., the number of the photo-electrons of energy $h\nu - K_{abs}$, $h\nu - L_{abs}$, etc., per disintegration will be $p_r{_K}\alpha_r$, $p_r{_L}\alpha_r$, etc., and the number of quanta of radiation emitted from the atom will be

$$p_r\,[1 - (_K\alpha_r + {_L}\alpha_r + \ldots)] = p_r\,(1 - \alpha_r),$$

where $$\alpha_r = {_K}\alpha_r + {_L}\alpha_r + \ldots.$$

The first estimation of the magnitude of the internal conversion coefficient $_K\alpha$ in the K level was made by Ellis and Skinner[*], who concluded that it must be at least as great as $0 \cdot 1$ for the main γ rays of radium B ($3 \cdot 0 \times 10^5$ volts). Gurney[†], by means of the apparatus described on p. 391, was able to measure the number of electrons contributed to the main groups of the radium B β ray spectra from each disintegrating atom, that is he was able to find $\Sigma p_r {_K}\alpha_r$ for the corresponding γ rays. Making the plausible assumption that Σp_r for these γ rays was approximately 1, he concluded that the average value of $_K\alpha$ was equal to about $0 \cdot 12$. Similar considerations for the main γ ray of thorium B led to a value at least as great as $0 \cdot 25$[‡]. Since this latter γ ray was of considerably lower frequency than those of radium B, it appeared likely that $_K\alpha$ decreased with decreasing wave-length.

Estimates have also been made of the value of the coefficient for internal conversion in the L level. Gray and O'Leary[§] investigated the case of radium D, which emits only one ray of energy $0 \cdot 472 \times 10^5$ volts. This is of too low a frequency to eject electrons from the K level but gives strong electronic groups from the L, M, N, ... levels. Using a source of radium D in equilibrium with radium E, they measured the relative ionisation produced in an electroscope by the β particles of radium E, the γ ray in question, and the characteristic L radiation of radium D excited by the internal conversion of the γ ray. The fast β rays of radium E were easily distinguished from the slow β rays of radium D and were a measure of the number of disintegrating atoms. From these measurements they were able to

[*] Ellis and Skinner, Proc. Roy. Soc. A, **105**, 185, 1924.
[†] Gurney, Proc. Roy. Soc. A, **109**, 540, 1925.
[‡] Gurney, Proc. Roy. Soc. A, **112**, 380, 1926.
[§] Gray and O'Leary, Nature, **123**, 568, 1929.

deduce the total energy emitted in these three different forms, and, knowing the average energy of each type, were able to calculate the relative frequency of occurrence of each process. They found that out of 43 disintegrations of radium D, eleven atoms emitted a quantum of L radiation, and five a quantum of the γ ray.

Hence
$$p\left[1 - (_L\alpha + _M\alpha + _N\alpha)\right] = 5/43,$$
$$p \cdot _L\alpha\left(1 - \alpha_x\right) = 11/43,$$

where α_x is the internal conversion coefficient of the L X radiation in the $M, N, \ldots$ levels. Further, they assumed from inspection of the β ray spectrum that

$$_M\alpha + _N\alpha = 0.7\,_L\alpha.$$

The value of $_L\alpha$ is insensitive to the particular value chosen for α_x, and we find $_L\alpha \sim 0.5$.

The value of the corresponding coefficient for the much higher frequency γ rays of radium B (about 3×10^5 volts) may be estimated as about one-fifth of the coefficient for conversion in the K level, that is about 0.02. This follows from the fact that the intensity of the photo-electric groups due to conversion in the L level ($p_{r\,L}\alpha_r$) is about one-fifth of those due to conversion in the K level ($p_{r\,K}\alpha_r$).

These results considered as a whole gave the order of magnitude of the effect and suggested that the coefficient of internal conversion in both the K and L levels decreased with increasing frequency. On this evidence Ellis and Wooster* assumed a simple power law for the dependence of α on the frequency, and made use of the intensities of the β ray lines ($p_{r\,K}\alpha_r$) to determine the relative intensities p_r of the γ rays. A similar attempt was made later by Stoner†, with a view to utilising the intensities of the γ rays to explain absorption experiments. Later measurements of Ellis and Aston‡ have, however, shown that the assumption that the internal conversion coefficient varies smoothly with the frequency is not correct. These experiments consisted in comparing the intensity of the electronic group due to internal conversion of a γ ray with that due to normal photo-electric conversion of the same γ ray in platinum. Figs. 4 and 5 of Plate XI show these two types of spectra. The intensity of the former groups can be written $p_r \cdot _K\alpha_r$, and it was indicated in § 115 how it was possible to deduce $p_r\left(1 - \alpha_r\right)\tau$ from the

* Ellis and Wooster, *Proc. Camb. Phil. Soc.* **23**, 717, 1927.

† Stoner, *Phil. Mag.* **7**, 841, 1929.

‡ Ellis and Aston, *Proc. Roy. Soc.* A, **129**, 180, 1930.

latter groups. The relative intensities of the groups therefore give the values of $\frac{{}_K\alpha_r}{1 - \alpha_r} \cdot \frac{1}{\tau}$. By photographing the spectra under comparable conditions, and knowing the relative number of disintegrations occurring in each experiment, it was possible to obtain approximately the absolute values of this quantity for several different γ rays. The results are shown in column II of the next table, and column III was obtained by multiplying the figures in column II by the values of τ used in § 115. While the dependence of τ on λ is not known with certainty, it is sufficient to assume that τ decreases smoothly with decreasing wave-length to show that the internal conversion coefficient cannot vary smoothly with the frequency.

Col. I	Col. II	Col. III
$h\nu$ of γ ray in volts $\times 10^{-5}$	$\dfrac{{}_K\alpha_r}{1 - \alpha_r} \cdot \dfrac{1}{\tau}$	$\dfrac{{}_K\alpha_r}{1 - \alpha_r}$
Radium B		
2·43	0·0405	0·36
2·97	0·0348	0·19
3·54	0·0340	0·12
Radium C		
6·12	0·0065	0·0061
7·73	0·0085	0·0048
9·41	0·0164	0·0061
11·30	0·0240	0·0062
12·48	0·0270	0·0057
13·90	0·0082	0·0014
14·26	—	—
17·78	0·0150	0·0016
22·19	0·0174	0·0013

The quantity α_r is equal to ${}_K\alpha_r + {}_L\alpha_r + \dots$ and the relative values of the intensities of the natural β ray lines due to the conversion of the same γ ray in the K, L, M states show that α_r is approximately equal to $1 \cdot 2\,{}_K\alpha_r$. We thus see that the values for radium B accord with the earlier estimates, and appear to show a distinct decrease with increasing frequency. The values for radium C however are peculiar. The first five γ rays appear to have a practically constant value of the internal conversion coefficient, and the coefficients for the remaining three γ rays are also constant but are much smaller in absolute magnitude.

We have already mentioned that the first attempt to interpret this phenomenon was to consider the radiation to be emitted and subsequently re-absorbed in the same atom. A detailed calculation of this effect, according to the quantum mechanics, was made by Swirles* previous to the publication of the results just described. It was supposed that the γ rays were emitted from a doublet, situated at the centre of the atom, and the perturbation of the electrons in the K and L levels of a hydrogen-like atom was calculated. As was to be expected from a radiation hypothesis, the coefficient of internal conversion was found to decrease rapidly with decreasing wave-length. The ratio α_L/α_K came to be about 1/5 in agreement with experiment and although the calculated value of α_K for radium B was only about 1/10 of that measured it yet seemed possible to attribute this to difficulties in approximating to the case of a heavy atom. It is clear, however, that the behaviour of the internal conversion coefficient shown in the last table indicates some type of coupling, other than radiative, between the nucleus and the outside electronic structure. An interesting application of this standpoint arises in connection with the γ ray of $14\cdot26 \times 10^5$ volts of radium C. Thibaud was the first to remark that while β ray groups were observed to be emitted from radium C whose energies corresponded to internal conversion of such a γ ray, no direct evidence could be obtained that radiation of this frequency was ever actually emitted. Ellis and Aston paid special attention to this point and confirmed Thibaud's conclusion. We can infer that a quantum switch of this magnitude occurs in the nucleus, but that owing to some type of selection rule this transition can only take place during interaction with the outer electronic system. In the analogous case of the electronic system of the atom, where an atom ionised in the K ring is found to be capable of disposing of its surplus energy either by emitting radiation or by emitting a photo-electron, it is always assumed that the latter phenomenon is the result of a direct interaction of the different electrons without the intermediary of any radiation. The measurements we have described would suggest a similar interpretation for the nucleus. This view was advocated on general grounds very early by Smekal†, but it did not receive great support, since at that time the separation between the nucleus and the electronic system appeared to be complete. The new views about the nucleus, involving the possibility of penetration of particles through a potential barrier, remove this difficulty. It is quite possible for a particle to be kept in the nucleus by a potential barrier, even to

* Swirles, *Proc. Roy. Soc.* A, **116**, 491, 1927; **121**, 447, 1928.
† Smekal, *loc. cit.*

have negative energy and to be permanently stable in the nucleus, but yet for its effect to penetrate out a sufficient distance into the potential barrier for it to affect the K and L electrons. It is doubtless in some such sense that an explanation of this coupling of the nucleus and the electronic system will be reached.

It is of interest to consider these results from another angle. If we consider an atom which is capable of making a nuclear quantum switch which liberates energy $h\nu$, we can state that it has several alternative modes of emitting this energy, either as a quantum of radiation, or as electrons of energy $h\nu - K_{abs}$, $h\nu - L_{abs}$, ..., the latter modes being associated with corresponding adjustments in the electronic system to maintain the energy balance. The branching probabilities between these different modes are proportional to $(1 - \alpha_r)$, $_K\alpha_r$, $_L\alpha_r$, ..., etc. We can however go further and give the absolute probabilities that any of these modes will be followed at a given disintegration. In the notation we have used these are expressed by $p_r(1 - \alpha_r)$, $p_r \cdot {_K\alpha_r}$, $p_r \cdot {_L\alpha_r}$, ..., etc. The values for $p_r(1 - \alpha_r)$, that is the average number of quanta emitted per disintegration, were given in the table in § 116, and it will be found that the corresponding absolute values for the average number of electrons per disintegration which go to form each of the β ray lines can be obtained from the relative intensities given on pp. 362 and 364 by multiplication by $5 \cdot 3 \times 10^{-4}$. Thus for example the β ray line $H\rho$ 1938 of radium B, whose relative intensity is given as 100, arises by an average emission of 0·053 electron per disintegration.

§ 118. **Energy levels in the nucleus.** It was pointed out by Ellis* that there appeared to be certain simple additive relations between the frequencies of the γ rays of radium B of the type $\nu_1 + \nu_2 = \nu_3$, and that the simplest interpretation of this fact was that the emission of the γ rays was due to transitions of the nucleus between various quantised energy states representable by a set of levels.

Subsequently attempts were made to set up level systems for radium B, and radium C by Ellis and Skinner†, and for meso-thorium 2 and thorium C″ by Black‡. With all these bodies so

---

* Ellis, *Proc. Roy. Soc.* A, **101**, 1, 1922.

† Ellis and Skinner, *Proc. Roy. Soc.* A, **105**, 185, 1924; Ellis, *Proc. Camb. Phil. Soc.* **22**, 369, 1924.

‡ Black, *Proc. Roy. Soc.* A, **109**, 166, 1925.

many examples of additive relation between the frequencies were found that there is good reason to believe in the essential correctness of the central idea. It is unnecessary to emphasise that it is plausible on theoretical grounds. It is by no means so certain, however, that the actual level systems which have been proposed are the correct ones. In general the frequencies are not known to more than 1 part in 200, and if a tolerance of this amount is allowed alternative solutions are possible. The additional principle which has been used by the above workers to distinguish between the different possibilities has been to use as few levels as possible to account for the γ rays. While this is reasonable it is not necessarily a true guide. A far sounder basis is likely to be provided by a study of the intensities of the γ rays. Since the possibility of the nucleus emitting γ rays is due to the departure of the disintegration particle, it is clear that while γ rays of several different frequencies may be emitted by one nucleus, not more than one quanta of any one frequency can be emitted. An obvious

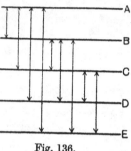

Fig. 136.

extension of this view is to state that any nucleus can only make the transition through its level system once. In Fig. 136 *A, B, C, D, E* represent a level system and we will assume for the purpose of the argument that γ rays according to the transitions shown are emitted. Suppose, in the ordinary terminology, that level *E* is ionised, then we may have a transition *A—E*, but in that case no other γ ray will be emitted; or the transitions might be *A—B, B—C, C—E*, and in this case it would be impossible, for example, to have from this particular nucleus the transition *B—E*.

If we denote by $\overset{C}{\underset{D}{\Sigma}} p$ the sum of the intensities of the γ rays whose transitions involve on the diagrams a passage over the region *C—D* [i.e. the γ rays *AD, AE, BD, BE, CD* and *CE*], then this "cross-sectional sum" can at no place of the level system be greater than unity, otherwise it would imply two excitations of the same nucleus. This is an exact and valuable criterion which might be used to distinguish between different level systems which were equally acceptable on energy grounds. Further, it will be seen that $\overset{B}{\underset{C}{\Sigma}} p - \overset{C}{\underset{D}{\Sigma}} p$ is a measure

of the number of nuclei which finally remain excited in the C level
after the whole disintegration is over. Again, this might furnish a type
of qualitative criterion, since on general physical grounds one would
not expect a large fraction of the nuclei to remain excited in a level
such as C if at the same time the total intensity of γ rays emitted by
transitions from C was at all large.

Little progress has as yet been made in investigating γ ray level
systems by this method, but there seems every reason to hope that a
solution along these lines is possible.

CHAPTER XVII

ATOMIC NUCLEI

§ 119. In previous chapters we have discussed some of the properties of radioactive nuclei and the types of radiation which accompany their transformations. The instability of these nuclei has given us important information on their structure, but unfortunately such information is not available in the case of the ordinary non-radioactive elements. Apart from their instability, there is no reason to believe that the nuclei of the radioactive elements differ in any marked way in their general type of structure from ordinary elements of high atomic weight. It is thus important to examine the properties of atomic nuclei in general to see whether we can obtain evidence to throw light on their structure and their connection with one another. In particular, it is of great interest to see whether any definite evidence can be obtained of the reasons why the property of radioactivity only manifests itself in any marked degree in the two elements of highest atomic weight, thorium and uranium, and their products of transformation.

In chapter VII an account has been given of the genesis of the nuclear theory of the atom and the evidence in its support. On this theory, the ordinary physical and chemical properties of the atom, excluding its mass, depend on the magnitude of the nuclear charge, for on this depends the number and distribution of the outer electrons. It has been shown that the nuclear charge of an atom can be measured directly from observations of the scattering of α particles, and the evidence obtained in this way suggested that the charge on a nucleus is equal to Ze, where Z is the atomic or ordinal number of the element when the elements are arranged in order of increasing atomic weight. In his classical researches on the X ray spectra of the elements, Moseley showed that the properties of an element depend on a number which varies by unity in passing from one element to the next, and concluded that the nuclear charge, in terms of the fundamental unit e, is given by the atomic number of the element. The lightest element hydrogen has a nuclear charge 1 and the heaviest element uranium 92, and with few exceptions all the values of Z between these limits are represented by known elements. He showed that the atomic number of an element was of more fundamental importance in defining its physical and chemical properties than its atomic weight, and that the well-known exceptions in the periodic classification of the elements were removed when atomic number was substituted for atomic weight. For example, the study of the X ray spectra of cobalt and nickel showed that the value of Z for cobalt was 27 and lower than that for nickel 28, although the atomic weight of cobalt, 58·97, was higher than that for nickel, 58·68. Similarly the atomic number found for argon and potassium, tellurium and iodine, reversed the order found from the atomic weights and was in accord with their chemical properties. The results of Moseley showed that between hydrogen and uranium only 90 elements of different nuclear charge were possible and of these only six were missing, of atomic numbers 43, 61, 72, 75, 85, 87. The method of X ray spectra gave a new and powerful method of attack for detecting the missing elements, even if present only in small quantity mixed with other elements. From measurements of the X ray spectrum of neighbouring elements, the wavelength of the main lines of a missing element could be predicted with considerable certainty. The first success of this new method was the discovery of *hafnium*, of number 72, by Coster and Hevesy. This element, of atomic weight 178·6, has chemical properties allied to zirconium and is the first element after the rare earth group of numbers 57 to 71. Later Noddack, Tacke and Berg found X ray evidence of the missing elements 43 and 75 in platinum minerals and suggested the names *masurium* and *rhenium* respectively. The element rhenium has been isolated in a pure state and its atomic weight found to be 188·7. While the evidence for the existence of masurium seems definite, the element has not yet been sufficiently concentrated to determine its atomic weight. Later Harris, Yntema and Hopkins found evidence of the missing element number 61 in the rare earths and named it *illinium*. Excluding number 43, there are thus only two missing elements of atomic number 85 and 87. Even if these elements were initially present in the earth, they may have been radioactive and reduced to so minute a quantity to-day as to be difficult of detection by chemical or radioactive methods.

§ 120. **Isotopes.** We have seen that if uranium is the element of highest atomic number existing in the earth, Moseley's classification permits of only 92 possible elements in the ordinary sense, of nuclear charge from 1 to 92. The discovery, however, that many of the ordinary elements consist of a number of isotopes, i.e. of atoms of the same nuclear charge but different atomic masses, shows that the number of species of atomic nuclei present in the earth is much greater than the number of elements and may amount to two hundred or more.

In chapter I we have seen that the first discovery of isotopes resulted from an examination of the chemical properties of radioactive bodies. In many cases these isotopes can be obtained in a pure state, although in very small amount and only detectable by the characteristic radiations emitted during their transformations. For example, radium B (214), thorium B (212), actinium B (211), and radium D (210) are radioactive isotopes of lead, while the end products of the three radioactive series, uranium lead (206), actinium lead (207), and thorium lead (208), are non-radioactive isotopes of lead. The nuclei of these isotopes, varying in mass from 206 to 214, exhibit a wide range of stability as measured by their rate of transformation. The non-radioactive isotopes are either permanently stable or have a life exceedingly long compared with uranium. The average life of transformation of the radioactive isotopes ranges between 25 years for radium D, and 38 minutes for radium B. It is thus clear that an identity of the nuclear charge may conceal marked differences of mass and constitution of the nucleus as revealed by their distinctive radioactive properties.

The proof of the existence of such isotopes in the radioactive bodies led to experiments to test whether some of the ordinary elements may not also consist of a mixture of isotopes. The methods employed have in general involved the production of charged atoms or compounds of the element under examination by the action of the electric discharge in a bulb at low pressure. The mass, or rather the ratio m/e, of the charged particles is determined by observations of their deflection by an electric and by a magnetic field. In the ingenious apparatus devised by Aston*, called the Mass Spectrograph, particles of the same charge and mass, even if differing individually in velocity, are brought to a focus by their passage first through an electric field and then through a magnetic field disposed in a special manner. In this way, a line spectrum is obtained on a photographic plate, each line corresponding to particles of different m/e. In the first apparatus, the masses of the isotopes could be measured with an

* Aston, *Proc. Roy. Soc.* A, **115**, 487, 1927; *Phil. Mag.* **49**, 1192, 1925; *Isotopes*, 2nd edn., Arnold, London, 1924.

accuracy of 1 in 1000. In a second modified apparatus, still greater resolving power was obtained and the masses of some of the isotopes could be determined with an accuracy of about 1 in 10,000.

The masses of the atoms or molecules are not directly measured but compared with the value found for the line due to oxygen which is assumed to have a mass 16. Oxygen was at first believed to be a simple element, and Aston[*] has given the evidence in favour of this assumption. He concluded that if an oxygen isotope of mass 18 exists, it cannot be present in quantity greater than 1/1000 of the main isotope. From an examination of the absorption bands of oxygen in the spectrum of the sun, Giauque and Johnston[†] showed that O = 18 must be present in the atmosphere and Babcock[‡] estimated from the intensity of the bands that it is present in about 1/1250 of the amount of O = 16. In a later communication Giauque and Johnston claim that definite evidence is also obtained of an isotope of oxygen of mass 17. It is of interest to note (§ 71) that Blackett deduced that an isotope of oxygen of mass 17 should be formed when nitrogen is disintegrated by α particles. It may be mentioned here that evidence of a similar character has been obtained to show also that carbon is not a simple element. According to King and Birge[§] the Swan spectrum of neutral C_2 obtained in the vacuum furnace indicates that carbon has an isotope C = 13, and later Birge[||] found additional evidence of this isotope in the absorption spectrum of CO and the emission spectrum of CN. No definite statement about the relative abundance of this isotope has yet been made, but it is probably very small. These results, if confirmed, are of great interest and indicate that the examination of band spectra affords in some cases a powerful means of detecting isotopes which are present in very small relative proportions. It should, however, be pointed out that the existence of an isotope in oxygen does not alter the relative masses of the nuclei found by Aston but lowers the actual mass of oxygen and other nuclei by about 1 in 10,000, assuming the average atomic weight of the atom of ordinary oxygen is taken as 16, a difference within the experimental error even of the latest measurements of Aston.

[*] Aston, *Nature*, **123**, 488, 1929.
[†] Giauque and Johnston, *Nature*, **123**, 318 and 831, 1929.
[‡] Babcock, *Nature*, **123**, 761, 1929; **124**, 467, 1929.
[§] King and Birge, *Nature*, **124**, 127, 1929.
[||] Birge, *Nature*, **124**, 182, 1929.

It was found that some of the elements were simple but others consisted of a number of isotopes. The elements, 57 in number, so far examined and the masses of the isotopes in order of intensity are included in the following table. With the exception of a few measurements by Dempster, and the results for carbon and oxygen given above, the data have been obtained by Aston.

Apart from the determination of isotopic constitution of a large number of the elements, the most important result from the measurements is that the masses of the isotopes follow the "whole number rule," i.e. the atomic mass of an isotope in terms of $O = 16$ is very nearly a whole number in all cases except hydrogen. This whole number is usually called the mass number of the element. While the rule is of great convenience in summarising the essential facts, it is only approximate, holding to an accuracy of about 1 in 1000. Subsequent accurate measurements by Aston with a more sensitive apparatus show that there are small but systematic departures from this rule over the whole range of the elements. It will be seen that these variations are of much importance in throwing light on the structure of nuclei.

Since the masses of many of the isotopes differ very nearly by unity, it is natural to suppose that there exists a unit of mass entering into the constitution of nuclei which is very nearly 1, in terms of $O = 16$. This unit, which is called the "proton," is to be identified with the nucleus of the lightest atom, hydrogen. The mass of the hydrogen nucleus in the free state is $1 \cdot 0072$ and is thus greater than the average mass of the proton in the nucleus. This difference of mass is ascribed to the so-called "packing effect." The electrons and protons making up a nucleus are so closely packed that their electric and magnetic fields closely interact. For these reasons, it is to be anticipated that the apparent mass of the proton and the electron in the nucleus should be less than in the free state. This change of mass may also be regarded from another general point of view. It is believed that mass and energy are closely related and that a decrease of mass δm in a system is to be ascribed to a loss of energy $c^2 \delta m$, where c is the velocity of light. On this view, the change of mass of the proton is to be ascribed to an emission of energy during the process of construction of the nucleus. This important question of the energy changes in the building up of nuclei will be discussed in more detail later.

We have seen that radioactive evidence indicates that the nuclei of the radioactive atoms are in part composed of α particles (helium

Table of elements and isotopes

Element	Atomic number	Atomic weight	Minimum number of isotopes	Mass numbers of isotopes in order of intensity
H	1	1·0078	1	1
He	2	4·0022	1	4
Li	3	6·94	2	7, 6
Be	4	9·02	1	9
B	5	10·83	2	11, 10
C	6	12·0036	2	12, 13
N	7	14·008	1	14
O	8	16·00	3	16, 18, 17
F	9	19·00	1	19
Ne	10	20·18	3	20, 22, 21
Na	11	23·00	1	23
Mg	12	24·30	3	24, 25, 26
Al	13	26·97	1	27
Si	14	28·08	3	28, 29, 30
P	15	30·98	1	31
S	16	32·06	3	32, 33, 34
Cl	17	35·46	2	35, 37
A	18	39·94	2	40, 36
K	19	39·10	2	39, 41
Ca	20	40·09	2	40, 44
Sc	21	45·1	1	45
Ti	22	47·9	1	48
V	23	51·0	1	51
Cr	24	52·0	4	52, 53, 50, 54
Mn	25	54·95	1	55
Fe	26	55·84	2	56, 54
Co	27	58·95	1	59
Ni	28	58·69	2	58, 60
Cu	29	63·55	2	63, 65
Zn	30	65·38	6	64, 66, 68, 67, 65, 70
Ga	31	69·72	2	69, 71
Ge	32	72·60	8	74, 72, 70, 73, 75, 76, 71, 77
As	33	74·93	1	75
Se	34	79·2	6	80, 78, 76, 82, 77, 74
Br	35	79·92	2	79, 81
Kr	36	82·9	6	84, 86, 82, 83, 80, 78
Rb	37	85·43	2	85, 87
Sr	38	87·63	2	88, 86
Y	39	88·9	1	89
Zr	40	91·2	3 (4)	90, 94, 92, (96)
Ag	47	107·88	2	107, 109
Cd	48	112·40	6	114, 112, 110, 113, 111, 116
In	49	114·8	1	115
Sn	50	118·70	11	120, 118, 116, 124, 119, 117, 122, 121, 112, 114, 115
Sb	51	121·76	2	121, 123
Te	52	127·5	3	128, 130, 126
I	53	126·93	1	127
X	54	130·2	9	129, 132, 131, 134, 136, 128, 130, 126, 124
Cs	55	132·81	1	133
Ba	56	137·36	?	138
La	57	138·90	1	139
Ce	58	140·2	2	140, 142
Pr	59	140·9	1	141
Nd	60	144·25	3 (4)	142, 144, 146, (145)
Hg	80	200·6	7	202, 200, 199, 198, 201, 204, 196
Pb	82	207·2	3 (4)	206, 207, 208, (209)
Bi	83	209·00	1	209

nuclei) and electrons. No evidence has been obtained of the emission of protons in the radioactive transformations. On the other hand, it is known that protons can be liberated from the nuclei of certain light atoms by bombardment of matter with swift α particles. It seems clear that while the proton can exist as an individual unit in the structure of some of the lighter atoms, the α particle is to be regarded as an important secondary unit in the building up of the heavier nuclei and probably of nuclei in general. The helium nucleus itself is believed to consist of a close combination of four protons and two electrons.

Assuming that all nuclei are composed of two fundamental units, the electron and the proton, the number of electrons and protons in any nucleus can at once be determined. If M be the mass of the nucleus to the nearest integer and Z its nuclear charge or atomic number, the nucleus contains M protons and $M - Z$ electrons. For example, radium of atomic number 88 and atomic weight 226 contains 226 protons and 138 electrons. The ratio of the number of electrons to the number of protons in a nucleus is either 1/2 or greater than 1/2 for all nuclei except that of hydrogen. The ratio tends to increase with increasing mass of the nucleus. It is very unlikely that the component electrons and protons can exist as separate and independent units in a complex nucleus. They no doubt tend to form aggregates like the α particles and other combinations of a simple type. Various suggestions of such combinations have been made from time to time, but in the absence of definite evidence they must be considered as hypothetical. In order to account for the observation that in the radioactive bodies an α ray change is in a number of cases followed by two successive β ray changes, Meitner suggested that a neutral α particle, consisting of a close combination of four protons and four electrons, has an independent existence in a radioactive nucleus. This question is mentioned in § 78. Similarly the existence of a neutron, i.e. a close combination of a proton and electron, has been suggested. From a consideration of the masses of the isotopes Harkins has proposed a building unit of three protons and two electrons, i.e. of mass 3 and charge 1.

On modern views, it appears difficult to account for the equilibrium of free electrons in a nucleus containing massive charged particles. It may be that the electrons in a nucleus are always bound to positively charged particles and thus have no independent existence within the nucleus. Unfortunately, it is very difficult to obtain convincing evidence on any of these points, for we have little if any definite information on the internal structure of nuclei to guide us.

When considering the table of isotopes (p. 522) it is significant that while odd-numbered elements in general have two isotopes differing in mass by two units, even-numbered elements may have a large number of isotopes. For example, for tin, atomic number 50, 11 isotopes have been found ranging in atomic mass between 112 and 124 with only 113 and 123 missing. Similarly xenon has nine isotopes and germanium eight. It may be significant that isotopes only become numerous for atomic numbers greater than 29.

The existence of elements representing nearly all nuclear charges between 1 and 92 is a striking and unexpected fact indicating that, whatever the nuclear charge, stable nuclei can be formed between these limits. The remarkable stability of nuclei and particularly even-numbered nuclei is even more emphasised by the existence of a whole series of isotopes over a considerable range of mass. The nucleus appears to survive the introduction of a certain number of additional protons and electrons without any apparent loss of stability. While so far the observed differences of mass between the lightest and heaviest isotope of an element is limited in general to about 10 per cent. or less, it is by no means certain that a more systematic search may not reveal the presence of additional isotopes existing in relatively small amount over a still greater range of mass.

Aston has observed a number of cases of *isobaric* atoms, i.e. atoms of the same mass but different atomic number. One of the most notable of these isobaric pairs is argon-calcium of mass 40, which consists of the main isotope of the constituent elements. A number of other such pairs have been found and no doubt more will be revealed when the detection of weaker isotopes becomes feasible. In the elements germanium-selenium, three isobaric pairs are present of masses 74, 76, 77. The atomic numbers of the elements forming the pairs so far noted are even and differ by two units and the common mass number is in most cases even.

Various rules have been suggested to fix the masses of the elements. One of the most interesting of these has been put forward by A. S. Russell*. He has pointed out that there is a close relation between the isotopes of the radioactive elements and those of the inactive elements, and it is possible on certain assumptions to deduce the

---

* A. S. Russell, *Phil. Mag.* **47**, 1121, 1924; **48**, 365, 1924. W. Widdowson and A. S. Russell, *Phil. Mag.* **48**, 293, 1924.

masses of the most abundant isotopes of the ordinary elements. The heads of three main series of radioactive elements, thorium, actinium and uranium, have mass numbers $4n,$ $4n + 1$, $4n + 2$, respectively, and the existence of a fourth group of mass $4n + 3$ is assumed. From a consideration of the modes of transformation of these series and certain general rules, Russell has been able to predict with some success the masses of the main isotopes of many of the elements. While there appears to be an undoubted analogy between the radio-active isotopes and the isotopes of inactive elements, there is no necessity to assume that the inactive elements have been derived from the transformations of the radio-elements.

§ 121. **Abundance of the elements.** Attention has already been drawn to marked differences in the number and masses of the isotopes of even- and odd-numbered elements. This distinction between even and odd elements is manifested also in other directions. For example, in the study of the artificial disintegration of light elements by a particles, it is found that the effect is most marked in odd-numbered elements. The even elements either show no effect at all or give rise to protons of lower average speed than from the odd elements. Another striking difference, first drawn attention to by Harkins*, is observed in connection with the relative abundance of the chemical elements in the material of the earth. Elements of even atomic number predominate and indeed form the greater part of the earth's crust. Various estimates have been made of the relative abundance of the elements both in stone meteorites and in the average material of our lithosphere. While it is of course difficult to be certain that the analyses represent the average composition of the whole material of the earth, certain general conclusions can be safely drawn. The following table, compiled by Harkins, shows the percentage composition by weight of the lithosphere taken from the data given by Clarke.

It is seen that the six even-numbered elements make up 85·7 per cent. of the weight of the lithosphere and the odd-numbered 12·8 per cent. Even if hydrogen is included, the proportion is not seriously altered. The two even-numbered elements oxygen and silicon together are responsible for 75 per cent. of the material of the earth, while less than 0·2 per cent. is contributed by elements of the higher atomic numbers 30 to 92.

---

* Harkins, *Journ. Amer. Chem. Soc.* **39**, 856, 1917.

If it be supposed that the elements are built up of protons and electrons, or of simple combinations of these units, it is to be anticipated that the lighter stable nuclei would be formed in much greater quantity than the heavier and more complex nuclei. The comparative rarity of the elements of atomic number greater than 30 is thus an indication that the nuclei arise from the building up of simple units into more complicated structures rather than from the disintegration of some hypothetical heavy element or elements.

Element	Atomic number	Percentage by weight	
Oxygen	8	47·33	—
Silicon	14	27·74	—
Aluminium	13	—	7·85
Iron	26	4·50	—
Calcium	20	3·47	—
Sodium	11	—	2·46
Potassium	19	—	2·46
Magnesium	12	2·24	—
Titanium	22	0·46	—
Totals	... ...	85·74	12·77

From analogy with the transformation of radioactive bodies there has been a tendency to assume that the relative abundance of an element is to be taken as a measure of its stability. No doubt if the processes of building up of nuclei of all possible types from simple units takes place in the stellar system, it is to be expected that any unstable nuclei which have a limited life of the order of 100 million years or less would be relatively rare in the earth. On the other hand, general evidence indicates that the great majority of the elements and their isotopes are either permanently stable or have an average life of transformation long compared with the estimates of the age of the sun (10^{12} to 10^{14} years). Under such conditions, the relative abundance of the elements must be mainly due to other factors, like the supply of the requisite building units and the probability of the formation of the nuclei in question. While the experiments on artificial disintegration of the elements by α particles certainly suggest that odd-numbered elements are more easily altered than even elements, and thus in a sense appear less stable, it does not follow that the odd-numbered elements would show any sign of radioactive instability under normal conditions in the earth. For example, aluminium, which is the most abundant of the odd-numbered elements and therefore presumably very stable, not only emits protons freely under α ray bombardment, but with an average individual energy greater than for any other element. It is seen that any deductions on the stability of elements based directly on their abundance must be interpreted with caution.

§ 122. **Masses of isotopes.** The original mass spectrograph of Aston gave the masses of the isotopes with an accuracy of about 1 in 1000, and sufficed to show that the masses were nearly integers in terms of $O = 16$. In order to examine the accuracy of the whole-number rule, and to resolve the isotopes of the heavier elements, a modified form of apparatus was constructed in which it was possible to compare the masses of atoms with an accuracy of about 1 in 10,000. The mass spectrum obtained on a photographic plate about 16 cm. long corresponded to rather more than one octave of mass. A change of mass of 1 per cent. corresponded to about 1·5 mm. on the photographic plate at the most deflected end and 3 mm. at the other. The dispersion of the new instrument was so great that the lines due to O^+ and CH_4^+ were clearly separated on the plate*.

The masses of the elements carefully examined by Aston with the new apparatus are given in the table below. The results for the two isotopes of lithium are due to Costa and are probably accurate to 1 in 3000.

Atom	Packing fraction $\times 10^4$	Mass $O = 16$	Atom	Packing fraction $\times 10^4$	Mass $O = 16$
H	$77\cdot8 \pm 1\cdot5$	$1\cdot00778$	Cl^{35}	$-4\cdot8 \pm 1\cdot5$	$34\cdot983$
He	$5\cdot4 \pm 1$	$4\cdot00216$	A^{36}	$-6\cdot6 \pm 1\cdot5$	$35\cdot976$
Li^6	$20\cdot0 \pm 3$	$6\cdot012$	Cl^{37}	$-5\cdot0 \pm 1\cdot5$	$36\cdot980$
Li^7	$17\cdot0 \pm 3$	$7\cdot012$	A^{40}	$-7\cdot2 \pm 1$	$39\cdot971$
B^{10}	$13\cdot5 \pm 1\cdot5$	$10\cdot0135$	As	$-8\cdot8 \pm 1\cdot5$	$74\cdot934$
B^{11}	$10\cdot0 \pm 1\cdot5$	$11\cdot0110$	Kr^{78}	$-9\cdot4 \pm 2$	$77\cdot926$
C	$3\cdot0 \pm 1$	$12\cdot0036$	Br^{79}	$-9\cdot0 \pm 1\cdot5$	$78\cdot929$
N	$5\cdot7 \pm 2$	$14\cdot008$	Kr^{80}	$-9\cdot1 \pm 2$	$79\cdot926$
O	$0\cdot0$	$16\cdot0000$	Br^{81}	$-8\cdot6 \pm 1\cdot5$	$80\cdot926$
F	$0\cdot0 \pm 1$	$19\cdot0000$	Kr^{82}	$-8\cdot8 \pm 1\cdot5$	$81\cdot927$
Ne^{20}	$0\cdot2 \pm 1$	$20\cdot0004$	Kr^{83}	$-8\cdot7 \pm 1\cdot5$	$82\cdot927$
Ne^{22}	$(2\cdot2 \quad ?$	$22\cdot0048)$	Kr^{84}	$-8\cdot5 \pm 1\cdot5$	$83\cdot928$
P	$-5\cdot6 \pm 1\cdot5$	$30\cdot9825$	Kr^{86}	$-8\cdot2 \pm 1\cdot5$	$85\cdot929$
			I	$-5\cdot3 \pm 2$	$126\cdot932$
Tin (eleven isotopes)			Sn^{120}	$-7\cdot3 \pm 2$	$119\cdot912$
Xenon (nine isotopes)			Xe^{134}	$-5\cdot3 \pm 2$	$133\cdot929$
Mercury (six isotopes)			Hg^{200}	$+0\cdot8 \pm 2$	$200\cdot016$

* Aston, Bakerian Lecture, *Proc. Roy. Soc.* A, **115**, 487, 1927.

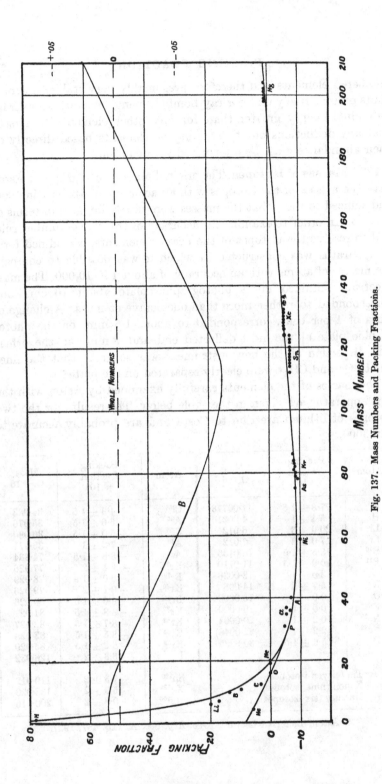

Fig. 137. Mass Numbers and Packing Fractions.

It is seen that there are systematic departures from the whole-number rule, showing that the average mass of the proton in different nuclei is subject to small variations which are ascribed to changes in the "packing effect" of the nucleus. To express this variation in convenient form, the packing fraction, shown in column 2, is calculated from the observed masses. This packing fraction, expressed in parts per 10,000, represents the mean gain or loss of mass per proton compared with the protons of oxygen taken as a standard. The variation of the packing fraction with the atomic mass is shown clearly in Fig. 137. It is seen that the packing fraction, initially positive for atoms lighter than oxygen, reaches a maximum negative value for atomic mass about 60 and then slowly rises again crossing the zero line for a mass about 200, and presumably rises steadily up to the heaviest element uranium. The actual variation of mass from an integer is shown in the same figure, curve B (scale on right). It is seen that the difference at first positive becomes zero for mass about 20 and reaches a maximum negative value for about mass 110, and becomes zero again for a mass about 190. No accurate measurements have been made for elements heavier than mercury, but the curve if extrapolated to include uranium should continue to rise well above the zero line.

In Fig. 138, the values of the packing fractions for the lighter elements are shown on a larger scale. It is seen that the packing fractions for the odd-numbered elements lie approximately on a curve hyperbolic in shape. On the other hand, the packing fractions for the even-numbered elements, helium, carbon and oxygen, fall well below the curve. It is clear from these results that the structure of the even-numbered elements differs in some way from the odd-numbered elements. It may be that carbon and oxygen are composed of α particles and that the odd-numbered elements consist of a sub-structure of α particles with additional protons of average mass larger than unity but considerably less than the mass of the free proton. The data available are not, however, sufficient to draw any very definite conclusions on this point.

The choice of $O = 16$ as a basis for comparison of the packing effects is of course quite arbitrary, and Strum[*] and St. Meyer[†] have taken $H = 1$ as a basis. The same general features in the variation of the packing fraction with atomic number must naturally appear in

[*] Strum, *Zeit. f. Phys.* **50**, 555, 1928.
[†] St. Meyer, *Naturwiss.* **15**, 623, 1927; *Wien. Ber.* **138**, 431, 1929.

these calculations. In addition, St. Meyer has pointed out that the packing fraction seems to show a periodic behaviour which is related to the well-known periods in the natural classification of the elements. The data are probably not yet sufficiently precise to establish such an unexpected relation.

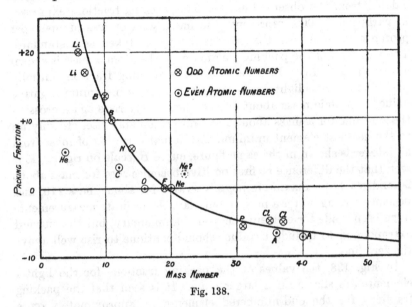

Fig. 138.

§ 123. Energy considerations.

On Einstein's theory, the energy E associated with a mass m is given by $E = mc^2$, where c is the velocity of light. If there is a change of mass δm in a system due to any rearrangement, the change of energy δE is given by $\delta E = c^2 \delta m$. In dealing with the changes of energy in atomic nuclei, it is convenient to express the energy in terms of the kinetic energy gained by an electron falling freely between a difference of potential V in volts. The energy per atom associated with a change of mass 1 on the atomic mass scale can be shown to be 933 million electron volts. This is a convenient number on which to base calculations of changes of energy involved in the structure of single nuclei. From Aston's data, the mass of the free proton is 1·0072 and the mass of the helium nucleus 4·0011 in terms of the oxygen *atom* = 16. If this helium nucleus be taken as a combination of four protons and two electrons, the loss of mass 4·0301 − 4·0011 = ·029, corresponding to an energy loss of 27 million volts. The energy involved in the change of mass of the proton 1·0072 in the free state to a mass of exactly 1 is 6·7 million volts. Similarly it can be deduced that the energy equivalent to the mass of a free electron and a free proton is about 511,000 volts and 940 million volts respectively.

If we suppose that all atomic nuclei have been formed by the combination of free protons and electrons, it is clear, since the average mass of the proton in most nuclei is nearly 1, that a large amount of energy must have been radiated in the process of their formation. For example, for the main isotope of mercury of mass 200·016, which is composed of 200 protons and 120 electrons, the loss of atomic mass in its formation from free electrons and protons is 1·45, corresponding to 1350 million volts. Similarly for the krypton isotope, mass 80, the loss is 600 million volts. On this hypothesis, the formation of all atomic nuclei from protons and electrons involves a loss, and often a considerable loss, of energy. If the nucleus of an atom were taken to pieces proton by proton and electron by electron, a large amount of work would have to be done. Consequently, the atom instead of being a source of energy is a sink of energy. It is difficult at first sight to reconcile such a conclusion with the observed fact that the uranium atom in its successive transformations spontaneously emits in the form of its characteristic radiation a total amount of energy corresponding to about 45 million volts. On the other hand, an emission of this amount of energy seems to be consistent with the rise of Aston's mass curve given in Fig. 137, if extrapolated to include uranium. In the transformation of uranium to lead, eight α particles and six β particles are liberated with a total emission of energy per atom of 45 million volts. The total mass of eight free helium nuclei and six electrons corresponds to 32·012, while the mass equivalent to an emission of energy of 45 million volts is ·048. Consequently the theoretical difference of mass to be expected between uranium and uranium lead should be 32·060, and the departure of the atomic mass of uranium from a whole number should be ·060 greater than for uranium lead. This is about the difference to be expected from an extrapolation of either of the curves given in Fig. 137, and is thus in general accord with observation. If the amount of energy liberated in each α and β ray transformation is known, it is clear that the change of mass from atom to atom can be calculated. If the mass of one atom in the series is known accurately, the masses of all the others can be deduced.

The difficulty on nuclear theories of accounting for the spontaneous emission of energy from radioactive atoms is largely removed if we assume that the main building unit entering into the structure of the heavier nuclei is not the proton but the α particle. This supposes that the emission of energy in the formation of an α particle from protons and electrons has occurred before the α particle enters the nucleus in question. In the table given by Aston the mass of Xe^{134} is 133·929 and Hg^{200} is 200·016. Making allowance for the mass of the extra-nuclear electrons, the total difference of the nuclear mass of xenon and mercury is 66·073. This corresponds on the average over this range to a change of mass per proton of 1·0011 and to a change of mass per α particle of 4·0044. In a similar way, from data already considered, it can be calculated that the average nuclear change of mass per α particle from xenon to uranium is about 4·0047. Now we have seen that the nuclear mass of the α particle in the free state is 4·0011. On the assumption therefore that the α particle is the main mass unit in building up the nuclei, it is seen that the average mass of the α particle successively added to the nucleus must be greater than in the free state. In other words, the α particle in the nucleus possesses greater total energy than in the free state. If the uranium nucleus were taken to pieces α particle by α particle, till xenon is reached, it is clear that a considerable amount of free energy would be liberated. On the other hand, we have seen that if the nucleus consisted mainly of protons, a large amount of energy would be expended in the corresponding process. If it be supposed that the average mass of the proton included in the structure of a nucleus is very nearly 1, its mass is much less in the nucleus than in the free state and a considerable amount of energy must be communicated to it by the other parts of the nucleus in order to give it sufficient energy to escape. We can thus see in a general way why protons, even if they are present, are not liberated in radioactive transformations.

The general evidence on nuclei strongly supports the view that the α particle is of primary importance as a unit of the structure of nuclei in general and particularly of the heavier elements. It seems very possible that the greater part of the mass of heavy nuclei is due to α particles which have an independent existence in the nuclear structure.

On these views, it is possible to form a general picture of the gradual building up of nuclei and of the energy changes involved. Probably in the lighter elements, the nucleus is composed of a combination of α particles, protons and electrons. The constituents of the nucleus attract one another strongly, partly no doubt due to the intense forces arising from the distortion or polarisation of the particles and possibly due to magnetic forces. It is difficult in the present state of our knowledge to define the relative importance of these various agencies. At first a concentrated and firmly bound nucleus is formed accompanied by the emission of energy, and for an atom about 120 a minimum mass is reached representing the closest binding of the constituent particles. With successive additions of mass, the a particle becomes less and less tightly bound until the heaviest element uranium is reached. The variation of mass from atom to atom may not be regular and is probably greater for a mass in the neighbourhood of 200 than for a mass 120.

We may thus suppose that the nucleus consists of a tightly packed structure near its centre, gradually becoming more diffuse towards the outside. This system is surrounded by a high-potential barrier which normally prevents the a particle from escaping. From the standpoint of the wave-mechanics the a particle is accompanied by a wave system which fills the whole volume inside the potential barrier. As we have seen in § 77 a, it is believed that the a particle, or rather its wave system, may penetrate this barrier under special conditions and escape from the nucleus.

For simplicity of explanation, we have considered above a type of static atom in which the total energy of each a particle increases from the centre outwards. It is more probable, however, that the constituents of the nucleus are in rapid motion, continuously interchanging positions with one another, and it may not be valid to suppose that on the average one a particle has more energy than another. There is, however, always a probability that the inner highly condensed nucleus may form a compact and definitely orientated structure surrounded by a number of a particles as satellites. In such a case the equilibrium of a nucleus would be analogous to the equilibrium of a sphere of liquid in the presence of a surrounding vapour near the critical point. On such a view some of the satellites would have a higher total energy than the average of all the a particles in a radioactive nucleus. If the energy on the average is equally distributed amongst all the a particles constituting a heavy nucleus, it is not obvious why a uranium nucleus should disintegrate at all, since the average mass of the a particle is little if any greater than its mass in the free state. However, our information is too indefinite to be dogmatic on these points.

G. Gamow* has examined the equilibrium of a nucleus supposed to consist of a collection of α particles surrounded by a potential barrier. It is assumed that the α particles are acted on by attractive forces varying rapidly with the distance and overbalancing at short distances the electrostatic forces of repulsion. Since the α particles carry even charges, they should obey the Bose-Einstein statistics. He concludes they must all be in the same quantum state with quantum number unity. The equilibrium of such a collection of α particles under the action of rapidly varying attractive forces is analogous to the behaviour of a small drop of water in which the particles are held together by the forces of surface tension. It is possible to calculate approximately the relation between the total energy of the particles contained in the nucleus and the number of α particles, i.e. the mass of the nucleus. The calculations have so far been applied only to elements of mass $4n$ where n is an integer. By taking account of the effect of the introduction of electrons in the nucleus for certain values of n†, Gamow‡ has been able in a general way to account for the peculiarities of the mass-defect curve shown in Fig. 137, curve B. While this curve gives the general trend of the variation in mass over the whole range of atomic weights, Gamow concludes that the gain or loss of mass due to the addition of an α particle may vary markedly from point to point. This theory while admittedly imperfect and speculative in character is of much interest as the first attempt to give an interpretation of the mass-defect curve of the elements. For further progress in the theory, the masses of many of the isotopes require to be known with great accuracy.

We have seen that on energy considerations we can understand why α particles and not protons are liberated from the radioactive bodies. It is also clear why elements much heavier than uranium have never been observed. With increasing mass, more and more energy is stored in the nucleus which should become highly radioactive. While no doubt the stability of a nucleus depends on its structure as well as its mass, probably in general the greater the mass of the nucleus, the more rapidly it would be transformed, and it is probably not an accident that no element heavier than uranium has

---

* Discussion on the structure of atomic nuclei. *Proc. Roy. Soc.* A, **123**, 373, 1929.

† G. Beck (*Zeit. f. Phys.* **50**, 548, 1928) has pointed out that for even elements of type $4n$, the number of free nuclear electrons is always even, and that a new pair of electrons appears at certain mass numbers.

‡ Gamow, *Proc. Roy. Soc.* A, **126**, 632, 1930.

so far been detected. Even if such heavy elements were present in the earth at the time of its formation, they would have largely disappeared if their radioactive life were small compared with the age of the earth. Uranium and thorium are apparently the sole survivors of such elements to-day for the reason that their average life is probably greater than the age of the earth.

In § 8 b, we have given the evidence in support of the view that the actinium series originates in an isotope of uranium of mass 235, which has a considerably shorter life than the main isotope of mass 238. Other isotopes may also have been formed but of too short a life to be observed to-day. For this reason, as pointed out in § 38, it is important to examine with great care pleochroic haloes in mica and other substances to see whether definite evidence can be obtained of the existence of unknown groups of α particles which may be ascribed to elements that have vanished long ago.

In this connection, it is of interest to refer to a suggestion made by Kirsch* that thorium originally arose from an isotope of uranium of which the radioactive life was short compared with that of thorium itself. This was based on the observation that thorium and uranium are found together in old unaltered minerals and that the amount of thorium relative to uranium increases with the age of the mineral. More recent analyses of uranium minerals, in particular of a uraninite from Keystone, South Dakota, which is the oldest mineral known and yet contains only a very small amount of thorium, have not supported this suggestion, and it seems necessary to abandon it†.

There seems to be little doubt that the processes of formation of complex atomic nuclei are still taking place in the sun and the hot stars. Evidence bearing on this point has recently been obtained from consideration of the origin of actinium. We have seen (§ 8 b) that actinium is probably derived from the transformation of an isotope of uranium (actino-uranium), of mass 235, which has an average life of the order of 400 million years. It is natural to suppose that the uranium in our earth had its origin in the sun and has been decaying since the separation of the earth from the sun. In order to account for the amount of actinium at present observed in uranium minerals, it can be estimated that the earth cannot have an age much greater than 4×10^9 years‡. If the age of the sun is of the

---

* Kirsch, *Wien. Ber.* II a, **131**, 551, 1922.
† Kirsch, *Geologie u. Radioaktivität*, Springer, Berlin, 1928.
‡ Rutherford, *Nature*, **123**, 313, 1929.

order of magnitude estimated by Jeans, namely 7×10^{12} years, it is clear that the actino-uranium observed in the earth must have been forming in the sun at a late period of its history, viz. about 4×10^9 years ago. If the actino-uranium could only be formed under special conditions in the early history of our sun, the actino-uranium, on account of the shortness of its life compared with uranium, should have practically disappeared long ago. It thus seems likely that the processes of production of elements like uranium were taking place in the sun about 4×10^9 years ago and probably still continue to-day. It is natural to suppose that this is true also for complex nuclei in general.

CHAPTER XVIII

MISCELLANEOUS

§ 124. The apparent radioactivity of ordinary matter. It was pointed out by Schuster in 1903 that every physical property discovered for one element had later been found to be shared by all the other elements in varying degrees. On such general grounds it might perhaps be expected that the instability shown by the elements uranium, thorium, actinium, and their products should be a property common to all matter. It is indeed true that every substance which has been examined shows a feeble radioactivity which can be detected by the ionisation method, but it seems probable that in all cases, with the two exceptions of potassium and rubidium, this activity is to be ascribed to the presence of traces of bodies belonging to the well-known radioactive families rather than to an instability of the element itself.

The early investigations of C. T. R. Wilson[*] on the rate of loss of charge of an insulated conductor in a closed vessel indicated that the ionisation was produced by a radiation proceeding from the walls of the vessel. This view was confirmed by the experiments of Strutt[†], who found that the rate of discharge of an electroscope depended on the material of which it was composed. In some cases markedly different rates of discharge were found for different specimens of the same metal, indicating that in these cases at least a large part of the effect was due to radioactive impurities. On the other hand, Campbell[‡] concluded from an extensive series of measurements that all metals showed a specific radioactivity and emitted characteristic radiations of the α ray type. Of more recent work there may be mentioned the experiments of Harkins and Guy[§], who examined a large number of substances and found no evidence of any specific activity; and those of Hoffmann[||], who concluded that the activity of some materials could not be attributed only to radioactive impurities.

[*] C. T. R. Wilson, *Proc. Camb. Phil. Soc.* **11**, 32, 1900; *Proc. Roy. Soc.* A, **68**, 151, 1901.

[†] Strutt, *Phil. Mag.* **5**, 680, 1903.
[‡] Campbell, *Phil. Mag.* **9**, 531, 1905.
[§] Harkins and Guy, *Proc. Nat. Acad. Sci.* **11**, 628, 1925.
[||] Hoffmann, *Ann. d. Phys.* **62**, 738, 1920.

In considering the evidence obtained in this way, it is necessary to examine the factors which influence the rate of discharge of a metal electroscope filled with air. The ionisation may be due to one or more of the following causes:

(1) the presence of a small quantity of radon mixed with the air;

(2) a superficial activity due to the exposure of the metal to the radioactive matter present in the atmosphere;

(3) the presence of traces of known types of radioactive matter throughout the volume of the metal;

(4) the escape of a radioactive emanation from the metal into the gas;

(5) the effect of γ rays from radioactive bodies in surrounding objects and in the earth and atmosphere;

(6) the effect of the "penetrating" radiation;

(7) a spontaneous ionisation of the gas; for this effect there is no evidence.

The effect of (1) is usually small and does not amount to more than 1 or 2 ions per c.c. per second in the air. The radon, if necessary, can be removed entirely by passing the air in the vessel through a charcoal tube immersed in liquid air. In an airtight vessel the effect will of course disappear in the lapse of a few weeks. The effect of (2) may be considerable, for the surface of all metals exposed in the open air becomes temporarily active, due to the presence of radium and thorium emanations in the atmosphere. McLennan and Burton*, for example, have observed that the ionisation due to a sheet of metal exposed in the open air always decreases for several hours after its introduction into a closed vessel. This is no doubt due to the active deposit of radium collected from the atmosphere. In addition, if the plate has been exposed long in the open air, radium $(D + E + F)$ collect upon it. It is thus to be expected that a metal would produce less ionisation when its surface is removed or thoroughly cleaned to get rid of the active matter. This has been shown to be the case by Cooke, who found that a brass electroscope showed only about one-third of its original activity after thorough cleaning. McLennan also found that the ionisation in large metal vessels, after an initial decrease, gradually rose for several days. The activity fell again when fresh air was substituted. This no doubt was due to the presence of a trace of radium in the metal and the release of a small quantity of radon into the air of the ionisation chamber. The greater part of the effect due to (5) will

* McLennan and Burton, *Phil. Mag.* **6**, 343, 1903.

under normal conditions be caused by the radioactive matter in the earth. The effect of the γ radiation from the earth can be eliminated by making measurements on the surface of a deep lake or ocean, provided that the water does not itself contain sufficient radium to give a detectable amount of γ rays*.

When adequate precautions are taken to eliminate the effects of factors (1) to (5), the rate of discharge of a metal electroscope filled with air at standard pressure corresponds to the production of not more than 5 or 6 ions per c.c. per second, when observed at sea level. Of this amount about 1·5 to 2 ions must be attributed to factor (6), the effect of the "penetrating" radiation (see § 112). The residual ionisation is therefore less than 4 ions per c.c. per second. Some observers† have found residual ionisations of less than 1 ion per c.c. per second.

It is difficult to decide with certainty whether this residual ionisation is due to an inherent radioactivity possessed by the metal of the electroscope or to the presence of a trace of known radioactive matter. Since the radioactive elements are widely disseminated throughout the surface of the earth‡, it is to be expected that traces of active matter would be separated with all metals. This view receives some support from experiments made with lead, the most active of all common materials. If any radium be present in the ores from which the lead was prepared, the product radium D will be separated with the lead; the α ray activity of such lead will be due to the subsequent product radium F or polonium, and owing to the decay of radium D (with a period of 25 years) the activity of the lead will decrease with its age. Both these points have been tested. Elster and Geitel§ extracted from lead a small quantity of active matter which they considered to be polonium, and it has often been observed, first by McLennan‖, that old lead is much less active than new lead.

An estimation of the amount of radioactive impurity necessary to account for the residual ionisation in an electroscope can be made from calculations due to v. Schweidler¶. For example, suppose that a spherical copper vessel of 2 litres volume gives a residual ionisation

* The radium content of sea water varies between 0·25 and 26 × 10^{-15} gm. per c.c.
† Cf. Bergwitz, *Elster-Geitel Festschr.* 585, 1915; Hoffmann, *Ann. d. Phys.* **82**, 413, 1927.
‡ The earth's crust contains on the average about 2 × 10^{-12} gm. Ra per gm. of rock.
§ Elster and Geitel, *Phys. Zeit.* **9**, 289, 1908.
‖ McLennan, *Phil. Mag.* **14**, 760, 1907.
¶ v. Schweidler, *Phys. Zeit.* **15**, 685, 1914.

of 4 ions per c.c. per second. The observed ionisation would be accounted for by the presence of radium as impurity to the extent of 4×10^{-13} gm. per gm. of copper. If, on the other hand, it be assumed that the ionisation is due to a true radioactivity of the copper, the calculations lead to a half-transformation period of about 3×10^{15} years.

In the experiments of Hoffmann[*] the current in the ionisation chamber was measured by means of a special electrometer and registered photographically. The electrometer was so sensitive that the ionisation due to each α particle emitted in the chamber could be measured with fair accuracy. Hoffmann was thus able to separate the α ionisation in the chamber from the β and γ ionisation. From a statistical examination of his results he concluded that platinum and brass gave a weak α activity which could not be attributed to the presence of known radioelements. In a later series of experiments Hoffmann[†] measured the γ activity of various metals, viz. lead, zinc, copper, iron, and aluminium, and he concluded that in each case the radium content was not more than 10^{-14} gm. Ra per gm. of metal. He considered that these results strengthened his former conclusion of a true α activity of brass. According to Ziegert[‡], who has investigated by the same method a number of metals including copper and zinc, this activity of brass is to be ascribed to its zinc content. While copper showed a very small α activity which could be ascribed to a radium content of between 1 and 2×10^{-14} gm. Ra per gm. Cu, all his specimens of zinc emitted α particles of unusually short ranges. By chemical means he separated from the zinc a small residue which gave, in addition to some of the known α particles of the uranium series, three groups of particles of very small total ionisation, corresponding to ranges in air of 0·55, 1·15, and 2·05 cm. He suggested a connection between these particles and the "X" haloes of Joly and the "Z" haloes of Iimori and Yoshimura (see § 38), in which occur rings of radius corresponding in the former case to an α particle of range just over 1 cm., and in the latter case to 1·2 and 2·1 cm. On the other hand, if the Geiger-Nuttall relation between the range of the α particle and the transformation constant of the elements of the known radioactive series may be extrapolated to these bodies, it would suggest that their rates of transformation should be slow

[*] Hoffmann, *Ann. d. Phys.* 62, 738, 1920; *Zeit. f. Phys.* 7, 254, 1921.
[†] Hoffmann, *Ann. d. Phys.* 82, 413, 1927.
[‡] Ziegert, *Zeit. f. Phys.* 46, 668, 1927.

compared even with uranium, and this is not consistent with the observed rate of emission of α particles from the small quantity of residue which was obtained. It is indeed difficult to offer any suggestions as to the source of these α particles, and further information must be awaited.

The question of the activity of the common elements is one of great interest, and the method of investigation developed by Hoffmann and Ziegert should afford a much more definite answer than can be obtained from the earlier electroscope measurements. Taking the evidence as a whole, and leaving Ziegert's observations for further investigation, we may say that there is no adequate proof that the ordinary chemical elements show an intrinsic activity such as is possessed by the members of the known radioactive families. The very small activity often observed is in all probability to be ascribed to the presence of traces of the known radioelements as impurities.

§ 125. **The radioactivity of potassium and rubidium.** It has already been mentioned that there are two exceptions to this general conclusion—potassium and rubidium both show a weak β ray activity which is specific to these elements. This fact was discovered by Campbell and Wood* in an examination of the activities of the alkali metals. They found that the β ray activity was shown by all potassium salts, whatever their origin, in proportion to their content of potassium. Tests for the presence of an emanation in the substances used gave a negative result, and they concluded that the observed emission of β rays was a property of the potassium atom. The β radiation was more easily absorbed than that from a uranium oxide preparation, giving a value for μ/ρ in tin varying from 27 to 11 as against a value of 10 with uranium. Similarly, the β radiation from rubidium was shown to be a property of the rubidium atom. Rubidium gave, weight for weight, about seven times the effect of potassium, but the radiation was much more easily absorbed. Subsequent investigations by Campbell†, Henriot‡, Biltz and Marcus§, Hahn and Rothenbach‖, Hoffmann¶, and others have confirmed the conclusion that the β radiations of these elements are not due to traces of radioactive

---

* Campbell and Wood, *Proc. Camb. Phil. Soc.* **14**, 15, 1906.

† Campbell, *Proc. Camb. Phil. Soc.* **14**, 211, 557, 1907.

‡ Henriot, *Le Radium*, **7**, 40, 1910.

§ Biltz and Marcus, *Zeit. f. anorg. Chem.* **81**, 369, 1913.

‖ Hahn and Rothenbach, *Phys. Zeit.* **20**, 194, 1919.

¶ Hoffmann, *Zeit. f. Phys.* **25**, 177, 1924.

impurity but to the elements themselves. All attempts to influence the activity of either element by chemical operations, and thus to prove that the activity is due to radioactive impurity, have been unsuccessful. Biltz and Marcus prepared specimens of K_2SO_4 from potassium minerals of different geological ages; the activity of the specimens per atom of potassium was always the same. Hahn and Rothenbach compared the activities of rubidium sulphate of different ages, 21 years old, 11 years, and new. No difference could be found.

In both cases the radiation is a β radiation; no α rays have been observed. Campbell showed that the rays from potassium were deflected in an electric field in the same direction as the β rays from a layer of uranium oxide. The deflection in a magnetic field was observed by Henriot and Vavon[*] for the rays from potassium, by Bergwitz[†] for the rays from rubidium. The magnetic spectra of the radiations have not been obtained for either element, and estimates of the average velocity of the rays have been derived almost entirely from absorption measurements. The early measurements of Campbell and Wood showed that the potassium rays were slightly less penetrating than those from a layer of uranium oxide, i.e. from uranium X_2, since the β rays of uranium X_1 are difficult to observe except with very thin layers. Further measurements have been made by Harkins and Guy[‡], and by Kuban[§]. The latter obtained, after certain corrections, a value for the absorption coefficient in aluminium of $\mu = 28$ (cm.)$^{-1}$. From these results the average velocity of the β rays of potassium may be taken to be approximately $2 \cdot 5 \times 10^{10}$ cm./sec., or $0 \cdot 85$ c.

From a comparison of the magnetic deflection of the β rays of rubidium with that of the rays from radium E, Bergwitz estimated the velocity of the rubidium rays to be about $1 \cdot 85 \times 10^{10}$ cm./sec. On the whole, the measurements of the absorption of the rays in metal foils are in agreement with this value. Henriot, and Hahn and Rothenbach, found that the penetrating power of the rubidium rays was about the same as that of the β rays of radium itself, which have velocities of roughly $0 \cdot 6$ c. Kuban obtained a value of μ in aluminium of 173 (cm.)$^{-1}$, which is rather less than the corresponding value for radium. Using a thin layer of rubidium salt, Hoffmann[||] found

* Henriot and Vavon, C.R. 149, 30, 1909.
† Bergwitz, Phys. Zeit. 14, 655, 1913.
‡ Harkins and Guy, Proc. Nat. Acad. Sci. 11, 628, 1925.
§ Kuban, Wien. Ber. 137, 241, 1928.
|| Hoffmann, Zeit. f. Phys. 25, 177, 1924.

evidence of a very soft component in the radiation with a value $\mu = 900$ (cm.)$^{-1}$ in aluminium.

The rates of transformation of potassium and rubidium can be deduced from measurements of the rate of emission of the β rays. These measurements are difficult on account of the weak activity shown by both these elements, and only very rough estimates of the half-transformation periods can be made. A further difficulty arises from the absorbability of the rays, particularly in the case of rubidium. A surface of a thick layer of either substance gives about one-thousandth of the effect of a surface of uranium oxide of the same area, when compared with a β ray electroscope. A general survey of the evidence bearing on this question has been made by Holmes and Lawson*, and they conclude that the half-value period of rubidium is probably about 10^{11} years, and of potassium about $1\cdot5 \times 10^{12}$ years.

These estimates were based on the assumption that all the atoms of the element are similar in their radioactive behaviour. On the other hand, it is known that both potassium and rubidium consist of two isotopes of masses 39 and 41, 85 and 87, respectively, and it may be that only one of these isotopes, or possibly a third unknown isotope, is responsible for the β radiation. In the case of potassium, a partial separation of the isotopes was obtained by Hevesy†. The heavy fraction was collected and an atomic weight determination by Hönigschmid gave a value 0·005 unit greater than that of ordinary potassium. The activity of this fraction was measured by Biltz and Ziegert‡ and found to be about 4 per cent. greater than that of common potassium. The results are in good agreement with the assumption that the isotope of mass 41 is mainly if not solely responsible for the β radiation of potassium. Since this isotope forms only one-twentieth of the element, it will have, if entirely responsible for the radioactivity, a half-value period of about $7\cdot5 \times 10^{10}$ years.

No transformation products either of potassium or of rubidium have so far been detected. Since the emission of a β particle raises the nuclear charge by one unit, potassium should change into calcium and rubidium into strontium. If in potassium only the isotope 41 is radioactive, the product will be a calcium isotope of mass 41; at present only two calcium isotopes are known, of masses 40 and 44. Using the above period for potassium 41, it can be calculated that,

---

* Holmes and Lawson, *Phil. Mag.* **2**, 1218, 1926.

† Hevesy, *Nature*, **120**, 838, 1927.

‡ Biltz and Ziegert, *Phys. Zeit.* **29**, 197, 1928.

since the consolidation of the earth's crust, about 2 per cent. of this isotope will have disintegrated. The maximum amount of calcium 41 which has accumulated in potassium minerals during all geological time will amount therefore to about 0·1 per cent. of their potassium content. Even if a potassium mineral could be found which contained only a trace of calcium, it would be exceedingly difficult to detect the difference in atomic weight between the calcium of the mineral and calcium of other origin. In the case of rubidium, the chemical difficulties in the way of establishing the difference in atomic weight between strontium produced by the decay of rubidium and ordinary strontium would not be so great. On the other hand, there is no real rubidium mineral; this element is widely disseminated over the earth's crust but always in small amounts.

In addition to the β radiation, Kolhörster* has observed the emission of a γ radiation from potassium. He estimated that the intensity of this radiation corresponded roughly to an amount of 10^{-11} gm. Ra per gm. of sylvin, KCl. Measurements of the absorption coefficient were made in iron and gave values varying from 1 (cm.)$^{-1}$ to 0·19 (cm.)$^{-1}$ as the thickness of absorber was increased from a few millimetres up to 6 cm. The γ radiation from potassium thus contains components which are more penetrating than the hard γ radiation of radium C, for which $\mu = 0.36$ (cm.)$^{-1}$ in iron. It is possible that a γ radiation may also be emitted by rubidium, but as yet it has not been observed.

§ 126. The counting of scintillations. The scintillation method of counting α particles has been described very briefly in § 12 b. Although the counting of scintillations has been widely used as an experimental method, little systematic work was done to investigate the method itself and its limitations until recently, when the experiments of Karlik and Kara-Michailova† and of Chariton and Lea‡ were able to furnish a clear account of some of the factors involved. Chariton and Lea investigated the two fundamental processes concerned in the production and observation of the scintillations, (1) the transformation of the energy of the α particle which strikes the zinc sulphide crystal into radiant energy, and (2) the detection of this radiant energy by the eye under the experimental conditions. It is

* Kolhörster, *Naturwiss.* **16**, 28, 1928.

† Karlik, *Wien. Ber.* **136**, 531, 1927; Karlik and Kara-Michailova, *Zeit. f. Phys.* **48**, 765, 1928; *Wien. Ber.* **137**, 363, 1928.

‡ Chariton and Lea, *Proc. Roy. Soc.* A, **122**, 304, 1929.

the extraordinary efficiency of these two processes which gives the scintillation method its great power.

It is well known that pure zinc sulphide does not scintillate when bombarded by α particles nor does it fluoresce under the action of light. In order to sensitise the zinc sulphide to the action of light or of α particles, the insertion of a minute quantity of impurity, such as copper, manganese, or other metal, is necessary. The response of such sensitised zinc sulphide depends on the nature and amount of impurity, and it appears that the best scintillations are given when the impurity is copper in amount about 0·01 per cent. of the zinc present. The method of preparation of phosphorescent zinc sulphides has been given by Tomaschek*. These zinc sulphides cannot be prepared in large crystals and it is usual to make a screen by dusting a layer of small crystals on a slip of glass. The size of the crystals is in general between $5\,\mu$ and $50\,\mu$, and crystals of fairly uniform size can be sorted out by centrifuging or by levitation. Since the range of the radium C' α particle in zinc sulphide is about $35\,\mu$, a preparation of crystals of about 20 to $30\,\mu$ will be suitable for most observations.

The duration of a scintillation is, according to Wood† and Herszfinkiel and Wertenstein‡, about 10^{-4} second. The spectral distribution of the light emitted in scintillations has been examined by various workers. For ZnS.Cu preparations the spectrum consists of a broad band between about $400\,\mu\mu$ and $600\,\mu\mu$, with two maxima; the first maximum is at about $450\,\mu\mu$ and the second, and chief, at about $520\,\mu\mu$. The detailed distribution varies appreciably for different preparations. A comparatively large proportion of the light emitted by ZnS.Cu crystals is thus in the region of wave-lengths to which the dark-adapted eye is most sensitive, for the scotopic luminosity curve has a maximum at about $505\,\mu\mu$. This fact is partly responsible for the remarkable suitability of these crystals for the purpose of observing scintillations.

The efficiency of zinc sulphide in transforming the energy of the α particle into radiant energy is surprisingly high. According to the measurements of Chariton and Lea, the radiant efficiency, i.e. the fraction of the energy of the α particle converted into visible light, is about 0·25 for the best specimens of ZnS.Cu. Knowing the spectral

* Tomaschek, *Ann. d. Phys.* **65**, 198, 1921; cf. also Lenard, *Handbuch der Experimental Physik*, vol. 23.

† Wood, *Phil. Mag.* **10**, 427, 1905.

‡ Herszfinkiel and Wertenstein, *J. de physique et le Rad.* **2**, 21, 1921.

distribution of the emitted light and evaluating the energy emitted as green light of wave-length $505\,\mu\mu$, they calculated that the luminous efficiency of these specimens was about $0\cdot12$. It appears, from experiments in which very slow α particles were used, that the efficiency of the surface layers of the crystal is not very different from that of the interior. There is no evidence that the efficiency depends to any marked degree on the velocity of the α particle. The above data may therefore be used to give an estimate of the total amount of luminous energy in a scintillation produced by an α particle of any given range.

Of the total light of a scintillation only a fraction, depending on the optical system used for observation, will be received on the retina of the eye. From the definition of numerical aperture (n.a.), it follows that the fraction of the light from a scintillation which enters the objective of the microscope is $\frac{1}{2}\,(1-\sqrt{1-(\text{n.a.})^2})$ as long as n.a. < 1. Provided that the exit-pupil of the system is not greater than the diameter, about $7\cdot5$ mm., of the pupil of the dark-adapted eye, the whole of this light will reach the retina of the eye, except for losses, which may amount to about 20 per cent., due to reflection and absorption in the optical system. The observations of Chariton and Lea show that for a scintillation to be perceptible through a microscope of $0\cdot45$ n.a. and magnification 50, the amount of luminous energy which reaches the eye must be equivalent to more than 300 quanta of green light $(505\,\mu\mu)$. On the other hand, scintillations can be perceived by the naked eye when the amount of energy reaching the retina exceeds 30 quanta of green light. This divergence was explained in the following way. Since the shape of the crystal is very irregular, the light produced in it will be reflected and refracted at the crystal surfaces with the result that the image formed in the microscope is not an image of the actual path of the α particle but of the illuminated crystal*. Assuming that the "size" of a scintillation is the same as the size of a crystal, about $25\,\mu$ in the above experiments, the size of the image on the retina when the scintillation is viewed by the naked eye will be about $2\,\mu$, and when viewed through a microscope of $\times\,50$ magnification about $110\,\mu$. Now the experiments of Reeves† showed

---

* This can be seen by viewing a zinc sulphide screen through a microscope of large magnification (about $\times\,800$). When an α particle strikes a crystal the whole, or a large part, of the crystal lights up. Only with larger or more regular crystals is it possible to see the actual path of the α particle in the crystal. Thus Geiger (*Zeit. f. Phys.* **8**, 191, 1921) was able to observe the tracks of α particles in a willemite crystal, and Chariton and Lea in a thin flake of diamond.

† Reeves, *Astrophys. J.* **47**, 141, 1918.

that the minimum amount of energy perceptible by the eye depends on the size of the retinal image of the stimulus, and the variation is roughly sufficient to explain why so much smaller a fraction of the light from a scintillation is necessary to produce a visual sensation when the eye is unaided, than when a microscope is used.

It is customary in scintillation counting to use a microscope of magnification between × 20 and × 50. Under these conditions a scintillation should just be visible when the amount of luminous energy which reaches the retina corresponds to about 300 quanta of green light. Only a very small fraction of such scintillations can be counted. For reliable counting the energy entering the eye must be more than 1200 quanta. With a microscope of numerical aperture 0·45, this corresponds to the scintillation produced by an α particle of velocity greater than $0·25 V_0$, where V_0 is the velocity of the α particle of radium C′, or a range in air of about 3·5 mm. With the same numerical aperture, the limit of visibility is reached when the velocity of the α particle is about $0·13 V_0$, or a range in air of about 0·2 mm.

The mechanism of the production of the scintillation in the zinc sulphide crystal is still obscure. Lenard has supposed that the foreign atoms incorporated in the lattice of the zinc sulphide crystal produce local disturbances which are responsible for the observed change of properties, but very little is known about the nature of these active centres or the kind of disturbance they produce in the crystal lattice. From the fact that a large part of the energy of the α particle is transformed into visible light, Chariton and Lea concluded that at least a large part of the volume of a zinc sulphide crystal must be occupied by active centres, for the rate at which the α particle loses energy cannot be appreciably different in the active and inactive parts of the crystal*. The disturbances caused by the introduction of copper atoms into the zinc sulphide lattice are therefore apparently distributed throughout the volume of the crystal in such a way as to allow the molecules of the crystal to respond to excitation by α particles. This is in agreement with the conclusion arrived at by Rutherford† from an analysis of Marsden's experimental data concerning the decay of luminosity under prolonged bombardment. Rutherford, assuming

---

* It might be assumed that the efficiency of transformation of the energy of the α particle was greater than 100 per cent. in the active parts of the crystal; against this must be noted the fact that the total energy developed in the crystal is equal to the energy of the α particle.

† Rutherford, *Proc. Roy. Soc.* A, **83**, 561, 1910.

that an active centre is destroyed once it has taken part in the emission of light, found that the radius of action of the α particle, i.e: the radius of the cylinder around the path of the α particle in which all the active centres are destroyed, is about 7×10^{-8} cm. in zinc sulphide. The dimensions of the active centres are thus of molecular order, in accordance with Chariton and Lea's suggestion that the whole crystal is active.

Chariton and Lea regarded the disturbance caused by the impurities as a network of slight strains which allow the molecules of the crystal to respond to excitation by α particles; the passage of the α particle releases the strains in a cylinder of the above radius around its path. Karlik has suggested that the active centres in the crystal are in an excited state and recombine, with the emission of light, by picking up electrons set free by the α particle in its passage through the crystal. Such a recombined centre can naturally take no further part in the production of a scintillation. This suggestion is supported by her observations of the scintillations produced by particles of different ranges in weakly scintillating specimens of zinc sulphides; provided that the α particles had sufficient range to pass through the crystal, the brightness of the scintillation was constant, i.e. independent of the amount of energy lost in the crystal. If the number of active centres in the cylinder of action of the α particle is less than the number of ions available for recombination, the amount of light emitted will, on the above hypothesis, be constant, and the observation receives an immediate explanation.

§ 126 a. **The efficiencies of observers.** A method of determining the efficiency of an observer in counting scintillations has been given by Geiger and Werner, and used by them to eliminate errors of counting caused by fatigue or lack of concentration. The method consists in the simultaneous observation of the scintillations on the same screen by two counters. The zinc sulphide screen can be observed through each of two microscopes. The area of the screen is so small that all the crystals are visible through each microscope. Each observer records the occurrence of a scintillation through an electrical contact key on the tape of a chronograph. Now every scintillation which occurs on the screen should be visible to both observers, and, if both are perfect counters, should be recorded at the same time by each, and all the marks recorded by one observer A should coincide with those of the second observer B. It is found, however, that all the marks do not coincide; some scintillations are seen by A which are not recorded by B, and vice versa. From the tape record of a large number of observations we can deduce both the actual number of scintillations which occurred and the efficiencies of the counters. If N is the number of scintillations which actually occurred on the screen and λ_1 is the probability that the counter A sees and records a scintillation, i.e. his efficiency in counting, then on the average he will make $N_1 = \lambda_1 N$ marks on his tape. Similarly, counter B will record $N_2 = \lambda_2 N$ scintillations, where λ_2 is B's efficiency. The number of coincidences on the tape, or the number of scintillations which both A and B see, will be $C = \lambda_1 \lambda_2 N$. The record on the tape gives N_1, N_2, and C at once, and thus N, the number of scintillations which actually occurred, and λ_1, λ_2, the efficiencies of the two counters, can be readily calculated.

Some objections may be raised against applying this method to give the true number of scintillations occurring on the screen. Perhaps the chief objection is that the dual registration of a scintillation is not instantaneous, but there is a time lag for each observer which may be subject to small variations; there is thus some latitude in estimating the coincidences and this latitude will depend on the frequency of the scintillations. The efficiency of an observer will depend on the brightness of the scintillations and on the rate at which they appear. It is, of course, essential that no scintillations are present which cannot be perceived by either observer.

As a general test of the efficiency of an observer the method is very useful. When the scintillations are produced by α particles of a few centimetres range and occur at a convenient rate, say about 20 per minute, even an untrained observer will count from 70 to 80 per cent. of the scintillations, and after some practice about 90 to 95 per cent. Under good conditions the efficiency of a trained counter is about 95 per cent.

To obtain information about the efficiency of an observer in counting very weak scintillations, two methods given by Chariton and Lea may be used. In the simpler method a zinc sulphide screen is exposed to α particles of good range from a constant source, e.g. polonium, and the apparent intensity of the scintillations is reduced by varying the numerical aperture of the microscope. The manner in which the number of scintillations observed falls off as the numerical aperture is reduced determines the efficiency of the observer in counting scintillations of different known intensities. The second method is the one used by Chariton and Lea to determine the smallest amount of light perceptible by the dark-adapted eye. Short flashes of light are produced under conditions resembling as closely as possible those in which scintillations are counted. The duration of the flash and the colour of the light are adjusted to correspond to a scintillation. Such artificial scintillations occur regularly in time and in position, and are therefore more easily observed than actual scintillations. Nevertheless, very useful tests of an observer may be made in this way. The minimum energy visually perceptible under these conditions is about 20 quanta of green light for a trained observer, about 30 quanta for an untrained observer.

It may be remarked that the superior efficiency of an experienced observer appears to be due to greater concentration, to control of spontaneous movements of the eye, and to practice in using the excentral portions of the retina, thereby avoiding the insensitive fovea-centralis.

§ 127. **The chemical properties of the radioactive elements.** It is outside the scope of this book to give a detailed account of the chemical properties of the radioactive elements or of the methods which may be adopted for separating them from each other or from inactive matter mixed with them. A general account of their chemical behaviour can be simply and, for the present purpose, adequately given by comparison with the properties of the inactive elements. A large number of the radioactive substances are isotopes of well-known elements and their chemical properties are thus completely defined by reference to the latter. Thus:

uranium I and uranium II are isotopes of the common element		uranium
thorium, radiothorium, uranium X_1, uranium Y, ionium, radioactinium	do	thorium
radium E, radium C, thorium C, actinium C	do	bismuth
radium B, radium D, thorium B, actinium B, and the end products, radium G, thorium D, and actinium D	do	lead
radium C″, thorium C″, actinium C″	do	thallium

The remaining radioactive elements are not isotopic with any of the common elements. They can, however, be divided into five sets, each containing a number of radioactive isotopes, and the properties of each set may be defined by reference to the nearest element in the same group of the periodic table. Thus radium, which belongs to a group of four radioactive isotopes, is the higher homologue of the common element barium, and resembles it closely in chemical properties; any reaction exhibited by barium or its salts is shown also by radium and its salts, but to a slightly different degree.

radium, mesothorium 1, ⎫ are isotopes and resemble in chemical
thorium X, and actinium X ⎭ properties the common element   barium
radon, thoron, actinon   do   xenon
actinium, mesothorium 2   do   lanthanum
protactinium, uranium X_2, ⎫
uranium Z   ⎭   do   tantalum
radium A, thorium A, actinium A, ⎫
radium C′, thorium C′, actinium C′, ⎬   do   tellurium
polonium (radium F)   ⎭

For a detailed account of the chemistry of the radioactive substances the reader may consult F. Soddy, *The Chemistry of the Radioelements* (Longmans, Green and Co.); A. S. Russell, *An Introduction to the Chemistry of Radioactive Substances* (Murray, 1922); and F. Henrich, *Chemie und chemische Technologie radioaktiver Stoffe* (Springer, 1918).

§ 127 a. **Methods of separation.** In the preparation of pure specimens of the radioactive elements certain difficulties arise which do not occur with the inactive elements. A preparation of an element may be said to be "chemically pure" when the presence of any other element cannot be detected by chemical means. A chemically pure preparation of an inactive element may, and often does, contain two or more isotopes, which are identical from the chemical point of view; it is sufficient if all the atoms have the same atomic number. In the radioactive series, however, atoms of the same atomic number may not be identical in their radioactive behaviour, and it is in general much more important to obtain a radioactive element free from admixture with its radioactive isotopes than from foreign inactive elements. To take an example, it will be for most radioactive investigations more desirable to prepare radium free from mesothorium 1 than from barium, although the former preparation may be "chemically pure." We have thus to distinguish between "chemically pure" preparations and "radioactively pure" preparations. It is obvious that a radioactively pure preparation will not remain so, for the products of disintegration will accumulate in course of time, at a rate depending on the constants of transformation; thus a preparation of radium will, after only a few hours, contain detectable amounts of radon and its active deposit of short life. It is therefore possible to prepare a radioactively pure specimen of a radioelement only for a limited period of time, before the products of transformation make themselves evident.

In addition to the relatively stable elements uranium and thorium, only radium, radon, radium D, and protactinium have as yet been prepared in visible quantity in a pure state. The difficulty lies in the fact that the majority of the radioactive elements are isotopes of either active or inactive elements with which they occur naturally. As an example the case of radium D may be cited. Radium D cannot be obtained in a pure state from uranium minerals, although it is present in comparatively fair quantity, for other isotopes of atomic number 82—the end products radium G, etc.—are generally present to a greater amount. It can, however, be obtained pure by the decay of radon, since the intermediate products are of short life. If a quantity of radon is introduced into a sealed tube and allowed to decay, radium D together with its transformation products radium E and polonium are found in a pure state on the walls of the tube. Old "radon tubes" of this kind afford very convenient sources from which to obtain concentrated preparations of radium D, radium E, or polonium*.

In the case of some of the radioelements a further difficulty may arise from the rapid rate of transformation; the life of the substance may be too short for the usual chemical methods of separation. In such cases a parent substance of suitably long life may be first separated and used as a source of supply for the desired product, which may then sometimes be obtained by other than chemical methods, e.g. by recoil.

A radioactive substance may be isolated from others not isotopic with it in one of the following ways: (1) by ordinary chemical methods, (2) by electrochemical methods, (3) by differences in volatility, and (4) by the recoil method. Only a very general discussion of these methods and their limits of application can be given here.

(1) *Chemical methods.* In the separation of a radioelement by chemical methods three cases may be distinguished. If the radioelement is present in visible quantity it may be precipitated from solution by reagents, according to the usual methods of inorganic

---

* Cf. v. Hevesy and Paneth, *Ber. d. D. Chem. Ges.* 47, 2784, 1914.

chemistry. Thus radium may be precipitated as sulphate by the addition of sulphuric acid or a soluble sulphate, and thorium may be precipitated as hydroxide by the addition of ammonia or as oxalate by the addition of oxalic acid.

In many cases the amount of the radioelement is much too small to give a visible precipitate; a visible quantity of an isotope of the substance may, however, be present. The radioelement will then be obtained mixed with this isotope, and cannot be separated from it by chemical means. In all precipitations the radioelement will be divided between precipitate and solution in exactly the same proportions as the isotopic element. As an example we may take the case of the separation of mesothorium 1 from a thorium mineral. According to the proportion of uranium in the thorium mineral, the mesothorium will be obtained mixed with a quantity of radium, with which it is isotopic. In the mesothorium preparations in commerce the radium impurity is responsible for about 20 to 30 per cent. of the γ ray activity; the weight of the radium impurity is roughly a hundred times the weight of the mesothorium.

If no isotope is present in the material to form a visible precipitate, a small quantity either of an isotope or of a substance of similar chemical properties may be added. Thus the radiothorium grown in a preparation of mesothorium may be precipitated as hydroxide by the addition of ammonia. To give a visible precipitate a few milligrams of its isotope thorium, or of iron, may be added. The addition of an isotope for this purpose has the disadvantage that the radioelement can never be separated from the added isotope, and it is often preferable to use a substance which is not isotopic but merely has convenient chemical properties.

(2) *Electrochemical methods.* The electrochemical methods of separating radioactive substances are of great importance since they permit a substance to be obtained free not only from other radioelements but also from all other elements. These methods are, however, of very limited application, for few of the radioactive substances are electrochemically noble. Their use is mainly restricted to the separation of bodies in the active deposits. We shall deal here only with the active deposit of radium, but the methods described will of course apply equally to the active deposits of thorium and actinium.

As is to be expected from the chemical behaviour of their isotopes, radium A, an isotope of polonium, is electrochemically more noble than radium C, an isotope of bismuth, and radium C is more noble than radium B, an isotope of lead. Consequently, if a solution of radium active deposit is electrolysed, radium A will separate out most easily and radium B least easily. If the electrolysis be carried out in such a way that the cathode potential does not rise above a certain amount, it will be possible to obtain the most noble of the radio-elements present in the solution in a pure state. Thus from a solution of radium (B + C) in weak HCl or HNO_3 pure radium C can be obtained on the cathode if its potential be kept between -0.08 and -0.5 volt with respect to the normal calomel electrode. The deposition of radium B can be further hindered by the addition of a soluble lead salt to the solution; the very small quantity of lead deposited under the above conditions will then consist of inactive lead atoms and only a small proportion will be radium B atoms. Radium B is obtained on the cathode when the potential falls below $E_c = -0.5$ volt.

A simpler method of obtaining pure radium C consists in dipping a polished plate of nickel into a solution of radium (B + C) in hot weak hydrochloric acid. Nickel is less noble than radium C and the latter deposits on the nickel surface in a practically pure state. Any radium A present in the solution will also be deposited on the nickel.

The most important application of electrochemical methods is in the preparation of sources of polonium (radium F) from a solution of radium D*. The radium D may be separated, with lead, from uranium minerals or, without lead, from an old radium salt or by the decay of radon in a closed vessel. In a weakly acid solution of radium (D + E + F) the F or polonium will deposit when the cathode potential is less than $E_c = +0.35$ volt. When the potential falls to $E_c = -0.08$ volt radium E will also be deposited and at $E_c = -0.5$ volt radium D will begin to deposit. It is not always practicable to measure the potential of the cathode with respect to the calomel electrode. As a rough working rule it may be taken that in $\frac{1}{10}$ normal nitric acid pure polonium will be obtained with a current density of 3×10^{-5} amps./sq. cm. and that radium E will deposit when the current density rises to about 1.6×10^{-4} amps./sq. cm. The electrode may be either gold or platinum according as it is desired to remove the polonium by solution or by distillation. Polonium cannot be entirely removed from a platinum surface by solution in acid, but it can be removed by

* Cf. Paneth and v. Hevesy, *Wien. Ber.* **122**, 1037 and 1049, 1913; v. Hevesy and Paneth, *Wien. Ber.* **123**, 1619, 1914.

distillation; on the other hand, in distilling polonium from a gold surface a trace of gold often distils with it.

It may often be convenient to obtain radium D, as superoxide, on the anode by electrolysis from a 10 per cent. nitric acid solution. Polonium can also be deposited as superoxide on the anode under suitable conditions, but this separation is not quantitative.

Polonium and radium E can also be obtained from a solution of radium (D + E + F) in weak hydrochloric acid by dipping into the solution a plate of nickel. If it is desired to prepare a source of pure radium E the polonium must first be removed either by electrolysis or by the nickel method. Owing to the great difference in the rates of transformation of the bodies a large fraction of the radium E will, after a few days, be present in the solution while the amount of polonium formed in the same time will be very small. As in the case of the separation of radium C from a solution of the active deposit, it may often be of advantage when preparing sources of E or F to add a small quantity of a soluble lead salt to the solution of radium D.

Polonium can also be obtained from a radium D solution without the application of current by deposition on silver, a method first given by Marckwald[*] and developed later by I. Curie[†]. If a silver button is rotated in a solution of pure radium D (free from lead) in $\frac{1}{2}n$ HCl, practically the whole of the polonium deposits on the silver. Very strong sources of polonium can be prepared by this method on a relatively small surface. This method has also been investigated by Erbacher and Philipp[‡], who separated from a radium D solution first the polonium by deposition on silver, the E by deposition on nickel, and finally the D by electrolysis.

The electrochemical method of separation can be applied also to the less noble radioactive substances, such as radium, thorium, mesothorium 2, and their isotopes. In these cases the separation is an indirect action; the electrical current transports the ions of the substance to the cathode, where they are precipitated chemically and under suitable conditions form a coherent deposit on the electrode. Since the chemical action at the cathode cannot in general be controlled, a separation of one substance from another of similar chemical properties is not possible in the case of the less noble radioelements. Nevertheless, the method may be of great value for such experiments

* Marckwald, *Ber. d. D. Chem. Ges.* **38**, 593, 1905.
† I. Curie, *Journ. Chim. phys.* **22**, 471, 1925.
‡ Erbacher and Philipp, *Zeit. f. Phys.* **51**, 309, 1928.

as those in which the magnetic spectrum of the β rays is to be examined, where it is desirable to have a line source of the radio-element. Electrochemical methods were first applied to the less noble radioelements by Hahn and Meitner* for the preparation of line sources of radium, uranium X_1 (a thorium isotope), and mesothorium 2 (an actinium isotope). The processes concerned in the electrolysis of these substances were examined systematically by Tödt†, who has defined more precisely the experimental conditions under which they can be obtained with a reasonable yield and in a coherent deposit.

(3) *Volatilisation*. A partial separation of the bodies comprising the active deposits can be obtained by the effects of temperature, for the B bodies volatilise more readily than the A bodies and the A bodies more readily than the C bodies. The values given by different workers for the volatilisation temperatures of the active deposit products vary considerably, and it is not always clear whether the experiments were concerned with the elements or with their chemical compounds. According to Russell‡ neither radium A, B, nor C volatilises from a quartz surface in the presence of oxygen at temperatures below 700° C., and radium C under these conditions volatilises at a temperature greater than 1200° C.; in the presence of hydrogen, however, all three products are completely volatilised below 650° C. In the latter case the products are presumably present as elements, and in the former as oxides. Loria§ showed that the amount of volatilisation depends on the nature of the surface on which the active matter is deposited and also on the way in which the deposit has been obtained; thus thorium C deposited electrolytically on a platinum foil does not volatilise as readily as when deposited by exposure to the emanation.

The volatilisation process is not a very effective method of separating one radioactive substance from another, though it may sometimes be useful in the case of products of short life. It is of more importance as a method of transferring a product from one support to another, and it has been much used for this purpose‖.

A very simple apparatus for volatilising polonium has been described by Rona and Schmidt¶, and employed by them for transferring

* Meitner, *Phys. Zeit.* 12, 1094, 1911; Hahn and Meitner, *Phys. Zeit.* 14, 758, 1913; 16, 6, 1915.

† Tödt, *Zeit. f. phys. Chem.* 113, 329, 1924.

‡ Russell, *Phil. Mag.* 24, 134, 1912.

§ Loria, *Wien. Ber.* 124, 567 and 1077, 1915.

‖ Cf. Russell and Chadwick, *Phil. Mag.* 27, 112, 1914.

¶ Rona and Schmidt, *Wien. Ber.* 137, 103, 1928.

polonium deposited electrolytically on a platinum foil of relatively large area to a small disk or the end of a thin rod. In this way the polonium from a number of electrolyses may be concentrated into one source, and very large amounts of polonium may be obtained on a small area.

(4) *The recoil method.* The importance of the recoil method in separating radioactive substances in a pure state has already been pointed out in § 31, where the main features of both α ray and β ray recoil have been discussed. From its nature this method is most effective when it is desired to obtain a body which results from an α ray transformation, and it has mainly been used in the analysis of the active deposits. Thus the product thorium C″ may be obtained by placing a negatively charged plate near a plate coated with the active deposit of thorium. The recoil atoms of thorium C″ resulting from the α ray transformation of thorium C carry, in air, a positive charge and collect on the negatively charged plate.

Thorium C″ prepared by this method should be free from any other matter, whether active or inactive. It is found, however, that a very small amount of the parent active deposit always appears on the collecting plate. This spontaneous transference of active matter from the surface of a plate to other surfaces close to it has often been observed. It was attributed by some workers to a slight volatility of active deposit at ordinary temperatures, by others to the recoil of a compact cluster of atoms of the active matter when one of the atoms contained in the cluster disintegrates with the ejection of an α particle. Lawson* showed that the latter explanation was the correct one, and he gave to the phenomenon the name of "aggregate recoil." He measured the amount of aggregate recoil when the active matter (polonium) was deposited on different materials, and he found that it was least from the surface of an easily oxidisable metal. He showed that aggregate recoil may under certain conditions interfere seriously with measurements of the half-value period of the active matter. Lawson also investigated the part played by "sputtering" of the metal on which the active matter was deposited, due to bombardment by the α particles. He found that the loss of active matter due to this cause was in general less than 1 per cent. of the loss due to aggregate recoil.

From his results Lawson concluded that many of the clusters or aggregates of atoms of the active matter deposited on a foil must consist of at least three atoms. The tendency of radioactive atoms to

* Lawson, *Wien. Ber.* **127**, 1, 1918; **128**, 795, 1919; *Nature*, **102**, 465, 1919.

form aggregates was observed early in the history of radioactivity by different workers. In recent experiments Chamié* has obtained evidence that the active matter, whether deposited by the action of the emanations or chemically from solution, is not uniformly distributed but consists largely of groups or aggregates which may contain a million or more atoms. Harrington† also has observed the presence of aggregates on surfaces which had been exposed to the emanation of radium. He suggests that most of the aggregates are actually formed in the gas itself by the combination of active deposit ions, possibly facilitated by the presence of molecules of water-vapour or of any other substance not able to leave a combination once formed. On the other hand, the presence of aggregates in solutions of active matter is attributed by Hahn and Werner‡ to the tendency of the active deposit bodies, notably the C products, to the formation of colloids. They consider that the aggregates occur in solutions only when the chemical natures of the radioelement and of the solution are such that hydrolysis can take place. The products of the hydrolysis are adsorbed on impurities, such as particles of dust or colloidal silica, which are always unavoidably present.

§ 128. The preparation of sources of radon, radium active deposit, and thorium active deposit. In most physical laboratories the greater part of the stock of radium is kept in solution and the radon grown by the radium is collected from time to time as it is required to provide sources of radon or radium active deposit. Various forms of apparatus have been devised§ for the collection of the radon and for its purification; we shall give here a very brief description of the methods adopted in the Cavendish Laboratory.

The radium solution, in weak hydrochloric acid, is contained in a glass bulb which may be connected through a stopcock to a Toepler pump. When it is desired to collect the radon the stopcock is opened and the radon, mixed with the gases formed by the action of the radiations on the solution, is allowed to expand into the pump, from which it is transferred to a burette or tube standing in mercury. The gases with which the radon is mixed consist mainly of hydrogen, oxygen, and ozone from the water of the solution, and carbon dioxide (with perhaps some hydrocarbons) from organic matter such as tap-

---

* Chamié, *C.R.* **184**, 1243, 1927; **185**, 770, 1277, 1927; **186**, 1838, 1928.

† Harrington, *Phil. Mag.* **6**, 685, 1928.

‡ Hahn and Werner, *Naturwiss.* **49**, 961, 1929.

§ Cf. Duane, *Phys. Rev.* **5**, 311, 1915; Hess, *Phil. Mag.* **47**, 713, 1924.

grease, with traces of chlorine and helium. The volume of the electrolytic gases will depend on the amount of radium present and on the time which has elapsed since the last collection of radon. With a solution containing 250 mg. Ra about 30 c.c. of mixed gases are produced in one week. The concentration of radon in the mixed gases is therefore less than 10^{-5}.

The mixed gases are first transferred to a gas pipette and exploded by sparking to remove the hydrogen and oxygen. It is usually of advantage to add oxygen in order to remove completely any excess hydrogen. Excess oxygen is easily removed by means of phosphorus. The residual gases, now about 0·5 c.c. in volume, are then introduced over mercury into a small tube containing a piece of caustic potash to absorb the carbon dioxide. If the radon is required for the purpose of preparing active deposit sources, no further purification will usually be required.

For the preparation of radon tubes it is generally sufficient to condense, by means of liquid air, the radon in the mixed gas partially purified by sparking and exposure to P_2O_5 and KOH, and to pump off the uncondensed gases. The radon is then allowed to expand and compressed by a column of mercury into a tube or bulb. It is, however, sometimes necessary to prepare very small strong sources of radon and in such cases a much higher degree of purity is required. The mixed gases, after sparking, are introduced into a purification apparatus*, in which they are brought into contact with heated copper and copper oxide to remove hydrogen, oxygen, and any hydrocarbons. The gaseous products of the combustion, carbon dioxide and water-vapour, are then absorbed by P_2O_5 and KOH. The radon can then be condensed in a side tube and any uncondensed gases removed by pumping. This method, if used with care, will yield radon of about 50 to 70 per cent. purity, and it is a comparatively simple operation to compress 100 millicuries of radon into a volume of less than 1 c.mm.

The main difficulty in purifying radon arises from the fact that the volume of radon to be purified is seldom more than 0·1 c.mm. The partial pressure of the radon in the purification apparatus is therefore very small; for example, the partial pressure due to 100 millicuries in a volume of 50 c.c. is about 1·3 microbar. While the permanent gases may be removed by pumping after condensation of the radon by means of liquid air, the condensable impurities must be removed by chemical means. Chemical agents are not efficient enough for these

* Cf. Wertenstein, *Phil. Mag.* 5, 1017, 1928.

conditions, and the reactions are rather slow at low pressures. The most troublesome impurity is carbon dioxide, which is strongly adsorbed by old glass surfaces and is also produced by the action of the radiations on organic matter. The use of greased taps must be avoided in the purification apparatus. The simplest method of removing the carbon dioxide is to leave the gas in contact with KOH for about 24 hours.

In hospitals and laboratories where large quantities of radon are frequently dealt with, the apparatus for removal and purification of the radon is designed to work as automatically as possible, in order to reduce to a minimum the exposure of the operator to the radiations. The physiological action of the radiations and the precautions to be taken when handling radon tubes and active sources have been mentioned in § 41.

Active deposit sources. Sources of radium active deposit are prepared by exposing a disk or wire to radon in a suitable vessel. Radium A, the first product of the disintegration of radon, is a solid and deposits on surfaces exposed to radon. Since the majority of the recoil atoms of radium A are positively charged (p. 156), the yield of active deposit can be increased by charging the exposed disk or wire negatively with respect to the surrounding surfaces. If the disk is exposed for a short time only, practically pure radium A will be obtained (cf. p. 12). To obtain A, B, and C in equilibrium the time of exposure to the radon should be, when no field is applied, about 4¼ hours. It is found that when the disk is negatively charged equilibrium is attained after about 2 hours' exposure. To obtain on the disk a satisfactory proportion of the available active matter it is important that the area of the disk should be as large a fraction as possible of the total area exposed to the radon.

When the source of active deposit is required the radon is pumped off and collected over mercury. The disk is removed, washed in alcohol, and heated to about 400° C. in an evacuated quartz tube to remove any radon which may be occluded in the surface of the disk.

In preparing wire sources of active deposit it is advantageous to surround the wire with a cylinder of iron or nickel charged positively with respect to the wire*. It may be noted that the active matter may not be deposited uniformly over the surface of the wire.

* Henderson, *Nature*, 114, 503, 1924.

Pettersson* has obtained a high yield of active deposit on a disk by condensing the radon on the surface of the disk. After 4 hours the active deposit reaches equilibrium, the radon is pumped off and the disk removed. When this method is used a plate or foil may be supported above the disk to catch the recoil atoms from the disintegrating radon, and thus a second source may be prepared at the same time.

Jedrzejowski† has prepared sources by compressing purified radon into a short capillary tube. After 4 hours the radon is pumped off and the active deposit is removed from the capillary by solution in hydrochloric acid. The active matter is then obtained by evaporating the solution on a suitable support.

Very strong wire sources of active deposit have been prepared by Rosenblum‡ by introducing a glass or platinum wire, a few tenths of a millimetre in diameter, into a capillary tube containing purified radon at atmospheric pressure. The radon is sealed from the atmosphere by a short column of mercury, through which the wire is introduced and removed. Using quantities of radon of about 500 millicuries, Rosenblum obtained half the total active deposit on the wire.

A very simple arrangement for collecting the radon from a radium preparation has been given recently by Hahn and Heidenhain§. They have succeeded in making solid radium preparations of very high emanating power which preserve this property for at least a few years. The radium preparation is contained in an evacuated vessel which, when radon is required, can be connected through a stopcock to a capillary tube. The radon is condensed in the capillary tube by means of liquid air and sealed off. Owing to the absence of water no purification is necessary.

Thorium active deposit. Owing to its short period thoron itself cannot be used for the preparation of strong sources of radiation, but in conjunction with a parent substance it will supply strong preparations of thorium active deposit. As a parent source of thoron it is general to use the substance radiothorium, which has a conveniently long period of nearly two years. Preparations of radiothorium can be obtained in a highly emanating condition by a method due to Hahn‖. A small quantity of iron chloride is added to the radio-

---

* Pettersson, *Wien. Ber.* **132**, 155, 1923; Ortner and Pettersson, *Wien. Ber.* **133**, 229, 1924.
† Jedrzejowski, *C.R.* **182**, 1536, 1926.
‡ Rosenblum, *C.R.* **188**, 1549, 1929.
§ Hahn and Heidenhain, *Ber. d. D. Chem. Ges.* **59**, 284, 1926.
‖ Hahn, *Ann. d. Chem.* **440**, 121, 1924.

RRS

thorium solution and precipitated by ammonia. The precipitate, the hydroxides of iron and radiothorium, is washed very carefully and dried at ordinary temperature. The preparation should have an emanating power of about 80 per cent. and should retain it for many months. The radiothorium preparation is spread in a thin layer at the bottom of a small metal vessel, and the active deposit is collected on an insulated plate or wire charged negatively with respect to the vessel.

§ 129. **The measurement of quantities of radium.** It is important in many experiments to measure accurately the amount of radium present in a preparation of a radium salt. Two general methods have been used: (1) the γ ray method, which is suitable for measuring quantities varying from 1/100 of a milligram to 1 gram, and (2) the emanation method, for quantities of radium of the order of one-millionth of a milligram.

The γ ray method. This method of measurement depends on the fact that the radium in a radium salt sealed in a tube so that no radon can escape will, after about one month, be in equilibrium with its products of short life, and will therefore emit penetrating γ rays with an intensity proportional to the amount of radium present. The intensity of the γ radiation from the preparation to be tested is compared with that due to a radium standard, a preparation containing a known quantity of radium.

The intensity of the γ rays may be measured by means of a simple electroscope or by use of a suitable ionisation chamber and some form of sensitive electrometer. The walls of the electroscope or ionisation chamber must be thick enough to absorb completely the β radiation from the radium preparation; then, provided that the electric field in the chamber is sufficient to produce saturation, the ionisation current is a measure of the intensity of the γ rays.

The form of electroscope used in the Cavendish Laboratory consists of a cubical box of aluminium with a spherical cavity, provided with the usual gold-leaf system. The electroscope is covered with a sheath of lead of thickness about 1 cm. The volume of the cavity is about one litre, the capacity of the insulated system is about one or two centimetres, and the sensitivity, as measured by a tele-microscope of magnification about × 20, is about one division of the eye-piece scale per volt. With such an electroscope, the rate of movement of the gold-leaf over a selected part of the scale is proportional to the ionisation over as wide a range as can be used effectively for accurate measurement. The procedure in making measurements with an electroscope is as follows. A standard radium preparation is placed some distance from the electroscope and the rate of movement of the gold-leaf is measured. The preparation to be compared with the standard is placed at exactly the same distance from the electroscope and the rate of leak again observed over the same portion of the scale. If the electric field is sufficient to give saturation, and if the rate of movement of the leaf is proportional to the ionisation current, the quantity of radium in each preparation will be proportional to the rate of leak corrected for the natural leak of the electroscope. If a number of observations are taken, the comparison should be accurate to within one-half per cent., provided that the quantities are not very different in amount. If this is not the case, the distances of the radium preparations from the electroscope may be adjusted so as to give about equal rates of leak. The rate of leak produced by a preparation will be inversely proportional to the square of its distance from the electroscope, provided this is large compared with the dimensions of the electroscope, but a correction must be made for the absorption of the γ rays in their path through the air to the electroscope. It may in some cases be necessary to apply a correction for the absorption of the γ rays in the tube containing the radium preparation and in the radium preparation itself. The preparations should be placed at such a distance from the electroscope that errors due to distribution of the radium in the tube are small.

An account of the development of the electroscope as an instrument of precision for γ ray measurements has been given by Bastings[*], who has considered the various factors which tend to limit the accuracy of measurement.

The combination of ionisation chamber and electrometer has been used in many different forms. A convenient arrangement is that in which the chamber is cylindrical in shape, with walls of lead about 1 or 2 cm. thick and an insulated plate electrode. The ionisation current is measured by means of a Compton electrometer, using the Townsend compensation method. Care must be taken to ensure saturation for the greatest currents to be measured. The procedure in comparing the γ ray activity of a radium preparation with that due to the radium standard is similar to the procedure described above when using an electroscope.

[*] Bastings, *Journ. Sci. Inst.* 5, 113, 1928.

A balance method for accurately comparing quantities of radium has been devised by Rutherford and Chadwick*. The radium preparation is placed at such a distance from a lead ionisation chamber hat the ionisation current in the latter due to the γ rays from the radium is balanced against an equal and opposite ionisation current supplied by a constant source of radiation, e.g. uranium oxide in another ionisation chamber. The balance distances are determined for the test preparation and for the radium standard with which it is to be compared. From these distances the quantity of radium in the test preparation can be accurately determined. A small correction is necessary for the absorption of the γ rays by the air. By this method the quantity of radium in a preparation may be determined to about 1 part in 400.

Another reliable method has been used by Mme Curie. The radium is placed on top of a large plate condenser consisting of two sheets of lead. The ionisation current between the lead plates when a large potential difference is applied is balanced by the use of the quartz piezo-electrique. St. Meyer has used, with satisfactory results, a similar ionisation chamber and a Wulf electrometer as measuring instrument.

It will be noted that in these methods of comparing quantities of radium, it is not necessary to open the tube containing the radium. Such a method of estimation is only possible in the case of a radioactive element like radium which emits a very penetrating radiation. It is important that the tube containing the radium preparation should be hermetically sealed for about a month before observations are made. In this way the possible error due to escape of emanation and consequent decrease of the γ radiation is avoided.

It should, however, be pointed out that these methods of comparing quantities of radium are only applicable when the preparation does not contain radiothorium or mesothorium. Both of these latter substances give rise to γ rays of about the same penetrating power as those emitted by radium, and their presence can only be detected by further examination of the preparation.

The methods above outlined can be used not only for comparing quantities of radium but also for comparing quantities of radon and of the product radium C. It is often necessary to obtain a measure of the quantity of radon contained in a sealed tube. About 4 hours after the emanation is introduced, the γ ray effect reaches a maximum,

* Rutherford and Chadwick, Proc. Phys. Soc. 24, 141, 1912.

and thereafter remains very nearly proportional to the amount of radon in the tube. The γ ray effect due to the radon is then compared with that due to the standard quantity of radium, and expressed in terms of milligrams of radium.

The γ ray activity of the radon, after reaching its maximum, decays exponentially with the time with a half-value period of 3·82 days. Consequently the amount of radon present at any time can be calculated from observations made some time after its separation. The reason that the γ ray method can be used both for radium and its emanation depends on the fact that the penetrating γ rays arise not from radium itself but from its products, radium B and radium C. For reasons discussed in § 4, the quantity of emanation is about 0·9 per cent. less than that determined by direct comparison of its γ ray effect with that of the radium standard. In a similar way, we can measure by the γ ray method the amount of the product radium C in terms of the amount in equilibrium with the radium standard. In this case, however, it is desirable to pass the rays through a lead screen about 1 cm. thick in order to absorb the less penetrating γ rays from radium B (see § 103). It may be mentioned here that Hahn has utilised the γ ray method to standardise very active preparations of mesothorium and radiothorium. Both of these substances give rise to γ rays of approximately the same penetrating power as those from radium. Consequently the amount of mesothorium present may conveniently be expressed in terms of the quantity of radium which would give the same γ ray effect. In § 103 an account is given of an absorption method devised by Bothe to estimate the quantity of mesothorium and radiothorium mixed with a sealed radium preparation.

Emanation method. The accurate determination of minute quantities of radium is often of such importance in radioactive work that it is desirable to give a fairly complete account of the most suitable methods adopted for this purpose. The emanation method depends on the fact that radium produces a characteristic emanation, radon, of a comparatively long period of transformation, which can be completely separated from radium solutions. If a radium solution is contained in a sealed vessel, the radon present reaches its equilibrium quantity in about one month, and the amount present is then proportional to the content of radium. Since the radon can be completely expelled from a solution by boiling, the amount Q_t of radon present at a time t after boiling the solution is given by $Q_t/Q = 1 - e^{-\lambda t}$, where Q is the maximum quantity and $\lambda = 0 \cdot 182$ (day)$^{-1}$; the interval of collection being expressed in days. The amount of radon in the specimen under examination can be compared directly with the amount of radon from a standard solution by introducing the radon mixed with air into a suitable electroscope, or into a suitable testing vessel connected with an electrometer. Under the proper conditions, the saturation currents due to the radon from the two solutions are proportional to the quantities of radium present. For comparisons of this kind, it is necessary to prepare a standard radium solution. This can be simply done in the following way. A quantity of radium salt is accurately determined by the γ ray method in terms of the radium standard. Suppose, for example, the radium salt contains 1 mg. of radium. This radium salt is dissolved in water and hydrochloric acid added to ensure complete solution. By adding distilled water, the volume is then increased to a known amount, say 1 litre. A definite fraction of this solution, about 1 c.c., determined by weight or by an accurate pipette, is removed, and this is again diluted to 1 litre. Under these conditions 1 c.c. of the last solution should contain 1/1,000,000 of a milligram of radium. For most experimental purposes, this will serve as a convenient standard radium solution, for it gives a reasonable rate of leak in an emanation electroscope.

This method of solution can obviously be used to obtain a standard solution of any strength desired. In preparing such standard solutions, it is essential to take great precautions that the radium is initially in solution, and that the solutions are thoroughly mixed. In addition, it is necessary to add hydrochloric acid to the solutions to ensure that the radium will remain permanently dissolved. The importance of this procedure is exemplified by the difficulties that arose in the case of the first standard solutions prepared by Rutherford, where no acid was added. Eve[*] found some years later that a large part of the radium had been precipitated from the solution in a non-emanating form on the surface of the glass vessels. The amount of radon to be obtained from such solutions was consequently much less than that to be expected from the amount of radium present. The method outlined above for the preparation of standard solutions was used by Rutherford and Boltwood[†], and has proved quite satisfactory.

[*] Eve, *Amer. Journ. Sci.* **22**, 4, 1906.
[†] Rutherford and Boltwood, *Amer. Journ. Sci.* **20**, 55, 1905; **22**, 1, 1906.

After introduction into a flask, the radium solution is boiled so as to expel all the emanation present; the containing flask is then sealed before the solution becomes cold. About one month later the radon approximately reaches its equilibrium amount, and is then removed. For this purpose the end of the tube is opened and air rushes into the partially exhausted vessel so that no radon can escape. The solution is boiled, and the radon mixed with air is collected over a surface of boiling water. The radon is then transferred into a partially exhausted electroscope. Boltwood has shown experimentally that the radon is completely removed from a solution by rapid boiling for several minutes. It is necessary to collect the radon over hot water in order to prevent its partial absorption by the water. If the water is nearly at boiling temperature, the absorption is very small. The rate of leak due to the radon in the electroscope increases rapidly at first and then more slowly, reaching a maximum at about 3 hours after the introduction of the radon. This increase is due to the production by the radon of its products, radium A, B, and C. For accurate work, it is desirable to determine the rate of leak when this maximum is reached. When the measurements are completed, the electroscope is completely freed from radon by aspiration or by exhaustion, and the instrument is again ready for use when the active deposit has decayed.

This emanation method is a very certain and delicate method of determining the quantity of radium in a solution. The radon from 1/1,000,000 of a milligram of radium gives a comparatively rapid discharge, and with care 1/100 of this quantity can be measured with certainty. In the earlier observations of the amount of radium in solutions, it was usual to aspirate air through the solutions, and to measure the ionisation produced by the released radon under definite conditions in the electroscope. Unless the greatest precautions are taken that the conditions of successive experiments are identical, it is difficult to obtain an accurate comparison by this procedure. The method of boiling the solution is undoubtedly far more certain and reliable.

§ 130. **International Radium Standard.** A number of magnitudes connected with radium are capable of measurement with considerable accuracy, for example, the rate of emission of α particles, the production of helium, the volume of the emanation, the heating effect, and the total ionisation. The values of all these quantities depend on the purity of the radium standard which is employed. In order to compare the values obtained by different workers, it is thus of great importance that they should be expressed in terms of the same radium standard. In order to meet this purpose, the Congress of Radiology and Electricity held in Brussels in 1910 appointed a Committee to make arrangements for the preparation of an International Radium Standard. In the course of the following year, Mme Curie prepared a standard containing, in August 1911, 21·99 mg. of pure radium chloride sealed up in a thin glass tube. In March 1912 the Committee met in Paris and compared the standard of Mme Curie with similar standards prepared by Hönigschmid from the material in the possession of the Academy of Sciences of Vienna, which had been purified for atomic weight determinations. The Vienna standards consisted of three tubes containing 10·11, 31·17, and 40·43 mg. of radium chloride. In all cases the preparations had been obtained from the uraninite at Joachimsthal, which contains only a small trace of thorium. The relative quantities of radium in the standards prepared in Paris and Vienna were determined by the two balance methods described on p. 564. The comparison showed that the standards agreed within the limits of error of measurement, and certainly within 1 part in 300.

The standard prepared by Mme Curie was accepted as the International Radium Standard, and it was arranged that it should be preserved in the Bureau International des poids et mesures at Sèvre, near Paris. It was also arranged that one of the Vienna preparations should be preserved as a secondary standard in Vienna. The preparation containing at that time 31·17 mg. of radium chloride was selected for this purpose by the Vienna Academy of Sciences.

The Committee made arrangements for the preparation of duplicate standards for Governments which desire them. These duplicates are prepared in the Institut für Radiumforschung, Vienna. They are compared by the γ ray method with the official secondary standard in Vienna and with the International Standard in Paris.

The English duplicate standard is preserved in the National Physical Laboratory. It contained, in October 1912, 21·13 mg. of radium chloride, $RaCl_2$. A radium preparation may be standardised by sending it to the National Physical Laboratory, where it will be compared, by the γ ray method, with this duplicate radium standard. The amount of radium in a standard diminishes with time and a correction is required to give the amount of the radium standard at any subsequent time. The diminution is at the rate of 4 parts in 10,000 per year.

APPENDIX

Constants

The following values for the fundamental constants have been used in the numerical calculations in this book. A comprehensive survey of the experimental material and a table of the most probable values of the constants (as on January 1st, 1929) has recently been given by Birge*. The accuracy of the radioactive data is not sufficient to justify a recalculation at this stage. The greatest uncertainty lies in the value of Q (see p. 63), which may be 0·5 per cent. too high.

Constant	Symbol	Value used	Birge
Electronic charge	e	$4\cdot774 \times 10^{-10}$ e.s.u.	$(4\cdot770 \pm 0\cdot005) \times 10^{-10}$ e.s.u.
Specific electronic charge	e/m	$1\cdot769 \times 10^7$ e.m.u./gram	$(1\cdot769 \pm 0\cdot002) \times 10^7$ e.m.u./gram
Velocity of light	c	$3\cdot000 \times 10^{10}$ cm./sec.	$(2\cdot99796 \pm 0\cdot00004) \times 10^{10}$ cm./sec.
Planck constant	h	$6\cdot55 \times 10^{-27}$ erg. sec.	$(6\cdot547 \pm 0\cdot008) \times 10^{-27}$ erg. sec.
Number of α particles emitted from 1 gram of radium per sec.	Q	$3\cdot70 \times 10^{10}$ per gram per sec.	—

* R. T. Birge, *Reviews of Modern Physics*, 1, 1, 1929.

List of the elements, their atomic numbers and atomic weights

A table of isotopes will be found on p. 522.

1	Hydrogen	1·0078	32	Germanium	72·60	63	Europium	152·0
2	Helium	4·0022	33	Arsenic	74·934	64	Gadolinium	157·0
3	Lithium	6·94	34	Selenium	79·2	65	Terbium	159·2
4	Beryllium	9·02	35	Bromine	79·915	66	Dysprosium	162·46
5	Boron	10·83	36	Krypton	82·9	67	Holmium	163·5
6	Carbon	12·0036	37	Rubidium	85·43	68	Erbium	167·6
7	Nitrogen	14·008	38	Strontium	87·63	69	Thulium	169·4
8	Oxygen	16·0000	39	Yttrium	88·93	70	Ytterbium	173·0
9	Fluorine	19·00	40	Zirconium	91·2	71	Lutecium	175·0
10	Neon	20·18	41	Niobium	93·3	72	Hafnium	178·6
11	Sodium	23·000	42	Molybdenum	96·0	73	Tantalum	181·3
12	Magnesium	24·30	43	Masurium	—	74	Tungsten	184·1
13	Aluminium	26·97	44	Ruthenium	101·65	75	Rhenium	188·7
14	Silicon	28·08	45	Rhodium	102·9	76	Osmium	191·0
15	Phosphorus	30·98	46	Palladium	106·7	77	Iridium	193·04
16	Sulphur	32·065	47	Silver	107·880	78	Platinum	195·2
17	Chlorine	35·457	48	Cadmium	112·40	79	Gold	197·21
18	Argon	39·94	49	Indium	114·8	80	Mercury	200·60
19	Potassium	39·105	50	Tin	118·70	81	Thallium	204·3
20	Calcium	40·09	51	Antimony	121·76	82	Lead	207·22
21	Scandium	45·15	52	Tellurium	127·5	83	Bismuth	209·00
22	Titanium	47·90	53	Iodine	126·932	84	Polonium	210
23	Vanadium	50·95	54	Xenon	130·2	85	—	—
24	Chromium	52·04	55	Caesium	132·81	86	Radon	222
25	Manganese	54·95	56	Barium	137·36	87	—	—
26	Iron	55·84	57	Lanthanum	138·90	88	Radium	225·97
27	Cobalt	58·95	58	Cerium	140·2	89	Actinium	227
28	Nickel	58·69	59	Praseodymium	140·9	90	Thorium	232·12
29	Copper	63·55	60	Neodymium	144·25	91	Protactinium	231
30	Zinc	65·38	61	Illinium	—	92	Uranium	238·18
31	Gallium	69·72	62	Samarium	150·43			

Periodic table of elements

	I	II	III	IV	V	VI	VII	O	VIII
1st period	1 H 1·0078						1 H 1·0078	2 He 4·0022	
2nd period	3 Li 6·94	4 Be 9·02	5 B 10·83	6 C 12·004	7 N 14·008	8 O 16·000	9 F 19·00	10 Ne 20·18	
3rd period	11 Na 23·00	12 Mg 24·30	13 Al 26·97	14 Si 28·08	15 P 30·98	16 S 32·065	17 Cl 35·457	18 A 39·94	
4th period	19 K 39·105	20 Ca 40·09	21 Sc 45·15	22 Ti 47·90	23 V 50·95	24 Cr 52·04	25 Mn 54·95		26 Fe 55·84 27 Co 58·95 28 Ni 58·69
	29 Cu 63·55	30 Zn 65·38	31 Ga 69·72	32 Ge 72·60	33 As 74·934	34 Se 79·2	35 Br 79·915	36 Kr 82·9	
5th period	37 Rb 85·43	38 Sr 87·63	39 Yt 88·93	40 Zr 91·2	41 Nb 93·3	42 Mo 96·0	43 Ma		44 Ru 101·65 45 Rh 102·9 46 Pd 106·7
	47 Ag 107·880	48 Cd 112·40	49 In 114·8	50 Sn 118·70	51 Sb 121·76	52 Te 127·5	53 I 126·932	54 Xe 130·2	
6th period	55 Cs 132·81	56 Ba 137·36	57–71 Rare Earths	72 Hf 178·6	73 Ta 181·3	74 W 184·1	75 Re 188·7		76 Os 191·0 77 Ir 193·04 78 Pt 195·2
	79 Au 197·21	80 Hg 200·60	81 Tl 204·3	82 Pb 207·22	83 Bi 209·00	84 Po 210	85–	86 Rn 222	
7th period	87–	88 Ra 225·97	89 Ac 227	90 Th 232·12	91 Pa 231	92 U 238·18			

Artificial disintegration by α particles.

The discussion (§ 72) of artificial disintegration by α particles needs some revision in the light of recent work. In § 72 the deduction was made from certain experiments that the energy change in a disintegration was not always the same, and it was suggested that perhaps all nuclei of the same type were not identical but could have masses differing by very small amounts. This would be the case if the energy levels of the protons or α particles in the nucleus were not well defined. Such a supposition is, however, difficult to reconcile

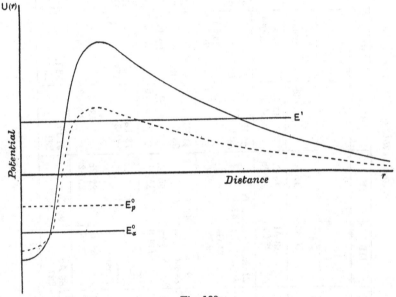

Fig. 139.

with the modern views of nuclear structure. It has been suggested by Chadwick and Gamow*, partly on general grounds and partly on the basis of experiments by Chadwick, Constable, and Pollard†, that the process of artificial disintegration of a nucleus by collision of an α particle may take place in two ways: (1) by the capture of the α particle by the atomic nucleus and the emission of a proton, (2) by the ejection of a proton without the capture of the α particle. In the

* Chadwick and Gamow, *Nature*, July 12, 1930.
† In course of publication.

former case the α particle must penetrate into the nuclear system, in the second case it seems likely that the disintegrations would arise mainly from collisions in which the α particle does not penetrate into the nucleus.

Consider a nucleus with a potential field of the type shown in Fig. 139, where the full line shows the potential barrier for the α particle and the dotted line that for the proton. Let the stable level on which the proton exists in the nucleus be $- E_p{}^\circ$, and the level on which the α particle remains after capture be $- E_a{}^\circ$. If an α particle of kinetic energy E_a penetrates into this nucleus and is captured, the energy of the proton emitted in the disintegration will be

$$E_p = E_a + E_a{}^\circ - E_p{}^\circ,$$

neglecting the small kinetic energy of the residual nucleus.

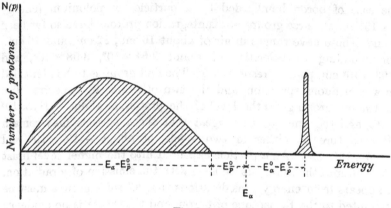

Fig. 140.

If the nucleus disintegrates without capture of the α particle, the initial kinetic energy of the α particle will be distributed between the ejected proton and the escaping α particle (again neglecting the recoiling nucleus). In this case the disintegration protons may have any energy between $E_p = 0$ and $E_p = E_a - E_p{}^\circ$.

If both these processes occur, the disintegration protons will consist of two groups: a continuous spectrum with a maximum energy less than that of the incident α particles, and a line spectrum with an energy greater or less than that of the original α particles according as $E_a{}^\circ > E_p{}^\circ$ or $E_a{}^\circ < E_p{}^\circ$, but in either case greater than the upper limit of the continuous spectrum. The energy spectrum of the protons will be roughly as represented in Fig. 140.

If there are two levels on either of which the α particle may remain after capture, a second line may appear in the energy distribution of the protons. If there are two proton levels from which a proton may be ejected, there may be both a second line and a second continuous spectrum.

It is to be expected that the two types of disintegration protons will differ in their angular distribution with respect to the direction of the initial α particles. It is probable that the protons of the line spectrum will be emitted nearly uniformly in all directions, while the protons of the continuous spectrum will be emitted mainly in the direction of the colliding α particles.*

Experimental evidence for the presence of groups of different ranges in the disintegration protons has been obtained by Bothe, by Pose, and by Fränz†, as well as in the experiments mentioned above. In the case of boron bombarded by α particles of polonium (energy 5×10^6 volts), three groups of disintegration protons have so far been found. These have ranges in air of about 16 cm., 32 cm. and 76 cm., corresponding to velocities of about $2 \cdot 54 \times 10^9$, $3 \cdot 08 \times 10^9$, and $3 \cdot 93 \times 10^9$ cm. per sec. respectively. The first group is to be identified as a continuous spectrum, and the two others as line spectra. The continuous group gives the level of the proton as about $- 1 \cdot 4 \times 10^6$ volts, and the line groups suggest that the α particle may remain after capture on either of two levels, of about $- 1 \cdot 6 \times 10^6$ and $- 4 \cdot 6 \times 10^6$ volts. An α particle captured into the higher level must later fall into the lower, stable level with the emission of γ radiation. It appears from energy considerations that all three groups must be attributed to the B_{10} isotope of boron, and that there is no evidence of the disintegration of the B_{11} isotope. The continuous spectrum of protons thus corresponds to the formation of the nucleus Be_9, the line group of greater energy to the formation of C_{13}, and the line group of smaller energy to the formation of an excited nucleus C_{13}, which passes after the proton emission to the stable C_{13}.‡

The disintegration of aluminium by polonium α particles shows similar features, but in addition Pose§ has obtained very clear

---

* See, however, Beck, Zeit. f. Phys. 64, 22, 1930, whose calculations suggest no pronounced difference in the angular distribution of the two types of protons.

† Bothe and Fränz, Zeit. f. Phys. 49, 1, 1928. Fränz, Zeit. f. Phys. 63, 370, 1930. Bothe, Zeit. f. Phys. 63, 381, 1930. Pose, Phys. Zeit. 30, 780, 1929; Zeit. f. Phys. 64, 1, 1930.

‡ A discussion of these questions is given by Chadwick, Constable, and Pollard, Proc. Roy. Soc. A (in course of publication).   § Pose, Zeit. f. Phys. 64, 1, 1930.

evidence of the occurrence of a resonance phenomenon between the α particle and the aluminium nucleus. The possibility of such a phenomenon was pointed out by Gurney and Condon*, and emphasised later by Gurney†. Suppose that the nucleus has a possible, but unstable, α particle level of positive energy E' (see Fig. 139). If the incident α particles have exactly energy E', their chance of penetrating into the nucleus will be unity and much greater than the chance for particles of larger or smaller energy. If a thick layer of nuclei is bombarded by α particles of initial energy greater than E', so that particles of energy exactly equal and very close to E' are present, there will appear a line group of protons corresponding to the increased chance of capture of α particles of energy E'. Pose's experiments show definitely the presence of two groups of protons which must be explained in this way. The measurements suggest two 'false' α levels in the aluminium nucleus of positive energies 3.9×10^6 and 4.8×10^6 volts. The normal proton groups can be explained on the assumption of one proton level and two α levels.

The experimental results are not yet sufficiently precise to permit their discussion in detail. Further work must be awaited, but it is already evident that the phenomenon of artificial disintegration now promises to reveal the intimate structure of the nuclei of the lighter elements.

* Gurney and Condon, *Phys. Rev.* **33**, 127, 1929.
† Gurney, *Nature*, **123**, 565, 1929; also Gamow, *Phys. Zeit.* **30**, 717, 1929.

SUBJECT INDEX

INDEX OF NAMES

Printed in the United States
By Bookmasters